D1107764

Between Pacific Tides

OCEAN 928

EDWARD F. RICKETTS AND JACK CALVIN

Between Pacific Tides

FOURTH EDITION

Revised by Joel W. Hedgpeth

STANFORD UNIVERSITY PRESS
STANFORD, CALIFORNIA

Between Pacific Tides was first published in 1939. A revised edition prepared by the original authors, with a Foreword by John Steinbeck, was published in 1948. Subsequent editions, with revisions by Joel W. Hedgpeth, appeared in 1952 and 1962. The present edition, incorporating further extensive revisions by Dr. Hedgpeth, has been completely redesigned and reset, and contains many new illustrations and photographs.

Stanford University Press
Stanford, California

Printed in the United States of America
First published 1939
FOURTH EDITION 1968
Last figure below indicates year of this printing:
78 77 76

Preface

In 1853 there appeared in England a book on the marvels of seashore life, illustrated directly on the lithographic stone by the author. This book, *A Naturalist's Rambles on the Devonshire Coast,* by Philip Henry Gosse, was the first of its kind in any language, and all others have been imitators and followers of this prototype. Yet only a few of the later books have been based on original observation done at the seashore. *Between Pacific Tides,* first published in 1939, is one of those few. It is also one of the few to survive an original edition and to remain in demand for still another generation of readers and tide-pool watchers. There are many reasons for this sustained popularity; but perhaps the chief one is that people want to know about seashore animals, and no one else has presented this information in terms of the way of life—the ecology, if you will—of the seashore in such a readable manner. Where Gosse was a particularizer, dealing with limited localities, E. F. Ricketts was a generalizer—with an eye for particulars, to be sure, yet interested in all that went on along a coast of more than a thousand miles.

The author of *Between Pacific Tides* and P. H. Gosse would have much in common if they could meet on some Elysian shore; but that is unlikely, since, according to their respective religious orientations, they could not both be destined for the place Gosse fervently believed in and into which he firmly anticipated admission. But they did have in common a love for the creatures of the seashore and an interest in telling people what they had learned from their own observation. Gosse, despite his narrowly doctrinaire concept of special creation and his resistance to the ideas of a certain Charles Darwin (whom he nevertheless graciously supplied with information on request), stimulated both popular and professional interest in the seashore life of Britain. His books were for some time the primary source of information, and even now, more than a century later, they are still cherished on the shelves of professional biologists. This book by Ed Ricketts has had very much the same place in the development of serious marine biology on the Pacific coast of North America. At the same time it has continued to introduce the interested bystander to a strange and complex way of life. As John Steinbeck wrote in his foreword to an earlier edition, "This is a book for laymen, for beginners, and as such, its main purpose is to stimulate curiosity, not to answer finally questions which are only temporarily answerable."

Many books about seashore life appeared in the 86 years between *A Naturalist's Rambles* and *Between Pacific Tides,* and in the last 30 years there have been as many more. (Quite a few of the more recent books, of course, have been about shores that were previously inadequately known.) Books of this sort cannot be written without some firm background of scientific knowledge of the marine life in the region concerned, and one of the unique features of *Between Pacific Tides* was (and is) an appendix that summarized the state of the literature. This appendix is in part an expression of the personality of Ed Ricketts: he evidently liked to do this sort of thing, although arid pedantry was one trait he did not have. But the preparation of *Between Pacific Tides* was a long, arduous process, which involved ransacking the scattered literature on marine invertebrates and corresponding with specialists, as well as observing in the field. It took a lot of hard work to find out all these things, and obviously it seemed best to share this work with others who would use the book. It must be remembered that Ed Ricketts was not, in one sense, a professional biologist, and he was not concerned with formal degrees or with conforming to what society might expect of him. For example, he did not bother to graduate from the University of Chicago after studying there for two or three years. Had he done so, he would probably have turned into an obscure instructor somewhere, and *Between Pacific Tides* would never have appeared. Instead, and in the ideal sense, Ed was a professional naturalist; and once started, he discovered that he wanted to communicate.

Ed was other things as well, as John Steinbeck has so eloquently said in the Preface to *Log of the Sea of Cortez.* It would seem that those of us who have some personal memory of Ed, who shared some private part of the world with him and knew Cannery Row before it was converted to less utilitarian purposes than canning fish, are fewer now. Indeed we are, not because most of us have disappeared, but because there are so many more people at Monterey, and everywhere on the Pacific coast. Perhaps it is a good thing that people no longer make pilgrimages to the old place on Cannery Row—or even a good thing that literary critics have ceased to worry about John Steinbeck's soul and which part of his mind might really have been Ed's, or about who really wrote some of those philosophical passages in *Sea of Cortez*. In his own way Ed was a born teacher, with the instincts of Aristotle, and John Steinbeck was his most devoted student. At least this is the feeling I now get when rereading Steinbeck's memoir of Ed.

From all this, the reader may wonder whether Jack Calvin, the junior author, really exists, and what he had to do with the making of this book. He exists indeed, and lives among the fjords of Alaska, where he operates a printing business. Although not a zoologist, he still has a lively interest in tide-pool life from his days at Pacific Grove. He took most of the photographs for the original edition, and helped with the wearisome business of putting things together in words. Many of his photographs were fine in their day, but photography has improved in recent years, and we have replaced quite a few of them for this edition; those familiar with earlier editions, however, will recognize some old friends. We have been able to revise the text completely for the first time, so that it is difficult to decide, here and there, who is now responsible for particular words. I am responsible for revisions of factual statements and for the annotated bibliographies in their present state. Nevertheless, the book is still essentially as Ed Ricketts intended it to be.

In the days when *Between Pacific Tides* was an idea in the heads of Ed Ricketts and the friends who encouraged him to write it, there were only three institutions on the Pacific coast that officially paid much attention to things of the seashore: the University of Washington, with its summer station at Friday Harbor; Hopkins Marine Station at Pacific Grove, a short walk from Cannery Row; and Scripps Institution of Oceanography at La Jolla, where not much of anything happened in those days. It is perhaps symptomatic of how things were that the only adequate book for professional and amateur alike was Johnson and Snook's *Seashore Animals of the Pacific Coast,* written by a professor at San Diego State College and a high school biology teacher from Stockton. This is still a fine book, although time has not dealt kindly with parts of it; but it was also symptomatic of those days that the publishers remaindered it.

In those days more was being done for the cause of what we now so glibly call marine biology by a somewhat old-fashioned professor at

Left: E. F. Ricketts

Right: S. F. Light

Berkeley, S. F. Light, who conducted field trips to such places as Moss Beach attired in his gray business suit, complete with vest and starched collar; a pair of rubber boots was his only concession to the environment. Dr. Light was interested in all sorts of invertebrates, and he welcomed students who would study any sort of seashore animal. He was, as one of my friends once remarked, an "inspired pedagogue," and, through his students, he has left his mark on virtually every institution of learning on the Pacific coast. It was he who persuaded the Stanford University Press, which was at first disconcerted by the annotated systematic index, to retain it as a valuable part of this book. He and his students produced the essential professional guide to central California, familiarly known as "Light's Manual."

It is pleasant to think of some implausible Elysian seashore, with Philip Henry Gosse, in an "immense wide-awake, loose black coat and trousers, and fisherman's boots, with a collecting basket in one hand, a staff or prod in the other," leading Ed and Sol Felty Light in search of zoophytes and corals in some still unexplored tide pool—Ed in one of his favorite wool shirts, a battered hat, and drooping rubber boots, and Light impeccable under his fedora, with boots neatly supported over his business suit. Surprisingly enough, it would be Gosse who would strip off his clothes to reach some almost inaccessible bright spot in a deep tide pool, as he did to find that *Balanophyllia* was still a living entity on the shores of Britain. Ed would probably not remove his clothes, but would wade in anyhow; and Light, interested but aloof, would stand on the rocks.

At least that is the way I prefer to think of them: Gosse, whose writings are still alive and fresh despite his extremely narrow theological Weltanschauung; Ed, the friend of many visits at Pacific Grove during those days I drove my mother down to see her old friend who lived where there is now a supermarket; and Light, my professor who perhaps never quite approved of me. I am not sure that Light completely accepted anyone, despite Ted Bullock's faithful reminiscence, since I do remember his startled glance when, as a vestryman, he ushered me to a seat in the church at Berkeley of which, following their chairman's lead, a number of zoologists were staunch members. There was indeed something more than words can express to that brief encounter on a Sunday morning—perhaps because it was never alluded to afterwards—the feeling that perhaps I had caught my professor out at something he was not sure his students should know about him, as well as amazement that I should turn up. He almost forgot to give me a program.

All of this seems so long ago. Dr. Light never lived to see the marine station of his dreams, but there are now marine stations up and down this coast, almost as close together as Father Junípero Serra wanted to have his missions. There are at least 10 such establishments open all year around, and almost as many more open only during the summer—counting only those associated with colleges and universities. The study of marine life is no longer something for the chosen few; it has become fashionable among both professors and students. Inevitably, all sorts of new and interesting things are being observed and recorded. Some of them now seem so obvious that it is a bit embarrassing to many of us that we didn't see them on our first trip to the beach. But, in retrospect, that is asking too much. There is still much to learn, and no one can be expected to learn it all by himself.

One of the disconcerting things to me, as the editor of this work, is the frequency with which it has been quoted in the scholarly literature as a source of something that subsequent research has demonstrated to be erroneous. It would be easy to say that many of these errors have been left in this edition in order to stimulate still further research, but it would be fairer to say that I just don't know all the mistakes that still lurk in these pages. I am especially aware that many statements of range are inaccurate from the viewpoint of the growing fraternity of divers; but since the book remains essentially a guide for those who do not dive but who approach the seashore from the land in boots or old tennis shoes, I have left most of the ranges as they were in earlier editions. Perhaps someone will produce the logical companion volume, *Below Pacific Tides,* for those who explore the shallow sea, as Rupert Riedl has done for the Mediterranean.

Well, that is for someone else to do, some time from now. There is still much to be learned about the shore, and my most pleasant memory of Ed Ricketts is a morning when he was sorting some pickled beasts into cans for shipment, in the basement of his place on Cannery Row. Ed had just taken someone from the East on his first field trip to the Pacific tide pools, and the visitor had almost immediately found some creature that Ed had never seen before. Ed was pleased that his guest had found this animal. He had no resentment, just pleasure in remembering someone who had made a discovery that Ed could share with him. Remember this when you use this book. It is still mostly Ed Ricketts, and he would have been pleased, as will I who have inherited it, to know that something new and more interesting has been learned about what is going on between Pacific tides.

J. W. H.

Acknowledgments

Over the years, many people have helped with this book—identifying material, correcting names, suggesting references, and supplying other information. From the very beginning, *Between Pacific Tides* could never have been written without the assistance of people at Hopkins Marine Station, as should be obvious to any alert reader: these included Walter K. Fisher, Harold Heath, and Tage Skogsberg, all no longer with us, and Rolf L. Bolin. The tradition still continues, and it is a pleasure to acknowledge the help of Donald and Isabella Abbott in preparing this edition. Josephine F. L. Hart, Olga Hartman, Paul Illg, Wheeler J. North, and many others have also contributed in many ways.

Much of the advice and assistance received for former editions is still part of the book; but it might be misleading to rethank everyone here, since the final version of the information supplied is the reviser's responsibility. I wish especially, however, to thank all those who have sent me reprints of their papers; whenever appropriate, these have been added to the Bibliographies. In these days of proliferating journals it is more difficult than ever to keep up with the world, and I am sure that I would be much further behind than I am were it not for this courtesy.

Except as indicated below, the illustrations in this book are those used in the original edition—the photographs by Jack Calvin, and the line drawings by the late Ritchie Lovejoy.

Nick Carter of San Francisco provided photographs for the following: jacket; endpapers; Figures 9, 11, 15, 18, 26, 28, 30, 37, 52, 54, 77, 81, 83,

86–88, 108, 143–44, 146, 150, 153, 160, 173, 175, 181, 198, 200, 203, 210–11, 215, 227, 231, 244, 245, 249, 272–73.

Robert Morris provided photos for Figures 24, 36, 55, 149, 154, 166, 170, 183, 197, 199, and 221.

We have retained from previous editions the pictures taken by Woody Williams for Figures 12, 76, 118, 192, 212, and 233.

Other photographs were contributed by the following persons: Figure 138 by Jack Nielsen; Figure 167 by Joel W. Hedgpeth; Figure 172 by Boyd Walker; Figure 176 by John LeBaron; Figure 223 by Wheeler North; Figure 252 by G. P. Wells; Figure 237 by C. B. Howland.

In the color section, Nick Carter provided the five habitat scenes and the photos of *Patiria miniata* and *Hermissenda crassicornis*; the rest of the pictures were taken by Robert Morris.

Drawings and charts were done by various people for the following figures. Darl Bowers: 139–41. Sam Hinton: 169, 242. Mrs. Carl Janish: 156, and drawings on pp. 407–15. M. W. Johnson: 238. Joel W. Hedgpeth: 4, 6–8, 31, 142, 145, 157, 174, 179, 239, 255, 258, 268. R. J. Menzies: 89, 156, 177, 180. Frank A. Pitelka and R. E. Paulson: 266, 267. Lynn Rudy: 97–100, 164. Gunnar Thorson: 53. Florence V. White: 260.

We also wish to thank the following for contributing illustrations to this edition: Willard Bascom for the photograph of E. F. Ricketts; Josephine F. L. Hart for the photograph of *Betaeus setosus* taken by C. B. Howland (Fig. 237); and C. Dale Snow of the Oregon Fish Commission for the pictures of *Cancer magister* taken by Jack Nielsen. We thank the Geological Society of America for permission to use figures from the *Treatise on Marine Ecology,* and the University of California Press for the use of various illustrations from *Seashore Life of the San Francisco Bay Area and Northern California.* Stanford University Press has allowed the inclusion of Mrs. Janish's drawings of marine algae, originally done for Gilbert M. Smith's *Marine Algae of the Monterey Peninsula*; most of the pictures by Robert Morris will appear in a pictorial guide to marine invertebrates now in preparation at Stanford by Robert Morris and Donald P. Abbott.

Other illustrations have been borrowed from various papers in the "open" literature, usually with some modifications. These include the drawings of *Velella,* originally by George O. Mackie and W. Garstang; diagrams from papers by Joseph Connell and Peter Glynn; various items from the CalCOFI reports; and illustrations from the work of J. L. Barnard, L. A. Zenkevich, and O. B. Mokievsky.

Contents

Between Pacific Tides

Introduction

THE SHORE topography of the Pacific coast differs considerably from that of the Atlantic coast, and this factor, since it determines the conditions of life of the shore animals, often produces animal communities that seem strange to students from depauperate eastern shores. More obviously on the Pacific coast than on the Atlantic coast, the three coordinate and interlocking factors that determine the distribution of shore invertebrates are: (*a*) the degree of wave shock, (*b*) the type of bottom (whether rock, sand, mud, or some combination of these), and (*c*) the tidal exposure.

Considering each in turn:

(*a*) On the Pacific coast the degree of wave shock is of particular importance. According to the physics of wave motion, the size of the unbroken water area (and to a limited extent the depth), the wind velocity, and the wind direction determine the size of waves. Assuming that wind velocities are the same in the Atlantic and the Pacific, the great unobstructed expanses of the latter ocean make for larger wave possibilities. When, in addition, there is a bluff, unprotected coast with few islands or high submarine ridges for thousands of miles, as from Cape Flattery to Point Conception, and with prevailing northwesterly to westerly winds, wave shock is probably more powerful than in any other part of the Northern Hemisphere.

Nevertheless, as will be noted later, various factors modify the force of the waves in particular regions and cause correlative changes in the animal assemblages. The extreme modification comes in closed bays, sounds, and estuaries, where there is almost no surf at all. Given all gradations in wave shock from the pounding surf of 20-foot groundswells to the quiet waters of Puget Sound, where the waves barely lap the shore, one could enumerate an indefinite number of gradient stages. For the purposes of this book, however, we use only three, although even in Puget Sound it is quite possible to recognize sheltered and exposed positions and to note corresponding faunal distinctions. Ours is a classification of convenience. It is as though we divided the numbers from 1 to 100 into three parts, calling the divisions I, II, and III. In spite of these arbitrary divisions, however, the collector will find that there is often a reasonably sharp dividing line, as where the surf beats against a jutting point that protects a relatively quiet bay; and that there is a fairly high degree of correlation between these divisions and the animal communities. There is often overlapping of the animals, of course, but if test counts show that 75 per cent of the total observed number of a certain starfish are found on surf-swept rocky points, then we feel no hesitation in classifying it as an animal that belongs predominantly in that environment, noting also that some individuals stray to more protected shores and to wharf piling.

These qualifying statements apply with equal force to the other two factors—type of bottom and tidal exposure—on which we base our classification. It must be understood that infinite variations exist, that few regions belong purely to one or the other of our divisions, and that any animal, even the most characteristic "horizon marker," may occasionally be found in totally unexpected associations.

(*b*) The intergradation in types of bottom is too obvious to stress beyond remarking that in the cases of the innumerable variations between sand, muddy sand, sandy mud, and mud we have begged the question somewhat by using only two headings—sand flats and mud flats—leaving it to the judgment of the collector to decide where one merges into the other.

(*c*) The third important aspect of habitat, tidal exposure, has to do with the zoning of animals according to the relative lengths of their exposure to air and water (bathymetrical zoning)—in other words, the level at which the animals occur. A glance at any beach will lend more weight than would much discussion to an understanding of the extreme variations that exist between the uppermost region at and above the line of high spring tides, where the animals are wetted only a few times in each month by waves and spray, and the line of low spring tides, where the animals are uncovered only a few hours in each month. Many animals adhere closely to one particular level, and all have their preferences; but again the overlapping of one zone with another should be taken for granted. Also, local conditions may affect the animals' level, as along the coast of Baja California, where the high summer temperatures force the animals down to a lower level than they would normally assume. Even there, however, the animals' relative positions remain the same. There is merely a compressing of the life zone into something less than the actual intertidal zone. With these provisos, then, our system of zonation is equally applicable to San Quintín Bay, where the extreme range of tides is less than eight feet, and to Juneau, where it is more than 23 feet.

These three aspects of habitat—wave shock, type of bottom, and tidal level—which so tremendously influence the types of animals that occur, suggest the scheme of topic organization adopted for this book, a scheme that may at first glance seem cumbersome but that is in fact quite convenient, and certainly more practical than a single taxonomic sequence of the animals themselves. The most unscientific shore visitor will know whether he is observing a surf-swept or a protected shore, and whether the substratum is predominantly rock or sand or mud, although he probably would not know the classification by which zoologists arrange the animals he is interested in. He will further know, or can easily find out, whether the tide at a given time is high or low; and when it is low he can easily visualize the four zones or levels of the bared area. The observer possessing this information should be able, with the aid of the illustrations and descriptions provided, to identify most of the animals that he is likely to find and to acquire considerable information about each.

Users of this book will accordingly find it profitable to fix in mind the following classification of shore habitats:

I. Protected Outer Coast. Under this division we treat the animals of the semi-sheltered coast and open bays, where the force of the surf is somewhat dissipated. These shores, which provide rich collecting, are generally concave, and are characteristically protected by a headland or a close-lying island (Fig. 1, *a*). Todos Santos Bay, Laguna, Santa Barbara, much of Monterey Bay, Half Moon Bay, Bodega Bay, and Point Arena are good examples. However, stretches of shore that appear at first glance to be entirely open to the sea may actually belong in this category. An offshore reef of submerged rocks or a long, gradually sloping strand is sufficient to break the full force of the waves and will greatly modify the character of the animal assemblages. Even distant headlands, outlying bars, or offshore islands, if they are in the direction of the prevailing winds, provide a degree of protection that makes the animals correspond to our protected-outer-coast division. An offshore kelp bed will serve the same purpose, smoothing out the water to such an extent that many years ago small coasting vessels often used to seek shelter inside the beds and take advantage of their protection whenever possible in landing goods through the surf. The 1889 edition of the *Coast Pilot* mentions many places where the kelp beds were frequently so utilized, and in recent years fishermen and an occasional yachtsman have kept the tradition alive. Today the beds are favorite places for venturesome divers and small-boat sailors.

In many places the protection is not provided by headlands or offshore rocks directly, but by the refraction of waves as they reach headlands or rocks (Fig. 2). Because the waves tend to approach parallel to the shore, they tend to bend around a headland, meeting it headlong from all sides, so to speak. Where there are two headlands close together, the energy of the waves along the shore between them is reduced. But this energy can be concentrated on the sides, and sometimes the apparently protected rear, of headlands and offshore rocks. Such places can be dangerous, and

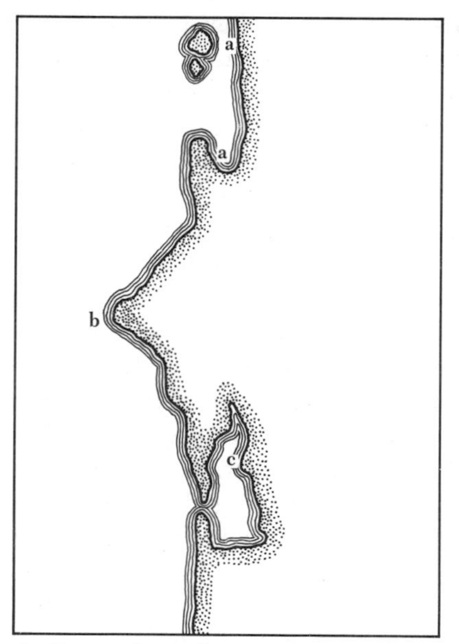

1. Types of coastline: *a—a* is a typical stretch of protected outer coast; *b*, a characteristic headland on the open coast; *c*, an enclosed bay and estuary.

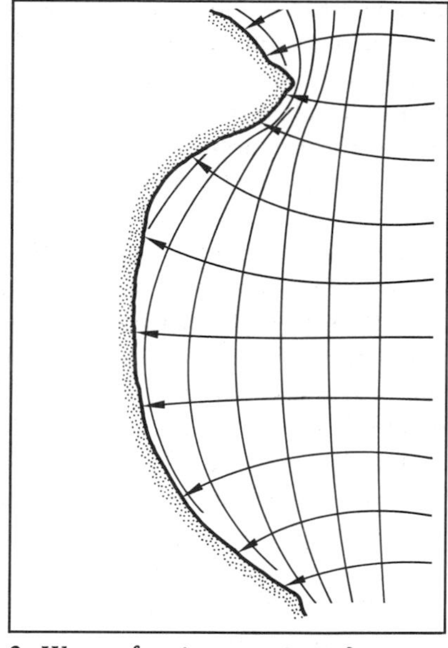

2. Wave refraction on an irregular shore. A beach need not be in the lee of an offshore feature to be protected from heavy wave action.

no one should venture on an unfamiliar shore without first observing for several minutes the patterns of waves against the shore. Many dangerous places along the coast are now marked by park services, and the warnings should be taken seriously. Often they indicate places where fishermen have been lost.

Here on the Pacific we have only two types of shore in this division, rocky shores and sandy beaches.

II. Open Coast. Entirely unprotected, surf-swept shores, though by no means as rich in animals as the partially protected shores of Division I, support a distinctive and characteristic assemblage of animals that either require surf or have learned to tolerate it. This type of shore is generally convex (Fig. 1, *b*), varying from bold headlands to gently bulging stretches, and there will be fairly deep water close offshore. Pismo Beach, the Point Sur and Point Lobos outer rocks, and the outer reefs of Cypress Point and Point Pinos are all good examples from the central California coast. And most of the coast of northern California and Oregon falls in this category. Obviously there are no muddy shores in this division; so again we have only rocky shores and sandy beaches.

III. Bay and Estuary. Animals of the sloughs, enclosed bays, sounds, and estuaries, where the rise and fall of the tides are not complicated by surf, enjoy the ultimate in wave protection and are commonly different species from the animals inhabiting the open coast and the protected outer coast. Where the same species occur, they frequently differ in habit and in habitat. The shores embraced in this division are sharply concave; that is, they have great protected area with a relatively small and often indirect opening to the sea (Fig. 1, *c*). San Quintín Bay in Baja California is an example, and moving northward we find other such areas in San Diego, Newport, Morro, San Francisco, Tomales, and Coos bays, in Puget Sound, and in all the inside waters of British Columbia and southeastern Alaska. This time we have a greater variety of types of shore, including: rocky shores, as in the San Juan Islands and north; sand flats, found in quiet waterways along the entire coastline; eelgrass, growing on many types of completely protected shore along the entire coast; and mud flats, found in nearly all protected waterways.

IV. Wharf Piling. In addition to many animals found elsewhere, wooden pilings support numerous species, such as the infamous *Teredo,* that will seldom or never be found in any other environment. The nature of piling fauna justifies its division (for convenience, and probably in point of fact) into exposed piling and protected piling.

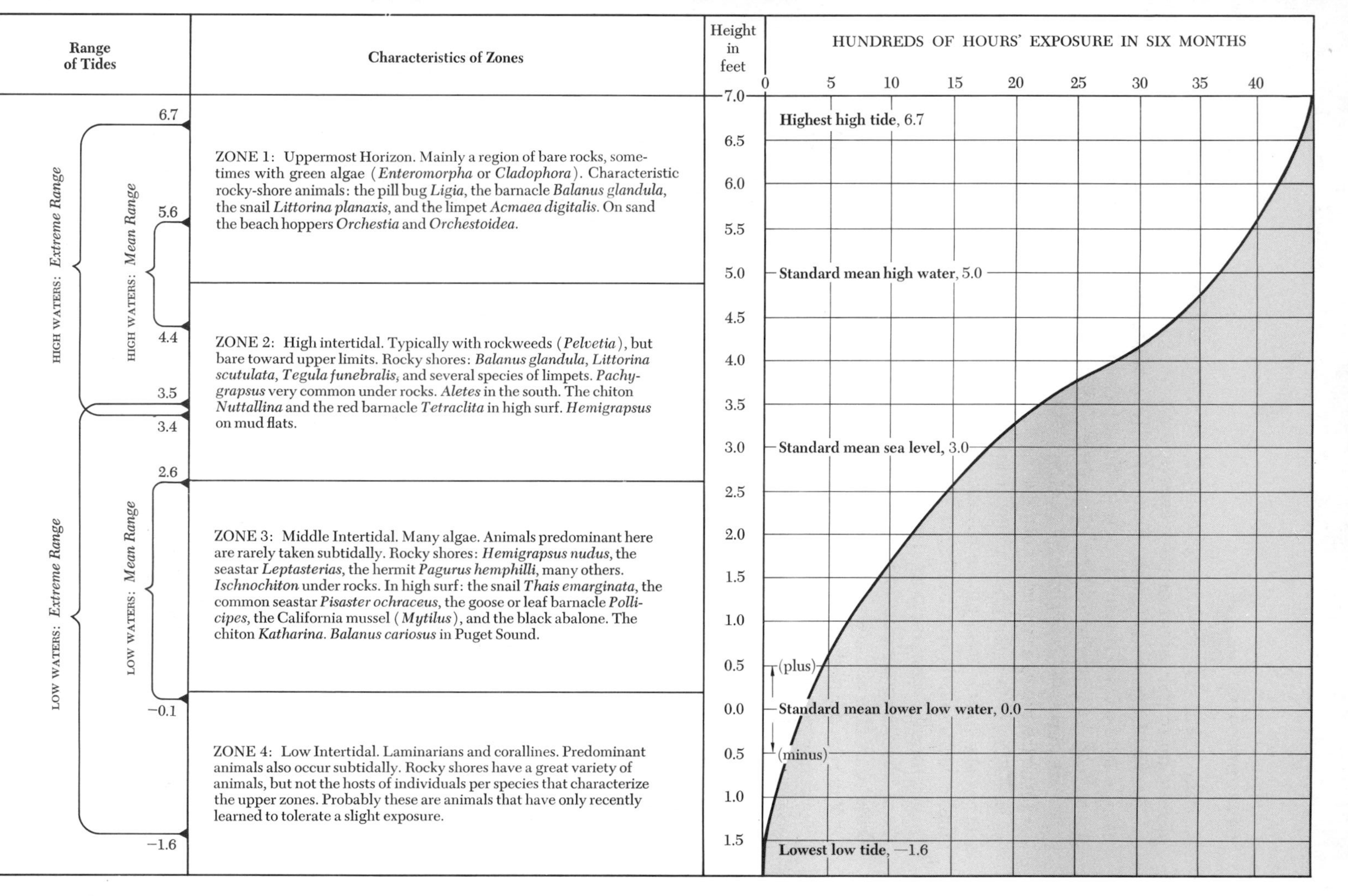

3. Vertical zoning of animals on the Pacific coast

Each of the three habitat types is also subdivided into zones (Fig. 3), according to the extent of tidal exposure, thus:

Zone 1. Uppermost beach: from the highest reach of spray and storm waves to about the mean of all high tides (which is plus 5 feet at Monterey). In this infrequently wetted zone, only organisms like the land-aspiring periwinkles and pill bugs occur. This is the splash, supralittoral, or Littorina zone, of various authors.

Zone 2. High intertidal, or upper horizon: from mean high water to about the mean flood of the higher of the two daily lows, a bit below mean sea level. This is the home of barnacles and other animals accustomed to tolerating more air than water. In some zonation schemes, our Zone 2 is the upper part of the balanoid, or midlittoral, zone. On the Pacific coast this is the zone above the mussel beds.

Zone 3. Middle intertidal, or middle horizon: from about mean higher low water to mean lower low water—the zero of the tide tables. Typically covered and uncovered twice each day, this belt extends from plus 2½ feet to 0 on the California coast, and corresponds to the lower part of the balanoid, or midlittoral, zone of many schemes of classification. Most of these schemes recognize one general midlittoral zone, but in practice we can recognize the two major subdivisions that we refer to as Zones 2 and 3. The animals found in Zone 3 have accustomed themselves to, and many require, the rhythm of the tides.

Zone 4. Low intertidal, or lower horizon: normally uncovered by "minus" tides only, extending from 0 to −1.8 feet or so at Pacific Grove, corresponding to the upper parts of the "Laminarian zone" or "infralittoral fringe" of some schemes of classification. This zone can be examined during only a few hours in each month and supports animals that work up as far as possible from deep water. Most of them remain in this zone, forgoing the advantages of the less crowded conditions higher up, because they are unable to tolerate more than the minimum of exposure incident to minus tides.

These zones, however, are not immutably fixed according to tidal levels, but tend to spread wider and higher toward the region of heavier wave action (Fig. 4). Often this can be seen on exposed coasts, along a vertical face set at right angles to the main direction of the waves. The conspicuous mussel zone will be wider toward the sea. What is not so obvious to the casual observer is that in very sheltered regions the whole zonal pattern may be lowered so that the uppermost zone is actually below the highest water mark of the year, as Figure 4 suggests. The generalized zonal pattern described here is essentially a summary of obser-

OUTLINE OF HABITAT AREAS TREATED

Protected Outer Coast, §§ 1–156

ROCKY SHORES, §§ 1–153. *Zone 1,* uppermost horizon, §§ 1–9. *Zone 2,* high intertidal, §§ 10–23. *Zone 3,* middle intertidal: exposed rock, §§ 24–25; protected rock, §§ 26–43; under-rock, §§ 44–51; substratum, §§ 52–58; pool, §§ 59–62. *Zone 4,* low intertidal: rock, §§ 63–120; under-rock, §§ 121–44; substratum, §§ 145–47; pool, §§ 148–52; root and holdfast, § 153.

SANDY BEACHES, §§ 154–56.

Open Coast, §§ 157–93

ROCKY SHORES, §§ 157–83. *Non-zoned animals,* §§ 157–60. *Zone 1,* uppermost horizon, §§ 161–62. *Zones 2 and 3,* high and middle intertidal, §§ 163–71. *Zone 4,* low intertidal, §§ 172–83.

SANDY BEACHES, §§ 184–93.

Bay and Estuary, §§ 194–314

ROCKY SHORES, §§ 194–245. *Zones 1 and 2,* uppermost horizon and high intertidal, §§ 194–98. *Zone 3,* middle intertidal, §§ 199–206. *Zone 4,* low intertidal: rock and rockweed, §§ 207–29; under-rock, §§ 230–40; substratum, §§ 241–45.

SAND FLATS, §§ 246–68.

EELGRASS FLATS, §§ 269–83.

MUD FLATS, §§ 284–314. *Zones 1 and 2,* uppermost horizon and middle intertidal, §§ 284–85. *Zone 3,* middle intertidal: surface, §§ 286–88; substratum, §§ 289–93. *Zone 4,* low intertidal: surface, §§ 294–98; substratum, §§ 299–313; pool, § 314.

Wharf Piling, §§ 315–38

EXPOSED PILES, §§ 315–30. *Zones 1 and 2,* uppermost horizon and middle intertidal, § 315. *Zone 3,* middle intertidal, §§ 316–20. *Zone 4,* low intertidal, §§ 321–30.

PROTECTED PILES, §§ 331–38.

vations of a semiprotected coastal situation; most specifically, of the zonation observed on the central California coast (see Chapter 14).

The outline on the facing page may be consulted as a summary of the Divisions and Zones within which the living creatures here described are classified.

Absolute beginners will do well to devote their primary attention to large, common, and spectacular animals, which may easily be identified merely by reference to the illustrations. And the more sophisticated will find that the book says a good deal also about the nondescript, the unusual, and the secretive. But the beginner, as well as the experienced, will often find that a creature he has observed is not mentioned in this book. There must be several thousand relatively common species of invertebrates along this coast. One hopeful ecologist, inspired by the Eltonian dictum that it would be best to learn as much as possible about a simple community rather than a little about a more complicated one, set himself to studying the apparently simple and limited group of creatures found among the little tufts of the red seaweed *Endocladia* at the higher tidal level. He soon found that he had more than 90 species—five of them undescribed, and several previously unreported from the Pacific coast!

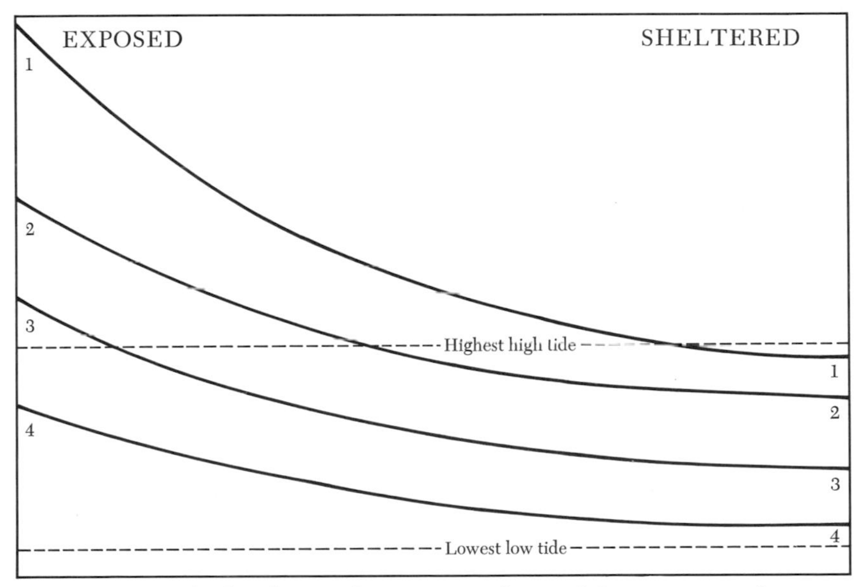

4. The zones are displaced upward as one proceeds toward the more exposed part of the coast; such a displacement is often seen on rocky headlands. This is a schematic diagram, and it is not intended to indicate actual height.

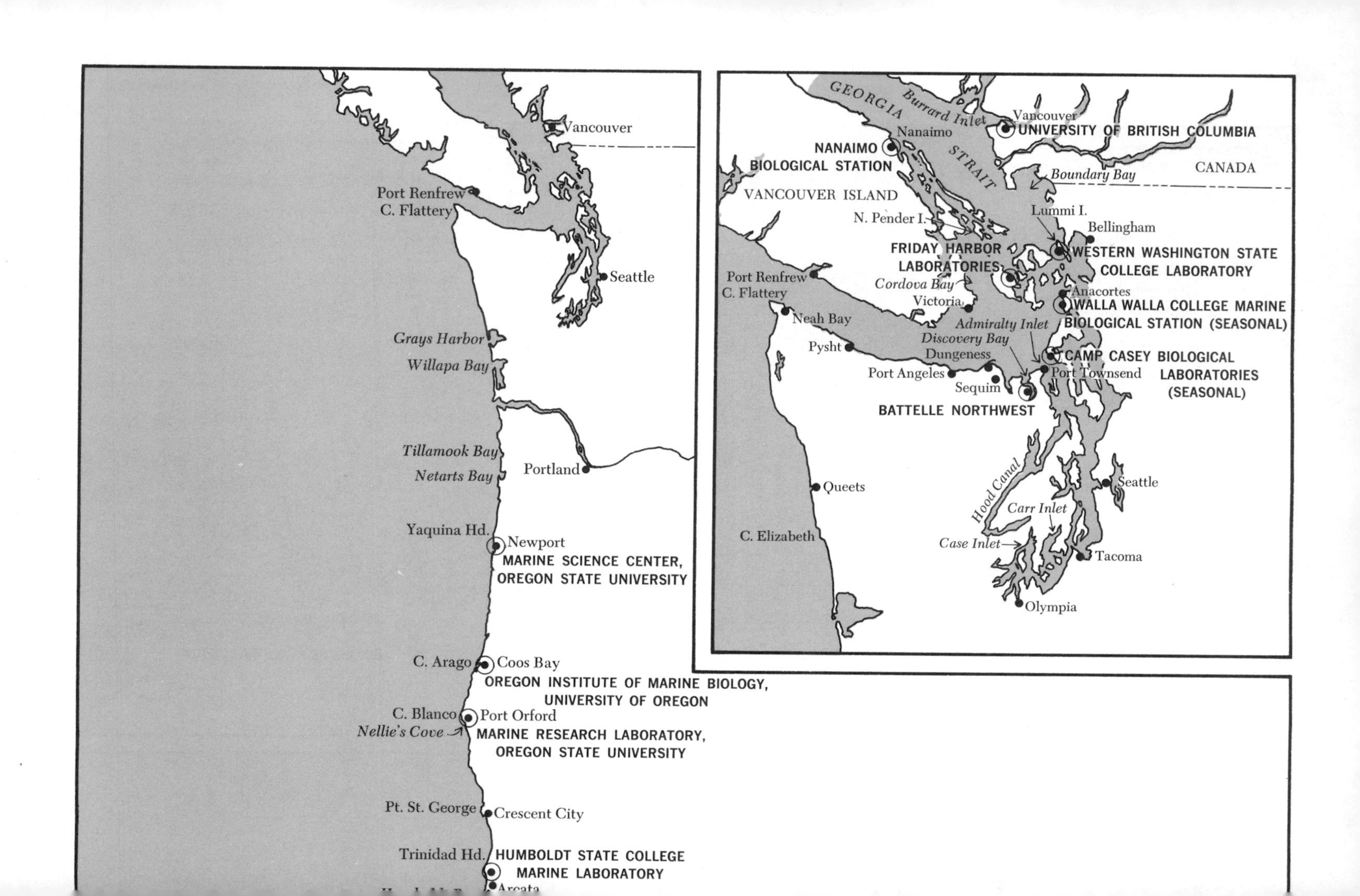

Vancouver
Port Renfrew
C. Flattery
Seattle
Grays Harbor
Willapa Bay
Tillamook Bay
Netarts Bay
Portland
Yaquina Hd.
Newport
MARINE SCIENCE CENTER,
OREGON STATE UNIVERSITY
C. Arago
Coos Bay
OREGON INSTITUTE OF MARINE BIOLOGY,
UNIVERSITY OF OREGON
C. Blanco
Port Orford
Nellie's Cove
MARINE RESEARCH LABORATORY,
OREGON STATE UNIVERSITY
Pt. St. George
Crescent City
Trinidad Hd.
HUMBOLDT STATE COLLEGE
MARINE LABORATORY
GEORGIA
Burrard Inlet
STRAIT
Vancouver
UNIVERSITY OF BRITISH COLUMBIA
Nanaimo
NANAIMO
BIOLOGICAL STATION
Boundary Bay
CANADA
VANCOUVER ISLAND
N. Pender I.
Lummi I.
Bellingham
FRIDAY HARBOR
LABORATORIES
WESTERN WASHINGTON STATE
COLLEGE LABORATORY
Port Renfrew
C. Flattery
Cordova Bay
Victoria
Anacortes
WALLA WALLA COLLEGE MARINE
BIOLOGICAL STATION (SEASONAL)
Neah Bay
Admiralty Inlet
Discovery Bay
Pysht
Dungeness
CAMP CASEY BIOLOGICAL
LABORATORIES
(SEASONAL)
Port Angeles
Sequim
Port Townsend
BATTELLE NORTHWEST
Queets
Hood Canal
Seattle
Carr Inlet
C. Elizabeth
Case Inlet
Tacoma
Olympia

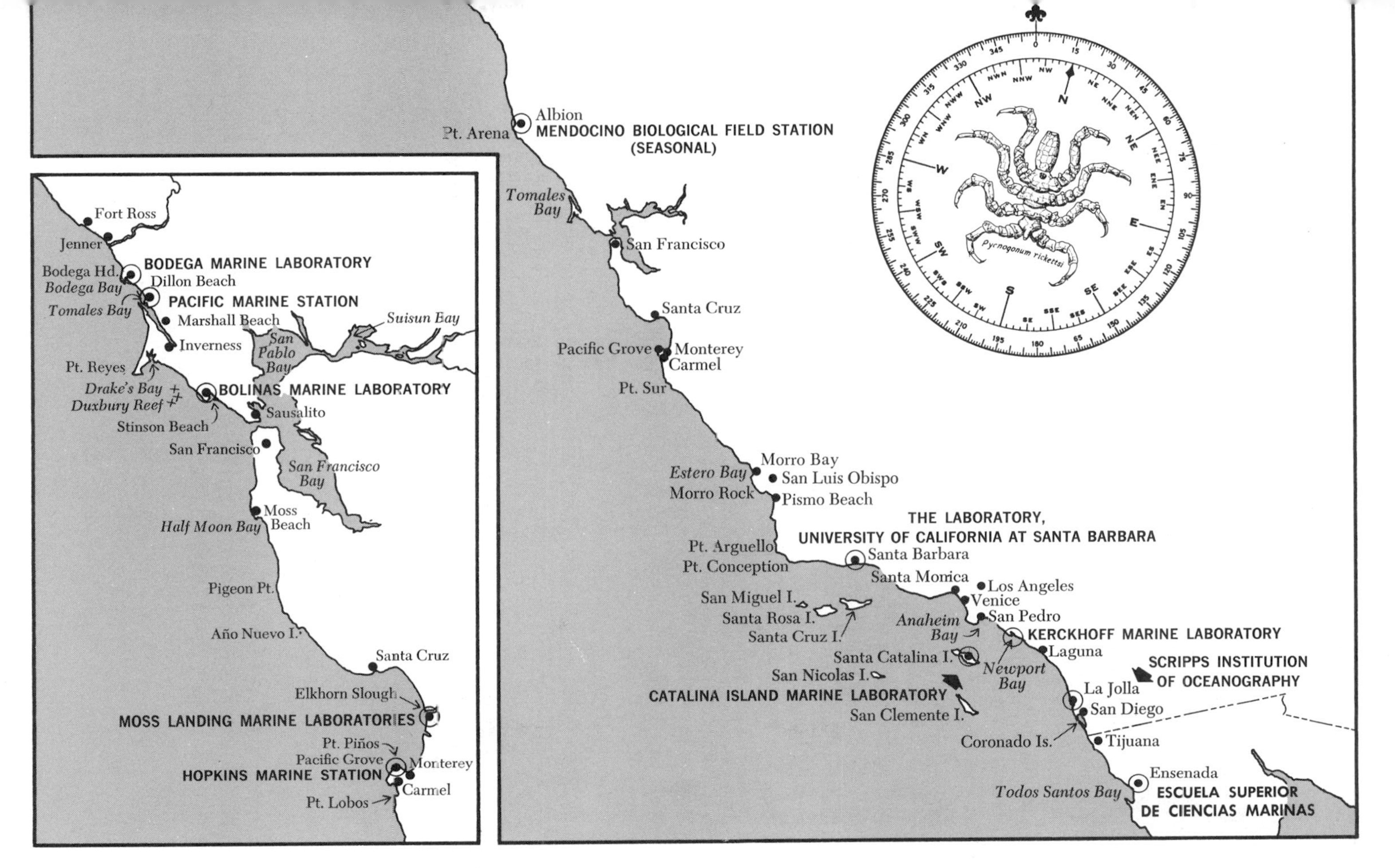

5. The Pacific Coast from Nanaimo to Ensenada, showing present marine stations. The insets show collecting localities in Puget Sound and central California.

The Systematic Index lists every species covered in the book according to its zoological classification; here the reader will be directed to books or papers that provide more information. We have cited, from the widely scattered literature, the following only: (1) the most recent general taxonomic papers covering the region; (2) smaller papers likely to be overlooked because of publication in obscure or foreign journals; (3) important references to the natural history, embryology, life history, and occasionally even the physiology, of common Pacific types or of comparable forms occurring elsewhere under supposedly similar circumstances. Papers concerned exclusively with pelagic, dredged, parasitic, or very minute species have rarely been cited unless they also carry information on related littoral forms.

This is a book for the observer or student who is limited to the shore and without equipment. The scope had to be limited in some way, and this choice accommodates the largest field. Though we have attempted to give a fair showing to the entire coast from Mexico to Alaska, as described by the maps in Figure 5, it may well be that the central California area has received undue attention. Expediency has dictated that this should be so, despite summer trips to Puget Sound, British Columbia, and southeastern Alaska, and frequent winter excursions south. The writers have themselves captured and observed practically all the animals listed—except as otherwise stated—but some we may not have identified correctly.

To the Bibliography for this edition we have added a brief review of all other books on seashore life, whether for young or old, concerning the Pacific coast of North America. One or two may have been missed, but none intentionally. Some of the comments on these books are based on a personal prejudice: some adult authors who write for children seem to forget that quite often the children may be more intelligent than the authors.

One more word of explanation is necessary before we consider the animals themselves. It would have been desirable in some respects to depend on scientific names less frequently than we have been forced to do in this account. Certainly it would make pleasanter reading for the layman if a common name were provided in each case, and if he did not have to read, or to skip over, long, tongue-twisting names compounded of Latin and Greek words and the latinized names of scientists—or even of boats that have carried scientific expeditions. Unfortunately, few of the marine animals have popular names that are of any diagnostic value. Popular names seldom have more than local significance, and often the same name is applied to different animals in different regions. Just as often a particular animal goes by a different name in each region where it occurs. We have accordingly given, for what they are worth, any popular names already in use, along with their scientific equivalents. To explain other scientific terms, we have added a brief Glossary to this edition (pp. 563–78). Persons who believe that magnetos, carburetors, cams, and valves are all "gadgets" are free to designate animals without common names, indifferently, as bugs, worms, critters, and beasties.

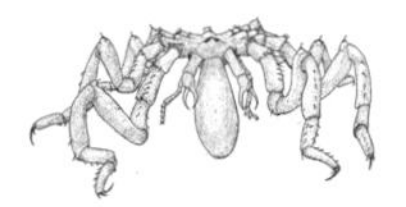  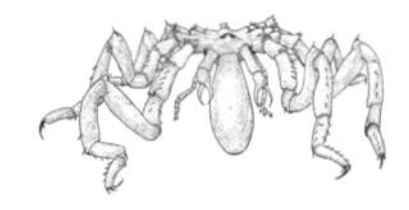

We are, alas, no longer in the halcyon days of carefree collecting and unspoiled abundance of life on our seashores. We must hope that those interested enough to obtain a copy of this book and use it for a guide will also be interested enough to watch intertidal life between the tides, rather than bringing it away to die in buckets or suffocate in poorly managed aquaria. We also hope that teachers will reconsider the utility of bringing brigades of students to some choice spot and turning them loose to trample paths among the seaweeds and the urchins. There is probably no hope for those teachers and counselors who stand placidly by while robust, undirected adolescents throw sea urchins at each other or stomp the gumboot chiton to death, but all should be aware that the life of the seashore was never adapted to withstand the pressure of hordes of people. Places that once abounded with urchins are barren of them now; the abalone, once a dominant animal of the intertidal regions, is now common only on inaccessible offshore rocks or islands (and these are accessible to the increasing shoals of divers); and the rock scallop *Hinnites*, once common in the lower intertidal, is now a rare animal.

This has happened before. Edmund Gosse wrote, in his moving story of life with his father, P. H. Gosse: "All this is long over and done with. The ring of living beauty drawn about our shores was a very thin and fragile one. It had existed all those centuries only in consequence of the indifference, the blissful ignorance, of man. These rock basins, fringed by corallines, filled with still water almost as pellucid as the upper air itself, thronged with beautiful, sensitive forms of life—they exist no longer, they are all profaned, and emptied, and vulgarized. An army of 'collectors' has passed over them and ravaged every corner of them. The fairy paradise has been violated; the exquisite product of centuries of natural selection has been crushed under the rough paw of well-meaning, idle-minded curiosity" (*Father and Son,* Ch. VI).

At least we seem to realize the danger more, nowadays. Perhaps Edmund Gosse sacrificed a bit of actuality to literary force, perhaps not. In any event, on the Pacific coast we are establishing more and more reserves, state parks, and national seashores where collecting is forbidden; and there are still inaccessible areas to serve as sources for repopulation. But many of our seashore animals do not move around very much, quite a few of them may lead extraordinarily long lives, and the process of recovery is slow; not everyone can be privileged to visit an unspoiled area. To be sure, it is necessary to take specimens for study and identification, but it is not necessary to take buckets of them; nor is it necessary for each of a group of 50 school children to collect "one of each" as part of his field trip. Especially is it unnecessary to remove the larger, conspicuous animals, which give our shore its unique character. Not only should you replace the rocks you overturn, you should leave the animals where they are unless you have a good reason for taking a very few of them.

PART I. PROTECTED OUTER COAST

From the standpoint of the shore collector, the rocky tide flats of the protected outer coast are the most important of all regions (see color section). Rocky-shore animals are abundant, easy to find, and frequently spectacular in their bright colors and unexpected shapes. So keen is the struggle for existence here that not only is every square inch of shore surface likely to be utilized but the holdfasts and stipes of kelp are also occupied; and such forms as sponges, tube worms, and barnacles often take up positions on the shells of larger animals.

Rocky-intertidal animals are characterized by interesting physiological processes, which offer methods of attachment, means of surviving wave shock or of coping with an alternate exposure to air and water, and techniques of offense and defense—all intensely specialized to meet the crowded environment. Obviously, there is no need for the elaborate devices developed by sand-flat and mud-flat animals to avoid suffocation or to retain orientation. But the rocky shore presents its own special problems. Perhaps the most interesting process, highly developed in this environment, is that by which a captured crab or brittle star deliberately breaks off an appendage in order to free itself. Losses occasioned by this process, called autotomy, are usually replaced by subsequent growth, or regeneration. This is a trait of great survival value to animals that may be imprisoned by loose rocks overturned by wave action.

1. Upper Zones

ROCKY-SHORE collecting is best managed by turning over small rocks by hand and lifting big ones with a bar. Moreover, it is highly important that the collector, in the interests of conservation, carefully replace all rocks in their depressions; otherwise many of the delicate bottom animals are exposed to fatal drying, sunlight, or wave action. Whoever doubts the necessity for this care should examine a familiar intertidal area directly after it has been combed by a biology class that has failed to observe the precaution of replacing overturned rocks, and should visit it again a few days later. At first the rocks will simply look strange and scarred. In a few days whole colonies of tunicates, solitary corals, and tube worms will be found dead, and a noticeable line of demarcation will set off the desolate area from its natural surroundings. It takes weeks or months for such a spot to recover.

Zone 1. Uppermost Horizon

This is a bare rock area in the main, but sometimes there are sparse growths of green algae. It is inhabited by hardy, semiterrestrial animals. At Pacific Grove the zone lies between plus 7 or 8 feet and plus 5 feet.

§1. The small, dingy snails that litter the highest rocks are not easy to see until one can distinguish their dry, dirty-gray shells from the rocky background of the same tone. Their habitat, higher on the occasionally wetted rocks than that of almost any other animal, gives them their Latin name of *Littorina,* "shore-dwellers"—a more diagnostic term than the popular "periwinkle." Marine animals though they are, the littorines keep as far as possible away from the sea, staying barely close enough to wet their gills occasionally. In regions of moderate surf they may be several feet above the high-tide line, scattered over the rock face by the thousands or clustered along crevices. Specimens placed in an aquarium show their distaste for seawater by immediately crawling up the sides until they reach the air; if forced to remain under water, they will ultimately drown.

It seems probable that some species of *Littorina* are well along in the process of changing from sea-dwellers to land-dwellers, and they have, in consequence, an extraordinary resistance to un-marine conditions. European littorines have been kept dry experimentally for 42 days without being damaged, and they can stand immersion in fresh water, which kills all true marine animals, for 11 days. We once subjected *Littorina planaxis* (Fig. 6), the commonest species on these high rocks, to a shorter but even more severe test: we fed it to a sea anemone (*Metridium*). There was no intention, however, of testing *Littorina*'s endurance; we merely wished to feed the anemone something to keep it alive in captivity, and assumed that its powerful digestive juices would circumvent the difficulty of the shell. The anemone swallowed the snail promptly, and our expectation was that in due time the empty shell would be disgorged. But it was an intact and healthy *Littorina* that emerged, like Jonah, after a residence of from 12 to 20 hours in the anemone's stomach. When first discovered, the disgorged snail was lying on the bottom of the dish, and since it must have been in some doubt as to just where it was, its shell was still tightly closed. When it was picked up for examination, however, it showed signs of life; and after being returned to the dish, it crawled away at its liveliest pace. It had apparently suffered no harm whatever, but its shell was beautifully cleaned and polished.

Such resistance to distinctly unfavorable conditions is made possible by a horny door, the operculum, with which *Littorina*, like other marine snails, closes its shell. This door serves the double purpose of retaining moisture in the gills and keeping out fresh water and desiccating wind. That it can also exclude the digestive enzymes of an anemone is a considerable tribute to its efficiency.

Like other marine snails, *Littorina* possesses a wicked food-getting instrument, the radula—a hard ribbon armed with rows of filelike teeth. The radula of this particular animal is unoffendingly used to scrape detritus and microscopic plants from almost bare rock—an operation that necessitates continual renewal by growth—but the radulae of several other snails are used to drill holes through the hard shells of oysters, mussels, and other snails.

The activity of littorines on coarse or loosely bound sandstone may in time abrade the surface. Wheeler North has estimated that the grazing of littorines at La Jolla erodes the rock at the rate of a centimeter a century.

The sexes of the littorines are separate. In central California, at least, some specimens may be found copulating at almost any time of the year; but reproduction in mass seems to be confined to the spring and summer months, when it is difficult to find, among thousands, a dozen specimens that are not copulating. The Pacific coast forms have not been studied specifically, but the female of a European species is known to lay in the neighborhood of 5,000 eggs in a month after one act of fertilization, the eggs being laid singly or in pairs in tiny transparent capsules shaped "like a soldier's tin hat." It is probable that the female of *L. planaxis* lays her eggs in a similar manner.

The extreme high-water form, *Littorina planaxis,* is said to range from Puget Sound to southern Baja California. It is most abundant on the California coast, and north of Cape Arago on the Oregon coast it is replaced by *Littorina sitkana* (Fig. 7). The slightly smaller, taller-spired *Littorina scutulata* (Fig. 8) is found all along the coast, usually at slightly lower levels. It appears to be more tolerant of reduced salinities than the other two, and is commonly found on rocks and pilings in bays; it is, for example, the littorine of such places as the sheltered parts of Tomales Bay, or Coyote Point in San Francisco Bay.

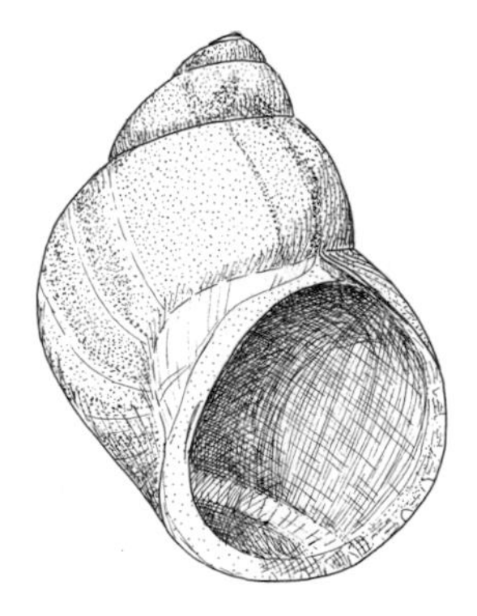

6. *Littorina planaxis* (§1).

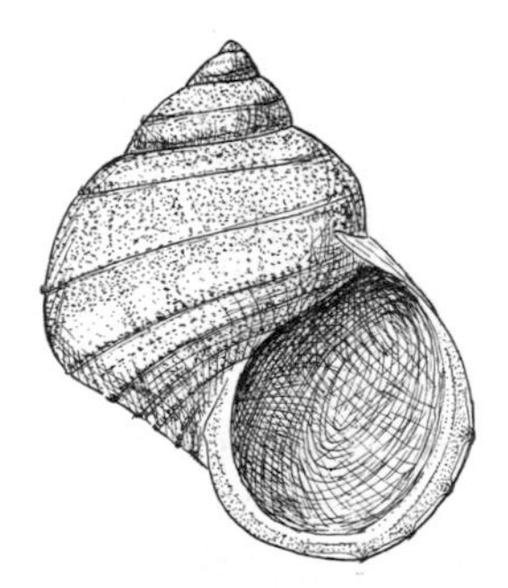

7. *L. sitkana* (§195).

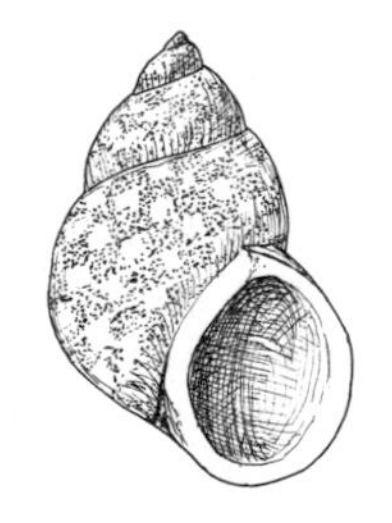

8. *L. scutulata* (§10).

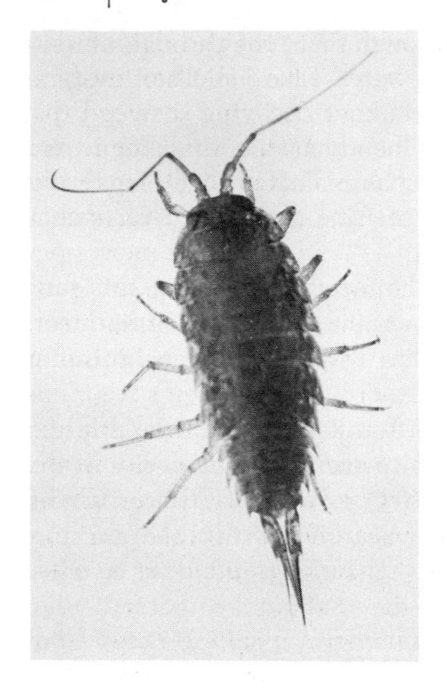

9. *Ligia occidentalis,* the rock louse (§2).

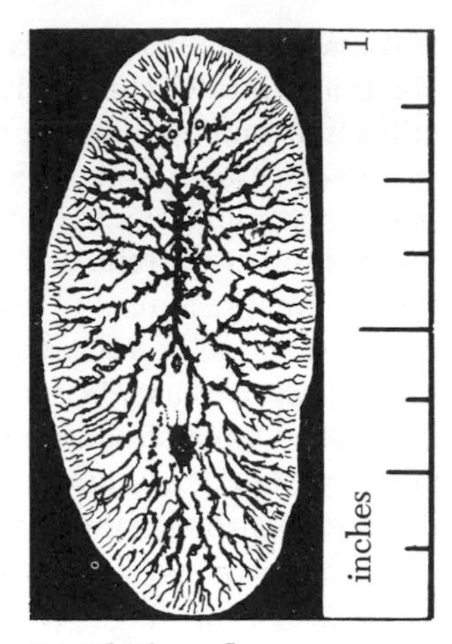

10. The large flatworm *Alloioplana californica* (§3).

§2. In some places specimens of a large isopod, the rock louse *Ligia occidentalis* (Fig. 9), up to 1⅓ inches long, can be found scurrying around on the rocks at distances so far removed from the water that one is led to believe it is another animal in the very act of changing from a marine to a terrestrial habitat. *Ligia* (including a more northern, open-coast species, §162) is very careful to avoid wetting its feet, and will drown in a short time if kept under water. When the tide is out it may venture down into the intertidal area, but when there is any surf it will be found just above the spray line. Specimens rarely occur singly; there will be hundreds or none. This species is recognized by the presence of two long forked spines projecting behind—which are used, apparently, as a rudder to help the animal steer its rapid and circuitous course over the rocks. We have taken specimens at Pacific Grove early in March that had the incubatory pouch on the underside of the abdomen turgid with developing eggs. The range is from at least Sonoma County to the Gulf of California. It is common on bay shores and occurs in the Sacramento River in almost fresh water.

Still another child of the ocean that has taken to living on land is the sand flea, or beach hopper, *Orchestia traskiana*. Along the steep cliffs west of Santa Cruz we have found it more than 20 feet above tidewater and above the usual spray line, living practically in the irrigated fields. It is found at Laguna in a brackish water slough that runs through marshy fields and rarely receives an influx of tidewater. Like the other hoppers, it is a scavenger, and may often be found about decaying seaweed that has been thrown high up on the shore. The animal's body color is dull green or gray-brown, with the legs slightly blue. Except for its small size (½ inch or a little more in length) and its short antennae, this form is similar in appearance to the large, handsome hoppers of the more open beaches. The hoppers are members of the family Talitridae; this family includes at least one strictly terrestrial species, which has spread from its native New Zealand and has become a resident of greenhouses in many parts of the world.

§3. A conspicuous rocky-shore animal that keeps a considerable distance above mean sea level is the large flatworm *Alloioplana californica* (Fig. 10), which is up to 1½ inches long. Thick-bodied and firm, almost round in outline, these animals are rather beautiful with their markings of blue-green, black, and white. They are relatively abundant, near Monterey at least, but are not often seen, since their habitat is restricted, when the tide is out, to the undersides of large boulders on gravel that is damp but seldom wet. The casual observer should feel properly thrilled at finding the several specimens that usually turn up, and will easily remember them from their size and beauty. As a matter of fact, they richly reward any layman who undertakes the labor involved in clearing the surroundings and rolling over one of these huge boulders.

Like most of the flatworms, *Alloioplana* is hermaphroditic, each animal developing first male and then female gonads. The egg masses, found almost throughout the year, form round encrusting bodies up to ¼ inch in diameter under boulders and in rock crevices. *Alloioplana* is known to feed on minute snails, since portions of snails' radulae, or rasping "tongues," have been found in its digestive tract. As far as we know, this flatworm has been recorded only from the Monterey Bay region, but it will surely be found elsewhere. We have taken an identical or similar form in Baja California.

§4. Small acorn barnacles, with small limpets interspersed among them, occur abundantly near the extreme upper limit of the intertidal zone. Except at extreme high tide, both have dried, dead-looking shells of the dingy gray color so common to high-tide rocks. The barnacles, usually *Balanus glandula* (Fig. 11), are actually crustaceans, more nearly related to the crabs and shrimps than to the shelled mollusks with which the tyro associates them. The sharp projecting shells, ½ inch or less in diameter, are sometimes crowded together like cells in a honeycomb on the

11. *Balanus glandula* (§4), the acorn barnacle, in a characteristic grouping; two limpets (*Acmaea digitalis*) are also visible.

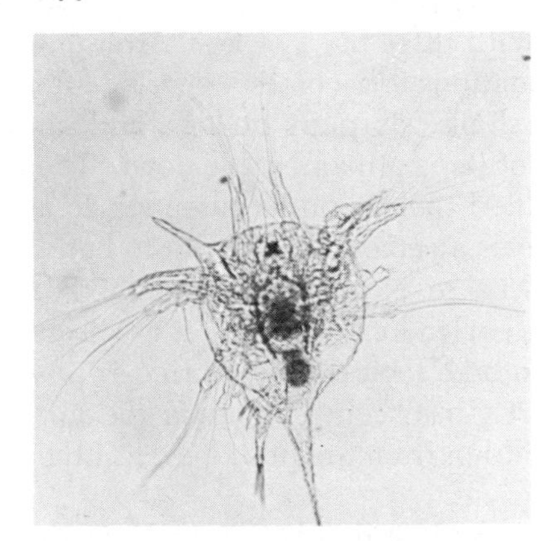

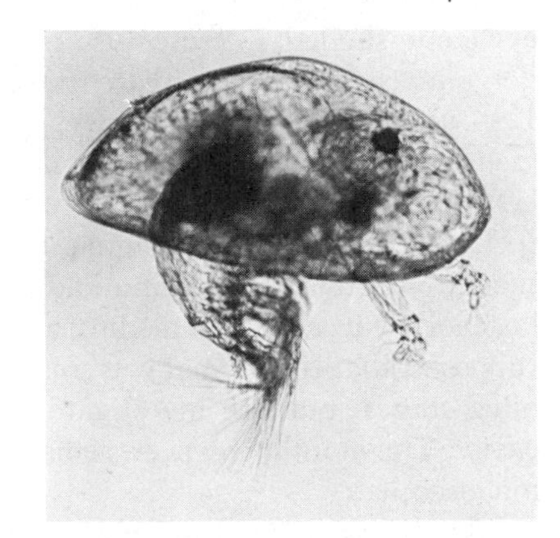

12. Larval stages of a barnacle (§4): left, nauplius; right, cypris. Both pictures are greatly magnified.

sloping or vertical rock faces. They are necessarily exposed a great deal to sun, rain, or wind; but at such times the operculum, a set of hinged plates, is kept tightly closed to protect the delicate internal anatomy. At extreme high tide, sometimes for only a few hours in a week, each animal throws open its operculum and rhythmically sweeps the water with its brightly colored appendages, respiring and searching for the minute animals that constitute its food. A dry cluster of barnacles will "come to" very quickly if immersed in a jar of fresh seawater, and they live well in aquaria.

Balanus glandula is one of the most abundant single animals on the coast, and is the Pacific's high-intertidal representative of a group of animals distributed throughout all oceans and in all depths. It is also one of the beautifully generalized (that is, unspecialized, hence able to meet unexpected conditions) animals that range all the way from the Aleutian Islands to Ensenada, Baja California, and probably still further south. It thrives not only in the constantly aerated ocean waters along the protected outer coast (and even on violently surf-swept points), but in the quiet waters of Puget Sound, where according to Shelford *et al.* it actually prefers estuarine conditions—enclosed bays, poor circulation of water, little wave action, and low salinity. On the quiet beaches of Tomales Bay there are individual barnacles perched on pebbles the size of walnuts; in such circumstances, even solid rock is obviously unnecessary.

The reproductive habits of barnacles are noteworthy. The eggs and embryos are retained within the shell of the parent, to be discharged as free-swimming larvae called nauplii. A nauplius (Fig. 12) is a one-eyed, one-shelled, microscopic animal with three pairs of legs. After undergoing several intermediate transformations it becomes a cypris (Fig. 12)—an animal with three eyes, two shells, six pairs of legs, and an inclination to give up the roving habits of its youth and settle down. The cypris attaches itself, by specially modified suckers on its antennae, to a rock or other object, whereupon it secretes a cement and begins to build the limy protective shell of an adult barnacle. Now it is a blind animal, fastened by its head and feeding with its curled, feathery legs. It has been supposed that the animal was unable to take food during its free-swimming stages; but one investigator, Hertz, has seen diatoms in the gut cavity. The mouth, however, remains hidden from the prying eyes of the microscope.

Like their relatives the crabs, barnacles molt at regular intervals, shedding the thin, skinlike covering of the animal itself rather than the familiar jagged shell. In the quiet waterways of Puget Sound and British Columbia these almost transparent casts may be seen floating on the water in incalculable numbers.

It should be mentioned in passing that occasionally we may have mistaken *Balanus hesperius* for *B. glandula,* although no specimen from our random shore collections has been so identified. We have taken undoubted specimens of *B. hesperius* from boat bottoms and from deep water only. In any case, it makes little difference; the two are so similar that only specialists are thoroughly competent to tell them apart.

§5. Ranging southward from the neighborhood of Santa Barbara is a much smaller gray or brown barnacle, *Chthamalus fissus,* whose shell has less than half the diameter of *Balanus*'s. Everything else being equal, this tiny barnacle will be found gradually but evenly replacing its larger relative as one moves southward, until, in northern Baja California, it becomes the predominant form. In some sheltered localities, or on gradually sloping beaches, it is the only barnacle to be found; but where there is a bit of surf the larger *Balanus* steps in again. There is one wave-pounded cliff on a jutting headland about 35 miles below the Mexican border that has *Balanus* only, although the smaller form is found within a few miles on both sides.

An Italian investigator, experimenting with another species of *Chthamalus*, found that they prefer atmospheric respiration, using the air dissolved in spray, and that consequently they thrive best on wave-battered reefs. While water remains on the rock, the cirri, or feeding legs, are frequently protruded into the air. This investigator kept 100 individuals on his laboratory table for 3 years, immersing them for 1 or 2 days about every 3 months. They were immersed a total of only 59 days out of the 1,036, and yet only 10 to 12 of them died each year. Other lots lived as long as 4 months continually immersed in fresh water, which kills nearly all marine animals. Some lived 2 months completely immersed in vaseline! Under either natural or artificial conditions, periods of drying are as necessary to them as periods of immersion.

§6. *Acmaea,* the limpet, is a gastropod ("stomach-foot"), and is related to the common marine snails, however little its lopsidedly conical shell resembles theirs. It occurs in great abundance, plastered so tightly against the rocks that specimens can be removed undamaged only by slipping a knife blade under them unawares. If the animal receives any warning of an impending attack, it draws its shell into such firm contact with the rock that a determined attempt to remove it by force will break the shell, hard as it is. So powerful is the suction foot, once it has taken hold, that an estimated 70-pound pull is required to remove a limpet with a basal area of 1 square inch. The sexes of limpets are separate, and fertilization occurs in the sea, since the eggs and sperm are shed freely into the sea.

Our limpets, except *Acmaea persona*, stay pretty much at home during the day, but species found elsewhere will move about in daylight as long as they are covered with water. Much can be learned, however, from a night trip to the shore. A single excursion into the tidelands with a flashlight will often reveal more of an animal's habits than a score of daylight trips. Seen in the daytime, a limpet seems as immobile and inactive as a dry barnacle; but at night the foot extends, the shell lifts from the rock, and the animal goes cruising in search of food, which it scrapes from the rock with a broad, filelike ribbon, the radula.

Studies of an Atlantic form indicate that there is a direct connection between a limpet's position in the intertidal zone and the height of its shell. Individuals inhabiting relatively low positions have low shells, whereas the higher (and hence the drier) the habitat the higher the shell. Presumably the animals that are least submerged must have the greatest amount of storage space for water and can afford the extra wave resistance. As usual, there are exceptions to every rule; the tallest *Acmaea, A. mitra,* frequents the lowest intertidal, but in general this relation applies.* Through the work of Orton, 1929, it is known also that shade and sunlight are the chief factors regulating the distribution of the British

* *Acmaea mitra* has several other characteristics that set it apart from the rest of the *Acmaeas*; some specialists place it in a separate subgenus.

limpet *Patella*; position in the littoral in relation to tidal levels is secondary. It seems that the needs of this creature are met in the following order: (*a*) shade, (*b*) moisture, (*c*) food (this least, since even such a slow-moving animal can work out into food-laden areas during the night or when the tide is high).

The commonest species on these high Pacific rocks is *Acmaea digitalis* (Figs. 13, 16), which ranges from the Aleutian Islands to San Diego but is abundant only from Monterey northward. This is the exceedingly common form that has been listed as *A. persona* in some older accounts. Great colonies occur even where (or especially where) the surf is fairly brisk, occasionally extending clear up to the splash line in areas sheltered from direct sunlight. The average small, dingy brown specimen found in these high colonies will measure under 25 mm., even under 15 mm.; but the larger, usually solitary, individuals found farther down may be half again as large and more brilliantly marked. *A. digitalis,* apparently preferring rough rocks, notably tolerates conditions such as swirling sand, mud, debris, and even sewage or industrial pollution (most of which are disastrous to other limpets), and is therefore possibly the most abundant *Acmaea* on the whole outer coast north of Point Conception. A related, heavily eroded type, the former subspecies *textilis* Gould, is now known to be simply an ordinary *A. digitalis* whose shell has been pitted by a parasitic fungus found also on other limpets inhabiting successively lower levels (§§ 11, 25); the reader should refer to accounts of these other forms, since some of them are occasionally found on the upper beach.

It has been recognized for some years that some of the limpets superficially resembling *Acmaea digitalis* are in fact another species. Though similar in external shape, and sometimes in color pattern, these limpets always lack the solid brown spot inside the apex of the shell (which can be seen only if the animal is removed) and the ridges on the outside of the shell that are characteristic of *A. digitalis.* The new species, which turns out to be common from Oregon to California, has been named *Acmaea paradigitalis* by Harry K. Fritchman II.

§7. The giant owl limpet, *Lottia gigantea* (Fig. 14), ranges from northern California to Cedros Island, Mexico; it is found in the high intertidal, especially in the southern part of its range. The largest specimens, however, more than 3 inches long, are found in the middle zone, especially on surf-swept rocks. Beside the animal in the photograph is the characteristic scar that marks the normal position of an absent specimen. Work done by European zoologists indicates that these animals range far

3. A group of limpets, *Acmaea digitalis* (§6), with periwinkles and barnacles.

4. *Lottia gigantea* (§7), the owl limpet, with its rock scar.

afield in search of food, presumably algae; each then returns to its own scar, even though a thousand scars in an area of a few square yards seem to be identical.

The Mexicans justly prize the owl limpet as food. When properly prepared, it is delicious, having finer meat and a more delicate flavor than abalone. Each animal provides one steak the size of a silver dollar, which must be pounded between two blocks of wood before it is rolled in egg and flour and fried. Unfortunately, *Lottia* is not protected under our fish and game regulations; a pity, for its edibility may be its downfall. It is much too interesting an animal to be carelessly tossed down a gourmet's gullet. Some observations made by John Stimson at Santa Barbara (in 1966 or 1967) indicate that our *Lottia* does not roam very far, but actively defends its home ground, bulldozing invaders off.

§8. In localities that have spray pools at or above the high-tide line, the vivid, yellow-green strands of the plant *Enteromorpha* are likely to appear, especially where there is an admixture of fresh water. In these brackish pools, which often have a high temperature because of long exposure to the sun, very tiny red "bugs"—crustaceans—may often be seen darting about. These animals, easily visible because of their color, are specimens of the copepod *Tigriopus californicus.* This hardy animal breeds easily in captivity and may be raised at home in old coffee cups or cheese glasses on a diet of pablum or the like. The eggs, which are carried in April in central California, are contained in a single sac. *Tigriopus* is one of the harpacticoid copepods, most of which live near the bottom, or among detritus and other accumulations everywhere on the seashore. *Tigriopus,* however, seems to be the most successful colonizer of the isolated, high pools. Some of these pools may contain only 1 or 2 quarts of water, and are subject to changes in salinity from rain and evaporation.

Where pools have an accumulation of sand, clumps of small algae, and the like, the waving tentacles of a spionid worm, *Boccardia proboscidea,* may be seen emerging from the bottom, although the worm is often difficult to find. There are two species of *Boccardia,* very much alike, and both have rather catholic habits on the seashore (§152).

§9. A few other animals may occur in the high intertidal zone: occasional small clusters of leaf, or goose, barnacles (§160) and even small mussels (§158). These, however, achieve their greatest development in other zones and are therefore treated elsewhere. All the above are abundant and characteristic; they are fairly well restricted to the uppermost zone, and we should say that all except *Alloioplana* (which takes searching for) and the uncertain *Ligia* can be found and identified by the neophyte. They should be a pleasant enough introduction for whoever cares to proceed further into the tidelands.

Zone 2. High Intertidal

In this belt there is a greater variety of species and a tremendously greater number of individuals of many of the species represented than in the highest zone. Here there are some more or less permanent tide pools, with their active fauna of hermit crabs, snails, and crabs. Barnacles occur here also, certainly larger and possibly in greater abundance. Plants begin to appear in the uppermost part of this zone, and grow lush

15. A high intertidal assembly: tufts of *Endocladia*; California mussels, *Mytilus californianus*; barnacles, *Balanus cariosus*; and scattered limpets, *Acmaea digitalis*. Some periwinkles (*Littorina* sp.) and chitons (*Nuttallina*) can be seen along the bottom of the picture.

in its lower part. Most of the blue-green algae and some of the greens begin here and extend on down into the lower zones. The slim brown rockweed *Pelvetia* begins at about the middle of Zone 2, and cuts off fairly sharply at the lower boundary. In mildly surf-beaten areas the first alga to appear may be the brown, crinkly *Endocladia*, sparsely covering the granite ledges with tufted clusters (Fig. 15). Below this the more massive *Fucus* appears. These plants not only provide bases of attachment for sessile animals but furnish shelter for others that adhere to the rock, and protect them from desiccation. (Several of the more common Pacific coast algae are illustrated on pp. 407–15.)

In the highest zone one can collect dry-shod, but in this zone and downwards wet feet are likely to penalize the un-rubber-booted observer. However, it is sometimes safer to wear sneakers, or even old leather shoes, in an area heavily overgrown with slippery seaweeds, since rubber boots are somewhat clumsy. Even the coldest ocean seems to become warmer after a bit of immersion. It is a good rule, when exploring an unfamiliar seashore, to use one hand and both feet to assure stability.

§10. *Littorina scutulata* (Fig. 8), one of the periwinkles, takes up the burden here, where *L. planaxis* leaves off, and covers the territory down to the lower reaches of the upper tide-pool zone, where *Tegula* becomes the dominant snail. The two species of *Littorina* may be found side by side; but of a thousand individuals, more than 80 per cent will indicate this depth zoning. *L. scutulata,* for instance, never occurs more than a few feet above high-tide line, whereas in surfy regions *L. planaxis* may be found 20 feet above the water. The two may be differentiated easily. *L. planaxis* has an aperture bounded at the inner margin by a large flattened area; it is dingy gray in color, chunky in outline, and comparatively large (up to ¾ inch high). The smaller, daintier *L. scutulata* is slim, has one more whorl to the shell, lacks the flattened area, and may be prettily checkered with white on shining brown or black. The range of *L. scutulata* is from Alaska, and that of *L. planaxis* from Washington, south to Baja California.

Certain of the rockweeds (*Pelvetia* or *Fucus*) serve the young periwinkles as a sort of nursery, for it is on their fronds and stems that the young will nearly always be found.

§11. A number of limpets find their optimum conditions in this zone, and scattered individuals may be found of others, like *Acmaea digitalis* (§6, Figs. 13, 16) or types more abundant lower down (§25). A large, high-peaked, spectacular form, *A. persona,* occurs high up, sometimes even in the uppermost zone; but in dark places, as in deep crevices or

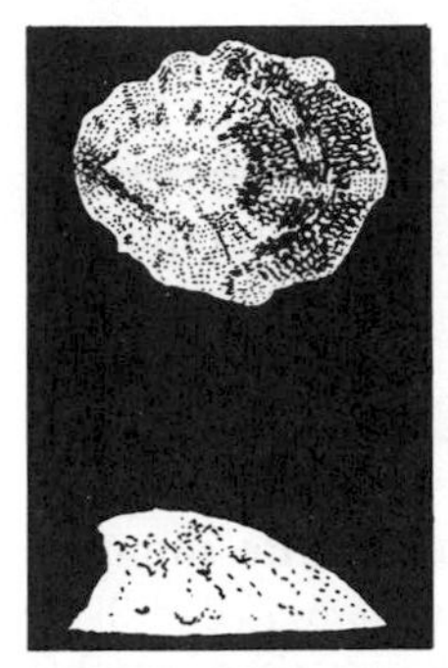
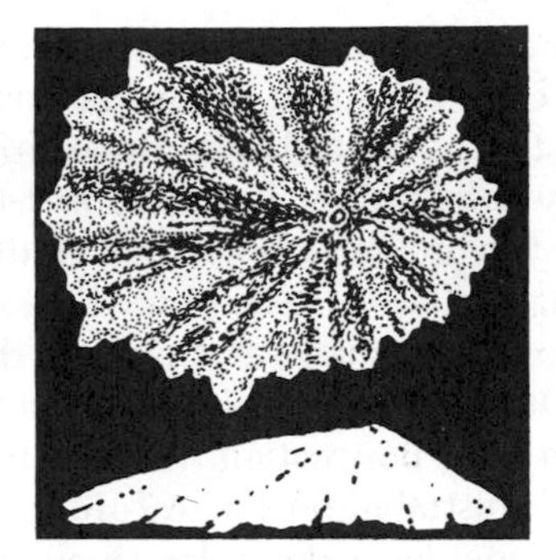
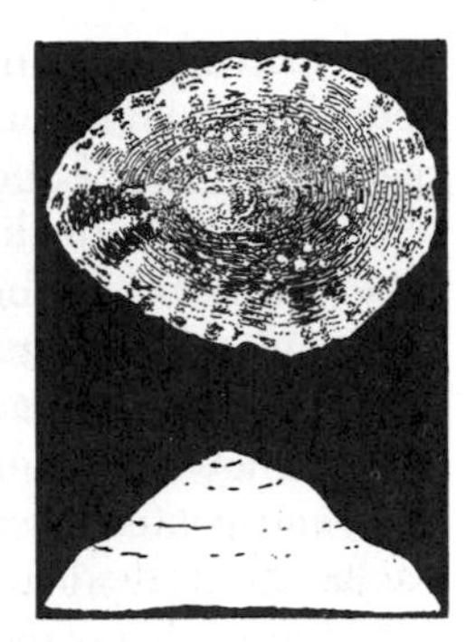

16. Three common limpets of the genus *Acmaea,* all shown close to natural size: left, *A. digitalis* (§11); center, *A. scabra* (§11); right, *A. pelta* (§11).

(particularly) on the roofs of caverns, it emerges to feed only at night. Mrs. Grant* says of it: "Exterior color olivaceous, variegated by numerous fine white dots sprinkled over surface. Larger white dots on anterolateral area make definite butterfly pattern in anterior view." The range is from the Aleutians to Monterey, and an average specimen is 34 × 25 × 16 mm.

The dainty and relatively flat *A. scabra* (Fig. 16), the ribbed limpet, ranging from north central California to Cape San Lucas, is regularly and deeply scalloped. It may be locally abundant on "rough rocks otherwise sparsely populated" (Grant); its average dimensions are 19 × 15 × 7 mm.

The brown and white shield limpet *A. pelta* (Fig. 16) occurs in this zone also, but around Monterey it may be taken far out. North of Crescent City it occurs high up with *A. digitalis.* Its distribution probably includes the entire littoral, but the submerged specimens are solitary, although larger than the average. *A. pelta*'s range is from the Aleutians south to Punta Santo Tomas, and a typical specimen measures 38 × 27 × 27 mm. The largest specimens are said to occur on the large, immovable rocks of the reefs, which are frequently covered by masses of *Ulva,* the sea lettuce, and on the surf-swept sea palm, *Postelsia.*

All the limpets illustrated are typical and representative specimens of their species, and show such marked characteristics of form and texture

* A. R. Grant, "The genus *Acmaea,*" Ph.D. thesis, Department of Zoology, University of California, 1938. Parts of this have also been published under the name A. R. (Grant) Test. Frederick Test (Mrs. Grant's second husband) also published a paper on *Acmaea.* The marriage has been a success.

that identification, from the illustrations, of equally typical specimens seen in life should be easy. However, the collector is quite likely to find occasional specimens that seem to have grown up in indecision about just which species they would favor. It is possible to arrange the shells of a large number of limpets so that it is difficult to tell where one variety ends and the other begins.

§12. The pleasant and absurd hermit crabs are the clowns of the tide pools. They rush about on the floors and sides of the rock pools, withdrawing instantly into their borrowed or stolen shells and dropping to the bottom at the least sign of danger. Picked up, they remain in the shell; but if allowed to rest quietly in the palm of one's hand, they quickly protrude and walk about. Among themselves, when they are not busy scavenging or love-making, the gregarious "hermits" fight with tireless enthusiasm tempered with caution. Despite the seeming viciousness of their battles, none, apparently, are ever injured. When the vanquished has been surprised or frightened into withdrawing his soft body from his shell, he is allowed to dart back into it, or at least to snap his hindquarters into the shell discarded by his conqueror.

It is a moot question whether or not hermit crabs have the grace to wait until a snail is overcome by some fatal calamity before making off with its shell. Many observers suppose that the house-hunting hermit may be the very calamity responsible for the snail's demise, in which case the hermit would obtain a meal and a home by one master stroke. The British naturalist J. H. Orton, however, determined conclusively that the common British hermits are unable to ingest large food, and that their fare consists entirely of microscopic, or at the most very small, particles of food and debris. In the Monterey Bay region a great many small hermits use the shells of *Tegula* of two species, snails that are strong, tough, and very ready to retreat inside their shells, closing the entrance with the horny operculum. It would seem that ingress to a *Tegula* shell occupied by a living animal determined to sell out as dearly as possible would be very difficult, even if the hermit ate its way in. Attacks have actually been observed in which the snail, by sawing the rough edge of its shell back and forth across the hermit's claw, convinced the hermit that that particular shell was not for rent.

Even when the hermit finds a suitable house, his troubles are far from over, for at intervals he must find larger shells to accommodate his growing body. It is a lifelong job and one in which he seems never to lose interest. A few hermits and some spare shells in an aquarium will provide the watcher with hours of amusement. Inspecting a new shell—and every shell or similar object is a prospect—involves an unvarying sacred ritual: touch it, grasp it, rotate it until the orifice is in position to be explored with the antennae, and then, if it seems satisfactory, move in. The actual move is made so quickly that only a sharp eye can follow it. One experimenter deprived some hermits of their shells and placed them near shells that had been stopped up with plaster. After much fruitless activity the animals finally ignored the useless shells. But when shells of another kind, likewise closed with plaster, were put in the aquarium, they recommenced their active and systematic efforts to obtain homes.

Hermits will eat any available animal matter, living or long dead. In aquaria we have watched them nip the tube feet from living sea urchins until the urchins were practically denuded.

Along the California coast the commonest hermit in the upper tide pools is *Pagurus samuelis* (see color section), a small crab with bright red antennae and brilliant blue bands around the tips of its feet. At Pacific Grove there are 10 of these to every one of the granular-clawed *P. granosimanus* (§196), but in the north the latter is larger and more abundant, apparently having an efficient or dominant range only in Puget Sound and to the northward, although its total range is from Alaska to Mexico. *P. samuelis,* on the other hand, having about the same total range, is most numerous along the California coast. We have taken ovigerous (egg-bearing) specimens of *P. samuelis* from early May through July, and of *P. granosimanus* in late May, both in California. Eggs, as shown in the photograph of another hermit (Fig. 17), are carried on the female's abdomen—a region always protected by the shell, but provided with a continuous current of water for aeration by the action of the abdominal appendages.

§13. The original residents of the homes so coveted by hermit crabs, the turban snails of the genus *Tegula,* often congregate in great clusters in crevices or on the sides of small boulders, especially at night or during still, cloudy days. The shells are pretty enough when wet; dry, they have a dingy gray color that blends very well with the dry rocks. The black turban, *Tegula funebralis* (Fig. 18), ranging from Sitka to Baja California, is the dominant type in this upper level. Lower down it is replaced by the brown turban (§29), which more often than the black turban carries specimens of the commensal slipper shell *Crepidula.* Both species frequently carry a small black limpet, *Acmaea asmi.*

Italians consider these snails fine food. The animals are cooked in oil and served in the shell, the bodies being removed from the shells with a pin as they are eaten. The lucky Americans who can overcome their food

17. An egg-bearing female hermit crab, *Pagurus hemphilli* (§12).

18. Black turban snails, *Tegula funebralis* (§13). The wet shells at the left are a brilliant black, the others an inconspicuous gray.

prejudices in this and other respects will at least achieve a greater gastronomic independence and may even develop an epicurean appreciation of many of the intertidal delicacies. They will also contribute to obliterating the fauna of such localities as Moss Beach. Studies by Peter Frank at Cape Arago suggest that *Tegula* may live 25 years; such a long-lived, slow-growing snail cannot withstand an intensive fishery!

§14. A third obvious and distinctly "visible" animal occurring in this zone is the pugnacious little rock crab *Pachygrapsus crassipes,* without which any rocky beach must seem lonesome and quiet (see color section). These dark red or green, square-shelled crabs run sideways or backward when alarmed, scurrying away before the intruder or rearing up and offering to fight all comers. They may be found on top of the rocks at night and hiding in crevices and pools during the day. To see a group of them attack a discarded apple core is to understand one method by which the rock pools are kept clear of all foreign matter that is to any degree edible. These crabs are common in the rocky intertidal from Coos Bay, Oregon (the northern limit seems to be Yaquina Head), to the Gulf of California. They and the beach hoppers are the most active scavengers in this particular ecological association.

Pachygrapsus females carry their eggs, attached to the underside of the recurved tail, until the late larval stages; just how long is not known. Egg-carrying specimens have been taken in the Monterey region in such divergent seasons as June and February.

§15. Collectors in the La Jolla region will find, on and under fairly bare rocks on the upper shore, the small, clean-cut, and rather beautifully marked volcano shell limpet *Fissurella volcano.* For some obscure reason this animal moves down into the low intertidal as it ranges northward. It has been reported from Crescent City to Panama. *Fissurella's* shell is pink between the red or purple ridges that radiate from the small opening at the apex.

§16. In the muddy or sandy layer between closely adjacent rocks, or in the substratum, the upper tide pools are characterized by the presence of the hairy-gilled worm, *Cirriformia luxuriosa* (formerly *Cirratulus,* Fig. 19), which may be found in bits of the blackest and foulest mud. The gills of this worm, which resemble thin, pink roundworms, will often be noticed on the substratum. They are withdrawn immediately when the animal is disturbed. This method of respiration is particularly efficient under the circumstances; it gives the worm an opportunity to burrow in the mud in search of food (and protection) while leaving its "lungs" behind, at the surface. *Cirriformia* is thus one of the few animals that can live in an environment of foul mud; the noxious substratum that repels others protects this animal from enemies and desiccation. An English species is known to feed on small food particles like one-celled plants and the spores of algae, which are selected from the mud by delicate sensory flaps on the mouth and carried into the digestive tract proper by minute waving hairs.

Our collecting records show that sexually mature specimens of *Cirriformia* have been observed in June under interesting conditions. We quote from a report dated June 21, 1927, from 9:30 to 10:30 P.M.: "For the past several evenings I have noticed that the *Cirriformia* were very active in mud-bottom pools of the upper tide flats. Some of them lie half exposed, whereas only the tentacles are protruded normally. Tonight I have noticed further that the shallow pools containing them were murky and milky. Then I noted some of the worms almost fully out of the substratum, exuding milky fluid or more solid white 'castings.' I took two worms back to the laboratory, examined the exudate, and found it to be spermatic fluid." Eggs were not specifically observed at this time.

§17. Under flat rocks in the upper tide-pool region (but possibly as abundant in the middle tide pools) may be found the polyclad flatworm *Notoplana acticola* (Fig. 20). It is tan in color, with darker markings around the midline. Large specimens may exceed an inch in length, although the average will be nearer ½ inch. By examining the pharynx contents Dr. Heath found that the food of many flatworms consists of one-celled plants, spores of algae, micro-crustacea, and worm larvae. Miss Boone found that captive *N. acticola* fed on the red nudibranch *Rostanga*.

These worms (and other species) crawl about on the damp undersides of freshly moved rocks, appearing to move much as a drop of glycerine flows down the side of a glass dish. This unique motion is achieved by means of cilia, whose grip is so effective that it is difficult to remove the soft-bodied worms from the rock without damage. We use a camel's-hair brush, to which they adhere as firmly as to the rock unless they are immediately cast into a jar of water.

§18. From British Columbia to the Gulf of California *Petrolisthes cinctipes*, the porcelain crab (Fig. 21), is also a common and characteristic inhabitant of the under-rock association, although, like *Notoplana*, it is almost as characteristic of the middle zone. A large specimen will measure about 9/16 inch across the carapace. It is these small, flat crabs that scurry about so feverishly for cover when a stone is upended. Their pronounced flatness makes it easy for them to creep into crevices, and they avoid imprisonment by throwing off, or autotomizing, a claw or walking leg at the slightest provocation. Indeed, *Petrolisthes* is one of

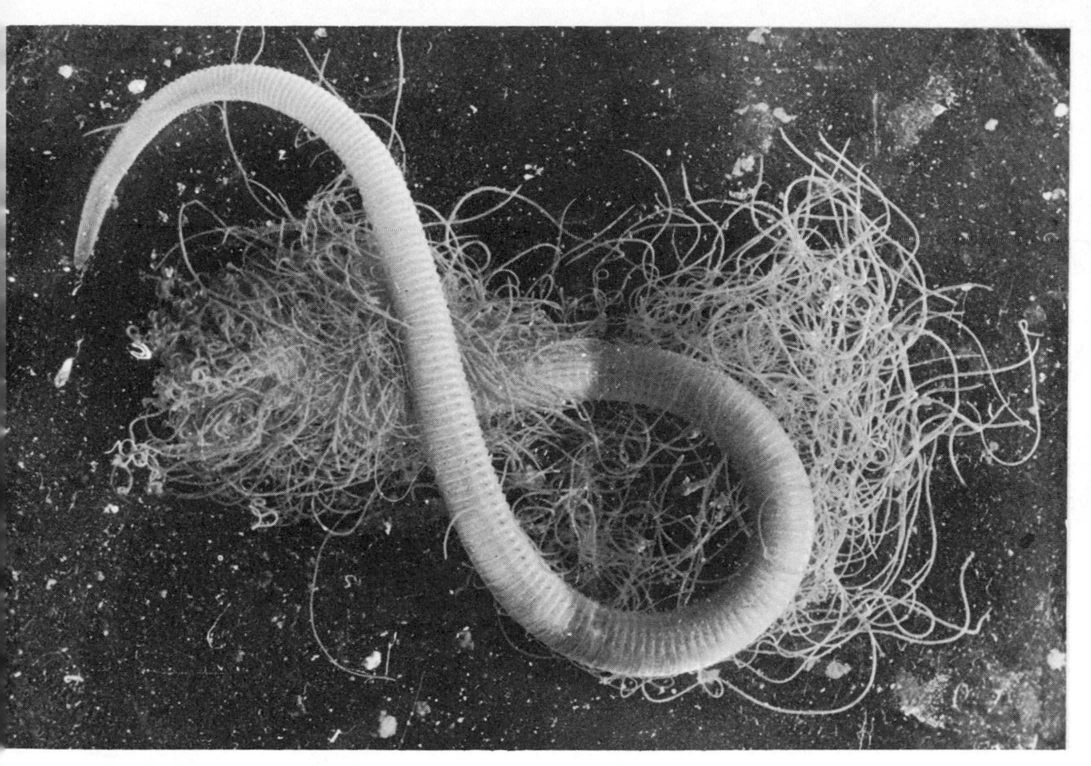

19. The hairy-gilled worm *Cirriformia luxuriosa* (§16).

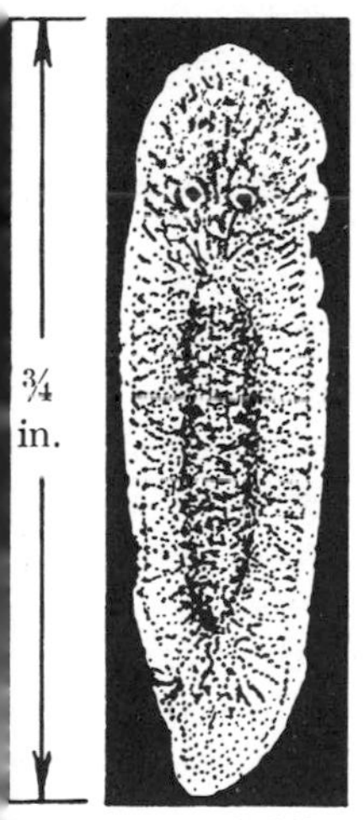

20. The polyclad flatworm *Notoplana acticola* (§17).

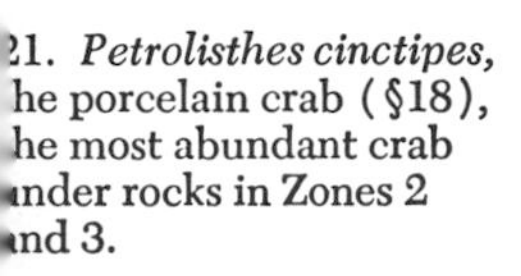

21. *Petrolisthes cinctipes,* he porcelain crab (§18), he most abundant crab under rocks in Zones 2 and 3.

the champion autotomizers of this coast—no mean distinction, for many animals have this peculiar ability.

In autotomizing, a crab casts off its limb voluntarily; an automatic reaction instantly closes the severed blood vessels to prevent bleeding to death. This process takes place at a prearranged breaking point near the base of the limb, marked externally by a ringlike groove. The muscles and tendons there are adapted to facilitate breakage; and after autotomy a membrane consisting of two flaps is forced across the stump by the congestion of blood at that point. Regeneration begins at once, and a miniature limb is soon formed, though the new limb undergoes no further growth until the next molt, when it grows rapidly to several times its previous size.

Female crabs will often be found with large egg masses under their curved and extended abdomens (we have taken ovigerous *Petrolisthes* in California in March, May, and June). These eggs do not hatch into young crabs, but into fantastic larval forms that the shore collector, as such, will never see. In the first larval stage, called zoea, the animals are minute, transparent organisms swimming at the surface of the sea. They seem so totally unrelated to their parents and to their own later stages that zoologists long mistook them for an independent species of animal. After casting its skin several times as it grows in size, the zoea changes to still another form, the megalops (also formerly regarded as a separate genus, *Megalopa*). The megalops, related to a crab in much the same way that a tadpole is related to a frog, casts its skin several times, finally emerging as a recognizable young crab. The pelagic zoea of *Petrolisthes*, somewhat resembling a preposterous unicorn, is a wondrous sight under a microscope. Projecting from the front of the carapace is an enormous spine, often longer than all the rest of the animal. If this spine has any function, it is, presumably, to make ingestion by the zoea's enemies difficult. An enterprising pup that had just attacked its first porcupine would understand the principle involved. There are also two long posterior spines on each side, which protect the larva from behind.

Petrolisthes has in recent years become a favorite in the curio business, since its flat shape makes it an attractive subject for embedding in blocks of plastic for key ring dingbats, pendants, and the like. The bright red color of these embalmed specimens is due to a preservative or other treatment.

§19. Of the two exceedingly common and shoreward-extending brittle stars, the dainty black-and-white *Amphipholis pugetana* (Fig. 22) is the higher, being characteristically an upper tide-pool animal. Since large adults measure only about ¾ inch in arm spread, it is further distinguished

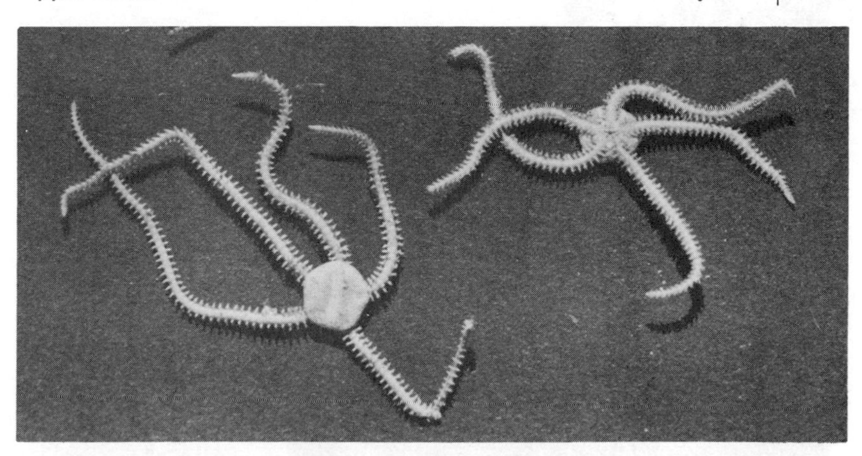

22. Dorsal and ventral views of *Amphipholis pugetana* (§19), the smallest brittle star of the Pacific coast.

by being the smallest of all Pacific coast ophiurans, although it is a rather hardy form that does not autotomize readily, unlike many of the other brittle stars. *Amphipholis* occurs in great beds, so commonly that it would be difficult to find a suitable rock in this zone that hid no specimens; but it is never found with intertwined arms, as is its larger relative *Amphiodia* (§44). The possible significance of this concentration of population is discussed in connection with the latter form. The recorded range of *A. pugetana* is from Alaska to San Diego; but it extends further south, for we have also taken it (and possibly the similar *A. squamata*) in northern Baja California.

§20. Representatives of two related groups of crustaceans, the amphipods and the isopods, are found under practically every embedded rock and boulder in the upper tide-pool zone, and further down also. The amphipods, a group containing the sand hoppers or beach fleas, are, as their popular names imply, lively animals. Being compressed laterally, so that they seem to stand on edge, they are structurally adapted to the jumping careers that they follow with such seeming enthusiasm. *Melita palmata* (Fig. 23) is almost certain to be present under most of the suitable rocks down as far as the middle tide pools. The amphipods of the Pacific coast have been worked over very little; and the collector who would identify his species has before him a weary search through scattered literature unless he enlists the aid of the National Museum.

§21. The isopods, being flattened bellywise, or dorso-ventrally, are crawlers, usually slow ones. These are the pill bugs, so called from their

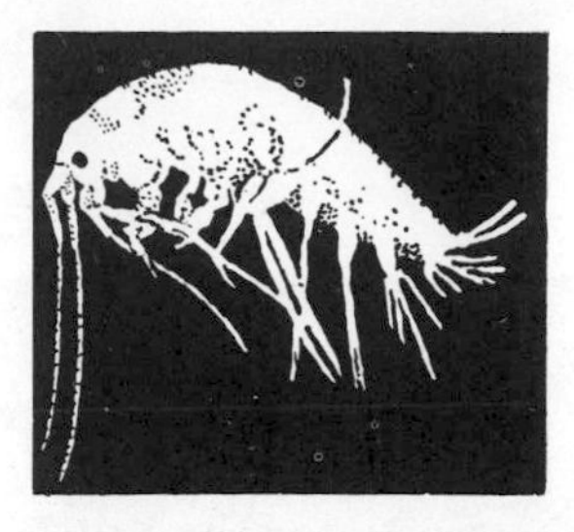

23. The beach hopper *Melita palmata* (§20), a common amphipod of the high intertidal zone.

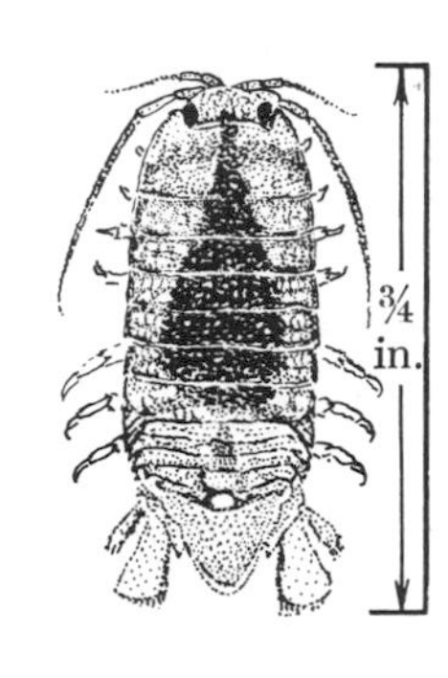

24. Photograph and drawing of *Cirolana harfordi* (§21), an isopod common along the entire Pacific coast.

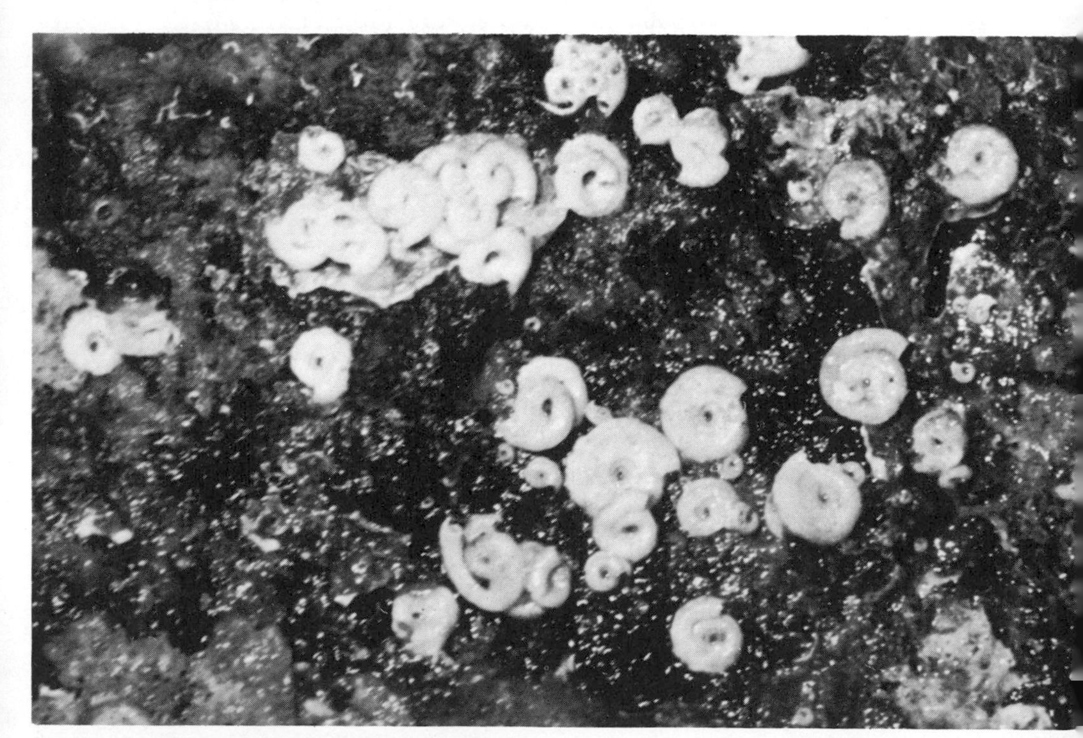

25. Tubes of the annelid worm *Spirorbis* (§22).

habit of curling up into little balls when disturbed; they are also known as sow bugs because of their resemblance to swine. Representatives of the group are common on land, where they are found in damp places—under boards lying on the ground, for example. The drab *Cirolana harfordi* (Fig. 24), a common inhabitant of the entire coastline from British Columbia to at least Ensenada, occurs most abundantly under rocks in the high intertidal but will be found in other zones as well. The isopods are fairly well known.

§22. Small round boulders that are always immersed in the pools are very likely to be covered with the tiny, calcareous tubes of a serpulid worm, the sinistral *Spirorbis* (Fig. 25). The almost microscopic red gills protrude from the tubes to form dots of color, which disappear instantly whenever the stone is disturbed. *Spirorbis* is not unique in being hermaphroditic, but its bisexuality takes an interesting form in that its anterior abdominal segments are female and its posterior segments male. It is thought that the worm's operculum may serve as a brood pouch. The eggs, which are extruded in February, develop into free-swimming larvae. In a related form (*Serpula,* §213) the sexes are separate.

§23. Other less common or less characteristic animals will sometimes be found in this zone. The purple snail *Thais* may be locally abundant in crevices and along the lower edges of rocks, but larger and more noticeable examples are found on the open coast (§182). Coincident with the start of algae beds, and so abundant as to be a feature of the tidal landscape, are extensive beds of a warty anemone, which is, however, more characteristic of the next lower zone (§24). A very small six-rayed starfish will now and then be seen in the pools (§30), together with small snails and hosts of red flatworms (§59). Even an occasional ribbon worm may be turned out of its home under some rock; but the ribbon worms, too, reach their maximum development in other zones and are treated elsewhere.

The animals of the uppermost beach and the upper tide pools are interesting in their resistance to some of the "un-marine" conditions that are a part of the environment. During rainy seasons they are drenched with fresh water. Violent temperature changes suddenly descend on them when the cool tide sweeps in over sun-heated rocks. When the sun beats on isolated pools at low tide, evaporation increases the salinity of the water. All this is in addition to the usual problems of wave shock, respiration, and drying that all the intertidal animals must meet. Animals successfully adjusting themselves to these conditions must be hard-boiled indeed, and it is not surprising that successful ones are found throughout ranges of literally thousands of miles of variable coastline.

2. *Middle Intertidal*

THIS area, Zone 3, is below mean sea level and is largely uncovered by most low tides and covered by most high tides. The higher zones seem relatively barren by comparison with the teeming life of this mid-region, which is exceeded in density only by that of the outer tidelands. Here plants grow in great profusion. There may be a bit of *Pelvetia* in the upper parts; *Fucus* is more likely to occur; *Gigartina* and *Porphyra* grow lush, with the sea lettuce *Ulva.* These plants, in providing protection from deadly sunlight and drying, are of importance to the animals, which are predominantly intertidal forms and rarely occur in deep water (Figs. 26, 27).

The variety of animals associated in this zone is so great that it seems desirable to consider them according to a further division of habitat, as follows: exposed rock; protected rock; under-rock; burrowing; pool; root and holdfast.

Exposed Rock Habitat

These are animals living on the tops and sides of rocks and ledges, exposed to waves, sun, and wind.

§24. Extensive beds of a small verrucose anemone, the aggregated *Anthopleura elegantissima* (often considered a size and color phase of the great green anemone, §64), occur on the exposed surfaces of rocks, especially where sand is being deposited. It is beds of these hardy animals, certainly the most abundant anemone on the coast, that make the

26. A view of the characteristic Zone 2 alga *Pelvetia,* with a dash of *Fucus.*

27. A rich area of protected outer coast at Carmel; Ed Ricketts collecting.

intertidal rocks so "squishy" underfoot, the pressure of a step causing myriad jets of water to issue from their mouths and from pores in their body walls. These anemones commonly cover themselves with bits of gravel or shell, so that when contracted they are part of the background. Many a weary collector has rested on a bit of innocent-looking rock ledge, only to discover that an inch or so of extreme and very contagious wetness separated him from the bare rock (Fig. 28).

Although this anemone reproduces sexually, from gametes released by male and female, it also reproduces by budding. Perhaps it is more successful without sex: areas of rocky surface are evidently colonized by budding, since the anemones occur in patches of one sex or the other.

These animals are not restricted to pure seawater, for they live well where there is industrial pollution or sewage, their hardihood belying their apparent delicacy. They can be kept in marine aquaria, even where there is no running seawater, provided only that the water is aerated or changed once or twice a day. In keeping with its exposed position high in the intertidal zone, this *Anthopleura* expands fully in the brightest sun-

28. A dense cover of the aggregated anemone *Anthopleura elegantissima* (§24) on a large rock.

shine, closing up at night. Contracted, it is rarely more than 1½ inches in diameter; expanded, the disk may be more than twice that size and may display rather dainty colors over a groundwork of solid green. The tentacles are often banded with pink or lavender, and lines of color frequently radiate from the mouth to the edges of the transparent disk, marking the position of the partitions (mesenteries) inside. We have found this form abundant from San Luis Obispo northward into British Columbia. South of central California its place is often, but not always, taken by small specimens of the giant green anemone, which typically attains its maximum development farther down in the intertidal zone.

Anemones are voracious feeders and when hungry will ingest almost anything offered, even apparently clean stones. Their tentacles are armed with minute stinging structures called nematocysts, which paralyze any small animals so unfortunate as to touch them (the nematocysts are often erroneously called nettle, or stinging, cells). We have many times seen small shore crabs disappearing down the anemone's gullet, and as often seen an anemone spew out bits of the crabs' shells. The surprising rapidity of an anemone's digestion was illustrated by specimens of another species, *Metridium* (§321), with which we experimented at Pender Harbor, B.C. One half of a small chiton that had been cut in two was placed on the disk of an anemone in a glass dish. The disk was immediately depressed at the center, so that the food disappeared into the body cavity. Within 15 minutes the cleaned shell of the chiton, all the meat completely digested away, was disgorged through the mouth—a striking tribute to the potency of *Metridium*'s enzymes.* And this action takes place at about the temperature of seawater! Those of us troubled with dyspepsia can very well envy animals so equipped, looking forward to the time when we may be able to help along an inadequate digestion with artificially extracted and properly administered anemone enzymes.

Not much is known about the anemone's rate of growth, which presumably varies directly with the food supply; but several examples of extreme longevity have been observed in this group. Professor Ashworth tells of several anemones that were donated to the University of Edinburgh in 1900 by a woman who had collected them, already fully grown, 30 years before. She had kept them throughout that period in a round glass aquarium, strictly observing a daily rite of aerating the water with a dipper and a weekly rite of feeding them on fresh liver, which she be-

* See Edge (1934, "Digestive Rates of Marine Invertebrates," *American Midland Naturalist*, Vol. 15: 187–89). Edge found that digestion required 2–12½ hours in most forms investigated, action taking place most rapidly in the actively voracious animals.

lieved they preferred to anything else. When they came into the possession of the University they were fed, possibly not so regularly, any scraps of protein that came to hand, such as shredded crab meat, or even beefsteak. Nevertheless, these anemones continued on in the best of health, annually producing clouds of sperm and eggs. Unfortunately, this experiment in longevity was brought to an untimely end after some 80 years by the ineptitude of (we understand) a botanist; but there seemed to be no reason why the animals might not have lived a century or more.

§25. Also occurring more or less exposed on the tops and sides of rocks are a number of forms more properly treated elsewhere. Occasional acorn barnacles extend their range down into this middle zone when competition is not too keen, and such open-coast forms as the red barnacle (§168) and the common tough-skinned seastar (§157) may be locally plentiful, depending on the exposure. There may also be clusters of mussels and goose barnacles, and the small chiton, or sea cradle, *Nuttallina* is occasionally found.

Three common limpets may be found at this level on the protected outer coast. The large *Acmaea scutum* (Fig. 29), formerly *A. patina*, the plate limpet, is the flattest of the tribe. It is frequently green, often with long trailers of the stringy green algae *Enteromorpha* and *Ulva* attached to its shell, and ranges north from Point Conception clear into the Bering Sea. *A. scutum* is one of the commonest large limpets in northern California. South of Monterey, however, according to Mrs. Grant, it becomes increasingly rare; and south of Point Conception it is replaced entirely by the *A. limatula* mentioned below, former literary references to the contrary notwithstanding (the southern *A. scutum* identifications were incorrect). Both species are thin and flat, but *A. scutum* particularly has a knife-edge shell easily broken in removal, and it lacks the black sides of the foot so characteristic of the other. An average specimen of *A. scutum* measures 33 × 25 × 9 mm.

A. limatula (Fig. 29), frequently called *A. scabra* in the past, is sculptured with coarse and imbricated ribs (hence the name "file limpet") that are sometimes obscured by algae. A diagnostic feature of this species is the characteristic deep black color around the side of the foot, the bottom being contrastingly white. It is reported to range from Vancouver Island to Central America, becoming increasingly common south of Monterey. The average specimen is 29 × 23 × 9 mm.

There is in the south a species related to *A. scabra*, described by Dr. A. R. (Grant) Test as *Acmaea conus*. It is ivory-colored to brown and, with *A. limatula*, replaces *A. scutum* on all reefs south of Point Conception. Its average dimensions are 15 × 23 × 6 mm.

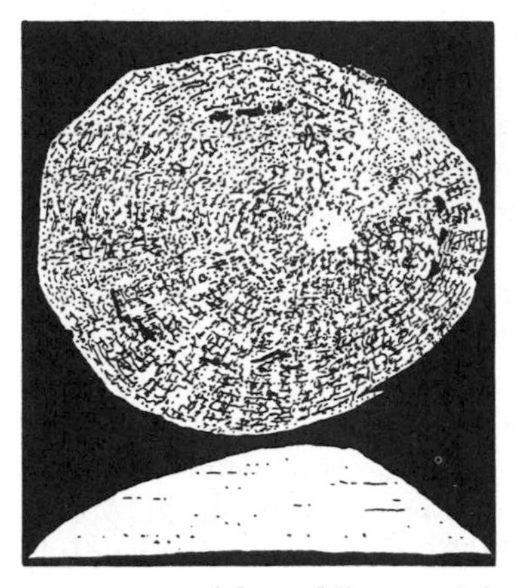

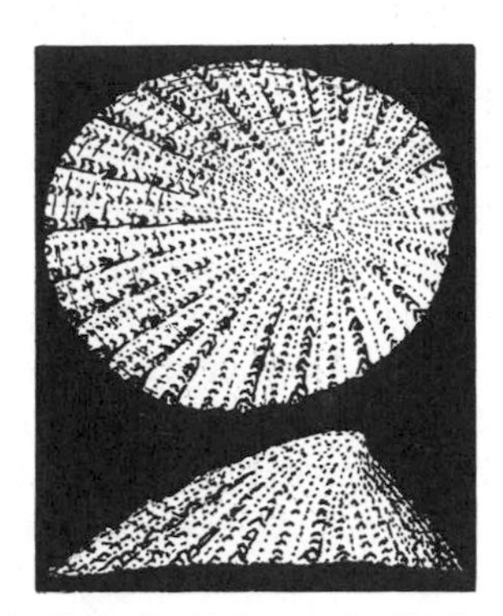

29. Limpets of the middle zone: left, *Acmaea scutum* (§25); right, *A. limatula* (§25).

Occasional specimens of any of the limpets previously mentioned (§§6 and 11) may be turned up here; *A. fenestrata* and *A. mitra* of the lowest horizon (§117) are also found.

Protected Rock Habitat

Animals here are attached to, or living on, protected rocks and rockweed. Many of the commonest inhabitants of this subregion can scarcely be considered obvious, since in order to see them advantageously one must almost lie down in the pools, looking up into the crevices and under ledges.

§26. The large red, yellow, or purple seastar *Patiria miniata* (color section, Fig. 30), locally called "sea bat" because of its webbed rays, is quite possibly the most obvious animal in this association, extending downward to deep water. Because it is readily available, because it has an unusually long breeding season, and because it extrudes its sexual products so obligingly, *Patiria* is used extensively for embryological experimentation. Most male and female specimens, when laid out on wet seaweed, will discharge their ripe sperm and eggs at almost any time of the year, but especially during the winter. With ordinary aseptic precautions, the ova can be fertilized by fresh sperm; and if spread out in scrupulously clean glass dishes, they will develop overnight into minute larvae, which swim by vibrating their cilia. The young seastars develop inside these larvae, gradually absorbing the larval form as they grow. After the adult specimens have shed their sexual products, they should be put back alive into the tide pools to prevent depletion of the race.

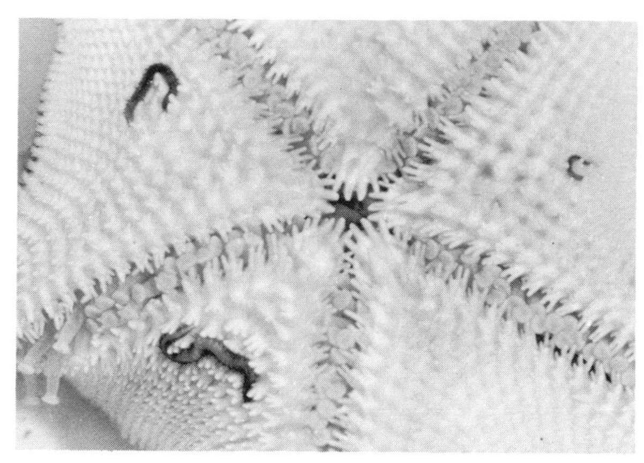

30. Ventral surface of *Patiria miniata* (§26), the bat star, showing the commensal polychaete *Podarke pugettensis.*

In the Monterey region, and probably because of collecting activities, this interesting animal is no longer found in the great congregations that formerly occupied the middle tide pools, although it still occurs in large numbers in the outermost pools and reefs. The recorded range is from Sitka to San Diego, and even to La Paz, Baja California; but south of Point Conception the specimens are few, small, and stunted, and are almost invariably found under rocks. The bat star is rare intertidally on the Oregon coast.

A polychaete, *Podarke pugettensis,* often lives commensally on the oral (or "under") surface of *Patiria.* It usually frequents the ambulacral grooves, but may crawl actively over the whole oral surface. There may be several worms on an individual seastar. In southern California the worm *Podarke* may live independently of its host. Davenport has shown that *Podarke* evidently finds a suitable host by detecting some essence of *Patiria* by chemosensory means.

§27. The little square-shelled rock crab *Pachygrapsus* is still abundant, but the purple shore crab *Hemigrapsus nudus* (see color section) is the dominant representative of this group in the middle tide-pool region, and it also extends into the next lower zone. *Hemigrapsus* is as characteristic of the rockweed zone as *Pachygrapsus* is of the naked rock zone higher up, or as *Cancer antennarius* (§105) is of the lower tide-pool zone. The purple spots on the claws of *Hemigrapsus,* particularly noticeable on their white undersides, differentiate it from the smaller and slimmer *Pachygrapsus.* Egg-bearing females of *Hemigrapsus* have been taken in November, December, and February in Monterey Bay, but in the Puget Sound area the females are ovigerous in early summer. The range is from Sitka to the Gulf of California.

§28. Hermit crabs are also abundant in this zone. Specimens of *Pagurus samuelis* larger than those encountered in the upper tide-pool region (§12) can be found, but even the large *P. samuelis* are small in comparison with the average *Pagurus hemphilli* (Fig. 17), a species characteristic of the middle and lower tide-pool regions. The stalwart and clean-cut *P. hemphilli,* up to 2 inches long and colored a fairly uniform straw tan, can be recognized by the laterally compressed wrist of its big claw, which subtends a sharp angle with the upper surface. Other hermits have the top of the wrist more or less rounded. Egg-bearing females of *P. hemphilli* have been taken in February and March. This hermit ranges from British Columbia to Monterey, occupying the shells of large *Tegula* almost to the exclusion of anything else.

Hermit crabs are sometimes afflicted with a parasite, *Peltogaster,* that produces a long white sac on the abdomen. This is actually a specialized barnacle, which gains access to its host as a free-swimming nauplius larva scarcely distinguishable from the nauplii of other barnacles. Once attached, the animal loses all its appendages and becomes a sac of generative organs devoted solely to the production of eggs and sperm, food and respiration being provided by the host. Hermit crabs so afflicted are not common on the Pacific coast of North America, and they are not easy to find, since the hermit's shell must be crushed in a vise to determine whether or not the animal is affected.

§29. In the middle zone the brown turban snail, *Tegula brunnea* (Fig. 31), ranging from Mendocino to Baja California, takes the place of *T. funebralis* (§13), its black brother in the upper zone, although each overlaps the territory of the other to some extent. Both species are definitely larger in the middle zone, however, and both are more often found

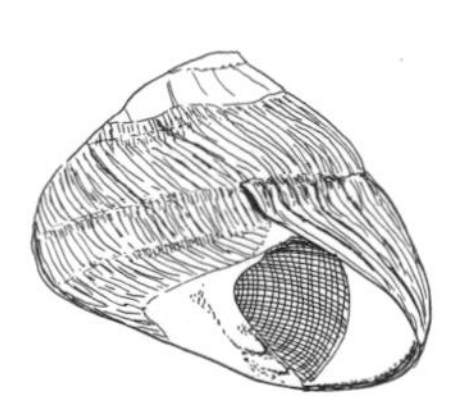

31. Turban snails of the genus *Tegula*: left, *T. funebralis* (§13), the black turban; right, *T. brunnea* (§29), the brown turban. *T. brunnea* is the more common in Zone 3.

carrying specimens of another snail, the horned slipper shell, or boat shell, *Crepidula adunca.* This small, dark brown form, with its sharply recurved hook, looks not unlike a limpet, and it affixes itself with limpet-like strength to the *Tegula* that carries it. The boat shell is known to occur from Vancouver to Cape San Lucas. Breeding individuals carry eggs in the shell cavity, where they can be seen if the animal is pried from its support. With a hand lens the embryos of *Crepidula* can sometimes be seen whirling around in their envelopes in the egg packets. We have taken specimens with the eggs at this stage throughout the spring months.

Atlantic coast forms of *Crepidula* have been used in the study of cell lineage, for the development of the animal from the two-celled stage follows, as Wilson has shown, a remarkable mathematical sequence. Ordinarily, it is impossible to say which part of a fertilized ovum will develop into a given organ. That is, development seems to be a matter of chance, the pattern being laid out as the form builds up. With *Crepidula,* however, it is possible to trace the ancestry of the cells and then, by reversing the process, to map out from the two-celled stage the regions that will develop into the various organs. In other words, the pattern has been laid out in advance, and the mathematical unfolding process can be predicted and observed. The throwing off of daughter cells is synchronous in the various regions, the cleavages being alternately to the right and to the left.*

An Eastern form, *Crepidula fornicata,* was introduced on the English coast about 1880, along with American oysters, and its habits and natural history have been studied there by Orton. It is considered an enemy of the oyster beds on the Essex coast because it uses the same kind of food in the same way. "Water is drawn in and expelled at the front end of the shell, the ingoing current entering on the left side, passing over the back of the animal and out at the right side. . . . Between the ingoing and outgoing current the gill of the animal acts as a strainer, which collects all the food material that occurs floating in the water." On its way toward the mouth the food is divided into two main batches, coarse and fine. The fine material, consisting of diatoms, etc., is collected in a cylindrical mass in a groove and ingested at intervals. The coarse particles are stored in a pouch in front of the mouth, so that the animal can feed whenever it wishes.

* Studies of cell lineage are no longer fashionable, and the beautiful monographs with colored plates that justified Ph.D.'s at the turn of the century now rest unread in our libraries while more up-to-date biologists talk learnedly of information units stored in molecules of DNA. But the riddle of morphogenesis is still with us.

32. *Leptasterias pusilla* (§30).

Crepidula fornicata is a protandric hermaphrodite.* It is at first male, then bisexual for a time, and finally female. In America it breeds in April and May, but in its new English environment breeding is in March. Special care is taken of the spawn. The animal constructs 50 to 60 membranous bags, into each of which it passes some 250 eggs, afterward closing each bag and tying it with a short cord. All the cords are then attached to the surface on which the slipper shell sits (our Pacific form carries its eggs in the mantle cavity, as previously noted), and the 13,000 or so eggs are carefully protected until, in about a month, they hatch. Thereafter the young are free-swimming for two weeks.

Orton has found the adults sticking together in long chains, one on the back of the other, with the shells partially interlocked by means of projections and grooves. Old individuals are sometimes permanently attached to their support or to each other by a calcareous secretion of the foot. There are as many as 13 individuals in a chain—the lower ones old and female, the middle ones hermaphroditic, and the upper ones young and hence male. Orton allows 1 year to the individual; thus in such a chain the oldest would be 14 years old.

§30. *Leptasterias pusilla* (see Fig. 32), a dainty little six-rayed seastar with a total arm spread usually under 1 inch, may be very numerous in this zone, crawling about in the sea lettuce or dotting the tide pools. At

* Orton (*Proc. Malacol. Soc. London,* Vol. 28 [1950]: 168–84) reports on the recent spread of this species in England, and suggests that self-fertilization may occur, thus facilitating its continued extension.

night, specimens are likely to be seen moving around on top of the rocks that have hidden them during the day. Unlike their larger relative *L. hexactis,* of the lower tide pools, these seastars are delicate and clean-cut, usually light gray in color. The breeding habits of *L. pusilla* are famous. The mother broods the eggs and larvae in clusters around her mouth region until the larvae have attained adult form. Ovigerous females may be found in January and February, and the very minute, liberated offspring are seen in the tide pools during February and March. This species is found in central California, including San Mateo, Santa Cruz, and Monterey counties, where it is very common.

§31. Encrusting the sides of rocks with gay colors are sponges and tunicates of several types. *Haliclona permollis* (Fig. 33), formerly *Reneira cineria,* is the vivid purple, fairly soft, but not slimy encrustation common in this association during some years. It is recognizable as a sponge by the regularly spaced, volcano-like "craters," the oscula, that produce the porousness characteristic of all the animals of this group, including the common bath sponge. *Haliclona* is a form almost cosmopolitan on north-temperate shores, and is known from central California, Europe, and the Atlantic coast of the United States. A similar species occurs in Puget Sound (§198).

§32. There are likely to be several types of encrusting red sponges growing in narrow crevices and on the undersides of overhanging ledges. *Ophlitaspongia pennata* (Fig. 34) is a beautifully coral-red form characterized, especially after drying, by starry oscula; its surface is velvety. De Laubenfels remarks that it occurs clear up to the half-tide mark (higher up than any other sponge), especially on vertical rocks under pendant seaweed, hence shaded from direct sunlight.

Plocamia karykina (Fig. 35), which grows in bright red layers up to ¾ inch thick, is firm and woody, with a smooth surface. The oscula are usually large, irregular, and rather far apart. De Laubenfels says it "emits copious quantities of a colorless slime not conspicuous before injury." A common neighbor is the slightly different-looking *Isociona lithophoenix,* which is more of a vermilion red, softer, and with a very lumpy, almost papillate, surface. Its oscula are very small and appear in depressions. The microscopic spicules of *I. lithophoenix* are straight, spiny rods, instead of short, thick, double-headed affairs like those of *P. karykina.* The brilliantly splashed colors of these red sponges are a feature of rocky caverns and granite rock faces in this zone, particularly in the Monterey Bay region. *P. karykina* is known to range from Vancouver to Laguna Beach. Other sponges occasionally occurring are mentioned in §140.

33. Above, left: *Haliclona permollis* (§31), a purple sponge.

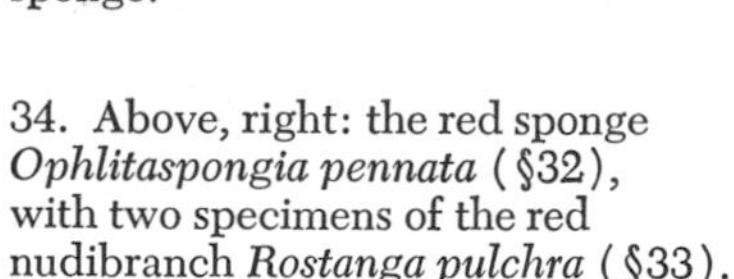

34. Above, right: the red sponge *Ophlitaspongia pennata* (§32), with two specimens of the red nudibranch *Rostanga pulchra* (§33).

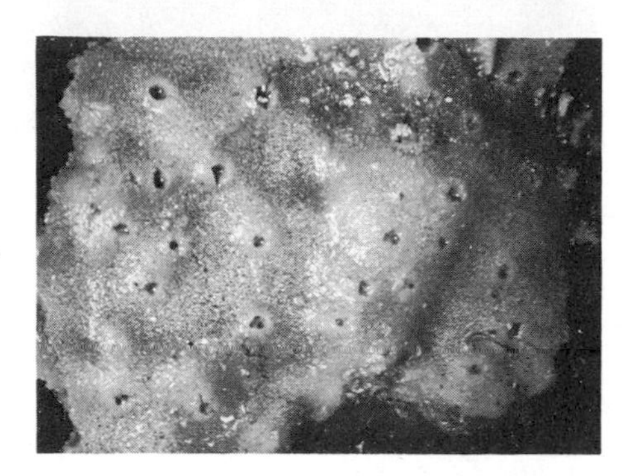

35. Right: the red sponge *Plocamia karykina* (§32).

§33. A vivid red nudibranch (shell-less snail), *Rostanga pulchra,* averaging ½ inch in length, is very commonly found feeding on red sponges —a good example of protective coloration if there is such a thing (two specimens are shown in the photograph of *Ophlitaspongia* (Fig. 34). Even *Rostanga's* egg ribbons, which may be found throughout the year attached to the sponge in a loose coil, are vividly red. *Rostanga pulchra* ranges from Vancouver Island to San Diego.

§34. Two other nudibranchs occur this high in the tide pools. One, *Triopha carpenteri* (Fig. 36), which is found in the Monterey Bay region and as far north as Oregon, can be seen crawling upside down suspended from the underside of the air-water surface film of pools. As it moves along the animal spins a path of slime, which prevents its weight from distorting and finally breaking the surface film. If the film is once broken, the animal sinks to the bottom of the pool, crawls up the side, and climbs to the surface film again—a long slow labor that may be watched to advantage in an aquarium. A good many nudibranchs, even the large ones, have this habit. *T. carpenteri* is seldom more than an inch

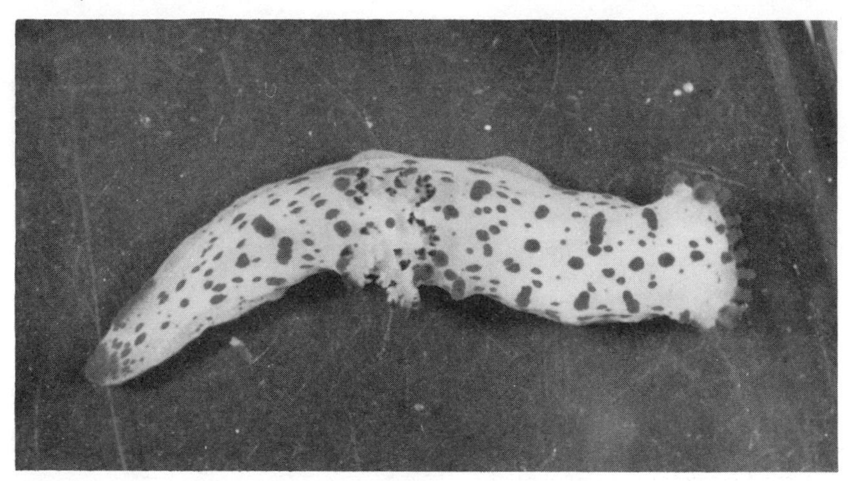

36. The nudibranch *Triopha carpenteri* (§34).

long, but the brilliant orange-red of its nipple-like appendages, contrasted with its white body, makes it very noticeable. Dumpy and sack-like in outline, it has none of the firmness of many other sea slugs and collapses very readily. It might be supposed that so tender-looking a morsel, apparently defenseless, would not last long among the voracious tide-pool animals, but for some reason as yet unknown the nudibranchs are avoided. Obviously this challenge to all enterprising human tasters could not go unanswered indefinitely. Professor Herdman took up the gauntlet by eating a vividly colored nudibranch alive. He reported that it had a pleasant oyster-like flavor, so the question remains open. McFarland says that the bright colors seem to serve as a warning. But more tasters are needed to find out why that warning is heeded. Certain special cases of inedibility among nudibranchs will be mentioned later.

§35. *Diaulula sandiegensis* (Fig. 37) occurs more frequently on the sides of boulders, or sometimes even under them. The colors are dainty rather than spectacular—consisting of as many as a dozen dark circles or spots on a light brown-gray background. Like other nudibranchs, *Diaulula* is hermaphroditic. At Monterey it has been observed depositing eggs, in broad white spiral bands, almost throughout the year, records being at hand for the summer, early October, January, and February. Apparently this is another of the Monterey Bay animals that has no restricted breeding season—probably a concomitant of the slight annual temperature variation at this point.

37. The nudibranch *Diaulula sandiegensis* (§35).

Diaulula and other dorids respire by means of a circlet of retractile gills at the posterior end; if the animal is left undisturbed, these flower out very beautifully. Never occurring in great numbers, but found steadily and regularly in the proper associations, *Diaulula* ranges from Unalaska to San Diego and probably further south. The specimens at Sitka are darker in color, gray brown, and larger than the average Monterey examples. Specimens 4 to 5 inches long are occasionally seen on the flats south of Dillon Beach.

§36. Some of the bolder chitons, or sea cradles, can be considered characteristic members of this zone. Most of them live under rocks, coming out to forage at night; but *Tonicella lineata* (formerly *Lepidochitona lineata*), the lined chiton (Fig. 38), is often in plain sight during the day. Although this is a small species, usually not much over an inch in length, it is the most strikingly beautiful sea cradle found in the Pacific coast intertidal. In semisheltered areas it maintains the high station (that is, high for a chiton) that characterizes *Nuttallina* (§169) on the surf-swept open coast. In southeastern Alaska *Tonicella lineata* is the most abundant shore chiton, but it occurs there with the leather chiton *Katharina*, which on the California coast is restricted to surf-swept areas. The lined chiton ranges from Alaska to Monterey and is reported to occur as far south as San Diego.

Another chiton, *Cyanoplax hartwegi* (§237), an oval, olive-green form, may be found almost at will, although not in great numbers, by lift-

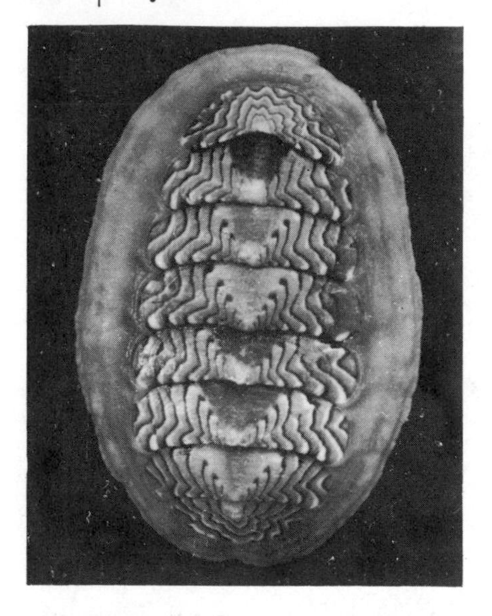

38. *Tonicella lineata* (§36), the lined chiton. This small species is one of the most distinctively marked on the Pacific coast.

ing up the clusters of *Pelvetia* on vertical rocks. In such places it spends its days, protected from sunlight and drying winds. It seems more at home, however, in quieter regions like Puget Sound.

The natural history of chitons is discussed in §76.

§37. The small anemone *Epiactis prolifera* (see color section) is locally abundant; it is usually found on the protected sides of small, smooth boulders and sometimes on the fronds of the sea lettuce, *Ulva*. This anemone is notable for its breeding habits. The eggs, instead of being discharged into the water, are retained in brood pits at the base of the body wall, where they develop into young anemones without leaving the parent, almost forming an accretionary colony. A fully expanded adult, with half a dozen flowerlike young, is a lovely thing to see. *Epiactis* is capable of considerable independent movement; if a stipe of alga bearing several specimens is moved to an aquarium, the anemones will leave the plant for the smooth walls of the tank. In the same way, the young, when they are sufficiently mature, glide away from the parent. Four offspring liberated themselves from one specimen that we observed within 3 days. Apparently their departure was hastened by the fact that the parent was dying. *Epiactis* ranges from Puget Sound to La Jolla. California specimens average perhaps ¾ inch in diameter; Puget Sound specimens are larger. The animal may be brown, green, or red with the stripes similar in color but darker.

§38. The solitary hydroid *Tubularia,* treated (in §80) as an outer tide-pool inhabitant, is occasionally found here also, as are small, fernlike colonies of *Abietinaria* (§83).

§39. The vividly orange-red solitary coral *Balanophyllia elegans* (Fig. 39) is abundant, but it must be well sheltered from desiccation and direct sunlight. In February and March, at Pacific Grove, the transparent and beautiful cadmium-yellow or orange polyps may be seen extending out of their stony, cuplike bases until the base comprises only a third of each animal's bulk. Large individuals may be ½ inch in diameter unexpanded. *Balanophyllia*'s range is from Puget Sound to Monterey Bay.

Another coral, *Caryophyllia alaskensis,* occurs intertidally in the Puget Sound region; according to Dr. Durham, it may be similar in appearance when alive to *B. elegans.* Perhaps some northern records for *B. elegans* should apply to *C. alaskensis.*

Astrangia lajollaensis, smaller and with light orange polyps, is said to occur from Point Conception south. Although we have never found it, reports (Johnson and Snook, 1927, p. 108) indicate its comparative abundance at La Jolla.

In fact as well as in appearance these corals are related to the anemones, and their method of food-getting is similar. Working with a related east coast species, *Astrangia danae,* Dr. Boschma found that in the natural state the coral fed on diatoms, small crustacean larvae, etc.; and that in captivity it would take almost anything offered, including coarse sand. In an average of less than an hour, however, the sand, covered with mucus, was regurgitated on the disk. The tentacles then bent down on one side, allowing the sand to slip off. Living copepods coming into contact with a tentacle remained affixed, harpooned and probably paralyzed by stinging nematocysts, and were then carried to the center of the mouth disk by the tentacle, the disk rising at the same time to meet the morsel. Larger animals usually struggled when they came in contact with the tentacles, but other tentacles then bent over to help hold the captive until it quieted.

§40. A form similar in appearance to *Balanophyllia,* but having white-tipped tentacles and of course lacking the hard skeleton, is the small red anemone *Corynactis californica* which often clusters under the overhang of rocks (Fig. 40). *Corynactis* does not like sunshine, hence the most fully expanded specimens are seen at night. It occurs intertidally from Marin County to Santa Barbara, and subtidally at least to the Coronado Islands near San Diego.

39. Unexpanded specimens of the solitary coral *Balanophyllia elegans* (§39). Each animal is in its own theca, separated from its neighbors.

40. The small anemone *Corynactis californica* (§40); the clubbed tentacles are clearly visible in these expanded specimens.

§41. The purple sea urchin is found frequently here, but not in the great beds that characterize the outer pools and reefs (see §174).

§42. Living on and in the rockweed itself, especially in *Pelvetia,* are several diminutive forms rarely found elsewhere. By shaking a mass of freshly gathered rockweed over a dishpan, one can easily dislodge and capture hosts of tiny amphipods (beach fleas) whose colors perfectly match the slaty green of the weed. *Atylopsis* (probably undescribed) is the commoner of two abundant forms. The other, *Ampithoë* (of unknown species), is noticeable because of its saddle-shaped marking in white. Other amphipods will sometimes be found in *Pelvetia,* but these seem to be characteristic. To enable them to cling to plants and other objects, both amphipods and isopods have the last joint of the legs incurved to form a claw. The shape of isopods (flattened horizontally) adapts them for crawling over rocks rather than over weeds, whereas amphipods (flattened vertically, or laterally) seem better adapted for balancing on vegetation. The leaps of amphipods are made with a sort of "flapping" motion. Driving power for the leap is obtained by suddenly snapping backward the bent posterior portion of the abdomen—a spring strong enough to send the animals several feet at a leap and make them adept at avoiding the specimen jar. In appearance they are scarcely distinguishable from *Melita* (§20).

§43. It is in this area also that one is likely to first see examples of the roundworms, a group of animals that include such well-known forms as the hookworm and the eelworm. Referring to the abundance and widespread distribution of these worms, Dr. Cobb has said that if everything else in the world were to be removed, there would still remain a ghost, or shadowy outline made up of roundworms, of mountains, rivers, plants, soil, and forests, and even of animals in which roundworms are parasitic. Their characteristically wiggling bodies are very noticeable to the keen-eyed collector who pulls loose a bit of alga, especially where decay has set in. Some forms writhe so rapidly that they seem almost to snap themselves back and forth.

Under-Rock Habitat

These are animals found on the underside of loosely embedded rocks (but not in the substratum) or between adjacent rocks and boulders where there is no accumulation of substratum. This area offers the maximum of protection from wave shock, sunlight, and desiccation, and a characteristic group of animals occurs here. Some of the brittle stars (*Ophioplocus,* §45) are diagnostic of this association and rarely found elsewhere, but such widely ranging and generalized forms as the flatworms (*Notoplana,* §17) may be the most abundant with respect to numbers of individuals. Members of the under-rock group are even less obvious than those of the protected rock habitat, but they are easily found by the collector who is willing to turn over a few rocks.

§44. The fragile and beautiful brittle stars, whose delicacy requires the protection of the under-rock habitat, are the most striking of all animals found there. They are bountifully represented, both in number of individuals and in number of species. In Monterey Bay six shore species occur in this habitat; and there are still others in the south, most of them restricted to the lower intertidal.

The tiny *Amphipholis pugetana,* already treated (§19) as an inhabitant of the upper tide-pool region, is almost as plentiful here in the middle zone. A much larger and hence more conspicuous form, ranging from Alaska to Monterey Bay, is *Amphiodia occidentalis* (Fig. 41), an annoyingly "brittle" brittle star with long snaky arms bearing spines that take off at right angles. An interested observer who places a specimen in the palm of his hand to examine it is likely to see the arms shed, piece by piece, until nothing is left but the disk, possibly bearing a few short stumps. Normally, regeneration takes place quickly; but experiments have shown that if the radial nerve is injured, little or no new growth will appear. During October the disk of the female, up to ½ inch in diameter, is greatly swollen with eggs; and at such times it is detached on the slightest provocation, possibly in a kind of autotomy. Whether or not the remaining part of the body, which is little more than the intersection of the arms, will regenerate a new disk is not known.

Although closely restricted to the underside of rocks (especially where the rocks are embedded in fine sand with detritus), and even occurring in the substratum, the brittle stars, like many other animals that hide away during the day, come out on calm nights and crawl about on top of the rocks at ebb tide. In their under-rock retreat they are almost invariably found in aggregations of from several to several dozen, so closely associated that their arms are intertwined; recent studies of this intertwining habit lead us to the border line of the metaphysical. Working with Atlantic brittle stars, isopods, and planarians, Dr. W. C. Allee found that aggregations have a distinct survival value for their members, bringing about a degree of resistance to untoward conditions that is not attainable by isolated individuals. By treating individual animals and also naturally and spontaneously formed aggregations with toxic substances, he found not only that the mass had greater resistance to the action of the poisons (partly because of absorption by secreted slime and the bodies of the outermost animals) but also that a specialized protective material was given off by the aggregations. This subtle material, which "once in solution passes through ordinary filter paper and persists after the filtrate is boiled," is apparently similar to an antibody, like those familiar to the general public in vaccine. Solutions containing the protective substance are capable of giving protection from poisons—fresh water or colloidal silver, for instance—to isolated animals that could not otherwise survive. Furthermore, certain animals can confer immunity on other taxonomically unrelated animals.

This rather astounding discovery opens new and unexplored vistas to students of biology. It will certainly throw some interesting light on animal communities in general, and may conceivably clarify the evolutionary background behind the gregariousness of animals, even behind that of human beings. At the moment, it explains a phenomenon regarding anemones that had puzzled us for some time. A single anemone may

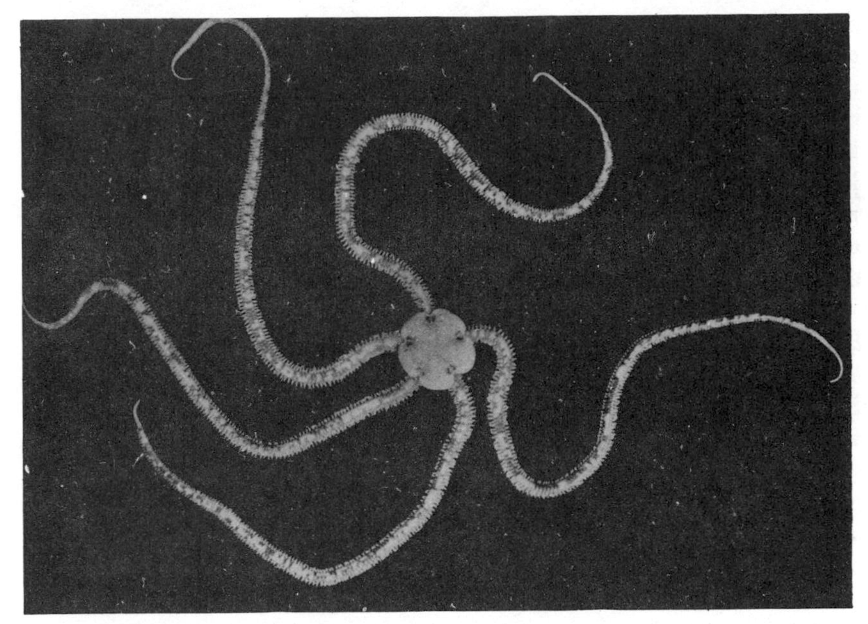

41. *Amphiodia occidentalis* (§44), a common brittle star of Zone 3.

very easily be anesthetized for preservation, but a large number in a tray show a tremendous degree of resistance, even though the amount of anesthetic employed per individual is greater for the group. Furthermore, the pans in which anemones are anesthetized become increasingly unsuitable; that is, each successive batch of animals is more resistant than the batch before. We finally took to scrubbing out the pans with hot water and soap powder and then revarnishing them.

Several animals, such as *Nereis* and *Amphitrite*, which do not form aggregations in their natural habitat, do so under artificial conditions, as when placed in trays. Since *Amphiodia* aggregates in its natural state, however, it may be assumed that it does so for protection against some usual environmental factor, probably oxygen suffocation. It is not known whether or not the aggregations disband when the tide is in, but this is probably the case.

Some more tangible aspects of brittle stars in general are their feeding habits and methods of locomotion. The locomotion is peculiar—quite different from that of seastars and very rapid as compared with the snail-like pace of even the fastest seastar. Most brittle stars that we have observed pull with two arms and push with the other three, apparently exerting the power in jerky muscular movements. Their very flexible arms take advantage of even slight inequalities in the surface.

Not a great deal is known of the brittle stars' feeding habits. The hard jaws surrounding the mouth can be rotated downward, thus greatly enlarging the mouth, and then rotated upward and inward, so that they form a strainer. The digestive tract is restricted to the disk and does not extend into the arms, as is the case with seastars. Moreover, a brittle star, again unlike the seastars, cannot project its stomach outside its body to envelop food. Some of them probably feed on detritus on the bottom. Others, which occur in deeper water in great numbers, hold their arms upward, and are thus suspension feeders. It has been observed that some brittle stars (if not all) feed by accumulating food particles on their tube feet and passing the material from foot to foot toward the mouth.

Any area colonized by animals is full of organic debris, which is continually being sorted over by animals of various capacities and with varying requirements, with the Victorian "economy of nature" factor always at work. The whole thing calls to mind the sorting effect of a set of sieves. Certain animals, like crabs, dispose of sizable chunks of food; others, like beach hoppers, eat minute particles and may, in time, reduce large pieces of edible matter; certain of the sea cucumbers, buried in the substratum, sweep the surface with their tentacles for adherent particles, an office that is performed for rock surfaces by limpets and periwinkles; and other animals, including the brittle stars, pass quantities of dirt and sand through their alimentary canals in order to extract whatever nourishment these may contain. Everything is accounted for: particles so small as to be overlooked by everything else are attacked and reduced by the last sieve of all, the bacteria.

The deep-water brittle stars (and presumably the shore forms as well) feed on the most superficial layer, which consists partly of microscopic algae and partly of decaying organic matter and is much more easily digested than the living animals on which the seastar preys. Thus "the simple structure of the alimentary canal seems to be correlated with the exceedingly simple character of the food." Deep-water forms under examination have frequently extruded a dirty-white fluid from the mouth, presumably a mixture of half-digested food, digestive fluid, and waste products, since there is no anus.

§45. *Ophioplocus esmarki* (Fig. 42) is a larger, sand-colored brittle star, common in some regions on a sandy-mud substratum under flat rocks, especially disintegrating granite. It averages ½ to ¾ inch in disk

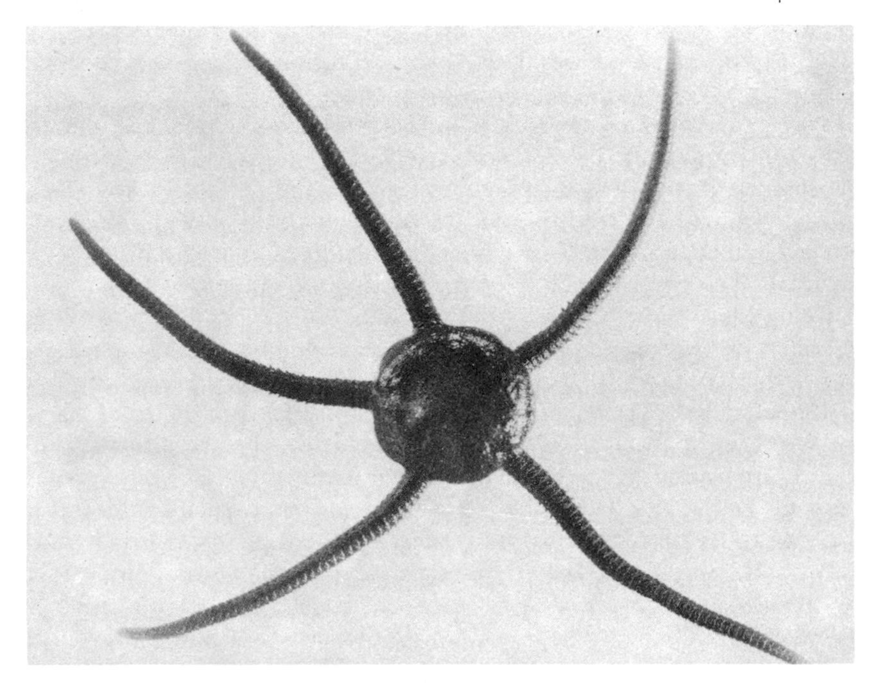

42. The brittle star *Ophioplocus esmarki* (§45).

diameter and has short stubby arms with a spread of 3 to 3½ inches. It is relatively hardy and maintains its form well when preserved. Specimens noted during January had the genital bursae, or sacs, swollen with eggs, and young are found in the bursae in July. *Ophioplocus* and *Ophioderma* (§121) differ from most brittle stars in having spines on the arms that extend outward at a sharp angle to the arms instead of being at right angles. *Ophioplocus* ranges from Pacific Grove to San Diego.

Two other common brittle stars of this southern range belong in the lower tide-pool zone but often stray up into the middle zone. Both (§121) are large and striking, and neither is found above San Pedro. One, *Ophioderma panamensis,* has a superficial resemblance to *Ophioplocus* but is larger, richer brown in color, and smoother in texture, with banded arms. The other, *Ophionereis annulata,* is the southern counterpart of the snaky-armed *Amphiodia.*

§46. The porcelain crab (*Petrolisthes,* §18), discussed with the upper tide-pool animals, is also a common and easily noticed member of the middle tide-pool under-rock association. The transparent tunicate

(*Clavelina,* §101) is fairly common here, but belongs more characteristically to the lower intertidal. Some serpulid worms occur (§§129, 213), but they are also more common farther down.

Belonging to this zone is a large, slow-crawling pill bug or isopod, *Idothea urotoma* (Fig. 43), which will be found under rocks, especially in the south. It may be recognized by its paddle-shaped tail and somewhat rectangular outline. It is brown in color and about ¾ inch in length.

Betaeus longidactylus (Fig. 44), the "long-fingered" shrimp, may be found under rocks or in pools. In adult males the nearly equal large claws are extraordinarily long, sometimes more than half as long as the animal's body. The elongated carapace projects somewhat over the animal's eyes, like a porch roof—a peculiarity that is still more exaggerated in a related shrimp that lives on wharf piling (§326). The legs of *B. longidactylus* are red, and there is a fringe of yellow hairs on the last segment of the abdomen. The body and the big claws vary in color from blue and blue-green to olive-brown. It occurs from Elkhorn Slough to San Diego at least, but is most plentiful in the south, where it occurs on the outer coast, being restricted in the north to quiet-water conditions.

§47. A common feature of the under-rock fauna below Santa Barbara is a great twisted mass of the intertwined tubes of a remarkable snail, *Aletes squamigerus* (Fig. 45), which occurs at times even on the sides of rocks. For some reason not readily apparent, this animal lost or gave up the power of independent movement and took to lying in wait for its food instead of going after it. Some adaptation to food-getting under these new conditions had to be made of course, and *Aletes* solved the problem by employing the mucus that normally lubricates a snail's path to entangle minute animals and bits of detritus. When a bit of passing food is thus captured, the animal draws the mucus back into its mouth and removes the food.

Aletes is ovoviviparous (that is, it lays eggs but retains them within the shell until the young are hatched), and produces offspring that have snail-like, coiled shells. It is only after the young have attached to rocks and started to grow that their shells lose all resemblance to a snail's. Calcareous matter is added to the mouth of the shell to produce a long, wormlike tube that is irregularly coiled or nearly straight. The hinder part of the body is black, the rest mottled with white. When disturbed, *Aletes* retracts slowly, like an anemone—a trait that distinguishes it instantly from the tube worms, which snap back in a flash, often with nothing more than a shadow to alarm them. Even experienced zoologists,

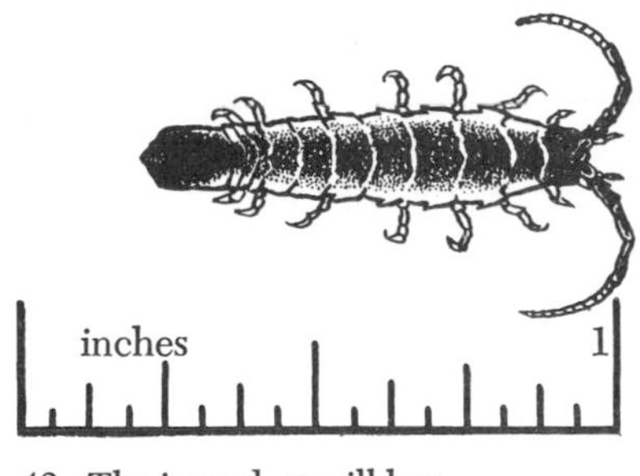

43. The isopod, or pill bug, *Idothea urotoma* (§46).

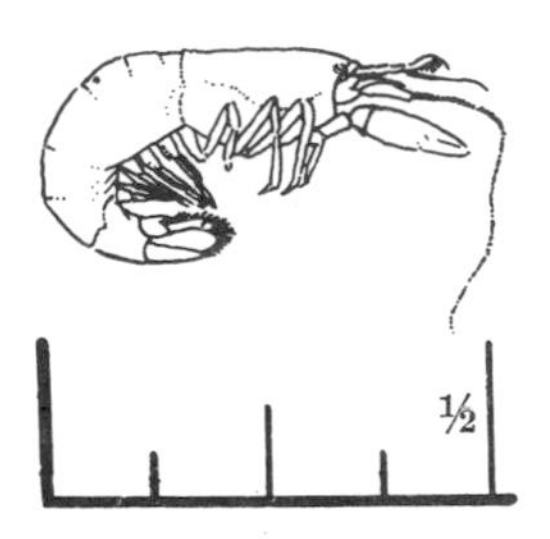

44. *Betaeus longidactylus* (§46), the long-fingered shrimp.

45. The fixed snail *Aletes squamigerus* (§47), with an encrusting sponge.

if unfamiliar with the Pacific fauna, may mistake this tubicolous mollusk for a serpulid worm. *Aletes* ranges from Monterey to Peru, but it is only occasional in the Monterey Bay region. The specimens shown were photographed in Newport Bay.

§48. More than a hundred species of ribbon worms (nemerteans) are found on the Pacific coast. One of them, *Emplectonema* (§165), is abundant here; but it is treated elsewhere, since it occurs even more commonly under mussel beds. The similar but larger *Amphiporus bimaculatus*, however, has possibly its chief center of distribution in this zone. It may be found occasionally as a twisted brown cord, capable of almost indefinite expansion and contraction, but averaging a few inches in length when undisturbed. As suggested by its specific name, *Amphiporus* has two dark spots on its head, not unlike great staring eyes but having no connection with the light-perceptive organs. The "real" eyes are the ocelli—dozens of minute black dots centered in two clusters that are marginal to the dark spots and extend forward from them. Otherwise the worm's head is light in color, a striking contrast to the rather beautiful dark orange of the rest of the body.

Nemerteans are ribbon worms, and they have a physical organization much more primitive than that of segmented worms like *Cirriformia* and the serpulids. They are unique in the possession of a remarkable evertible organ, the proboscis, which, in the case of some mud-living forms (*Cerebratulus*, §312), may be longer than the animal bearing it. The proboscis of *Amphiporus* is armed with a poisonous barb or stylet, not large enough to penetrate human skin but very effective when directed against smaller worms. This form is abundant throughout its range, from Sitka to Monterey Bay.

It is almost impossible to preserve some of the nemerteans, because of their habit of breaking up on the slightest provocation (*Amphiporus*, however, is only moderately fragile). Often they cannot even be picked up whole. If the head fragment of an autotomized specimen is left in the animal's natural environment, or undisturbed in a well-aerated aquarium, it will regenerate a complete new animal. Some interesting experiments in this connection are discussed in §312.

§49. A very common, insignificant-looking inhabitant of this region is the scale worm, or polynoid. *Halosydna brevisetosa* (Fig. 46) is the most common form in the Alaska to Monterey region, being replaced south of that point by *H. californica*, which occurs at least to San Diego. Free-living specimens seldom exceed an inch or so in length. Many, however, exchange their freedom for a parasitic but doubtless comfortable residence in the tubes of other worms. In this case, contrary to the usual run of things, the dependent form is considerably larger and more richly colored than its self-supporting relatives.

Halosydna's breeding habits are noteworthy. To quote Essenberg: "The eggs, after leaving the body cavity of the worm, are attached by a mucous secretion to each other and to the dorsal surface of the parent's body beneath the scales. There they develop until a preoral band of cilia is formed, when the larvae escape as the well-known trochophores, swimming freely near the surface of the water. The larvae finally settle to the bottom of the ocean, undergoing there further metamorphosis and assuming gradually the shape of the adult worm."

The polynoid worms are voracious feeders; and in captivity they will attack one another with their strongly developed, four-jawed probosces, displacing scales or removing entire segments from the after ends of their companions' bodies. *H. brevisetosa*, like allied annelids, bears

6. The common scale worm *Halosydna brevisetosa* (§49), often
›und as a parasite or in commensal association.

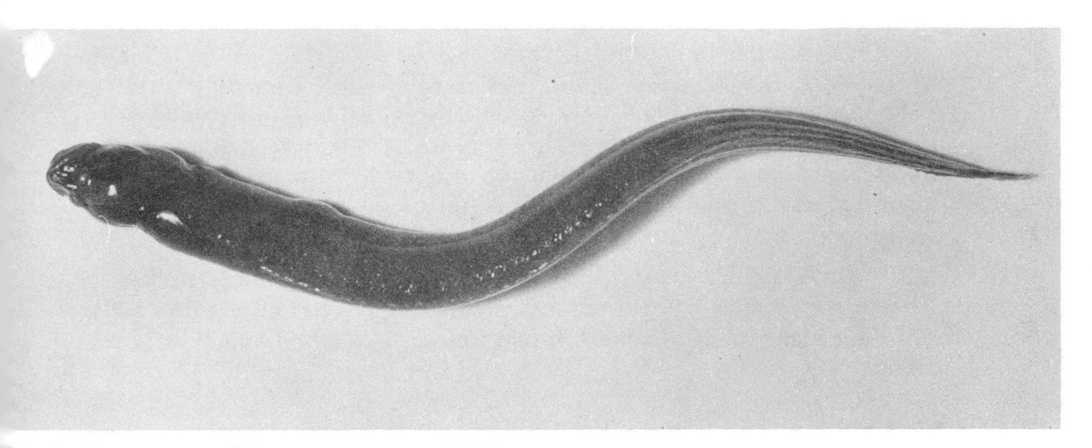

7. The blenny *Epigeichthys atro-purpureus* (§51),
representative tide-pool fish.

its scales, called elytra, in two longitudinal rows on the upper (dorsal) surface and may autotomize them when roughly handled. This animal has 18 pairs of scales, which are mottled, with white centers.

Many other worms, such as *Neanthes virens* (§311), are occasional in the under-rock habitat.

§50. Here, as in the upper tide pools, small isopods and amphipods, pill bugs and beach hoppers, are more numerous than any other animals found under rocks. *Melita palmata* (§20) is the commonest hopper in this zone, but others may occur, including the very active, lustrous slaty-green *Orchestia* (of undetermined species), which is closely related to the tiny fresh-water "shrimps." It should be mentioned in extenuation of the hoppers that their variant name of "beach flea" is a libelous misnomer. None of the beach hoppers has the slightest use for man or other warm-blooded animals.

The pill bug *Cirolana harfordi* (§21), likewise not restricted to any one habitat, occurs also, as do several others.

§51. A long, eel-like fish, one of the blennies (Fig. 47), is very common under boulders in this association from southeastern Alaska to Point Arguello. The young are greenish black; the adults (up to 12 inches long) are nearly black, with traces of dusky yellowish or whitish mottling toward the tail fin. This species, *Epigeichthys atro-purpureus,* lives even as an adult in the under-rock habitat, frequently in places that are merely damp when the tide is out. Most specimens have three stripes of light brown or white, bordered by darker colors, radiating backward and upward from the eyes. Not all *Epigeichthys* have this character, and other blennies may have it also; but the condition is fairly diagnostic. Another common species is the larger *Xiphister mucosus,* only the young of which are found in the intertidal zone. The young are a pale, translucent olive, and range from Alaska to Baja California. A third, less common, blenny represents an entirely different group, which is usually tropical. It is more compressed laterally than those described above and is not so elongate, but it is often brilliantly striped. Two subspecies of this form, *Gibbonsia elegans,* range from British Columbia to Baja California and probably below. Many people regard the blennies as excellent eating. However, some that we tried had green flesh, whose cooking smell was so reminiscent of defunct kelp that our research was discontinued.

With the blennies, or in similar situations, the little clingfish, *Gobiesox meandrica,* is very common, slithering over smooth, damp surfaces or clinging to a rock with limpetlike strength.

Burrowing Habitat

Animals of the burrowing habitat occur in the mud, sand, or gravel substratum underlying tide-pool rocks. A small stout trowel is a good tool here, for the surface boulders are often resting on a matrix of muddy gravel bound together by chunks of rock of varying sizes. Ordinarily, however, the prying up of a large boulder will disturb the substratum, and any animals present will be visible. The normal expectancy would be that the sand or mud under rocks would harbor forms similar to those burrowing in the sand or mud of estuaries. This proves true to some extent, although the under-rock substratum animals actually seem to form a fairly specific assemblage.

§52. Foremost of these substratum animals is the sipunculid worm *Phascolosoma agassizi* (Fig. 48), which ranges at least from Sitka to Monterey Bay, and is replaced in the south by the more beautifully flowering *Dendrostoma* (§145). These rough-skinned worms are related to the annelids, although themselves devoid of segmentation. They have been popularly termed peanut worms, a name that really describes the specimens only in their contracted state. *Phascolosoma* is abundant in the muddy crevices between rocks, and is often associated with terebellids (see next paragraph). A contracted specimen 2 inches long (larger than the average) is capable of extending its introvert, which bears a mouth surrounded by stubs of tentacles, to an extreme length of 6 inches, much as one would blow out the inverted finger of a rubber glove. Under anesthesia (used for museum preparations), it is a common thing for specimens taken in the spring to extrude eggs and sperm. In the quiet channels of upper British Columbia large representatives of this form are among the most characteristic substratum animals.

§53. The brown or flesh-colored terebellid worms (see color section), up to 6 inches long, commonly build their parchment-like tubes under rocks or in the substratum between rocks. Several species are involved. *Thelepus crispus* (= *plagiostoma*), *Eupolymnia heterobranchia* (formerly *Lanice*), and possibly some species of *Terebella* seem to prefer the cool, oceanic waters of open shores but may be taken also in the quiet waters of Puget Sound. Another form, *Amphitrite robusta* (§232), which reaches great size in Puget Sound, may also occur on outer coasts. All of these worms superficially resemble *Cirriformia* (Fig. 19), but the cirratulids have gills extending almost to the middle of the body, which is tapered at both ends, whereas the terebellids have long, delicate tentacles that are restricted to the enlarged head region.

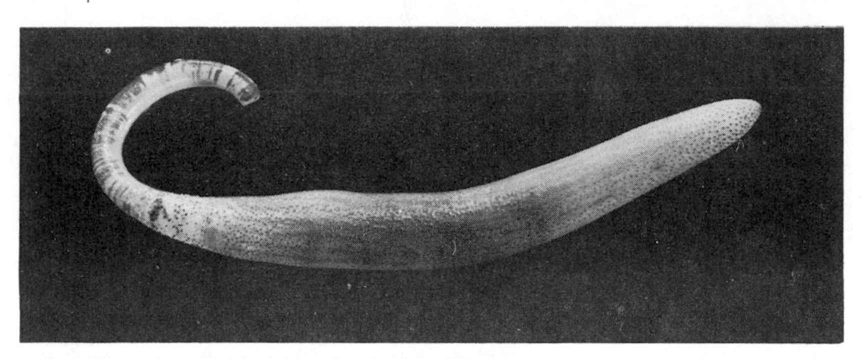

48. *Phascolosoma agassizi* (§52), a peanut worm.

Differentiating the many species of terebellid worms is a task for the specialist. We have had little success in identifying them by observations in the field, but the following suggestions supplied by Dr. Olga Hartman will be found useful: *Thelepus* has threadlike and unbranched (cirriform) tentacles, with gills of about the same length. The tentacles of *Amphitrite* and *Terebella* are branching (dendritic), dark, and longer than the white gills; but *Amphitrite* has only 17 thoracic segments, whereas *Terebella* has more—23 to 28.

A specimen of any of these tube worms, if deprived of its home and placed in an aquarium with bits of gravel, will quickly set about remedying the deficiency. In a few hours a makeshift burrow will have been constructed of the gravel, cemented together with mucus and detritus, and the worm's transparent, stringy tentacles will be fully extended on the bottom of the container, writhing slowly in search of food, which is transferred to the mouth by cilia.

In Bermuda, Welsh (1934), primarily interested in nerve structure, investigated the autonomy of the tentacles of a similar form. He found that they would live, squirm about, and react normally in most ways up to 12 days after they had been removed from the body. He also determined that new tentacles (in experimentally denuded worms) grew at the rate of 1 to 1½ cm. per day; and that the fully developed tentacle, up to 50 cm. long, acted as a food-getting organ, collecting the minute particles of detritus in a ciliated groove and transmitting them to the body.

Most of the tubes of our local forms, when carefully examined, will be found to contain presumably uninvited guests who share the substantial protection of the tubes, profiting from the food-laden water. Commensal polynoid worms (§49) are often present, and with them a minute pea crab, *Pinnixa tubicola*, which is never found living free.

§54. *Amphitrite* is one of several animals whose breeding time is known to be correlated with the spring tides and hence bears some sort of relationship to the phases of the moon. In this case one investigator, Scott, believes that both moon and tides are entirely secondary factors; and that the egg-laying is induced by the higher temperatures at the time of spring tides, caused by long exposure to sunlight, and possibly by the changes of pressure brought about by the alternate increase and decrease in the the depth of water. The increased food supply may be another inciting factor.

At any rate, both males and females discharge their ripe sexual products into their tubes during spring lows and expel them from the tubes into the water via the respiratory currents. Another possible reason for discharging the eggs and sperm at this time is that these products have a much greater opportunity to mix than if they were voided into the tons of water moving at flood tide—a reasoning that applies equally well, of course, to other animals. At the time of egg-laying the female's wavelike body contractions become stronger and faster as the ripe eggs that are to be released are separated from the immature eggs that are to be retained.

§55. In sandy and gravelly substrata, extending downward from the middle tide pools, the dirty-white sea cucumber *Leptosynapta albicans* (formerly *inhaerens*) is common. This and similar forms (the synaptids) lack the respiratory trees that are common to other cucumbers; instead they have a delicate, semitransparent skin that lets through enough dissolved oxygen for purposes of respiration. The skin, through which the internal anatomy can be seen vaguely, is slippery, but it is armed with numerous white particles, calcareous anchors that stiffen the body and give the animal a better purchase in burrowing. The anchors often come off on one's fingers. Although it lacks the tube feet that are characteristic of most of the other cucumbers and of echinoderms in general, this remarkably efficient little wormlike form is at home along temperate shores almost all over the world. When disturbed, it has the habit, also wormlike, of autotomizing by constricting itself into several portions—a trait that makes it difficult to collect and almost impossible to keep in aquaria. It is not known whether all the autotomized portions will regenerate complete new animals or whether only the head portion will.

Under feeding conditions (that is, when the tide is in) the animal lies fairly well buried but stays close enough to the surface to protrude its 10 or 12 feathery tentacles, which capture small living animals, minute floating plants, and other bits of organic matter and then transfer the food to the mouth.

Here on the Pacific coast this species is rarely more than an inch or two in length, but specimens up to 10 inches long have been found in other parts of the world; and a related form from southeastern Alaska, mostly from quiet water, may be 6 inches long or more (*Chiridota*, §245).

§56. The brittle star *Amphiodia* (§44) is possibly more commonly found buried in the substratum than crawling on the underside of rocks, but it is certainly not so easy to see there. In little pockets under rocks, where the bottom is soft and oozy but with coarse particles, one may sometimes sift out with the fingers a dozen or more of these snaky, long-armed specimens.

§57. In southern California and northern Mexico the muddy bottom under tide-pool rocks is often riddled with the burrows of a small, white, crayfish-like animal, the ghost shrimp *Callianassa affinis* (Fig. 49). A large specimen may be 2½ inches long. Although not so feeble and delicate as its relatives of the mud flats, *C. affinis* is one of the softest animals found in the tide pools. There it uses the protection of rocks and then burrows deeply into the substratum so as to obtain the maximum shelter. The ghost shrimp's recorded range is from Santa Monica to San Diego, but we have taken specimens (identified by the National Museum) from Todos Santos Bay. Like its larger mud-flat relatives (in connection with which the natural history of the ghost shrimps is discussed in more detail,

49. ***Callianassa affinis*** (§57), the ghost shrimp.

50. The blind goby *Typhlogobius californiensis* (§58).

inches 1 2

§292), this *Callianassa* has the crayfish habit of flipping its tail suddenly so as to swim backward in a rapid and disconcerting manner. Ovigerous specimens have been taken in early May.

§58. Associated with the above is the blind goby, or pinkfish, *Typhlogobius californiensis* (Fig. 50). It lives in the burrows made by the ghost shrimp and is about 2½ inches long, somewhat resembling a very stubby eel. Its eyes are rudimentary and apparently functionless.

Pool Habitat

§59. A minute red flatworm, *Polychoerus carmelensis,* only slightly longer than it is wide, may be so abundant locally during spring and summer as to dot the Monterey tide pools with color. This form (a turbellarian) is interesting in that it is one of the few free-living flatworms to have no intestine and to possess a tail. These animals are hermaphroditic, and the male organs of one fertilize the female organs of the other. (Self-fertilization rarely takes place in hermaphroditic animals except possibly by accident.) At Woods Hole, Massachusetts, it has been found that *Polychoerus* eggs are deposited at night in transparent, gelatinous capsules. *Polychoerus* is a member of the order Acoela, whose other members are reported to feed on copepods, other flatworms, or diatoms; little is known of the habits of our local form.

§60. Those who examine tide pools bordered with red algae in the early evening, about the time total darkness sets in, will probably be rewarded with an exhibition of phosphorescence by another flatworm, *Monocelis* (a rhabdocoele). These little dots of light can be seen crawling rapidly along the underside of the air-water film or swimming about at the surface. Certain small annelid worms, likewise luminescent, may also be found under similar circumstances, but they swim about more rapidly, change direction quickly in darting movements, and are generally somewhat larger. Like *Polychoerus, Monocelis* is hermaphroditic, with cross-fertilization. It forms hard-shelled eggs.

§61. During late winter or early spring (in February 1930, at Laguna Beach, for example), but unfortunately not every year, the middle and upper tide pools are temporarily inhabited by hordes of a very small crustacean, *Acanthomysis costata.* These are mysids, called "opossum shrimps" because the eggs are carried in thoracic pouches; they are very transparent and delicate, with a slender body up to ½ inch long, enormous eyes, and long, delicate legs on the thorax. The thoracic legs have outer branches, or exopodites. *Acanthomysis costata* seems to be a neritic (near-shore) species restricted to southern California, but is overlapped by the similar *A. sculpta,* which is often very abundant in Puget Sound.

§62. Especially at night, the transparent shrimp *Spirontocaris picta* (pale green, often with red bands) is likely to be found, darting about in the pools or quiescent in bits of sea lettuce or in rocky crevices. There is such a fairylike beauty to this ephemeral creature that the inexperienced collector will be certain that he has taken a rare form. The elusive animal is likely to lead him a lively chase, too: when at rest, it is difficult to see against the vivid colors of a tide pool; and when discovered, it darts backward in a disconcerting manner by suddenly flexing its tail. Once captured, the living specimen should be confined in a glass vial not much larger than itself and examined with a hand lens. The beating heart and all the other internal organs can be seen very plainly through the transparent body.

About 95 per cent of the transparent shrimps found in rock pools will be of the species named (*S. picta*). The very similar S *paludicola* (Fig. 51) is much more common on mud flats and in eelgrass beds, but it does occur in rock pools.

In the lower tide pools other species of *Spirontocaris* will be found; these are not transparent, but all are striking and all have the tail bent under the body in a characteristic manner.

51. The transparent shrimp *Spirontocaris paludicola* (§62); S. *picta* (§62) is similar in appearance.

3. *Low Intertidal Rocks*

It is here that the collector finds his most treasured prizes, since this zone, bared only by the minus tides (from 0 to −1.8 feet at Pacific Grove), is not often available for examination. In its lower ranges, the region of the outer tide pools is uncovered only a few times each month, sometimes not at all. Many of the middle-horizon animals occur here also, but much of the population consists of animals unable to exist higher up and not so accustomed to the rhythm of the tides. This tremendously crowded environment contributes a greater number of species than all the other tidal levels. Triton is a fruitful deity, and here, if anywhere, is his shrine. Not only do the waters teem, but there is no rock too small to harbor some living thing, and no single cluster of algae without its inhabitants. Since these creatures live and thrive in an environment that seems utterly strange to us, it is no wonder that we find interest in their ways of feeding, of breathing, of holding on, of ensuring the continuity of their kind—or in their strangely different weapons and methods of attack and escape.

In this zone, very few animals are exposed to wave action, sun, and wind on unprotected rock faces, except on open coasts where the surf is high. (The animals often found on bare rocks—purple seastars, mussels, and goose barnacles—are treated in Part II with the fauna of the surf-swept open coast.) On the contrary, almost every square inch in the lower tide-pool area is covered with protecting rockweeds, corallines, or laminarians. Though few of the animals are completely exposed, a rich and varied population can be seen without disturbing the rocks.

§63. Where there are limestone reefs, a black, tube-building cirratulid worm may colonize. This is *Dodecaceria fistulicola,* originally described, on account of its appearance in tubed masses, as a sabellid. Compact colonies of these worms may occasionally be found on the shells of the red abalone (§74). In the south there is another bare-rock form, *Chama pellucida* (§126). In some regions this clam is practically a reef builder, but it achieves its maximum development in quiet waters.

§64. The other zones exhibit a few characteristic and striking animals that occur in great numbers, but it would be hard to pick any highly dominant forms out of the lower tide-pool life. Perhaps the solitary great green anemone, called *Anthopleura xanthogrammica* (Fig. 52), is the most obvious; certainly it is the form most frequently observed by the layman. This surf-loving animal might have been included as justly with the open-coast fauna, but since it is here in the tide-pool region that the casual observer is first likely to see it, the inductive method of this book justifies its treatment here. So far as is known, the huge specimens of *A. xanthogrammica* that are found in the deep pools and channels are exceeded in size only by the giant anemones of the Australian Barrier Reef, specimens of which have been reported as over a foot in diameter,* with stinging capacities as great as the stinging nettle. Although our form sometimes attains a diameter of 10 inches, a bare hand placed in contact with its "petals" will feel no more than a slight tingling sensation and a highly disagreeable stickiness. This stickiness is possibly caused in part by suction. At La Jolla, Parker (1917) recorded a tentacular suction in *A. xanthogrammica* that was powerful enough to enable the animal to capture small fish, and also found that the tubercles had a suction of 15.6 pounds per square inch in retaining pieces of shell and stone.

Typically a west coast animal, this magnificent anemone has the enormous range of Unalaska and Sitka to Panama, and it has been the subject of observation and experimentation since Brandt first observed Alaskan specimens in 1835. It is a more or less solitary animal, typically of great size and with a uniform green color, and it is restricted to the lowest tide zone, or lower in areas where surf and currents continually provide a fresh supply of water. Unlike the smaller colonial species *A. elegantissima,* which occurs in great beds high in the intertidal zone, *A. xanthogrammica* cannot survive where there is sewage, industrial pollution, or sludgy water. The vivid green color of specimens living in

* MacGinitie writes (March 12, 1947), "There are some anemones on the little rock island in Carmel Bay that will measure a good 16 inches when expanded."

2. The giant green anemone, *nthopleura xanthogrammica* §64); the specimen at bottom enter is contracted.

3. The snail *Opalia crenimarginata* §64), a wentletrap, feeding on the iant green anemone; Gunnar horson's Christmas card for 1956.

sunlight is produced by symbiotic green algae, which live in the tissues of the animal in a mutually beneficial arrangement, as mosses and algae live together to form lichens. Specimens found in caves and other places sheltered from direct sunlight are naturally paler.

In its preference for brilliant light, as in many other physiological traits, this giant green anemone is similar to the colonial species of *Anthopleura* (treated in §24, to which the reader is referred for an account of anemones in general), and the two have often been considered different phases of the same species.

A sea spider, *Pycnogonum stearnsi* (§319), will often be found associated with the giant green anemone in central California, apparently forming some such relationship as most of the other sea spiders form with the hydroids on which their young live. Nor are sea spiders the only animals appreciative of *Anthopleura*'s society; a giant amoeba, *Trichamoeba schaefferi,* which is visible to the unaided eye if the specimen is in a strong light, has been taken from the base of the anemone.

Around the bases of the green anemone at La Jolla (and doubtless at other localities in southern California) may be found a predatory or quasi-parasitic snail, *Opalia crenimarginata* (Fig. 53). This snail occupies at San Diego the ecological position that *Pycnogonum stearnsi* does at places like Carmel and Duxbury Reef. A somewhat desultory study of this matter, lasting several years, has failed to reveal the presence of *Pycnogonum* and *Opalia* living together on the same anemones in southern California; but at Pacific Grove and at Shell Beach (Sonoma County) there were several finds of *Pycnogonum stearnsi* and various small *Epitonium* together on the same anemones. *Epitonium* is another genus of wentletraps, as the small, long-spired snails like *Opalia* are called. *Opalia* has been found in association with *Anthopleura* as far north as Bodega. In Pliocene times, where La Jolla is today, there were *Opalia* almost 3 inches long, which would suggest, if proportions mean anything, that there could have been anemones the size of washtubs in those days.

The wentletraps, many of whom have been identified as feeders upon coelenterates, produce a violet or purple toxic fluid that is thought to be an anesthetic permitting the animal to pierce the side of an otherwise resisting anemone with its proboscis. *Pycnogonum,* with its strong claws, need only dig and poke its way in, and requires no tranquilizer (if that indeed is the function of the wentletrap's purple juice).

§65. *Tealia crassicornis* is another large anemone, beautifully red in color, usually with green blotches, and up to 4 inches in diameter, that may be found only at the most extreme of the low tides. It attaches to rocks, but is often half buried in the sand and gravel and, like *Anthopleura,* has bits of shell and gravel attached to its body wall. *Tealia* (also known as *Urticina*) is also common on the coasts of Europe and, through circumpolar distribution, occurs on the east coast of America as far south as Maine, where it is called the thick-petaled rose anemone. Thus we have, as far south as La Jolla at least, another cosmopolitan species to keep company with the sponge *Haliclona,* the cucumber *Leptosynapta,* and the flatworm *Polychoerus.* Another species of *Tealia, T. lofotensis,* a bright cherry-red anemone with white warts on the column, occurs from Marin County northward. It is also a circumpolar species.

Other anemones, *Epiactis prolifera* and *Corynactis* (§§37, 40), and the solitary coral *Balanophyllia* (§39) may be found abundantly in this zone.

§66. Seastars are likely to be common and noticeable in this prolific zone. In most well-known regions on the Pacific (except Oregon), *Patiria* (§26) may be larger and more plentiful here than in the middle tide-pool region, but an examination of virgin areas indicates that the center of distribution is probably subtidal. Where surf keeps the rocks bare of large algae, *Pisaster* (§157), the common seastar, may also be abundant. There are a number of seastars, however, that are characteristic inhabitants of the low intertidal.

The many-rayed sunflower star, *Pycnopodia helianthoides* (Fig. 54), is possibly the largest seastar known, examples having a spread of more than 2 feet being not uncommon. Certainly it is the most active of our Pacific asteroids; and though all seastars can and do move, they are scarcely to be classed in general as active animals. Probably they possess the most highly specialized form of locomotion to be found in the intertidal zones. Locomotion is effected, in some debatable manner, by the tube feet, which cling so strongly that many will be broken when a seastar is forcibly removed from a rock. The formerly accepted theory was that the tube feet were swung in the direction of motion and took hold of the substratum as vacuum cups (the vacuum being created by the animal's highly effective water-vascular system); thereupon the seastar drew itself forward, destroyed the vacuum by readmitting water to the tube feet, and repeated the cycle. Jennings, however, has demonstrated that the animal can walk equally well on a greased surface or on sand, and concludes that the tube feet act very much like the legs of higher animals, since they could not exert an effective suction pull on either grease or sand. He believes that the adhesive action of the tube feet

54. Above: the under (oral) surface of the sunflower star, *Pycnopodia helianthoides* (§66), showing the tube feet. Below: *Pycnopodia* righting itself.

ordinarily serves merely to give the animal a firm foothold and becomes of primary importance only when the seastar is climbing a steep surface or hanging inverted. Of course, it also secures the animal against wave shock.

Pycnopodia is readily distinguished from all other seastars in this zone by its soft, delicate skin, often colored in lively pinks and purples, and the bunches of minute pincers (pedicelleriae) on its upper surface. It also has the greatest number of rays—up to 24. It is often popularly called "the 21-pointer," but the number of its rays is apparently dependent on its age, since it starts out in life with only six. Unless carefully supported when lifted from its tide pool, *Pycnopodia* indicates its attitude toward the human race by shedding an arm or two, and persistent ill-treatment will leave it in a very dilapidated condition. The animal ranges from Unalaska to Monterey Bay, and is found less frequently between Monterey and San Diego.

A favorite item in the sunflower star's diet is the sea urchin, and many of the nicely cleaned tests of *Strongylocentrotus* to be found in tide pools indicate that *Pycnopodia* has recently fed well. It evidently engulfs its prey, spines and all, with its ample stomach, and leaves the well-cleaned remains behind. It would appear that nowadays the sunflower star is urchin enemy number two—first place now goes to the developmental biologists.

§67. A less common multi-rayed seastar, rare in the intertidal of central California but slightly more common to the northward, is the sun star, *Solaster dawsoni.* The number of arms in this species is usually 12 but varies from 8 to 15. A similar form, *S. stimpsoni,* usually has 10 rays, very slim and uniformly tapering to a point. Both have a less webbed appearance than *Pycnopodia,* having disks smaller in proportion to the size of the animal, and both are found only at or below the extreme low-tide mark. The colors range from purple-gray to orange, and the skin, which is harder than *Pycnopodia*'s and softer than *Pisaster*'s, is rough to the touch. *S. dawsoni,* ranging from the Aleutian Islands to Monterey (at the latter place found in deep water only), is larger than *S. stimpsoni* (up to 14 inches) but very rarely occurs inshore. *S. stimpsoni,* ranging from the Bering Sea to Oregon, is not infrequent in the low intertidal.

§68. Two ovigerous seastars occur in this zone, and one of them, *Leptasterias hexactis,* is very common. This little six-rayed form, up to 2½ inches in spread and hence considerably larger than the delicate *L. pusilla* (Fig. 32) of the middle pools, is a feature of the low intertidal fauna throughout its range of Puget Sound to Newport Bay. The madreporite (a perforated plate through which water is admitted to the water-vascular system) is conspicuous in this species, and the rays are slightly swollen at the base, giving this dull pink, green, or olive species a bulky, unwieldy appearance. Eggs are carried from February on, and are discharged as minute seastars in April or May. Incubating females are likely to be humped up over the eggs and to keep fairly well hidden in rock crevices.

§69. *Henricia leviuscula* (Fig. 55), not nearly as common as the above, resembles *Leptasterias* in its usually small size (although 5-inch individuals have been seen) but differs in its usual vivid, blood-red color. The rays are long and sharply tapering. The females of this species brood their eggs in January, keeping absolutely in the dark, and hence well hidden, during this period. *Henricia* ranges from the Aleutian Islands to San Diego, and is related to an east coast species (*H. sanguinolenta*) that is circumpolar in distribution but does not range as far south as Monterey. The distinctive gawky shape and vivid color of this animal make it a form that will stand out in the mind of the collector who finds it.

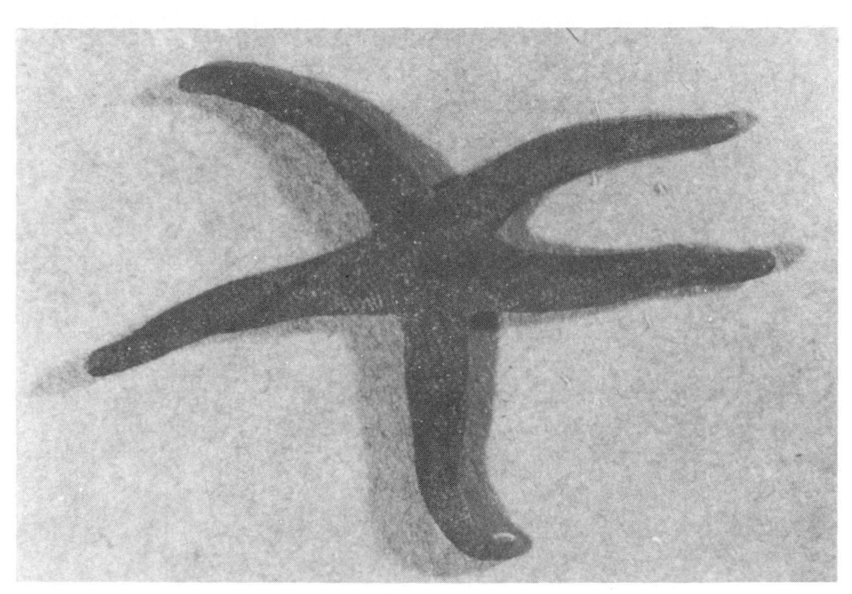

55. A red seastar, *Henricia leviuscula* (§§69, 211).

§70. The leather star, *Dermasterias imbricata* (§211), may be found occasionally in the lowest rock pools of the protected outer coast, but it is far more common in completely sheltered bays and sounds.

§71. Southern California has two characteristic rocky-shore seastars, both semitropical forms: *Linckia columbiae* (Fig. 56) and *Astrometis.* A symmetrical *Linckia* might be mistaken for *Henricia* were it not for the former's mottled gray and red colors and for the difference in the range. *Linckia,* however, a typical nonconformist, rarely falls into the bourgeois error of remaining symmetrical. This grotesque little species, which may have a spread of 4 inches, is famous for autotomy and variability. Dr. S. P. Monks says of them: "In over 400 specimens examined not more than four were symmetrical, and no two were alike. . . . The normal number of rays is five, but some specimens have only one, while others have four, six, seven, or even nine. . . .

"Single living rays without any external sign of disk are not uncommon. . . . I have been fortunate in a series of experiments . . . in having a number of single rays, cut at various places, regenerate the disk and other rays."

Referring to autotomy she says: "The cause of breaking is obscure. . . .

56. Three specimens of the highly variable *Linckia columbiae* (§71).

If any external force bears a part in breaking the animal, it is probably that the creature is surprised when limp and relaxed, but I am inclined to think that *Phataria* [*Linckia*] . . . always breaks itself, no matter what may be the impulse. They may break when conditions are changed, sometimes within a few hours after being placed in jars. . . . Some never break, but stand all kinds of inconvenience. . . . Whatever may be the stimulus, the animal can and does break of itself. . . . The ordinary method is for the main portion of the starfish to remain fixed and passive with the tube feet set on the side of the departing ray, and for this ray to walk slowly away at right angles to the body, to change position, twist, and do all the active labor necessary to the breakage. . . . I have found that rays cut at various distances from the disk make disks, mouths and new rays in about six months."

Other anomalies recorded are: the animals may have from one to several mouths; the usual number of madreporites is two, but there may be from one to five; some of the rays may be young, whereas others are sexually mature; "comet" forms, in which there is one large ray surmounted by a minute disk with from one to six tiny rays, are of common occurrence. Here is polyvitality (if one is permitted to use such a word) with a vengeance. How the single arm of an animal that normally has a disk and five arms can live, regenerate, and grow offers a striking example of the flexibility and persistence of what might have been termed "vital purpose" a generation ago. In the matter of autotomy, it would seem that in an animal that deliberately pulls itself apart we have the very acme of something or other.

Linckia columbiae reaches its furthest northward extension in the southernmost part of our territory, at San Pedro and San Diego, extending from there to the Galápagos Islands.

§72. *Astrometis sertulifera* (Fig. 57), compensating in symmetry for *Linckia*'s temperamental ways, still clings to a mild autotomy. Specimens that are roughly handled reward their collector by making him the custodian of a flexible three- or four-rayed animal. The rich ground color of the slightly slimy skin is brown or green-brown. The spines are purple, orange, or blue, with red tips, and the conspicuous tube feet are white or yellow. This active and noticeable seastar, having an arm spread of up to 5 or 6 inches, ranges from Santa Barbara to the Gulf of California but never occurs in central or northern California.

§73. *Strongylocentrotus franciscanus* (Fig. 58), radiating spines like a porcupine, is the largest sea urchin found in this section, which is fitting in view of the length of its name. This giant brick-red or purple form

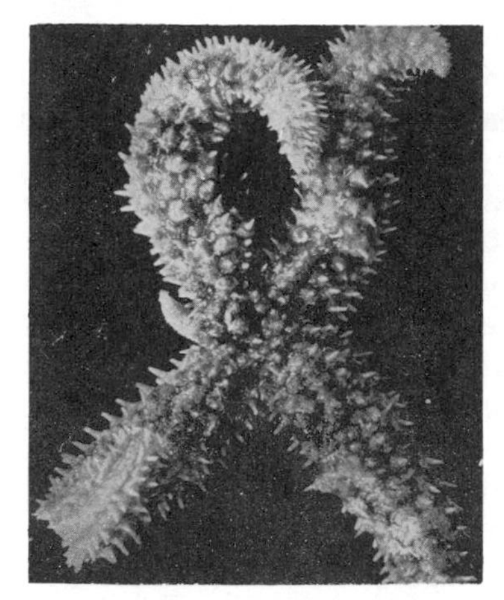
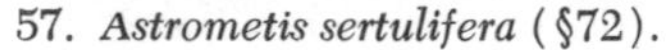

57. *Astrometis sertulifera* (§72).

58. *Strongylocentrotus franciscanus* (§73), the giant red urchin.

often has a total diameter exceeding 7 inches. Its gonads are considered a delicacy by the Italians, who come from a land where such seafood is highly valued (a smaller urchin being commonly marketed for food at Naples). We have sampled these gonads, eaten à l'Italienne (raw) with French bread, and found them very good—extremely rich, and possibly more subtle than caviar. If the race were not already being depleted by the appreciative Italians, and more recently by biologists, urchins could be highly recommended as a table delicacy.

Sea urchins are closely related to seastars and brittle stars. Whoever examines the test or shell of an urchin that has been denuded of spines will recognize this relationship in the pentamerous design and in the holes through which the sensory tube feet protrude. Urchins and seastars have three kinds of projecting appendages: spines, tube feet, and pedicellariae. In the urchins the movable spines are the most conspicuous of the three. When an urchin's test or shell is prodded with a sharp instrument, the spines converge toward the point touched so as to offer a strictly mechanical defense. However, if a blunt instrument is used, the spines turn away from the point of attack in order to give the pedicellariae free play. These peculiar appendages are thin, flexible stalks armed with three jaws apiece; each jaw is provided with a poison gland and a stiff sensory hair. The urchin can maintain a stout defense with its pedicellariae against the attack of, say, a predatory seastar; but if the attack is continued for long enough, the urchin is likely to succumb because of the loss of its weapons, since each pedicellaria is sacrificed after inflicting one wound.

Like the seastars, urchins make auxiliary use of their movable spines, crawling about by depressing the forward spines and pushing with the backward spines. Locomotion is relatively slow, since the fixed algae (such as sea lettuce) on which the urchin feeds are not famous for their speed of retreat. Movable jaw parts, forming a structure called, from its design, "Aristotle's Lantern," cut the seaweed into portions small enough for ingestion in the urchin's huge, coiled intestine. Urchins are also scavengers and will apparently eat almost anything they can manage, including dead fish.

Urchins reproduce much like the seastar *Patiria* (§26), by pouring out eggs and sperm into the water for chance fertilization; the embryonic development of the free-swimming stages is also very similar. The sexes are separate, and the extrusion of eggs and sperm has been observed in February and March along the California coast. Like *Patiria, Strongylocentrotus* has become an important object of embryological research. The larval development was investigated at Puget Sound by Johnson (1930), who found that the metamorphosis, similar in form to that of the other urchins, is not complete even after 62 days. Johnson also reported sexual maturity in March and April both in Puget Sound and at Monterey. The range of the species S. *franciscanus* is from Alaska to Cedros Island off the coast of Baja California, and it is abundant at Santo Tomas. It characteristically inhabits only the deeper pools and the rocky shores extending downward from the low-tide line.

An interesting feature of the digestive tract of S. *franciscanus* is the almost invariable presence of a small rhabdocoele flatworm, *Syndesmis franciscanus.* This commensal or parasitic worm, up to ¼ inch in length, usually reveals itself by sluggish movements when the intestine of its host is opened. Sometime in the past few thousand years *Syndesmis* presumably discovered that food-getting conditions on the inside of urchins were much superior to those on the outside and forthwith gave up free-living habits for good. How it gets inside is still a puzzle, however.

Often occurring with the giant red urchin, but preferring strongly aerated waters with violent surf (and hence treated in that habitat), is a smaller urchin, *Strongylocentrotus purpuratus* (§174). This species has a larger number of spines, which are vividly purple and are more closely cropped than those in the bristling panoply of S. *franciscanus.*

§74. The very word "abalone" conjures up a host of associations for the Californian who has accustomed himself to steaks from this delicious shellfish. Still more vivid memories of foggy dawns will be recalled by the sportsman who has captured this huge, limpetlike snail on the low-tide rocks and reefs. By prohibiting the shipping of even the shells to points outside the state, an often wise Fish and Game Commission has decreed that only California residents and visiting tourists may enjoy this finest of all mollusks. Even with only local consumption, however, abalones of legal size are becoming increasingly rare, and probably 99 per cent of those that the casual collector sees will be undersized and therefore protected by law. Abalones for market are taken by divers in 20 to 50 feet of water, but their distribution on isolated islands indicates that they were once abundant intertidally. Even in deep water, divers tell us, the abalone supply has become greatly depleted. Years ago, when the price was 50 cents a dozen, it would not have been worthwhile to bring them in, according to one diver—except that a man could sit down on a ledge and, without changing his position, fill his basket from a single rock. Now a diver must walk along the bottom; and, if fortunate, he may fill his basket in half an hour.

Haliotis rufescens, the red abalone (Fig. 59), may grow to be 10 or 11 inches long, but 6½ inches is the minimum size that may legally be taken, and larger specimens are rare near shore. Abalone shells have been valued since their discovery by human beings with an eye for iridescent colors; and abalone pearls, formed when the animal secretes a covering of concentric layers of pearly shell over parasites or irritating particles of gravel, have in times past made fashionable jewelry. One of the common instigators of abalone pearls is the small boring clam, *Pholadidea conradi* (a relative of the shipworm *Teredo*), which, except for the great difference in size, looks much like *Penitella penita* (Fig. 202).

Haliotis occurs most frequently on the undersides of rock ledges, where it clings, limpet-fashion, with its great muscular foot. If taken unawares, specimens may be loosened from their support easily; but once they have taken hold, it requires the leverage of a pinch bar to dislodge them. Some of this clinging force, according to Crofts, is probably the result of a vacuum. The copious secretion of mucus on which the animal "slides" would certainly be an aid in perfecting a vacuum beneath the foot. Stories of abalones that hold people until the incoming tide drowns them are probably fictitious, but it is nevertheless inadvisable to try to take them by slipping one's fingers under the shell and giving a sudden pull. We have captured them in this manner when no bar was available, but there is always danger of a severe pinch.

59. A red abalone, *Haliotis rufescens* (§74); this specimen has been turned on its back and is extending its great foot preparatory to righting itself.

The life history of the abalone has been carefully investigated, chiefly in connection with commercial fishery legislation. The red abalone, which is the only one that is marketed extensively, reaches breeding age at 6 years, at which time it is about 4 inches in diameter. One specimen known to be 13 years old was 8 inches long. During her first breeding season a female produces about 100,000 eggs, but a 7-inch specimen will produce nearly 2½ million. Crofts says, "The genital products are shot out in successive clouds," the sperm white and the eggs gray-green. "All five naturally spawning *Haliotis* observed were stationary during spawning, which occurred as the tide commenced to flow, and on a warm sunny day." Spawning takes place between the middle of February and the first of April. The animal's diet is strictly vegetarian. It crawls about slowly in the forests of algae, filling up an enormous gut with such fixed plants as sea lettuce and kelps.

Abalones are known to have existed as long ago as the Upper Cretaceous Age. They now occur in the Mediterranean, and even on the coast of England, but it is only in the Pacific that they attain great size, most being found off California, Japan, New Zealand, and Australia. The red abalone is distinguished by its great size, by the (usually) three or four open and elevated apertures in its shell (through which water used in the gills is discharged), and by the whole forests of hydroids, bryozoans, and plants that frequently grow on its shell. It ranges from Cape Arago, Oregon, to northern Baja California. It is most abundant on the central California coast between Mendocino County and San Luis Obispo, but specimens are now rare in the intertidal zone, and most of them are taken in deeper water by professional divers or near the shore by amateurs. The red abalones at Cape Arago are scarce (and becoming scarcer) and appear to be members of a nonreproducing population; all of them are large, and no young ones have been found. Recently (1967) the Oregon Fish Commission has planted several hundred young *Haliotis rufescens* just north of Newport, Oregon. The stock came from a commercial shellfish hatchery near Pigeon Point in California.

The abalones everywhere in the world need all the help they can get. The lovely *ama* of Japan, who used to dive almost nude for the *awabi,* have found that neoprene wet suits enable them to stay under water longer and to dive more frequently in cold water. Alas, the short Japanese figure is most unlovely in a black or white neoprene diving suit; aesthetics have been sacrificed for more *awabi.* There is talk of aiding the commercial divers on the coast of California with underwater television. Tests have been made with a TV camera suspended from a small boat with a screen on deck, so that concentrations of abalones may be spotted and the divers directed to them.

In the days before man learned to dive in black rubber suits with bright orange trimmings and yellow or international-orange air tanks (an outfit resembling nothing so much as a large bass lure), the abalones of shallow coastal waters were the principal food of the sea otter. The otters, once abundant, were hunted almost to extinction for their valuable fur more than a hundred years ago. Ever since a small herd of sea otters was observed south of Carmel in the 1930's there has been great concern for these interesting and valuable animals, and they have been rigidly protected. Now it seems that the sea otter may have been protected too well, as far as those interested in eating abalone are concerned. The herd now contains more than 500 otters, and their predilection for abalone has not gone unnoticed. The herd is invading the commercial abalone grounds to the south of the sea otter refuge. So there is the usual cry: the predators must go; man must have all the prey. The fish and game people wonder if the sea otters can be encouraged to stay within their reservation, perhaps by feeding them abalone scraps from the commercial abalone plants, or perhaps by some sort of sonic barrier. Somehow, this sounds too much like what we did to the Indians—chase them off the choice land and feed them beef culls in place of buffalo. Two hundred years ago there were evidently vast herds of sea otters, and the abalones survived. Today, even with the comparatively small herd of 500, the abalones are getting scarcer. The inference is obvious: instead of moving the sea otters away or "harvesting" them to keep the population down, we ought to "harvest" ourselves, if we really want to eat more abalone, especially on a per capita basis.* Finishing off the sea otters will not solve the problem of human population increase, and less popular alternatives will have to be considered —such as eating fewer abalones.

One enterprising group decided that establishing some artificial reefs where abalones might be planted and nurtured would be a profitable venture. There was even talk of an independent country, to be called "Abalonia," just outside the 12-mile limit near San Diego. Evidently, everyone involved was to become rich, as well as politically independent, with the aid of the luscious abalone to be raised in this venture. Alas, the hulk being towed out to start this empire sank in the wrong place; and when last heard of, the entrepreneurs of Abalonia were in difficulties with the Coast Guard for creating a menace to navigation.

When an abalone is to be eaten, its shell is removed, the entrails are cut away, and the huge foot is cut into several steaks about ⅜ inch thick. Commercial fish houses pound these steaks with heavy wooden mallets; amateurs use rolling pins or short pieces of two-by-four. Whatever the instrument, the pounding must be thorough, so as to break up the tough muscle fibers. Preferably, the meat should then be kept in a cool place for 24 hours before cooking. When properly prepared and cooked, abalone steak can be cut with a fork; otherwise it is scarcely fit to eat.

On the outside Alaska coast, as at Sitka, the small and dainty *H. kamtschatkana* may sometimes be taken in abundance, since it, like *H. rufescens,* occurs gregariously. It has rarely been taken as far south as Monterey. Our specimens from Sitka Sound were up to 4½ inches long, with butterscotch-colored flesh and brown-and-gray sides. Steaks from this small abalone are said to require no pounding, but it is probably well to work over any abalone before frying it.

There are eight species of abalone on the Pacific coast. For details concerning their ranges and habits, as well as a series of excellent color

* Harvesting the sea otter for its pelt could become a very lucrative affair; in 1968 four sea-otter pelts from Alaska were auctioned in Seattle for $2,300 apiece for a coat to be sold by a certain mercantile establishment in Dallas. The average price per pelt for all 826 skins at this auction was slightly more than $170, however. A return to the sea-otter trade can be expected to bring with it all the complications of biopolitics and might well cost the state of California more than it would be worth in extra wardens, red tape, and all.

plates of the shells, the reader is referred to Keith Cox's bulletin in *California Abalone* (Fish Bulletin No. 118 of the California Department of Fish and Game).

§75. Below Point Conception, and as far south as we have collected along the coast of Baja California, *H. fulgens*, the smaller green abalone, is the common form. It may be distinguished by its flat, clean shell, which has fine corrugations and six open holes with the rims only slightly elevated. Some green abalones are taken commercially, and we have often seen Mexicans gathering them for their own use.

Still another *Haliotis*, the black abalone (§179), may occur with either of the above, but it is more common in surf-swept regions.

§76. The collector who, from seeing *Tonicella* (§36) and an occasional *Nuttallina*, has become familiar with the chitons for which the Pacific is famous will surprise himself some day by turning up a perfectly enormous brick-red sea cradle with no apparent shell. This giant, *Cryptochiton stelleri* (see color section), sometimes called the "gum boot," is the largest chiton in the world, growing up to 13 inches long. It is reputed to have been used for food by the coast Indians and was eaten by the Russian settlers in southeastern Alaska. After one experiment we decided to reserve the animals for times of famine; one tough, paper-thin steak was all that could be obtained from a large *Cryptochiton*, and it radiated such a penetrating fishy odor that it was discarded before it reached the frying pan.

Although this giant form is sensitive to light and feeds mostly at night, as do the other chitons, it may remain out in the tide pools and on the rocks all day when there is fog. *Cryptochiton* feeds on fixed algae, rasping its food into small particles, snail-fashion, with a large radula, which may be examined by reaching into the chiton's mouth with a pair of blunt forceps and drawing out the filelike ribbon.

Chitons, although usually placed in a separate class, are closely related to snails. Like snails they have a long, flat foot, with the internal organs between it and the shell; but chitons have symmetrical double gills, one along each side of the foot, whereas most snails have their anatomy so convoluted and twisted that only one gill remains. A chiton's shell is formed of eight articulated plates, which allow the animal to curl up almost into a ball when disturbed. In *Cryptochiton*, however, the plates are not visible externally, the fleshy girdle having completely overgrown them. The hard, white, disarticulated plates of dead specimens are often cast up on the beach as "butterfly shells."

Cryptochiton ranges from Alaska westward to Japan and southward to San Nicolas Island off southern California, though specimens are not numerous below Monterey Bay. Sometimes in the spring great congregations of them gather on rocky beaches, having presumably come in from deep water to spawn. An account of the breeding habits of a related chiton (*Stenoplax heathiana*) is given in §127.

More than 25 per cent of the cryptochitons examined will have a commensal scale worm, *Arctonoë vittata,* formerly *Halosydna* (*Polynoë*) *lordi,* living on the gills. These worms, with 25 or more pairs of scales, sometimes shed them when disturbed. *Arctonoë* grows up to 3 or 4 inches long and is light yellow in color. The same worm may be found in the gill groove of the keyhole limpet *Diodora aspersa* (§181).

Another guest that may be found clinging to the gills of *Cryptochiton* is the pea crab, *Opisthopus transversus* (Fig. 60, inset). This tiny crab, likewise never free-living, takes to a variety of homes, for it is found regularly in the California cucumber (§77), frequently in the mantle cavities of the quiet-water mussel and the gaper (*Tresus*), occasionally in the giant keyhole limpet (*Megathura*), and even in the siphon tubes of rock-boring clams.

§77. Tourists at Monterey who go out in the glass-bottomed sightseeing boats are often shown "sea slugs," which are actually sea cucumbers. These rather spectacular animals, *Stichopus californicus* (Fig. 60), are usually subtidal, but they will often come to the attention of the shore collector. In the south—from Laguna Beach to Baja California—there is a similar-looking *Stichopus, S. parvimensis,* which is definitely intertidal, living as high as the middle tide pools. The cylindrical, highly contractile bodies of these cucumbers, black, dull brown, or red, are up to 18 inches long. The animals are covered above with elongated warts and below with tube feet with which they attach themselves and crawl. Cucumbers may be flaccid when unmolested; but when annoyed, they immediately become stiff and turgid, shorter in length, and very thick.

We have pointed out that the lay observer must tax his credulity before grasping the fact that urchins are closely related to seastars and brittle stars. Sea cucumbers tax one's credulity still further, for they are related to all three of these, despite their wormlike bodies, their possession of tentacles, and their apparent lack of a skeleton. Their pentamerous symmetry is much less obvious than in urchins. It can be seen readily, however, in *Cucumaria miniata* (Fig. 200). The water-vascular system characteristic of echinoderms is manifest in the cucumber's tube feet, which, with wormlike wrigglings of the body wall, serve for locomotion. The tentacles around the mouth are actually modified tube feet. Like other echinoderms ("spiny-skins"), cucumbers have a cal-

60. The California sea cucumber *Stichopus californicus* (§77), compared with a 6-inch rule; these animals are usually found in deep water, but sometimes move up into the intertidal. Inset: the pea crab *Opisthopus transversus* (§76), a common commensal of the cucumber and the gumboot chiton.

careous skeleton; but in their case it is only vestigial, composed of plates and spicules of lime buried in the skin and serving merely to stiffen the body wall.

Cucumbers in general have a specialized form of respiration that is unique among the echinoderms. Water is pumped in and out of the anus, distending two great water lungs, the respiratory trees, that extend almost the full length of the body. As would be expected, this hollow space, protected from the inclemencies of a survival-of-the-fittest habitat, attracts commensals and parasites. The respiratory tree is the common home of two microscopic, one-celled animals, *Licnophora* and *Boveria,* each of which clings by means of a ciliated sucking disk at the end of a posterior fleshy stalk. In aquaria the same pea crab that occurs with *Cryptochiton, Opisthopus transversus,* may be found in and about the posterior end of *Stichopus.* When cucumbers are being relaxed for preservation, the crab will often leave his heretofore dependable home in the interior to cling to the outside. A scale worm, *Arctonoë* (formerly *Acholoë, Halosydna,* or *Polynoë*) *pulchra,* distinguishable by the fact that the scales have dark centers that are lacking in *A. vittata,* may often occur with the crab.

Cucumbers usually lie half buried in the soft substratum, passing through the intestinal tract quantities of sand and mud from which their food is extracted. *Stichopus,* however, is a rock-loving form. Considering its habitat and the nature of its tentacles, it seems likely that this cucumber brushes its stumpy appendages along the surface of the substratum as a related English form is known to do, sweeping minute organisms and bits of debris into its mouth. A New Zealand holothurian has been observed to emerge from its hiding places at night to feed, returning sometime before dawn.

The collector who insists on taking living specimens of these animals away from the tide pools for subsequent observation is sure to be provided, if the water gets stale, with firsthand information on the general subject of evisceration, a cucumber trait that has been termed "disgusting" by unacclimated collectors. When annoyed, *Stichopus* spews out its internal anatomy in a kind of autotomy. The organs thus lost are regenerated if the animal is put back into the ocean. In the usual evisceration the anus is ruptured by the pressure of water caused by a sudden contraction of body-wall muscles. This contraction voids first the respiratory trees and subsequently the remainder of the internal organs. In one or two well-known examples the trait has protective value. An English form is called "the cotton spinner" from its habit of spewing out tubes covered with a mucus so sticky when mixed with seawater that it entangles predatory animals as large as lobsters.

The sexes are separate. As with many other marine animals, hosts of ova and sperm are discharged into the water, fertilization being pretty much a matter of chance. In Puget Sound spawning takes place in late July and August, and the animals come into shallow water to spawn (or at least their migration into shallow water is coincident with the ripening of the sexual products). However, it has been found that temperatures of 14° C. (about 59° F.) or above—temperatures often encountered in Puget Sound shallow water—mean death to the animals. The safe temperature for developing eggs is only 10° C.—a temperature found only where there is oceanic water with strong currents.

Stichopus californicus and its related species have a wide distribution on the Pacific coast, ranging down into Mexico.

Visitors to San Francisco's Chinatown (and probably visitors to other American Chinatowns) have seen *Stichopus* and a similar cucumber in the food shops. The body wall, first boiled and then dried, is a great delicacy. In the South Seas the production of this "trepang" or "bêche-de-mer" is an important fisheries industry. The very cheapest grades of trepang are sold by Chinese merchants in this country at more than a dollar a pound.* It has the reputation, along with ginseng and bird's-nest soup, of being an efficient aphrodisiac.

* At last report (1967), the price was up to $2.30 for a single dried animal packed neatly in a plastic bag.

§78. In rocky clefts and tide pools where currents of pure water sweep by, delicate sprays of hydroids will be found, often in conspicuous abundance. Until their peculiarities are pointed out to him, the uninitiated collector is quite likely to consider them seaweeds, for only on very close inspection, preferably with a microscope, does their animal nature become apparent. They may occur, like *Eudendrium californicum,* in great bushy colonies, sometimes 6 inches tall, or, like *Eucopella everta* and various species of *Halecium* and *Campanularia,* as solitary animals about 1/16 inch high, forming a fuzz about the stems of larger hydroids. (See Figs. 61–70.)

Whether large or small, the hydroids are likely to be first noticed for their delicate beauty and often exquisite design. Few things in the exotic tide-pool regions will bring forth more ecstatic "oh's and "ah's" when first examined than these hydroids, and one cannot but wonder why their fragile patterns have never been used as the motif for conventional designs. *Plumularia setacea* (Fig. 61), one of the plume hydroids, is possibly the most delicate of all intertidal animals. Under water, and against many backgrounds, it positively cannot be seen, so perfect is the glassy consistency of the living specimen. This is all the more remarkable when it is remembered that there is a covering of skeletal material over the stems and over the budlike individual animals, or hydranths. When removed from the water, or seen against certain backgrounds, the tiny sprays, ½ inch or so long, are plainly visible and well worth examination. *P. setacea* extends from Vancouver Island to California. In the north, the larger and more robust *P. lagenifera,* which occurs sparsely on the central California coast, largely replaces *P. setacea,* until, north of Vancouver Island and clear up at least to Sitka, it becomes the commonest plume hydroid.

Hydroids have a rather startling life history. It is a grotesque business, as bewildering to the average man as if he were asked to believe that rosebushes give birth to hummingbirds, and that the hummingbirds' progeny become rosebushes again. The plantlike hydroid that the shore collector sees gives rise by budding to male and female jellyfish, whose united sexual products develop into free-swimming larvae, the planulae, which attach and become hydroids, like their grandparents. The life cycle, then, is hydroid-jellyfish-larva-hydroid.

Unfortunately for the efficiency of this account, few or none of the rocky tide-pool hydroids produce free-swimming jellyfish. The reason is apparent: a jellyfish budded off from a rock-pool hydroid would begin its independent life in such a dangerous neighborhood that its chances of survival would be almost nil as compared to its chances of being dashed against the rocks and destroyed. Accordingly, the jellyfish generation, among the rock-pool species, is usually passed in sacklike gonangia distributed about the hydroid stem. In some cases, as with the sertularians and the plumularians, the gonangia have degenerated into little more than testes and ovaries; but the medusae of *Tubularia marina,* although attached, actually have tentacles and look like the tiny jellyfish that they are. Species like *Obelia* (§316), which inhabit more protected waters or the pilings of wharves, commonly have a conspicuous, free-swimming jellyfish generation. It should be mentioned that not all jellyfish are born of hydroids. All of the hydromedusae come from hydroids, however; and most of the scyphomedusae, or large jellyfish, are budded from a stalked, attached, vaselike animal (the scyphistoma), which is the equivalent of the hydroid stage. The hydroid stage of one of our most common hydromedusae, *Polyorchis,* is still unknown, however.

Hydroids are interesting for another reason—they provide a protective forest for many other small animals, some of them, like the skeleton shrimp, absurd beyond belief. These dependents will be mentioned in detail later.

With the aid of the accompanying photographs and drawings the collector should be able to identify the hydroids listed in the following paragraphs. These are, we believe, the commonest forms, but there are innumerable others, and an attempt to list them would only lead us into confusion worse confounded.

§79. *Eudendrium californicum* (Fig. 62), mentioned above, is one of the most conspicuous of the hydroids, with bushy colonies that are sometimes 6 inches tall. The brown, slender stem by which the spirally branching colony attaches to the rock is stiff and surprisingly strong, and the roots are very firmly attached. The zooids—the individual animals in the composite group—look like bits of pink cotton on some lovely sea plant. The minute white tentacles surmounting them can be seen readily if the colony is held to the light in a jar of seawater. They retract slowly if touched, and lack the protective hydranth cups (thecae) into which so many of the hydroids snap back their tentacles.

Hydroids belong in the same great group as jellyfish, anemones, and corals; all of these are characterized by their lack of a specialized digestive tract, food simply being dumped indiscriminately into the body cavity, where it is digested. The stem must be regarded as part of the colony, since through it runs part of the system of minute tubes that connects with all the zooids and transmits dissolved food. A good-sized

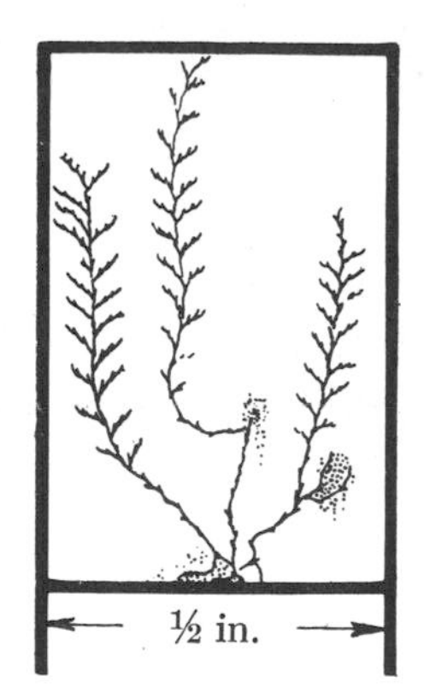

1. A small plume hydroid, *Plumularia setacea* (§78).

62. The hydroid *Eudendrium californicum* (§79).

3. The orange-colored hydroid *Garveia annulata* (§81).

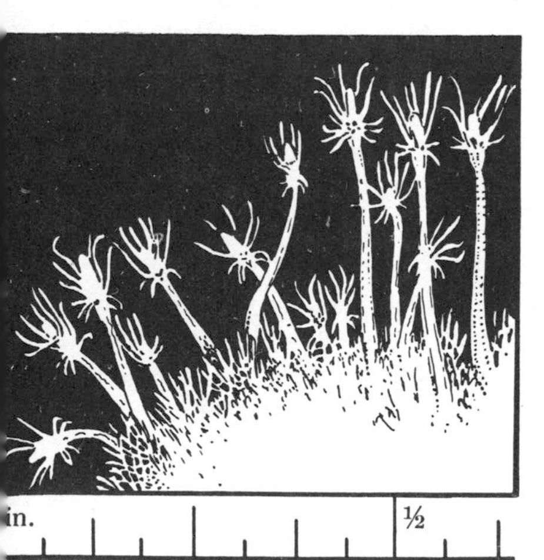

64. A colony of *Hydractinia milleri* (§82).

65. *Abietinaria anguina* (§83).

cluster of *Eudendrium* may have thousands of living zooids, all with tentacles outstretched for the capture of passersby smaller than themselves. The prey is stung by the nematocysts and carried into the body cavity through the central mouth. During the winter this hydroid bears fixed jellyfish in the form of orange (female) and green-centered pink (male) gonophores. The female gonangia, borne at the base of the cuplike hydranths, may be turgid, and may invest the zooid so completely as to hide all but the tips of the tentacles.

Except on very low spring tides, colonies of *Eudendrium* are rarely exposed. They occur, even in regions of high surf, from Sitka to San Diego.

§80. *Tubularia marina,* differing from its common and clustered brother *T. crocea,* of boat bottoms and wharf pilings, is strictly solitary. Individuals may be spaced evenly a few inches apart in rocky crevices or under ledges. The stem, usually not more than an inch long, bears a relatively large but dainty hydranth, often vividly orange. The medusoids are red, with pink centers, and have two very noticeable long tentacles. Specimens that are sexually mature, that is, with sessile medusae present, have been taken at Pacific Grove early in February. The range is from Trinidad (near Eureka) to Pacific Grove.

§81. The hydroid *Garveia annulata* (Fig. 63), sometimes 2 inches high, with 20 or 30 zooids similar in size and appearance to those of *Eudendrium,* is recognizable by its brilliant and uniform color—orange root, stem, and hydranth, with lighter orange tentacles. The colonies are likely to be seen at their best during the winter and spring months, when they commonly appear growing through or on a sponge. They also grow on coralline algae or at the base of other hydroids, and they themselves are likely to be overgrown with still other hydroids. *Garveia* ranges from Alaska to southern California.

§82. *Hydractinia milleri* (Fig. 64), which occurs in pink, fuzzy masses on the sides of rocks, is interesting because it illustrates one of the most primitive divisions of labor—nutrition, defense, and reproduction are each carried on by specialized zooids. *H. milleri* is known from the outside coast of Vancouver Island, and it ranges at least as far south as Carmel.

§83. The hydroids considered above are characterized by naked hydranths. Those that follow have hydranths housed in tough skeletal cups (the thecae), and most of them withdraw their tentacles into these protective cups very rapidly at the slightest sign of danger. The fernlike colonies of *Abietinaria anguina* (Fig. 65) have well-developed thecae, and the obvious skeleton is all that one ordinarily sees. During the winter or spring months one of the clean-cut sprays should be placed in a dish of seawater, left unmolested for a few minutes, and then observed. The transparent zooids will be seen popping out of the hard cups, forming a total pattern that is amazingly beautiful. At other times the dead branches will be furred with other, smaller hydroids and with the suctorian protozoan *Ephelota gemmipara,* and will be covered with brown particles, which are diatoms.

A. anguina sometimes carpets the undersides of ledges or loosely embedded rocks, together with a smaller, undetermined relative, possibly the *A. amphora* also reported from the outer shores of Vancouver Island. The bushy-colonied *A. greenei* (Fig. 66), another of the numerous Pacific intertidal species of *Abietinaria,* will be taken for a totally different animal. The various species range well up the coast and south as far as San Diego, *A. anguina* and *A. greenei* being fairly well restricted to oceanic waters.

§84. The Pacific coast is famous for its ostrich-plume hydroids, which belong to the genus *Aglaophenia.* Giant specimens (*A. struthionides*) are more common in violently surf-swept clefts of rock (§173) or in deep water, but they may be found here also. In the habitat considered here, however, there are at least two species so similar in appearance that only the larger form is illustrated. *A. inconspicua* occurs as 1-inch bristling plumes foxtailing over eelgrass or corallines, and is sometimes found at the base of larger species of the same genus. *A. latirostris* (Fig. 67) is a clustered form with dainty, dark brown plumes as long as 3 inches. Surprisingly enough, *A. latirostris* was originally described from Brazil; but it has since been recorded from Puget Sound. *A. inconspicua* is restricted to central and southern California.

§85. Several other skeletal-cupped forms, all beautiful, turn up very frequently in the lower tide pools. *Eucopella caliculata* (formerly *Orthopyxis,* Fig. 68) forms a creeping network on the "stems" and fronds of algae that grow just at or below the extreme low-tide line. This network and the stalked solitary zooids that arise from it are opaque white and are easily seen; even the extended tentacles are visible against the proper background. Authorities disagree on whether or not *Eucopella* liberates a free-swimming medusa stage. Certainly, conditions of surf in this habitat are such that free-swimming jellyfish would at least be at a disadvantage. Several species of *Eucopella* are widely distributed on the Pacific coast, and *E. caliculata* is almost cosmopolitan in both hemispheres.

§86. *Sertularia furcata* (Fig. 69) is a small hydroid, known from the outer coast of British Columbia to San Diego, that forms a furry growth on the blades of eelgrass, and occasionally on algae. The related *Sertu-*

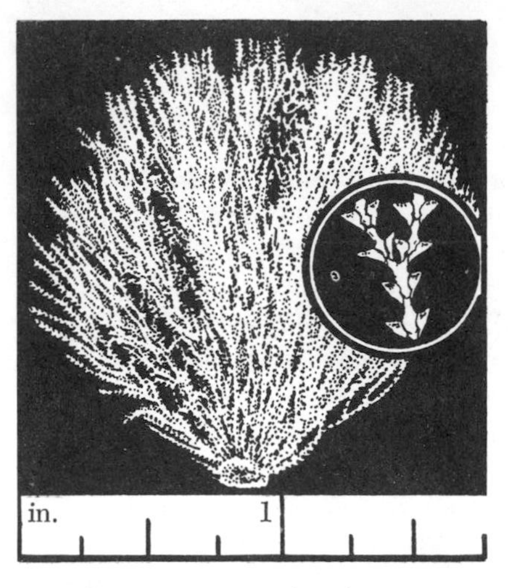

66. *Abietinaria greenei* (§83).

67. Right: an ostrich-plume hydroid, *Aglaophenia latirostris* (§84).

69. *Sertularia furcata* (§86).

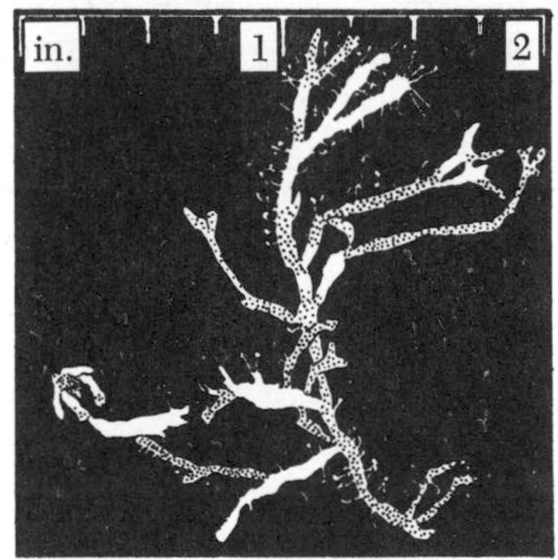

68. The hydroid *Eucopella caliculata* (§85).

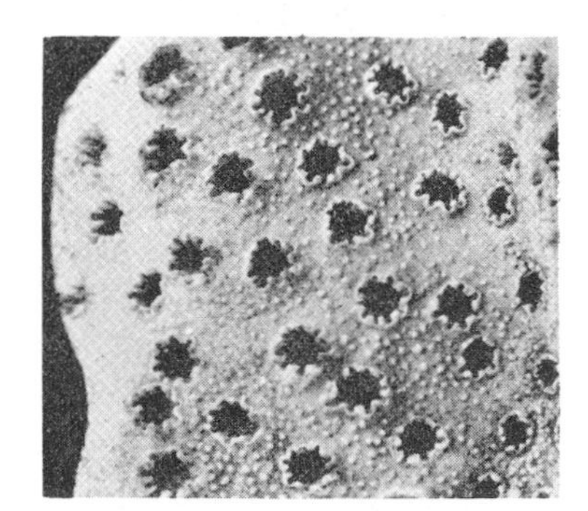

70. *Allopora porphyra* (§87), a hydrocoral.

larella turgida is one of the most common and widespread hydroids on the coast. It is recorded from many points between Vancouver Island and San Diego, from both the protected outer coast and the fully protected inner coast, and from deep water as well as intertidal. The stout and robust stalks of *S. turgida* have the zooids, which are swollen to turgidity, arranged alternately on the stem, rather than opposite each other, as in *Sertularia furcata.*

The coppinia masses (compact, fuzzy encrustations of gonangia concentrated about the stems of the hydranths) and the strangely irregular creeping or mildly erect hydroid colonies of *Lafoea* species (*L. dumosa* and *L. gracillima*) will often be seen in the north; these are very characteristic, and, once recognized, they are not likely to be mistaken for anything else. Both species range from Alaska to San Diego, but will be found intertidally only in the north.

§87. The hydrocoral *Allopora porphyra* (Fig. 70) is related to the hydroids, and not to the corals it resembles. Distinct and vividly purple colonies may be found encrusting rocky ledges at very low tide levels where the surf is fairly powerful but where no sediment occurs. In some of the semiprotected regions south of Monterey, and on the Oregon coast, *Allopora* colonies cover many square feet of continuous rock surface. Deep-water colonies of similar forms, on the Pacific coast and elsewhere, form huge erect masses, bristling with branches and the sharp points of the colorful lime skeleton. In the Caribbean they form some of the "coral" reefs.

Many hydrocorals expand only at night, but our California shore form is even more conservative—it never expands at all. Occasionally, however, one gets a glimpse of a white polyp down in the tiny craters that protect the feeding zooids.

§88. As would be expected in a haunt so prolific as the lower tide-pool zone, the shelter of the hydroid forests attracts a great many smaller animals, both sessile and active. In this work, it is difficult to say at just what point animals become too inconspicuous to be considered, for in the tidelands it is almost literally true that

> Great fleas have little fleas
> Upon their backs to bite 'em,
> And little fleas have lesser fleas,
> And so *ad infinitum.*

A hydroid colony no bigger than can be contained in one's cupped hands may be almost a whole universe in itself—a complete unit of life, with possibly dozens of units in one tide pool. Each little universe includes amphipods, isopods, sea spiders (grotesque arthropods that are spiders only in appearance), roundworms, other hydroids, bryozoans, and attached protozoa, with a possible number of individuals that is almost uncountable. Though all of these animals are at least visible to the naked eye, and are abundant, characteristic, and certainly not lacking in interest, most of them are too small to be seen in detail without a hand lens or microscope and hence cannot be included in this handbook. It is impossible, however, to pass over some of the largest of these fantastic creatures without more detailed mention.

§89. Caprellids, amphipods very aptly known as skeleton shrimps, may be present in such multitudes as to transform a hydroid colony into a writhing mass, and the tyro will insist that it is the hydroids themselves that do the wiggling, so perfect is the resemblance between the thin, gangly crustacea and the stems they inhabit. If caprellids were a few feet tall instead of around an inch, as in the case of the relatively gigantic *Metacaprella kennerlyi* (Fig. 71), no zoo would be without them, and their quarters would surpass those of the monkeys in popularity. Specimens seen under a hand lens actually seem to be bowing slowly, with ceremonial dignity; clasping their palmlike claws, they strike an attitude of prayer. Often they sway from side to side without any apparent reason, attached to the hydroid stem by the clinging hooks that terminate their bodies, scraping off diatoms and bits of debris, or possibly eating the living zooids of the host. As a counterbalance to the animal's preposterous appearance, it evidences great maternal solicitude; the female carries her eggs and larvae in a brood pouch on her thorax. This prettily pink-banded amphipod, most commonly found in *Abietinaria* colonies, ranges from Alaska to southern California. A number of smaller species (e.g., *Caprella equilibra* and several now incorrectly grouped as one species, *C. "acutifrons"*) will be found in *Aglaophenia* colonies and elsewhere.

Caprellids can, when they wish, climb rather actively about on the branches of their hydroid homes in a manner suggestive of measuring worms. They take hold with their front appendages and, releasing the hold of their hind legs, bend the body into a loop, moving the clinging hind legs forward for a new hold. The body is extended forward again and the motion repeated until the desired destination is reached.

§90. Among the pycnogonids, or sea spiders, the male carries the eggs on an extra pair of appendages lacking in many females. Sometimes the thin, gangly-legged animal will be seen weighted down with a white

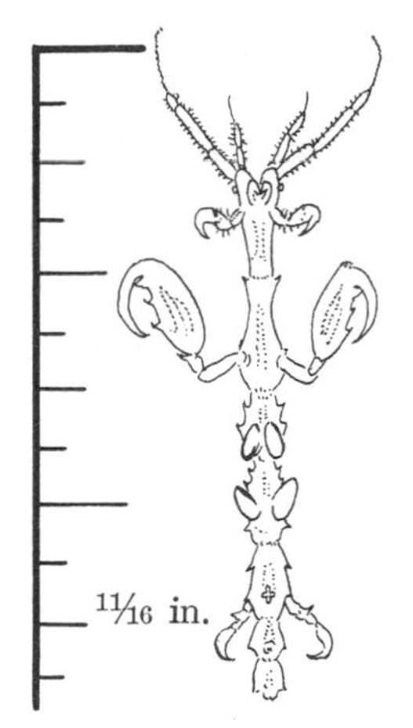

71. *Metacaprella kennerlyi* (§89), a skeleton shrimp.

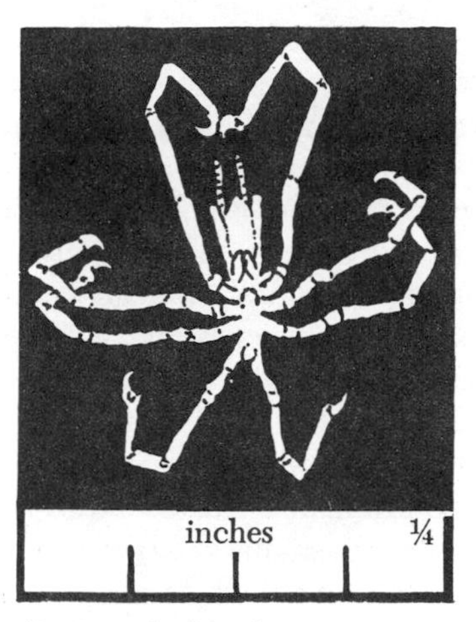

72. *Lecythorhynchus marginatus* (§90), a sea spider.

mass of eggs that must be as heavy as he is. The bodies of sea spiders are not large enough to contain all of the internal anatomy, so that, as with seastars, the stomach extends into the legs. Few large clusters of *Aglaophenia latirostris* are without these weird creatures, and usually the comparatively large *Lecythorhynchus marginatus* (Fig. 72), with a leg spread up to ¾ inch, is the most conspicuous species. It is well disguised by its color—amber, with darker bands on the legs—and by its sluggish habits, so that specimens often may not be seen until the hydroids on which they live are placed in a tray and carefully examined. One larval stage of *L. marginatus,* not even faintly resembling the adult (if the life history of this form resembles that of similar types that have been investigated), actually lives within the hydroid, entering by means of sharply pointed appendages and feeding on the body juices of the host. The legs of sea spiders end in recurved hooks, and are admirably adapted to crawling about on vegetation and hydroids. *L. marginatus* has been reported from the vicinity of Vladivostok (on *Zostera*) and it occurs on the California coast southward to Cedros Island.

Tanystylum californicum, a much smaller sea spider, occurs on *Aglaophenia.* It is rounder and more symmetrical than its larger fellow lodger, and its legs are banded with white instead of light yellow, in a color pattern that matches its host.

A. latirostris is for some reason a favored pycnogonid habitat, and most of the sea spiders mentioned here have been collected at one time or another on this hydroid. Particularly common along the California coast during some years (possibly alternating in abundance with *L. marginatus*) is the thin and gangly *Phoxichilidium femoratum*, reminiscent of the *Anoplodactylus* (§333) taken with *Tubularia* and *Corymorpha*. *P. femoratum* is known to range from Dutch Harbor, Alaska, to Laguna Beach.

§91. Sponges of many kinds are common in the low intertidal landscape. Far simpler in organization than any of the hydroids, and near the bottom of the scale in the animal kingdom, some of the sponges seem to be more or less loose aggregations of one-celled animals—colonies of protozoa, one might say—banded together for mutual advantages. This loose organization has been the subject of recent experiments with West Indies sponges. Bits of carefully dissociated tissue were strained through silk mesh and allowed to stand in dishes of seawater under proper conditions. In a few days the separated pieces were found to have united to form new individuals. Different-colored species of sponges have united in this manner to form a mosaic colony.

Though the sponges have no mouths, stomachs, or other specialized internal organs, they do have flagella, or lashing "tails," which pull a continual current of water through the colony, bringing in food particles and expelling waste products. The incoming, food-bearing current passes through innumerable fine pores, and the outgoing, waste-bearing current through characteristic crater-like vents (oscula). The observer who would see for himself the currents of water that mark these plantlike encrustations as animals may do so by transferring one of the large-pored sponges from its tide pool to a clean dish containing fresh seawater and adding a little carmine to the water.

Notwithstanding the ease of making this simple experiment, there has long been a popular belief that sponges are plants. English observers once supposed, the *Cambridge Natural History* notes, that sponge colonies were the homes of worms, which built them as wasps build nests, and as mud wasps build crater-like holes. The sharp-eyed and sharper-witted Ellis disproved this theory when he stated, in 1765, that sponges must be alive, since they sucked in and threw out currents of water. It is pleasant to note that the mighty intellect of Aristotle, far on the other side of the foggy Middle Ages, knew them for animals.

A modern investigator, Parker, remarks that in the vicinity of large tropical sponges, the "water often wells up so abundantly from the sponges as to deform the surface of the sea much as a vigorous spring deforms that of a pool into which it issues." Laboratory tests indicate that a single such sponge colony, with a score of "fingers," will circulate more than 400 gallons of water in a day.

The reproduction of sponges is sometimes by eggs and sperm (the latter being liberated into the water) and sometimes by asexual budding. In any case, free-swimming larvae emerge from the oscula of the parent and swim away, propelled by numerous flagella on the *outside* of the body. When these larvae attach to rocks, the cells move around so that those with flagella are on the *inside* of the animal, where they begin the function, already mentioned, of creating currents of water.

§92. The giant *Spheciospongia confoederata* is related to a West Indies sponge that is the largest in the world. The Pacific intertidal form occurs at the lowest tidal level in great, slaty clusters on which several people could be seated. On close inspection these clusters are recognizable as sponges by the presence of numerous pores and oscula. At a distance of a few feet, however, there is nothing to distinguish them from a hundred boulders that may be scattered about in the same acre of tide flat. *Spheciospongia* is known from Monterey Bay, and we have seen a similar or identical form along the shore in northern Baja California.

§93. Masses of the cream-colored *Leucosolenia eleanor* (Fig. 73), one of the most primitive of sponges, are common at the lowest intertidal

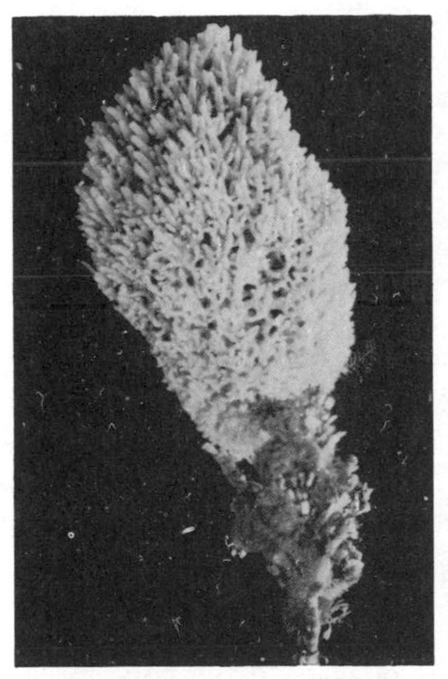

73. *Leucosolenia eleanor* (§93).

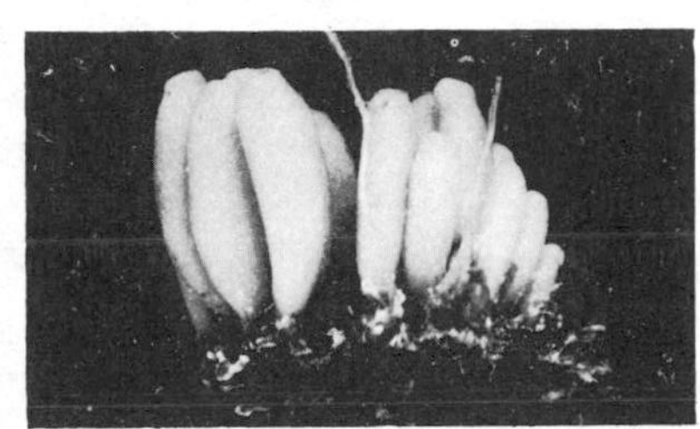

74. *Rhabdodermella nuttingi* (§94).

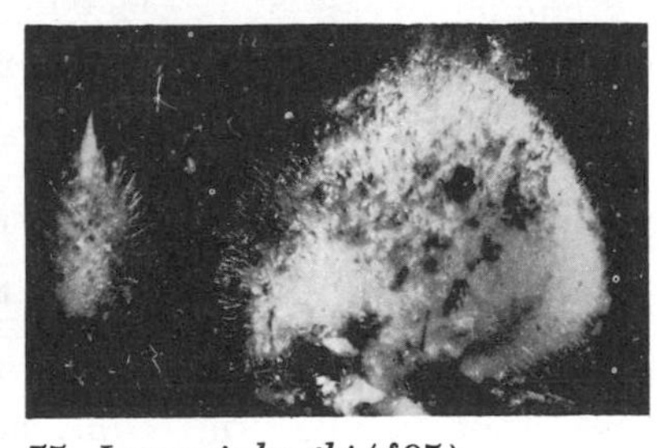

75. *Leuconia heathi* (§95).

horizon in crevices and at the bases of rocks. A similar "antler sponge" at Plymouth, England, has been observed to attain its complete growth as a massive cluster, several inches in diameter, in 6 or 8 months. Growth of the Pacific species seems to start in the spring with delicate colonies of branching tubes, which become clusters of rank, connecting growth by late fall; hence it is presumed to be an annual here also. It is common in Monterey Bay, and a similar species occurs in Puget Sound (§229).

§94. The urn-shaped *Rhabdodermella nuttingi* (Fig. 74) may be found growing suspended in crevices and under ledges at the extreme low-tide mark, where it is seldom exposed to the air. Large specimens, creamy white in color, may reach a length of 2 inches. Visitors from other regions commonly mistake this form for *Grantia,* a simple sponge that it closely resembles but to which it is not intimately related. Like *Leucosolenia,* the *Grantia* at Plymouth, England, is known to become adult in 6–8 months. At Pacific Grove large specimens of *Rhabdodermella* have been taken at all times of the year; only a few miles north of San Francisco, however, this sponge appears to be more seasonal, being most obvious in early spring and scarce by late summer.

§95. *Leuconia* (*Leucandra*) *heathi* (Fig. 75), a very sharp-spined, cream-colored sponge with a large, volcano-like, central osculum, typically occurs as a dome-shaped colony, but it accommodates its shape to that of the crevice it occupies. The average size is small, but large specimens may be several inches in diameter. The large osculum, which acts as a centralized excurrent pore, is fringed with spines longer than those that bristle formidably over the rest of the body. *Leuconia*'s dangerous-looking spicules are calcareous, however, and so crumble easily —quite the opposite of the really dangerous silicious spicules found in *Stelletta* (§96).

§96. It is amusing to note the controversies that centered on sponges a few hundred years ago, when their structure was still imperfectly known. The herbalist Gerard wrote in 1636: "There is found growing upon the rocks neer unto the sea, a certaine matter wrought togither, of the forme [foam] or froth of the sea, which we call spunges . . . whereof to write at large woulde greatly increase our volume, and little profite the Reader." The tactful Gerard might have been describing some of our white sponges. Possibly the most obvious of these are species of *Stelletta,* a feltlike, stinging sponge that encrusts the sides of low-tide caves. Certainly, this form is the most obvious to the collector who handles it carelessly and spends the rest of the day extracting the spicules from his hands with a pair of fine forceps. The spicules are glassy and have about the same effect on one's fingers as so much finely splintered glass. There are other white sponges, most of them smooth and harmless, but any uniformly white sponge with bristling spicules had best be pried from its support with knife and forceps. On one of the harmless varieties a small white nudibranch, *Aegires albopunctatus,* may be found, matching the color of its home just as the red *Rostanga* matches its red sponge.

§97. A good many other sponges occur in the low rocky intertidal; but we cannot consider them all in detail. Several others, however, are common enough to deserve mention.

Lissodendoryx firma, a strong and foul-smelling cluster up to 8 cm. thick and 15 cm. in diameter, has a lumpy yellow surface like a dried bath sponge, and is malodorously familiar to all prowlers in the Monterey Bay tide pools. When the sponge is broken open, its interior is found to be semicavernous and to harbor hosts of nematodes, annelids, and amphipods. This is one of the few sponges that may be found actually in contact with the substratum, on the undersides of rocks.

The famous and cosmopolitan "crumb of bread" sponge, *Halichondria panicea,* may also occur spottily up and down the coast; occasionally it is abundant in central California. Encrustations are orange to green, amorphous, up to 6 mm. thick and 3 cm. in diameter, of fragile consistency, and have the oscula raised 1 mm. above the superficially smooth to tuberculate surface.

Purple colonies of *Haliclona* (§31) occasionally extend into the low-tide zone from the middle tide pools, but these are rare compared to the red sponges. *Plocamia, Ophlitaspongia* (§32), and other red encrusting forms are often common here, splashing the sides of rocks with delightful bits of color. (See also §140, under-rock forms, in this connection.) Near relatives of the familiar bath sponges are represented here only by small and rare specimens; the bath sponges grow luxuriantly only in warm to tropic waters, since their habitat is correlated in some poorly understood way with their food supply.

§98. Under overhanging ledges in the La Jolla–Laguna region, and on the sides of rocks in the completely sheltered Newport Bay, occur clusters of the coarse-textured yellow sponge *Verongia thiona,* known as the sulphur sponge. It is shown in Figure 45 growing pretty well up in the tide pools, with the tubed snail *Aletes.* Clusters that fit one's cupped hands have from two to four raised oscula. This animal turns purple or black after being removed from the water, even if immediately preserved. This trait distinguishes it from *Hymeniacidon* (§183), another sponge occasionally abundant in the same habitat and similar superficially, although very different internally.

Living on *Verongia* clusters is another animal, actually related to the sea hares but more nearly resembling a soft-shelled limpet. This little-known creature, *Tylodina fungina* (Fig. 76), is colored the same yellow as the sponge—another of the many examples of apparent protective coloration. Such examples ought never to be accepted without reservations, for very often animals that are "protectively colored" in one environment turn up in other habitats where their coloring makes them conspicuous. In this case the color may be due to the diet and nothing more.

§99. Encrusting colonies of compound tunicates can be distinguished from sponges by their texture. Most sponge colonies, even the soft ones, have a gritty feeling caused by the presence of lime or glass spicules; most tunicates are slippery to the touch. Colonies of compound tunicates (*Amaroucium californicum* and similar species), which encrust the sides of rocks in yellows, reds, and browns, are actually the specialized adults of free-swimming "tadpole larvae" that have a bit of notochord, the structure around which the backbone of the vertebrates is formed. These tunicate colonies are a shining example of the error of judging the position of an animal in the evolutionary scale by its external appearance. Although they belong to the group that contains the vertebrates—most specialized of all animals—they most nearly resemble sponges, one of the lowest and least organized groups. The tunicates are highly specialized in their own right, to exploit the abundant food resources presented by the fine suspended and floating material in seawater—as are the sponges. The great abundance of filter- and suspension-feeding animals along the seashore is testimony enough to the quantity of this fine material.

Tunicate species are difficult to differentiate, or at least they have been for us. The following notes, based on identifications by Dr. N. J. Berrill of McGill University, may serve as field characters to differentiate these often side-by-side colonies, among the commonest sessile forms of the outer coast.

Eudistoma psammion (Fig. 77) is a compact, sand-covered, very dark, hard, and firm-textured colonial form, with zooids often raised slightly above the surface. It grows in coherent sheets, which may be an inch thick. It is probably the commonest Monterey Bay compound tunicate, especially in such regions as the cave area near Santa Cruz, where great beds carpet the ledges within the cave.

6. *Tylodina fungina* (§98), crawling n a sponge; greatly enlarged.

7. Colonial tunicates in a bed of the eelgrass *Phyllospadix* at Pacific Grove. The darker, coarser mass *Eudistoma psammion* (§99); the white species at the top is *E. molle*.

Amaroucium californicum forms amorphous but usually sheetlike masses, with a gelatinous, light-colored (creamy, yellow, or pink) cover through which the long, pinheaded, white zooids may be seen. The colony is exceedingly soft, almost mushy; its base may be sandy, but its surface is slippery and gives a feeling of flabbiness. This is one of the commonest tunicates of the region.

Eudistoma diaphanes resembles *Amaroucium,* except that the whole colony is much firmer, both in texture and on the surface. Although *Amaroucium* cannot easily be pried away unbroken from its support, sheets of *E. diaphanes* can be pulled off without breaking. *Eudistoma's* tunic is more opaque, and whiter, with a blue tinge; and the zooids are more closely packed together and less noticeable. Colonies may be thin or (usually) thick sheets, or pedunculated. A cross section of the colony shows cavelike areas, with the rest stiff and springy.

Distaplia occidentalis forms mushroomlike, flat, or pedunculated colonies. Its closely packed zooids loom up more clearly at the top of the colony, forming (in cross section) a superficial layer of differentiated tissue concentrated at the top surface. The richly brown zooids are arranged in rosettes around a common, raised, crater-like pore.

§100. The peculiar habits of the almost sessile amphipod *Polycheria antarctica* (Fig. 78), which lives in *Amaroucium* colonies, have been noted by several observers. *Polycheria* will usually be found in its burrow; but, whether swimming, walking awkwardly, or lying in its burrow, it always stays on its back. Only females have been observed, and they are about 3⁄16 inch long. Nothing but *Amaroucium* will satisfy them for a home. When placed for observation in an aquarium lacking the tunicate, they hunt it until they die, apparently from exhaustion. Finding it, they burrow in by grasping the surface of the tunicate and slowly pressing their backs into it. Once satisfactorily settled, *Polycheria* lies there, with antennae protruding, and waits for the currents to bring it food; its utmost exertion consists in creating currents with its legs and in bending its antennae down occasionally to scrape off and eat their accumulation of diatoms and other food particles. This mode of life is reminiscent of that permanently adopted by the barnacles. When danger impends, *Polycheria* draws the edges of its burrow together with its legs; it leaves the burrow only when temperature or water conditions become unfavorable.

The female carries her eggs (as many as 80) and young in a brood pouch throughout the summer, forcing the young out a few at a time, although they cling tenaciously to her hairs. The young, already moving

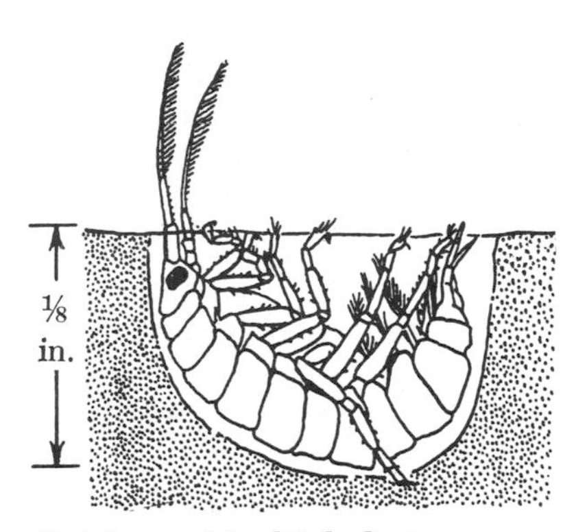

78. The amphipod *Polycheria antarctica* (§100), which lives in a tunicate.

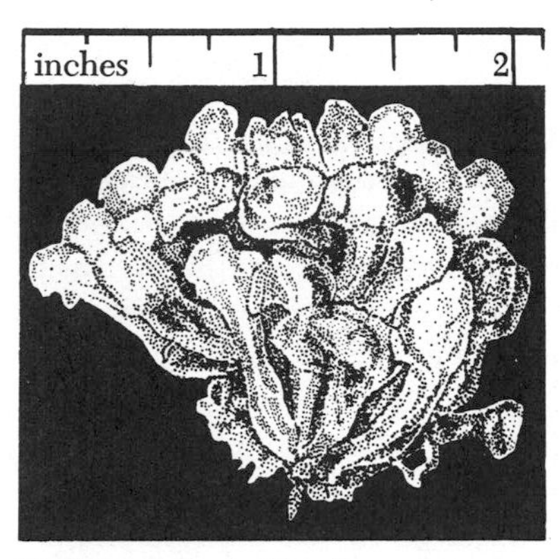

79. The tunicate *Clavelina huntsmani* (§101).

on their backs, immediately begin to search for a lodging place, preferring for their first burrow a slight depression in the surface of the tunicate.

Polycheria has been reported from Puget Sound; and it is fairly common in Monterey Bay, where it has been known as *P. osborni*. Careful search will probably reveal it elsewhere.

§101. A solitary stalked tunicate, *Styela montereyensis,* may be found in this zone, in crevices or on the protected sides of rocks exposed to fairly heavy surf, but more and larger specimens are features of the wharf-piling fauna (§322).* Other obvious solitary and semicolonial tunicates are characteristic of the lower rocky tide pools, notably great colonies of the club-shaped zooids of *Clavelina huntsmani* (Fig. 79). The individuals are ½ inch to 1½ inches long, and are closely connected to one another at the base only. The tunic is transparent and never heavily encrusted with sand, so that the pink internal organs can be seen plainly. In adaptation to their sessile lives, tunicates have incurrent and excurrent siphons not unlike those of clams, and in *Clavelina* these siphons are fairly visible, although they are not conspicuously erect, as in *Cnemidocarpa* (§216) of the Puget Sound region. *Clavelina* is apparently an annual species, since fresh, sparkling colonies are found during the summer; later in the fall these become dirty and degenerate, finally disappearing completely in the late fall. *Clavelina* is tremendously

* The large *Halocynthia johnsoni* will occur occasionally in these situations south of Point Conception, but it seems to be more highly developed in sheltered bay waters and will be considered in §216.

abundant at Monterey Bay, where next to *Amaroucium* it is the most common tunicate, but it is less abundant elsewhere on the California coast. In some years it is common in Marin and Sonoma Counties.

§102. The semicompound *Euherdmania claviformis* (Fig. 80) resembles *Clavelina,* but with longer, slimmer individuals that are more loosely adherent in the completely sand-covered colony. We have found this tunicate to be very plentiful in the sea caves west of Santa Cruz, and it is common at La Jolla. The dull-green zooids of *Euherdmania* are useful for embryological work, since the long oviduct is swollen during the summer with a series of embryos, the largest of them nearest the attached end of the parent.

§103. A tunicate not uncommon south of Point Conception is the stalked *Polyclinum* (formerly *Glossophorum*) *planum* (Fig. 81). This animal consists of a definite, usually somewhat flattened, bulb some 2 or 3 inches long by two-thirds as wide, mounted on a thick stalk, or peduncle. The zooids give the bulbous colony a flowery wallpaper effect.

As was the case with hydroids and sponges, many more species of tunicates are likely to be turned up in the lower tide pools than can profitably be treated here. At the risk of being misunderstood, we might quote the excellent Gerard and say, "to write at large woulde greatly increase our volume, and little profite the Reader." Another semicompound tunicate, *Perophora annectens,* must be considered, however, because it illustrates another stage in the evolutionary series from the simple, completely separated individuals, such as *Styela,* to the completely colonial *Amaroucium.* The dull-green matrix of the colony adheres closely to the rocks. When the animal is seen unmolested, the bright orange zooids will be well extended; but *Perophora* has powers of retraction most amazing for a tunicate. If disturbed, it pulls its delicate zooids away from danger, withdrawing them into the tough tunic of the matrix and almost instantly changing the color of the whole colony. The covering of the extended zooids is even more transparent than that of *Clavelina,* and through a good hand lens the internal anatomy is quite apparent. *Perophora* ranges from British Columbia to San Diego.

§104. The sea hare *Aplysia californica* (Fig. 82) occurs in various environments—from completely sheltered mud flats to fairly exposed rocky shores. It occurs below the low-tide line and is also a common feature of the intertidal zone almost up to the upper pools. We have found great numbers of small specimens with the under-rock fauna in northern Baja California in February, and hosts of medium-sized specimens—about 5 inches long—at Laguna Beach in May. These tectibranch (covered-gill) mollusks belong most characteristically, however, in the

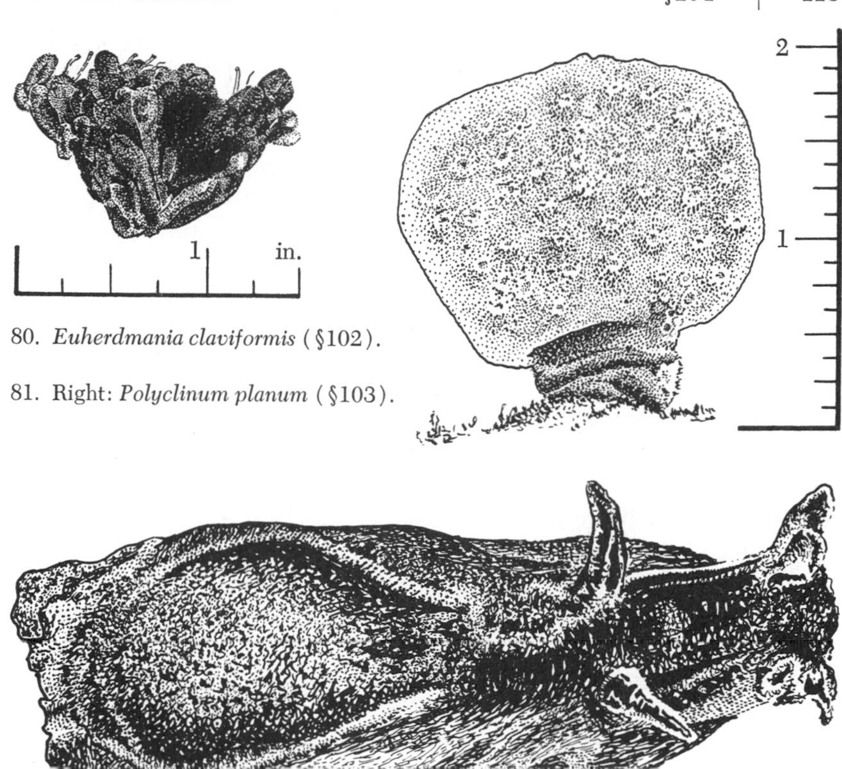

80. *Euherdmania claviformis* (§102).

81. Right: *Polyclinum planum* (§103).

82. The sea hare, *Aplysia californica* (§104), a tectibranch.

low rocky intertidal. Throughout the year at Pacific Grove adults may be found depositing their eggs in yellow, stringy masses larger than one's two fists. From a test count of a portion of such a mass MacGinitie estimated that the total number of eggs in the single mass was in the neighborhood of 86 million.*

This continuous breeding, common among central Pacific coast inver-

* MacGinitie (1934) also records a ±5% computation to the effect that a captive sea hare weighing under 6 pounds laid 478 million eggs in less than 5 months. The total egg mass, when untangled, was 60,565 cm. in length—about ⅓ mile! The largest of the 27 layings amounted to 17,520 cm., at an average rate of 5.9 cm. or 41,000 eggs per minute; and each egg averaged 55 microns in diameter (a micron is 1/1000 mm.). The larvae become free-swimming in 12 days, to contribute their mite of food toward appeasing a marine world of hungry, plankton-seeking forms. Presumably, few more than the biblical two from one of these litters can run the gauntlet into adulthood, long before which their flesh will have become distasteful to predacious neighbors.

tebrates, is very possibly a result of the even temperature throughout the year—especially likely when it is remembered that on the northern Atlantic coast, where there is great seasonal temperature change, the animals have definite and short reproductive periods.

Aplysia is hermaphroditic, having both testes and ovaries and both male and female organs of copulation. It may thus play female during one copulation and male during the next, as its whims happen to dictate; or it may play both roles at the same time. At Elkhorn Slough in the spring of 1931, seven or eight *Aplysia* were seen copulating in a "Roman circle," each animal having its penis inserted in the vaginal orifice of the animal ahead. Copulation lasts from several hours to several days.

There are other miscellaneous points of interest. *Aplysia* has an internal shell, which it has tremendously outgrown, and a more complicated digestive system than almost any other algae-feeding invertebrate. The food is first cut up by the radula and then passed through a series of three stomachs, the second and third of which are lined with teeth that continue the grinding process begun by the radula. Breathing is assisted by two flaps—extensions of the fleshy mantle—that extend up over the back and are used to create currents of water for the gills, so that the animal may be said to breathe by "flapping its wings." When disturbed, *Aplysia* extrudes a fluid comparable to that of the octopus, except that it is deep purple instead of sepia. A handkerchief dipped into a diluted solution of this fluid will be dyed a beautiful purple, but the color rinses out readily. Possibly the addition of lemon juice or some other fixative would make the color fast. The color of the sea hare itself is an inconspicuous olive-green or olive-brown, often with darker blotches. *Aplysia* is common as far north as Monterey Bay, and occasional specimens show up at Bodega in Sonoma County.

§105. Another obvious animal in this zone, with hosts of less obvious relatives, is the large crab *Cancer antennarius* (Fig. 83), which snaps and bubbles at the collector who disturbs it. The carapace of *C. antennarius* is rarely more than 5 inches wide, but its large claws provide food quite as good from the human point of view as those of its big brother *C. magister,* the "edible," or Dungeness, crab of the Pacific fish markets. (*C. magister* rarely occurs in the intertidal in California as an adult, but it is common on the Oregon coast in summer, when it comes into shallow water to molt.) *C. antennarius* can be identified easily by the red spots on its light undersurface, especially in front.

Here another word anent popular names is necessary, for to many people *C. antennarius* is known as "the rock crab," although the present

83. *Cancer antennarius* (§105).

writers have given that name to the bustling little *Pachygrapsus*. The reason is that the vastly more numerous and obvious *Pachygrapsus* seemed to be in dire need of a nontechnical name. It has been called "the striped shore crab," but except in occasional specimens the stripes are not particularly conspicuous. *Pachygrapsus*'s predominant green color is a more striking characteristic, but the name "green shore crab" designates a European form.

To return to *Cancer antennarius*: female specimens carrying eggs are seen from November to January, with the embryos well advanced by the latter month. The range is from British Columbia to Baja California. *C. antennarius* is a rather delicate animal that does not live well in aquaria—a direct corollary of its habitat. It is interesting to contrast it in this respect with the much hardier purple shore crab (*Hemigrapsus*) of the middle zone and with the extremely tough and resistant *Pachygrapsus* of the upper zone. It is an almost invariable rule that the lower down in the tidal area an animal lives, the less resistance it has to unfavorable conditions.

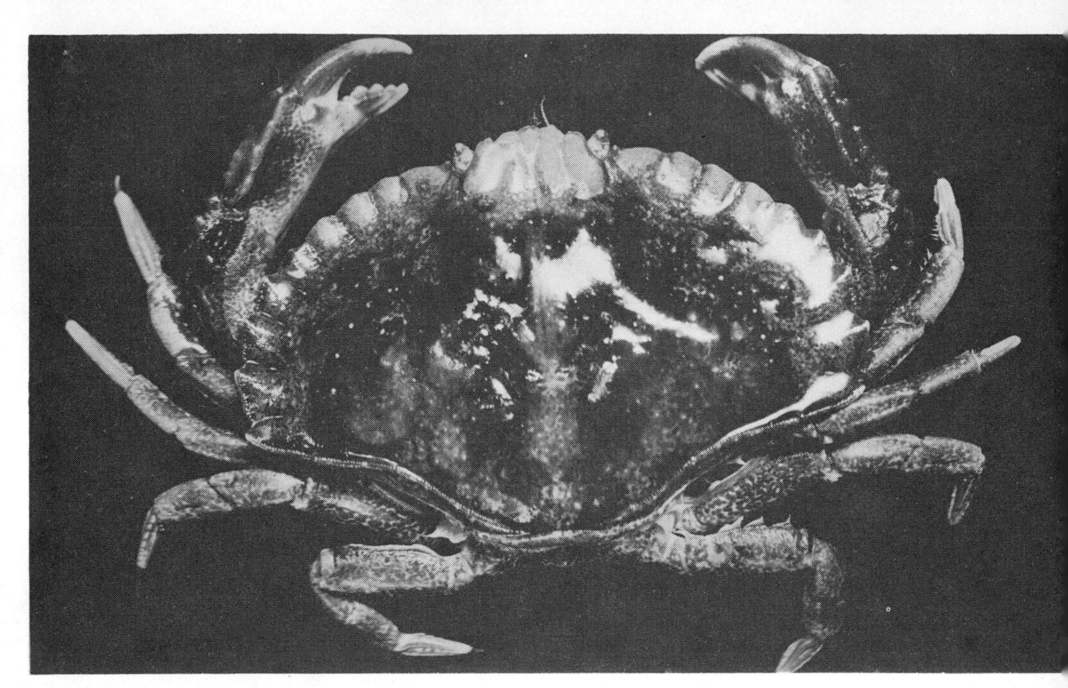

84. *Cancer productus* (§106).

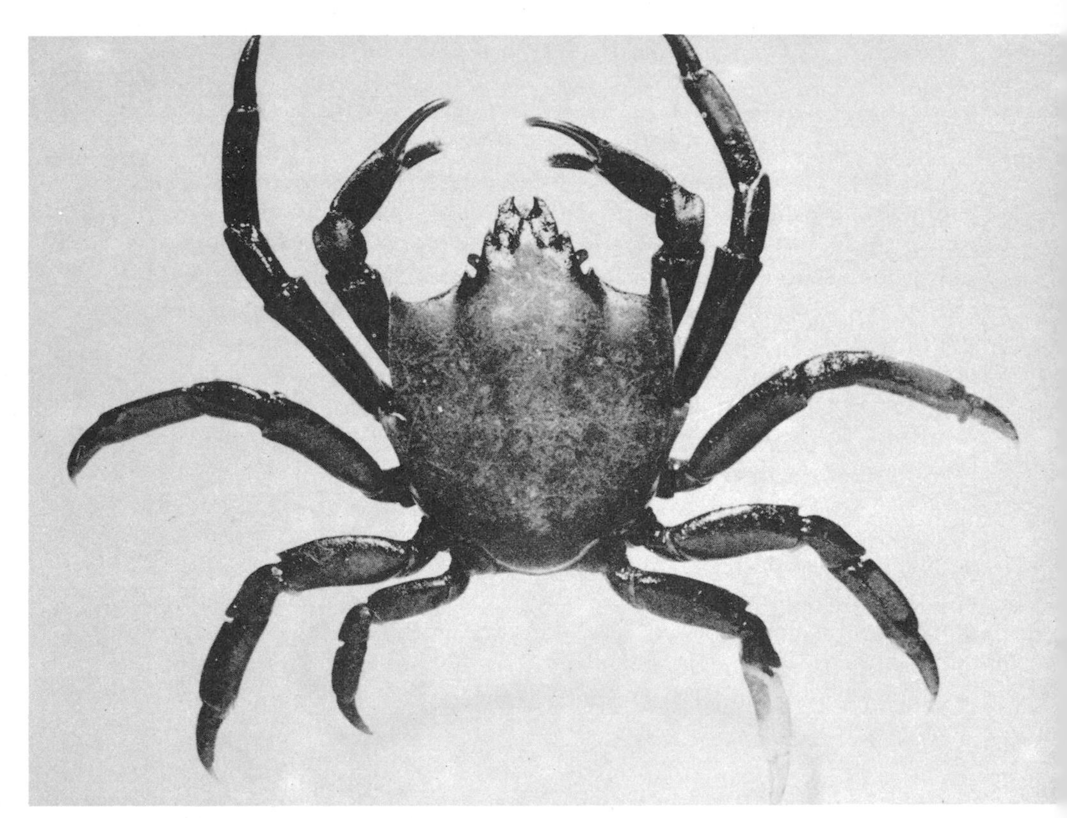

85. *Pugettia producta* (§107), the kelp crab.

§106. *Cancer productus* (Fig. 84), related to *C. antennarius,* may be considerably larger and therefore more useful for food, but it is not plentiful enough to be commercially important. Adults are uniformly brick-red above. The young exhibit striking color patterns, in which stripes are very common. Like *C. antennarius, C. productus* may often be found half buried in the sandy substratum under rocks. At night it stalks about in the tide pools, so large and powerful that it dominates its immediate vicinity unless predatory man happens along. *C. productus* is a good example of an animal that is pretty well restricted by its physiological makeup to this rocky habitat, for it lacks equipment for straining the fine debris encountered on bottoms of mud or pure sand out of its respiratory water. It ranges from Kodiak Island to Baja California.

§107. A dark olive-green spider crab, *Pugettia producta* (Fig. 85), occurs so frequently on strands of the seaweed *Egregia* and others that it is commonly called the kelp crab. The points on *Pugettia*'s carapace and the spines on its legs are sharp (the latter adapt the animal to holding to seaweed in the face of wave shock), and the strong claws are rather versatile. The collector who is experienced in the ways of kelp crabs will catch and hold them very cautiously, and the inexperienced collector is likely to find that he has caught a Tartar. Even a little fellow will cling to one's fingers so strongly as to puncture the skin, and a kelp crab large enough to wrap itself about a bare forearm had best be left to follow the normal course of its life.

Most spider crabs, being sluggish, pile their carapaces with bits of sponges, tunicates, and hydroids that effectively hide them, but *Pugettia* is moderately active and keeps its carapace relatively clean. Limited observation on this point suggests that specimens from rocky tide pools are more intolerant of foreign growth than specimens from wharf piling, where the animals are very common. We have examined many wharf-piling crabs that had barnacles and anemones growing on their backs, or even on their legs and claws.

Pugettia, which is often parasitized by a sacculinid (§327), is recorded from British Columbia to Baja California, but below Point Conception it is pretty largely replaced by the southern kelp crab *Taliepus* (formerly *Epialtus*) *nuttallii,* which ranges from there to Ballenas Bay, Baja California. The carapace of *Taliepus,* usually fairly clean, is rounder than that of *Pugettia,* and is often purple with blotches of lighter colors.

§108. Our friends *Hemigrapsus nudus* and *Pagurus hemphilli,* the pugnacious purple shore crab and the retiring hermit, are frequent visitors in the lower tide-pool region, *P. hemphilli* being the only hermit found this far down, at least in the California area. Many other crabs, some of which are almost certain to be turned up by the ambitious collector, are not sufficiently common here to justify their inclusion in this account.

§109. The northern Pacific coast is noted for its nudibranchs, for there are probably greater numbers here, of both species and individuals, than in any other several regions combined. Several nudibranchs have already been mentioned, but it is in the lower horizon that these brilliantly colored naked snails come into their own. It would be difficult for a careful observer to visit any good rocky area, especially in central California, without seeing several specimens. Their reds, purples, and golds belie their prosaic name, "sea slugs," so often imposed by people who have seen, presumably, only the dead and dingy remains. No animals will retain their color when exposed to the light after preservation, and a colorless, collapsed nudibranch is certainly not an inspiring object. Their appearance "before and after" is one of the strongest arguments against preserving animals unless they are to be used specifically for study or dissection. Every once in a while some student comes along from a teacher's college (we know they are no longer called this, but they still exist), who thinks that a suitable thesis topic would be the development of means to preserve nudibranchs in their original, living colors. Though we would hesitate to say "it can't be done" in these days of chemical marvels, the people who usually propose this do not know an atom from a molecule, and we advise them to get interested in something harmless, like color photography.

Nudibranchs breathe through "crepe-paper" appendages at the rear end (actually a rosette of gills around the anus), or through delicate, fingerlike protuberances along the back (called cerata; the central, darker part is a branch of the "liver," or digestive gland). Nudibranchs with centralized gills in a single retractable cluster are generally grouped together as dorids. Two common dorids have already been mentioned: the red *Rostanga pulchra,* found on sponges (§33); and *Diaulula sandiegensis,* which characteristically has ringlike markings on the back (§35). *Anisodoris nobilis,* the sea lemon, is a bright yellow nudibranch with a white gill plume, sometimes mottled with light brown (§178); a similar dorid is *Archidoris montereyensis,* which is paler yellow and has dark, almost black, spots scattered irregularly on its back (Fig. 86). There are several species of *Cadlina,* characterized by short, rounded tubercles scattered over the back (Fig. 87). Many nudibranchs, but especially the dorids, have a penetrating fruity odor that is pleasant

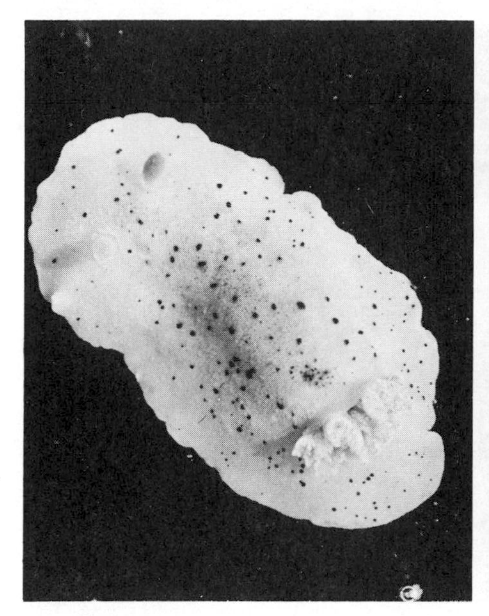

86. A dorid, *Archidoris montereyensis* (§109).

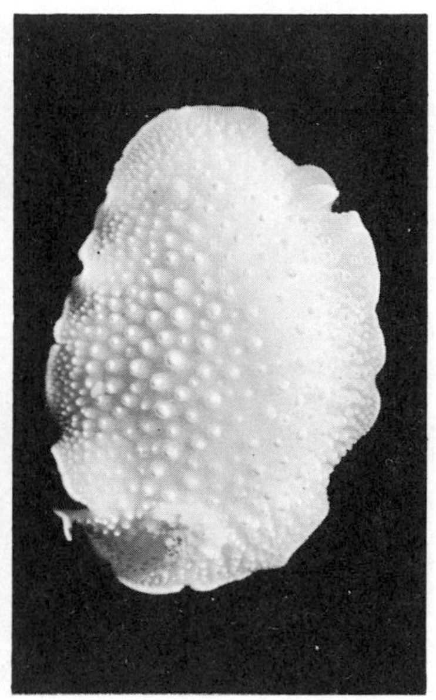

87. A dorid, *Cadlina* sp. (§109).

when mild but nauseating when concentrated. Undoubtedly, this odor is one of the reasons why nudibranchs seem to be let strictly alone by predatory animals: recent studies have indicated that fish will reject a nudibranch if it is accidentally swallowed.

Most of the dorids so far mentioned are prettily but not brilliantly colored (except for *Rostanga*). One of the brightest dorids on the central California coast, sometimes found as far north as Sonoma County, is *Hopkinsia rosacea* (see color section), which is a bright, uniform magenta or cerise ("rose pink," according to the MacFarlands). More characteristic of southern California waters are such gaudy blue-and-yellow creatures as *Glossodoris californiensis* (with separated yellow spots) and *G. porterae* (with a continuous yellow band around the margin of the back), and the lovely violet *Glossodoris macfarlandi.* Two brightly colored dorids are common in central California: *Triopha carpenteri,* with a snow-white body color and brilliant orange protuberances, occurs as far north as Oregon; *T. maculata,* which has a brown body speckled with blue spots and reddish-orange processes and gill plume, ranges from San Diego to Oregon.

Another group of nudibranchs commonly encountered is the eolids,

88. *Dirona albolineata* (§109), an eolid.

characterized by rows of cerata on the back. Perhaps the commonest of these is *Hermissenda crassicornis* ("most lovely of the eolids," according to the MacFarlands). Although variable in color, *Hermissenda* always has a clear blue line around the sides (see color section). This eolid is more or less common throughout its range from Alaska to San Diego. According to MacFarland, it is "voracious and irritable, attacking other nudibranchs and members of its own species indiscriminately." A striking contrast in coloration to *Hermissenda* is provided by the uniform dull grayish-mauve or pinkish-gray *Aeolidia papillosa,* which is found from Alaska to Monterey. Another contrast occurs in the arrangement of the cerata—clustered in *Hermissenda,* more or less uniform along both sides in *Aeolidia.*

One of the more widely distributed North Pacific eolids, found from San Diego to Alaska, and also in Japan, is the uniformly gray *Dirona albolineata.* As the specific name implies, this nudibranch has white lines: they are vividly white, adding a conspicuous trim to the cerata (Fig. 88).

The eolids characteristically prey on coelenterates, and their cerata may be packed with nematocysts from the hydroids and anemones upon which they feed. Since the nematocysts are "independent effectors," actually little mechanical devices that work by a trigger or exposure to some chemical, it is possible that they may protect the nudibranch; but it seems more likely that they are stored in the cerata as indigestible by-products. If so, the question is, why aren't the nematocysts discharged when their original owner is being eaten?

Many of the nudibranchs are sporadic or seasonal in occurrence; some apparently come near shore only during the summer. In some years certain species may not be seen at all. A good sampling of our rich opisthobranch fauna (nudibranchs, sea slugs, and their allies) is now available in the long-awaited MacFarland monograph, with its color plates. However, the publication of this work by a master systematist was so long delayed that some of the species named as "new" had already been described elsewhere.

§110. On the "leaves" and "stems" of outer tide-pool kelps, one almost always finds an encrusting white tracery delicate enough to be attributable to our childhood friend Jack Frost. But a hand lens reveals a beauty of design more intricate than any ever etched on frosty window panes. These encrustations are usually formed by colonies of the bryozoan, or "moss animal," *Membranipora membranacea* (Fig. 89), so named in the middle of the eighteenth century by Linnaeus, the founder of modern classification. *M. membranacea* is found in temperate regions all over the world and is common on this coast at least as far south as

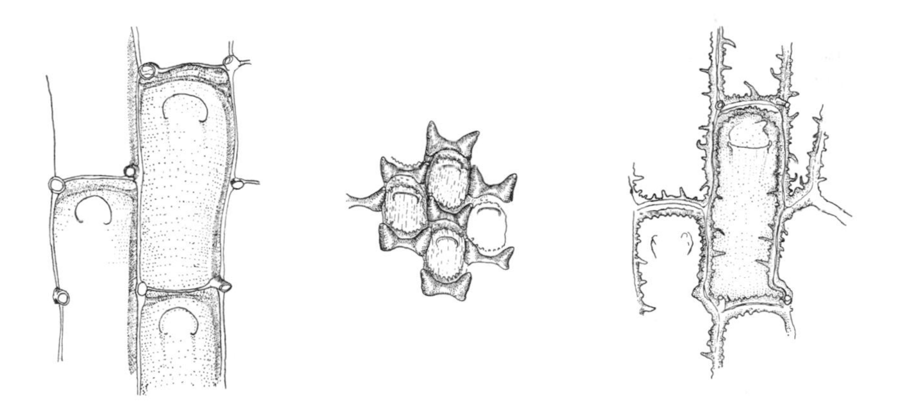

89. Three species of the bryozoan genus *Membranipora* (§110), greatly magnified to show the cells housing the individual zooids. Left to right: *M. membranacea, M. tuberculata, M. serrilamella.*

northern Baja California. The minute, calcareous cells, visible to the keen naked eye but seen to better advantage with a lens, radiate in irregular rows from the center of the colony. The colony may be irregular and may cover several square inches, or it may be fairly small and round. The minute crown of retractile tentacles (not unlike those of hydroids) by which the animals feed can be seen only with a microscope. As is usually the case with sessile animals, bryozoans have free-swimming larvae. The larvae were described nearly a hundred years ago as rotifers, later as worm larvae. Robertson mentions finding the larvae of a related species settling on the kelp at La Jolla during July. When a larva comes in contact with a bit of kelp, it opens its shell and settles down; the shell flattens over the larva; and the bit of tissue enlarges rapidly to form a colony of several cells, which soon reaches visible size. This whole process has been watched, and can be watched again by a careful observer with a good lens.

Living on *Membranipora* colonies, and feeding on them, is a small nudibranch, *Corambe pacifica.* This animal reaches a maximum length of ½ inch, but is nearly as wide as it is long. *Corambe*'s ground color is a pale, translucent gray that makes it almost indistinguishable from its bryozoan background; its yellowish liver shows through the center of its back, however, and surrounding this area the white of the foot may be seen. The rest of the animal's back, especially toward the margins, is flecked with irregular, sometimes broken, lines of yellow. *Corambe* occurs from Nanaimo, British Columbia, to Monterey Bay.

§111. The lower tide pools contain many other bryozoans. Possibly the most obvious, though certainly not the most common, is *Flustrella cervicornis* (Fig. 90), reported from the coast of Alaska and from Van-

90. *Flustrella cervicornis* (§111), an encrusting bryozoan, here shown growing on the fronds of a branching seaweed; natural size.

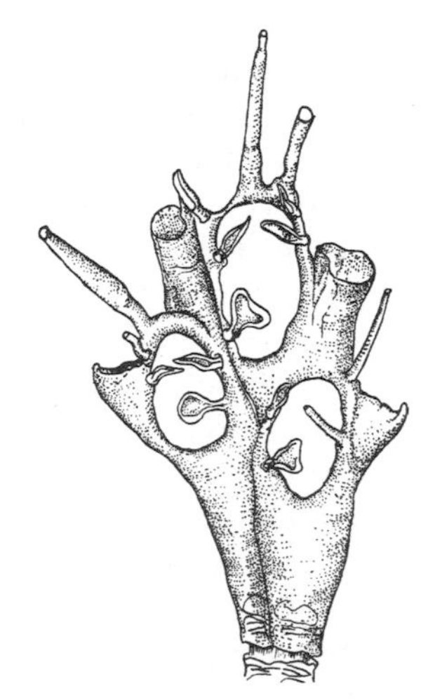

91. The branching bryozoan *Tricellaria occidentalis* (§112): left, a characteristic clump growing at the base of eelgrass, about natural size; right, a greatly enlarged view of one branch, with the individual zooids retracted.

couver, and common enough in sheltered areas along the cliffs west of Santa Cruz. The colonies are encrusting, but are soft rather than calcareous; they form great, dull, gray-brown masses over the erect seaweeds. Their surface is covered by erect, horny, branching spines more than ¼ inch tall.

§112. Other bryozoans are certain to be mistaken for hydroids—the delicate *Tricellaria occidentalis* (Fig. 91), for example, whose erect, branching colonies resemble those of *Obelia*, although the two belong to entirely unrelated groups. Probably some lusty forefather of *Tricellaria* made the happy discovery, back in an age geologically remote, that the branching pattern of growth was successful in resisting wave shock and provided plenty of area for waving tentacles—a discovery that the hydroids made somewhat earlier. Sprays of the white *Tricellaria* will often be found in rock crevices, with all the minute, tentacle-bearing colonies facing in one direction so that the "stem" will be curved slightly downward. The animal ranges south from the Queen Charlotte Islands (and probably Alaska) at least as far as San Diego.

§113. Branching bryozoans of the genus *Bugula* are found all over the world, and several species occur on the Pacific. A particularly handsome form is *Bugula californica* (Fig. 92). The colonies are large and definite, 3 inches or more in height, and have a pleasantly fresh purple tint. The individual branches are arranged in a distinctly spiral fashion around the axis. This species has heretofore been recorded only between Dillon Beach and Pacific Grove, but lush growths have been seen at Fort Ross and will probably be found to the northward (occurring, however, only at the extreme low-tide line).

The avicularia ("bird beaks") of *Bugula,* thought to be defensive in function, are classic objects of interest to the invertebrate zoologist. It is a pity that these, like so many other structural features of marine animals, can be seen only with a microscope. If the movable beaks of avicularia were a foot or so long, instead of a fraction of a millimeter, newspaper photographers and reporters would flock to see them. The snapping process would be observed excitedly, some enterprising cub would certainly have one of his fingers snipped off, and the crowds would amuse themselves by feeding stray puppies into the pincers. Avicularia and similar appendages, situated around the stems that support the tentacled zooids, probably have a function similar to that of the pedicellariae of urchins and seastars. Whatever else they do, they certainly keep bryozoan stems clean, as anyone will grant who has observed their vicious action under the microscope.

Working with a related Atlantic coast form, Grave (1930) found that during the summer and fall breeding season the embryos emerged at dawn and swam about freely for 4–6 hours before attaching; they became half-grown but sexually mature colonies in about one month, and

92. *Bugula californica* (§113), both life-size and greatly enlarged views.

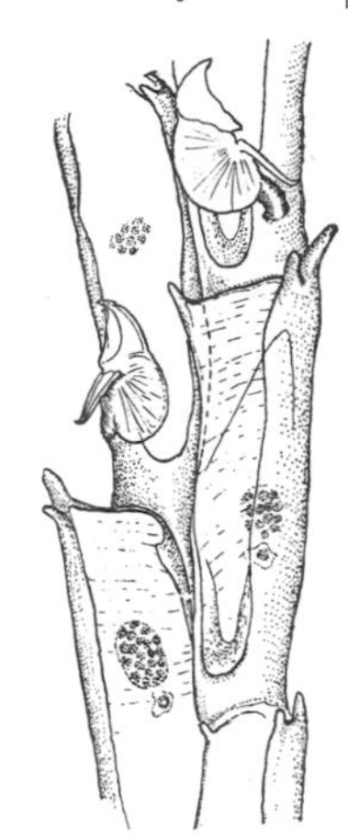

93. *Bugula pacifica* (§114), growing on a cluster of the hydroid *Aglaophenia*; natural size and enlarged views. The inset shows a "bird beak."

were senescent in three. Only the young colonies, that is, those produced toward the end of the breeding season, were able to tide over the winter by hibernation, resuming growth in early May.

§114. A more common *Bugula,* occurring along the whole Pacific coast from Bering Sea at least to Monterey Bay, where it is one of the most common bryozoans, is *B. pacifica* (Fig. 93). It is slightly spiral in form and 3 inches tall at its best, the largest specimens being found in the northern part of the range. The northern specimens are usually purple or yellowish in color, whereas those taken at Monterey Bay are light-colored—almost white—and rarely more than an inch high. At Monterey, however, they occur in profusion at a tide level much higher than that chosen by *B. californica.* Like hydroid colonies, bryozoans of this sort may provide shelter to hosts of different animals, notably skeleton shrimps and beach hoppers. Bits of *B. pacifica* will often be found attached to the sponge *Rhabdodermella* (§94).

§115. In the rocky caves west of Santa Cruz, rank clusters of *Barentsia ramosa* (Fig. 94), an entoproct, form thick-matted colonies with head-like bunches at the end of ½-to-1-inch stalks. *Barentsia's* tentacles contract unlike those of any hydroids or bryozoans herein considered, being pulled in toward the center like the diaphragm of a camera shutter. The similar *Pedicellina cernua* may also be found furring the stems of plants,

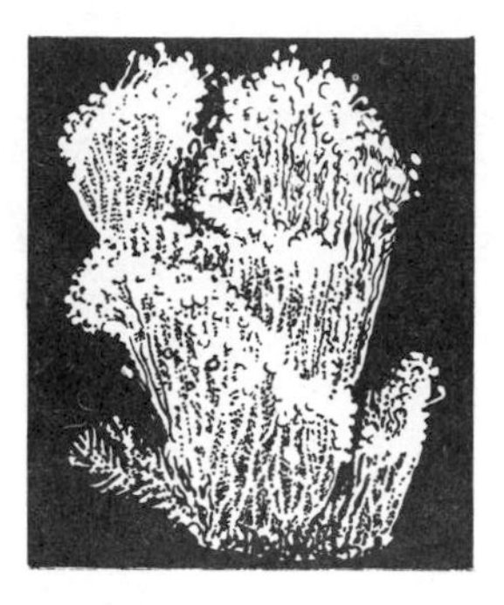

94. *Barentsia ramosa* (§115), an entoproct.

hydroids, and bryozoans. The entoprocts are superficially similar to bryozoans, and in older classifications were considered to be closely related; they are now classified as a distinct phylum.

§116. There are a good many small crustaceans in this zone, mostly amphipods and isopods. For the most part, they look pretty much alike superficially, and there seems to be no need for enumerating them here, for anyone interested in their differentiation will already have passed beyond the unpretentious scope of this handbook. We should mention, however, the hosts of quarter-inch beach hoppers that populate each cluster of the coralline alga *Amphiroa.* Whoever likes to deal with numbers of astronomical proportions should estimate the number of these amphipods in a handful of corallines, multiply by a factor large enough to account for the specimens present in one particular tide pool, estimate the probable number of tide pools per mile of coast, and, finally, consider that this association extends for something more than a thousand miles. One who has done this will have great respect, numerically speaking, for the humble amphipod, and will be willing to agree with Verrill that "these small crustacea are of great importance in connection with our fisheries, for we have found that they, together with the shrimps, constitute a very large part of the food of most of our more valuable fishes"; and that they "occur in such immense numbers in their favorite localities that they can nearly always be obtained by the fishes who eat them, for even the smallest of them are by no means despised or overlooked, even by large and powerful fishes that could easily capture larger game."

§117. There are a number of shelled snails in this zone, *Thais* and *Tegula* (§§182, 29) being fairly common. Larger and more spectacular than either, but not so common, is the brick-red top shell *Astraea gibberosa,* which ranges from Vancouver to San Diego. It may be 3 inches or more in diameter, and the larger *A. undosa* (Fig. 95) of southern California may reach a diameter of 6 inches. Large shells of this type are quite likely to be overgrown by algae or hydroids, so that they are not conspicuous unless the living animal is present to move them along. The cone-shaped shell of *Astraea* is so squat, and the whorls are so wide and regular, that identification is very easy.

A murex, known as *Ceratostoma foliatum* (Fig. 96) and related to the richly ornamented tropical shells so familiar to the conchologist, occurs on this coast. This large and active snail is carnivorous, and is closely related to the oyster drill that plays such havoc with commercial oyster beds on the east coast. Specimens 3 to 4 inches long have been taken. Its recorded range is from Sitka to San Diego, but it is seldom found below Pacific Grove. This avoidance of the southern range is interesting when it is remembered that the large and famous *Murex* is a strictly tropical animal. Neither *Astraea* nor *C. foliatum* will

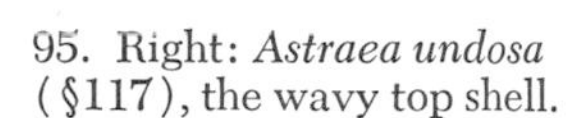

95. Right: *Astraea undosa* (§117), the wavy top shell.

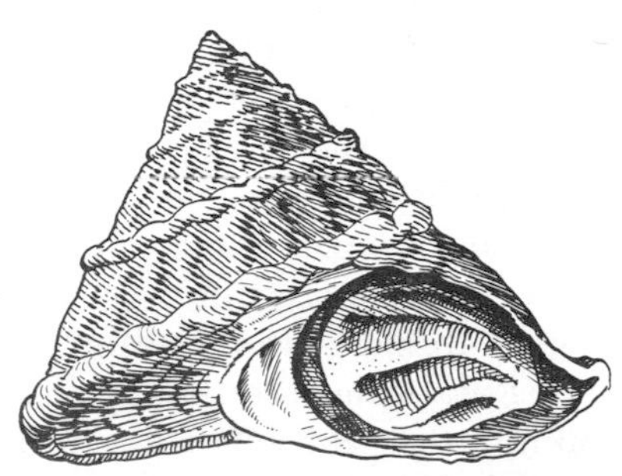

96. Below: the active and carnivorous snail *Ceratostoma foliatum* (§117).

be found frequently, and both occur only in the lowest of low-tide areas.

Some of the limpets occur here, and at least two are characteristic. The dunce-cap limpet *Acmaea mitra* (Fig. 97) is rarely found living above the line of extreme low tides, however common its empty shells may be on the beach. Its height and size are diagnostic; the average specimen will be 26 × 23 × 16 mm., making it the tallest of the limpets. Individuals up to 43 mm. long have been taken; in bulk these are probably the largest limpets, although some of the flat species may cover more surface. The shell is thick and white, but it is frequently obscured by growths of bryozoans and red corallines. *A. mitra*'s range is from Bering Sea to Santo Tomás, and the animals are ordinarily solitary and fairly uncommon; but Mrs. Grant has observed thousands of specimens south of the Point Arena lighthouse, where they occupy pits of the purple urchin.*

The flatter *A. fenestrata* (Fig. 98), formerly considered to be a subspecies of *A. scutum* (Fig. 99, §25), is now a separate species. It occurs from Half Moon Bay to the Aleutians, and is said to be the commonest limpet at Jenner. The apex of *A. scutum*'s shell is lower than in *A. mitra*, the faces are convex, and the aperture is more nearly circular than in any other limpet. Specimens may be identified readily by their fairly consistent size (25 × 23 × 11 mm. for a standard shell) and by the ecological distribution recorded by Grant—smooth, bare rocks in sand. When the tide recedes, the animal usually crawls down to sand and buries itself, so that one must dig to find them. At Dillon Beach, however, *A. fenestrata* is known to occur fairly high in the intertidal. A subspecies, *A. fenestrata cribrarius*, occurs south of Moss Beach to Santo Tomás.

As in previous zones, limpets more characteristic of other levels may be found occasionally, and reference should be made to §§11 and 25.

§118. The last form to be mentioned as belonging strictly to the protected rock habitat is a tiny colonial chaetopterid worm, whose membranous tubes mat large areas of vertical or slightly overhanging rock faces, especially where the boulders rest on a sandy substratum. The tubes are light gray, about 1/32 inch in diameter, and 2 to 2½ inches long; they are intertwined at the base, and are encrusted with sand. A cluster the size of one's fist will contain, at a guess, several hundred tubes. As is so often the case, this form is easily overlooked until the

* This form is also fairly common far down in the quiet waters of Puget Sound (§195).

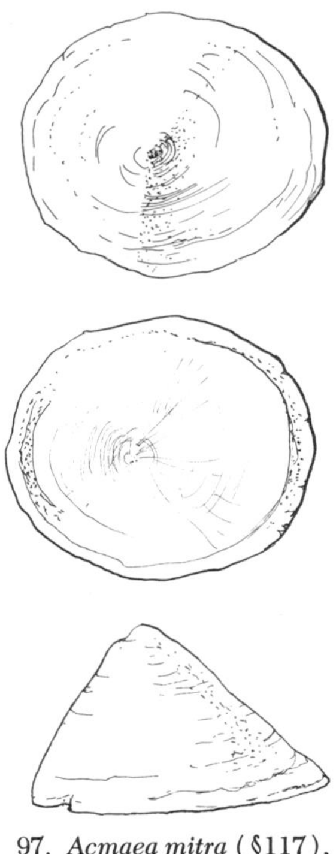

97. *Acmaea mitra* (§117), the dunce-cap limpet.

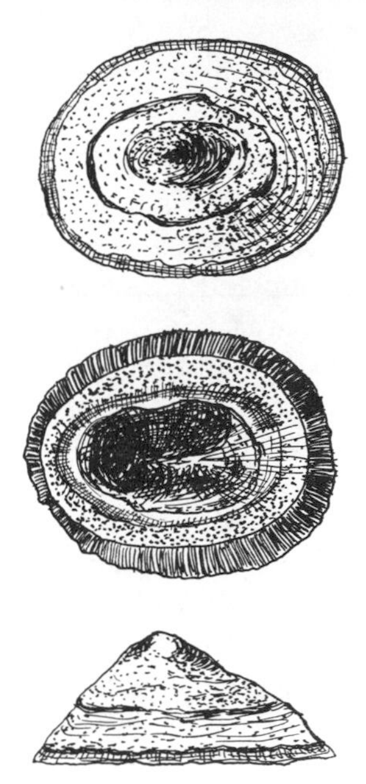

98. *Acmaea fenestrata* (§117).

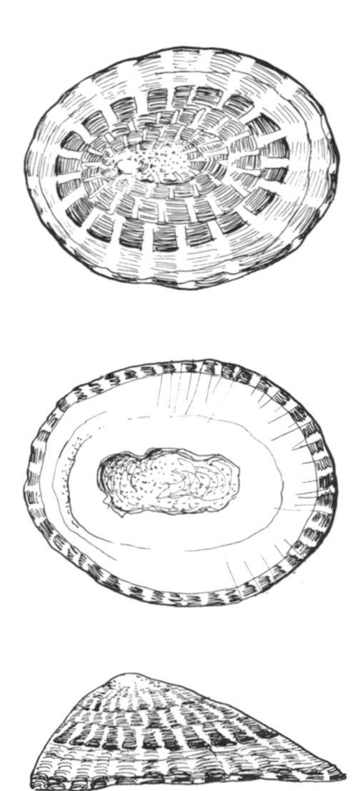

99. *Acmaea scutum* (§117).

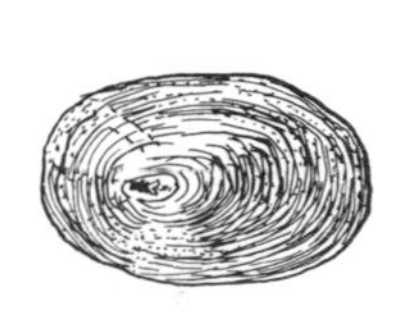

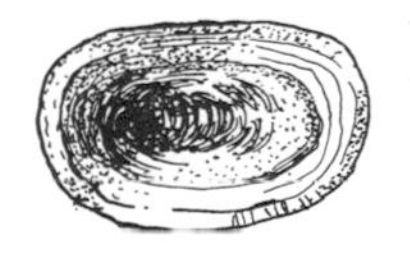

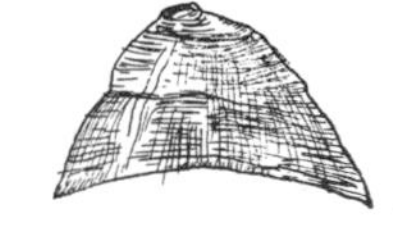

100. *Acmaea insessa* (§119).

collector has it brought to his attention and identifies it. After that it is obvious and seems to occur everywhere. (This species has not been encountered "everywhere" for many years, and it is still unknown to specialists.)

§119. A variety of eelgrass (*Phyllospadix*) grows in this habitat on the protected outer coast; on its delicate stalks occurs a limpet, ill-adapted as limpets would seem to be to such an attachment site. Even in the face of considerable surf, *Acmaea paleacea,* called the chaffy limpet, clings to its blade of eelgrass. Perhaps the feat is not as difficult as might be supposed, since the flexible grass streams out in the water, offering a minimum of resistance. Also, the size and shape of the limpet adapt it to this environment. A large specimen, usually less than ½ inch long, is higher than it is wide, and its width (about ⅕ of its length) is less than that of a blade of eelgrass. An average specimen is 6 × 1 × 3 mm. The ground color of *A. paleacea*'s shell is brown, with white at the apex where the brown, ridged covering has been worn away. This eelgrass limpet is fairly common in its recorded range of Trinidad to Baja California. It is abundant at Carmel.

Another common marine plant, *Egregia,* provides on the midrib of its elongated leathery stipe (sometimes 20 feet long) attachment sites for still another limpet, *Acmaea insessa,* apparently limited specifically to this habitat (Fig. 100). Grant records a range that pretty well includes the whole coast of California. The shell is brown, translucent, and smooth, or nearly so; but no description or illustration should be needed to identify an animal so obvious and so limited in its habitat. A typical shell is 16 × 14 × 9 mm.

§120. A good many small or obscure animals, listed elsewhere as referable to other environments, may turn up here occasionally. Common in the north, and rare to occasional at Monterey, is the sessile jellyfish *Haliclystus,* thought to be more characteristic of quiet waters (§273). The hydroid jellyfish *Gonionemus* (§283) may be locally abundant in the Sitka semiprotected areas (but never so far south as the Washington outer coast), in coves off the open coast or on the lee shore of protecting islands. The wave-shock factor, although possibly regulating the distribution of both *Gonionemus* and *Haliclystus,* is thought to be secondary to that of pure and clear oceanic water. Both species may be successful competitors in semisheltered and entirely quiet waters; but, especially in the case of *Haliclystus,* there must be strong currents of oceanic water from neighboring channels. They apparently tolerate no stagnation and very little fresh water.

4. *Low Intertidal: Under-Rock and Substratum*

Under-Rock Habitat

THE FIXED clams, the great tubed worms, the brittle stars, and the crabs are the most noticeable under-rock animals. The newly ordained collector will perhaps be attracted first of all by the brilliant and active brittle stars. He is likely to find most of the brittle stars that occur under the middle tide-pool rocks, and, especially in the south, he will surely find others.

§121. Two of the large and striking brittle stars not found north of San Pedro stray with considerable frequency into the middle zone, but never occur there in the great numbers that characterize this low zone. One of them, *Ophioderma panamensis* (Fig. 101), has a superficial resemblance to *Ophioplocus* (Fig. 42), but it is larger, richer brown in color, smoother in texture, and has banded arms. We have specimens as large as an inch in disk diameter with an arm spread of 7 inches from the Laguna region, which is near the upper limit of its recorded range, San Pedro to Panama. This species rarely autotomizes, and should prove to be a fair aquarium inhabitant. As in *Ophioplocus*, *Ophioderma*'s short spines extend outward at an acute angle to its arms.

The other species commonly found here, *Ophionereis annulata* (Fig. 102), is the southern counterpart of *Amphiodia* (Fig. 41), although it does not autotomize so readily. This very snaky-armed echinoderm, having a disk ½ inch or less in diameter and arms up to 4 inches in spread, ranges from San Pedro to Central America.

§122. *Ophiopholis aculeata* (Fig. 103, *a*), a dainty, variably colored, and well-proportioned animal—according to human aesthetic standards—is notable for an astonishing range, both geographically and bathymetrically. The same species not only extends throughout the world in northern temperate regions but occurs at various depths from the low intertidal zone to the sea bottom at 600 fathoms. The physical conditions in the tide pools at Pacific Grove may not differ markedly from those at Sitka or on the coasts of Maine, England, or Spitsbergen, but there is certainly a marked difference between conditions at the surface and those on the ocean bottom 7/10 mile below. In the intertidal zone there are wave shock, sudden temperature change, alternate exposure to air and water, and bright light at least some of the time. Under 3,600 feet of water there are absolute stillness, unchanging low temperature, light sufficient to affect a photographic plate only after long exposure, and pressure so great that only reinforced steel cylinders can resist it. Yet varieties of the delicate-looking *O. aculeata* bridge this gap and are equally at home in both environments.

The disk of this species has a granular appearance, and its predominant color is rusty red, although green may form a part of the mottled pattern. *O. aculeata* is found on this coast in Alaska, British Columbia, and Puget Sound, and we have taken it as far south as Point Sur.

§123. The ultra-spiny brittle star *Ophiothrix spiculata* (Fig. 103, *b*), has its disk so thickly covered with minute spicules that it appears fuzzy. The arm spines are longer, thinner, and more numerous than those of *Ophiopholis,* with which it is sometimes found. *O. spiculata*'s color is so highly variable that it furnishes little clue to the animal's identity, but it is often greenish-brown, with orange bands on the arms and orange specks on the disk. The disk may be 2/3 inch in diameter, and the arm spread several inches. The egg sacs of ovigerous specimens may be seen protruding from the disk between the arms. Although this is a southern form, ranging from Monterey Bay to Central America, specimens are not at all uncommon at the northern extremity of the range. Associated with *Ophiothrix,* but treated with the more open coast forms, may be found any of several "sea scorpions" (isopods), the large *Idothea stenops* at Monterey and *I. schmitti* there and elsewhere (§172).

§124. *Ophiopteris papillosa* (Fig. 103, *c*) occurs with other brittle stars in the south, where it is not uncommon; but there it seldom reaches the great size and spectacular appearance that characterize the rare Pacific Grove specimens—a strange situation, since Pacific

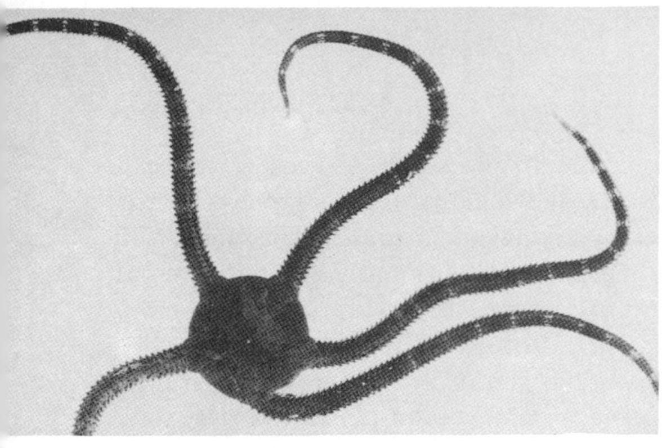

01. *Ophioderma panamensis* (§121).

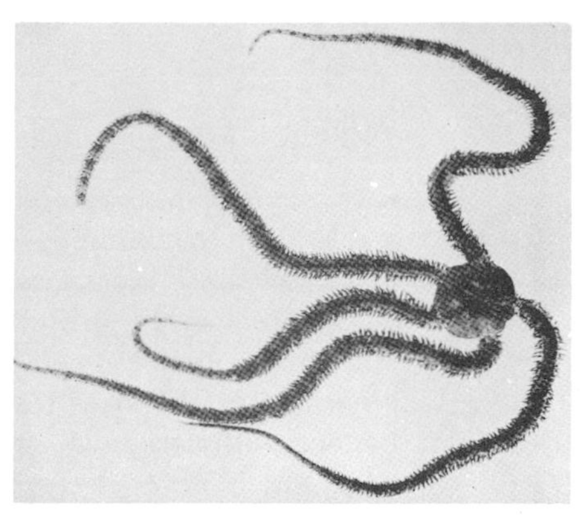

102. *Ophionereis annulata* (§121).

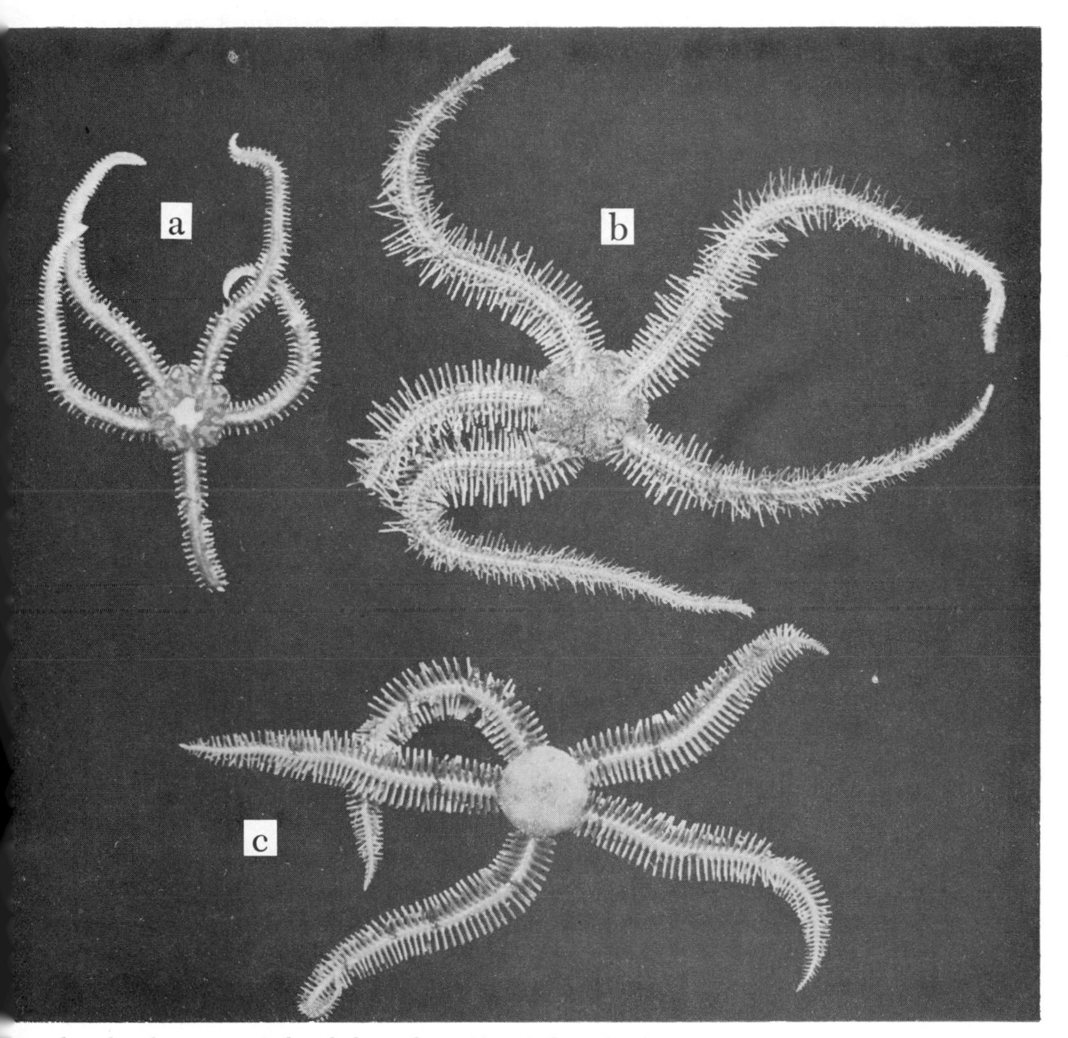

3. Three brittle stars: *a, Ophiopholis aculeata* (§122); *b, Ophiothrix iculata* (§123); *c, Ophiopteris papillosa* (§124).

Grove is the animal's northern limit. In the south, too, *O. papillosa* is a relatively hardy animal, whereas at Pacific Grove it is so fragile that only once have we succeeded in taking an unbroken specimen. The spines are large and blunt, almost square-cut, and relatively few, although the temperamental animal is nonetheless distinctly bristly in appearance. The arms reach a spread of 6 to 7 inches and are brown, banded with darker colors. This brittle star is never found in great numbers, and is never accessible except on the lowest of low tides, for it requires water that is pure and well aerated.

§125. The avid turner of rocks will see many such fixed bivalves as *Chama* (§126) and the rock oyster *Hinnites giganteus* (Fig. 104). *Hinnites,* when young, is not to be distinguished from a scallop, and it swims about by flapping its two equal shells. But the ways of placid old age creep on it rapidly. The half-buried undersurface of some great rock offers the appeal of a fireside nook to a sedentary scholar, and there the scallop settles. One valve attaches to the rock and becomes distorted to fit it; the other valve adapts itself to the shape of the attached one, and there *Hinnites* lives to an alert old age, with its half-open shell ready to snap shut and send a spurt of water into the face of some imprudent collector. The largest specimens of this species, which ranges from Alaska to Baja California, may have a diameter of more than 6 inches.

Hinnites was apparently much more abundant, perhaps at somewhat higher levels than we now find it, not so long ago. It was a popular animal for seafood banquets on the Oregon coast eighty years back, and the editor remembers his mother speaking of the rock oysters of Purissima Creek (below Half Moon Bay). Now it is eagerly sought after by the growing hordes of divers, who are harvesting the subtidal populations. It is possible that it may take 25 years for a *Hinnites* to reach full size, and even medium-sized specimens may be several years or a decade old. Bag limits of 30 or 40 a day are too generous for this slow-growing animal.

§126. *Hinnites* is fairly flat, without any sudden curves, but *Chama pellucida* (Fig. 105) is often taller than it is long. Nearly hemispherical, and only an inch or so high, the translucent white shell of this species is like nothing else to be found in the tide pools. Except that it snaps its shell shut at times most inconvenient for the observer, we have nothing to record of its natural history; and we suspect that little or nothing is known directly of its breeding or feeding habits, its growth rate, or its life-span. In the Baja California region *Chama* grows

104. The rock oyster, *Hinnites giganteus* (§125).

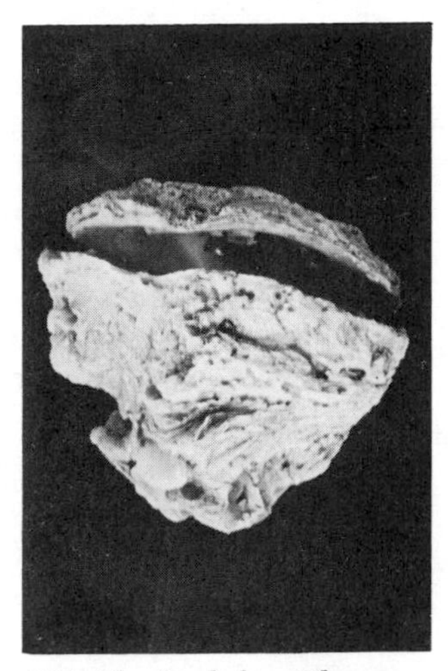

105. The fixed clam *Chama pellucida* (§126).

so thickly on the tops and sides of rocks as to be practically a reef-building form; in the Pacific Grove region, however, it is found only under rocks at the lowest of low tides. It is reported to range from Oregon to South America, but is rarely if ever seen in Oregon.

Most of the species of *Chama,* which share with the unrelated *Pododesmus* the name "rock oyster," are tropical forms, and are found attached to coral reefs. The few species in the world today are only a fraction of the number of extinct forms. Many are highly frilled and ornamented—an indication that the species is reaching its old age and is likely to become extinct soon.

The rock oyster, or "jingle," *Pododesmus* is occasional here, but it is more common on rocks in quiet bays, or on wharf piling (§217).

§127. The other molluscan inhabitants of the under-rock areas are chitons ("sea cradles"), several species of which have already been mentioned (see Fig. 106 for illustrations of various specimens). Most of the 110–25 species of chitons that inhabit the west coast are light-sensitive and will be found underneath the rocks in this zone, carefully protected from the sun.

Most chitons are vegetable feeders; but some are actively carnivorous, and others seem to prefer a mixed diet. In most cases the sexes are separate. According to observations made by Dr. Heath, the females will never release their eggs until the males have liberated their sperm into the water. Spawning takes place on May and June days when low tides occur in the early morning—not, apparently, because of the influence of the moon or merely because the tide is low, but because the water in the pools is then undisturbed. Egg-laying may continue for several hours, but it ceases the moment the returning tide sends the first wave breaking into the pool. Another investigator, however (Grave, 1922), finds that moonlight is the chief influence on the sexual activity of an east coast form similar to *Nuttallina.*

Stenoplax heathiana (up to 3 inches long) is a rather beautifully marbled, elongated, light gray to white chiton, ranging from Oregon to Magdalena Bay in Baja California; it is particularly common in Monterey Bay. Along a considerable stretch of coast, all the *S. heathiana* (Fig. 106) will spawn on the same day, and almost at the same hour. Usually each female lays two long, jellylike, spirals of eggs, which average 31 inches long and contain, together, from 100,000 to nearly 200,000 eggs. The larvae begin moving within 24 hours, and in 6 days the young break through, to swim freely for from 15 minutes to two hours before they settle down. After settling they remain inactive for about two days, but within 10–12 days they undergo several metamorphoses and become miniature chitons. The same sequence of events appears to apply to the other species of *Stenoplax,* to *Mopalia muscosa,* and to *Katharina.*

In the La Jolla tide pools the most abundant large chiton to be found at present is *Stenoplax conspicua* (Fig. 107), whose girdle is covered with fine, velvety bristles. Four-inch specimens are common there, in the San Diego area, and in northern Baja California, the extreme range being Monterey to the Gulf. Both of these large and noticeable chitons occur under fairly smooth and round rocks buried in types of substratum in which sand is the dominant constituent.

§128. Other common chitons (Fig. 106) are readily distinguished by their shape or color. *Ischnochiton regularis,* reported from Mendocino to Monterey only, is a uniform and beautiful slaty blue, and is up to 2 inches long. Another *Ischnochiton,* of unknown species, is also small and resembles *I. regularis* except that it is colored a brilliant turquoise. Specimens of *Callistochiton crassicostatus* (not illustrated), green or brown, have a noticeable median keel to the plates, are less than an inch long, and range from Alaska to San Diego. *Placiphorella velata* is recognized by its almost circular outline; it is 2 inches or less in length. James McLean has observed that *Placiphorella* is an active predator, capturing such agile prey as amphipods and worms by lift-

)6. Under-rock chitons (§§127, 128) of Zone 4. Upper row: *Stenoplax heathiana, Ischnochiton gularis, Lepidozona mertensi, L. cooperi, Placiphorella velata*. Lower row: *Mopalia ciliata, . muscosa, M. ciliata*.

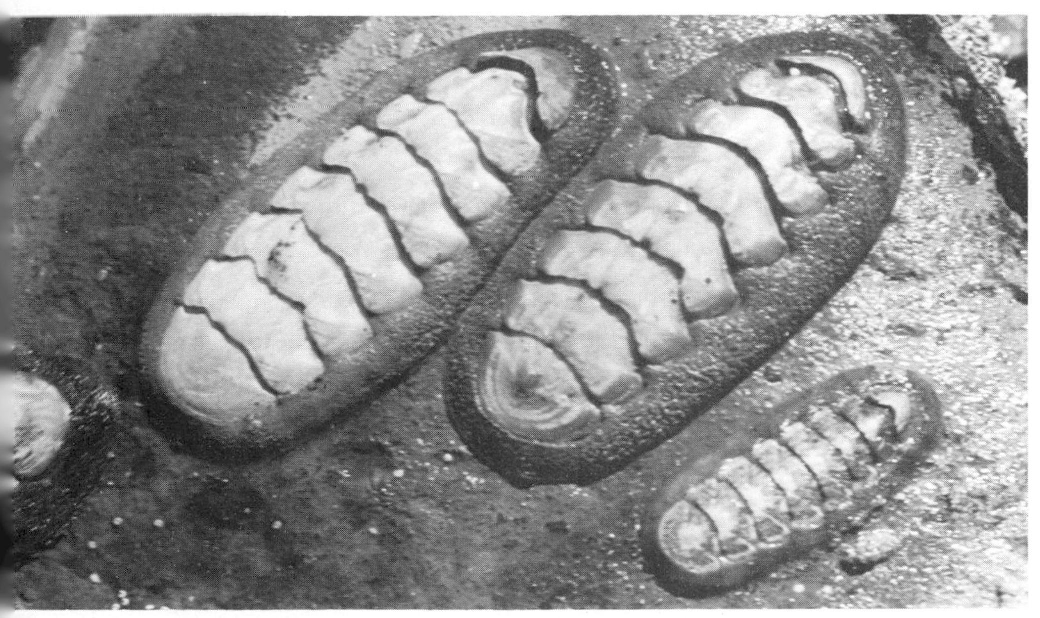

7. *Stenoplax conspicua* (§127).

ing up its head flap and trapping its victims. *Lepidozona mertensi,* up to 1½ inches long, with angular ridges and straight sides, is usually a mottled red, and ranges from Alaska to San Pedro. The similar but dull brown *L. cooperi* occurs also. All of these are exceedingly shy animals, presumably because they find sunlight highly toxic. But the several species of *Mopalia*—broad, chunky chitons that have the mantle haired to the point of furriness—are more tolerant of daylight. We have found fine, healthy specimens living on the walls and ceilings of caves, and during foggy weather they may often be found on rocks. Ordinarily, however, they are under-rock animals like their relatives. Some *Mopalia,* however, live in sheltered bays; and *Mopalia muscosa* is evidently tolerant of surprisingly muddy conditions on scattered rocks in such locations as Tomales Bay and Yaquina Bay.

The three common species of *Mopalia* may be identified according to the following field characters. *M. muscosa* (§221) is unmistakable; the thick, stiff hair is diagnostic. Of the other two, the more common *M. ciliata* has a definite split notch at the posterior end of the girdle, and the sculpturing is very evident; *M. lignosa* lacks both, but has prominent color lines, and its posterior valve has a decided central beak. This information has not so far been correlated with habitat; probably the various species are ecologically restricted. *M. ciliata* and *M. lignosa,* especially the former, occur between tides on open shores; both range from Alaska to Baja California.

§129. The large, tubed worms are of two sorts: serpulids, in which the tube is white, hard, and calcareous; and sabellids, which are larger, plumed worms with membranous tubes. The sabellid *Eudistylia polymorpha* (Fig. 108) is one of the vivid sights of the North Pacific tide pools. The dull yellow or gray, parchment-like tube is very tough and may be 18 inches long, extending far down into the crevices of rocks; nothing short of prolonged labor with a pick will dislodge the complete animal. When the worm is undisturbed, its lovely gills are protruded out of the tube into the waters of the tide pool, where they look like a delicate flower or like the tentacles of an anemone. Touch them, and they snap back into the tube with such rapidity as to leave the observer rubbing his eyes and wondering if there really could have been such a thing. So sensitive are the light-perceiving spots on the tentacles that the mere shadow of one's hand passing over them is enough to cause the tentacles to be snapped back. This worm ranges from Alaska to San Pedro but is not common south of Pacific Grove.

Eudistylia vancouveri, of the northwest, is gregarious, forming large,

108. Clustered tubes of the sabellid worm *Eudistylia polymorpha* (§129).

shrublike masses on piling (but it appears to be susceptible to reduced salinity); surprisingly enough, it is also found on vertical rock faces in heavy surf, where it would seem in danger of being washed away. It is possibly the same species as *E. polymorpha*.

A solitary serpulid worm, *Serpula vermicularis* (§213), may be locally common. It has a sinuous to nearly straight stony tube several inches in length, which is stoppered up with a red operculum when the tentacles are withdrawn. When removed from its tube, this worm obligingly extrudes red eggs or yellowish-white sperm. The sexes are separate, but there is no apparent external difference between the male and the female, although probably, as with Roland Young's fleas, "she can tell, and so can he." Spawning has been observed at Pacific Grove during the spring.

A smaller tubed worm, *Marphysa stylobranchiata*, occurs in under-rock chitinized tubes much like those of *Eudistylia* but considerably smaller, and has two big head palps projecting from the tube in place of the great sabellid's tentacle crown. *Marphysa* is a eunicid, a very near relative of the palolo worm of lunar spawning habits (§151); but it apparently lacks that spectacular trait on this coast.

§130. The worm *Glycera americana* should be mentioned in passing; although commoner in sloughs, particularly in eelgrass (§280), it will sometimes be found under rocks on the protected outer coast. These slender worms may be mistaken for *Nereis* (*Neanthes virens,* §311, is occasional here also), but differ in that the head tapers to a point, and in that the animal can shoot out a startling introvert. The long, slim, and fragile *Lumbrineris zonata* will be taken here occasionally, as well as the ubiquitous scale worm *Halosydna* (§49) and a good many smaller forms. Other worms frequently found are mentioned in the root and holdfast habitat (§153), but of the enumerating of worms there is no end: certainly 20 species could be taken in this environment at Monterey alone. Their differentiation offers a challenge even to the specialist; field determinations are rarely conclusive.

§131. Of the masking crabs, *Loxorhynchus crispatus* (Fig. 109) is easily the champion. Although it occurs so far down in the intertidal as to be more justly considered a subtidal form, it will be found often enough in the intertidal between Point Reyes and San Diego to justify mention, and it is certain to occasion interest whenever found. It is perhaps the most inactive of all the crabs, moving seldom, and then sluggishly. Until it does move one never suspects that it is a crab.

To some animals the accidental growth of algae or sessile animals on their shells seems to be a source of danger, presumably because the weight and the water resistance of this growth might hamper their movements. These animals keep their shells scrupulously clean. Others tolerate foreign growths; and still others, notably some of the spider crabs, go to the extreme of augmenting the natural growths by planting algae, hydroids, sponges, etc., on their backs. This masking may serve a double purpose: first, to make the animal inconspicuous to its enemies; and, second, to enable it to stalk its prey without detection. Masking crabs placed in a new environment will head for areas containing the same kind of growth that they carry on their backs if such areas are available; if placed in a totally different environment, they will often remove the existing growths from their backs and replace them with forms that are common in the new locality.

Experimenting with a European masking crab (*Dromia*), Dembowska found that if the animal were deprived of its sponge covering and

109. *Loxorhynchus crispatus* (§131), a masking crab.

placed in an aquarium with a piece of writing paper, it would tear the paper into a pattern roughly corresponding to the shape of its back and put it on. This was an emergency measure, however; when a sponge attached by a wire hook was hung within reach, the crab went toward it at once, dropping its paper shield on the way. It pulled itself up to the sponge, cut the sponge loose from the hook, and rolled with it to the bottom of the aquarium. Immediately afterward the crab placed the sponge on its back, holding it there with the upcurved fourth and fifth pairs of legs, which are modified for that purpose.

The crab showed considerable adaptability to circumstances and some capacity for learning, for when the sponge covering was removed time and again and buried in gravel, deeper each time, the crab learned to uncover it and finally to dig at the right spot even when the sponge was entirely out of sight. These experiments confirm the very complex nature of crab reactions, a matter that is further discussed in connection with the fiddler crab (§284). Observing such behavior, one is tempted to credit the animals with something akin to intelligence, but there is reason for believing that their actions are almost entirely automatic.

A more common and more truly intertidal masking crab is *Scyra acutifrons*. It is not more than half the size of *Loxorhynchus*, and it is not so efficient a masker. Although reported from Kodiak to San Diego, it is uncommon below Pacific Grove.

§132. Under-rock collecting will turn up a number of other crabs, which may be differentiated by reference to the illustrations. *Pachycheles* (§318), a common wharf-piling inhabitant, will occasionally be routed out of an unaccustomed home here.

Lophopanopeus frontalis is a common southern California form, its place being taken at Monterey Bay by *L. leucomanus* ssp. *heathi* (Fig. 110).

Mimulus foliatus (Fig. 111), of which we have seen ovigerous females during June in central California, may have a carapace of bright red or yellow, caused, in some cases at least, by growths of sponges. An extreme range of Unalaska to Mazatlán is recorded.

The queer-looking *Cryptolithodes sitchensis* (Fig. 112) has so large a carapace that none of its appendages are visible from above, and it resembles nothing so much as a timeworn fossil. This species ranges from Alaska to Monterey, and full-sized specimens are not uncommon even in the southern part of the range.

The fuzziest of the crabs is *Hapalogaster cavicauda* (Fig. 113), which ranges from Cape Mendocino to the Channel Islands off Santa Barbara. In life the animal presents an interesting sight, especially when, as during the winter months, the already swollen tail is distended with eggs. The male also has a characteristic redundance of tail.

Not so much under rocks as in shallow crevices, abandoned sea-urchin pits, and other holes will be found *Oedignathus inermis*, a slow-moving, stout, clawed anomuran that superficially resembles a fiddler crab because of its large claw. It has a large, somewhat bulbous abdomen, which outwardly suggests the presence of a sacculinid parasite. This crab occurs from Alaska to Monterey, but it is rare in central California. It is often encountered on the Oregon coast.

§133. Conditions of semitropical heat and aridity being what they are, it is quite in the order of things for the intertidal animals of the south to be restricted to the under-rock habitat. From Santa Barbara to as far down into Baja California as we have collected there is a prolific and interesting fauna, but only the upturning of rocks will reveal it.

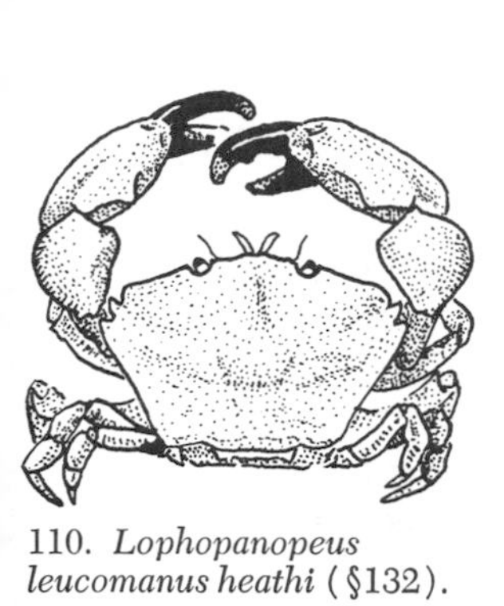

110. *Lophopanopeus leucomanus heathi* (§132).

111. *Mimulus foliatus* (§132).

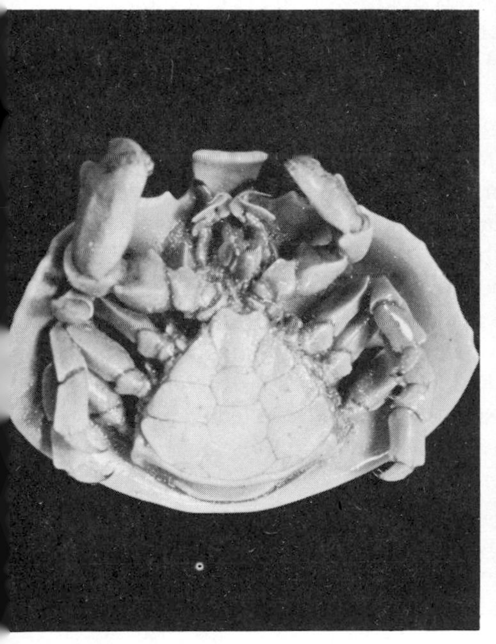

2. Underside of *Cryptolithodes sitchensis* 132), the umbrella-backed crab.

113. The furry *Hapalogaster cavicauda* (§132).

Such rock-beach inhabitants as *Ischnochiton,* octopuses, brittle stars, ghost shrimps, and blind gobies, the two last restricted to the substratum, form an association in which a common top shell and several crabs also figure.

The crab *Cycloxanthops novemdentatus* (Fig. 114), a dull reddish-brown animal, will be taken for a small *Cancer.* And the dark red *Paraxanthias taylori* (Fig. 115) is identified by the obvious bumps on its big claws. Both of these crabs extend down into Mexico from Monterey; but though they are found in the intertidal at La Jolla, for instance, neither is common in the northern part of their range.

Pilumnus spinohirsutus (Fig. 116), recognizable immediately by its hairiness, is another of the retiring, probably light-avoiding crabs of the southern coast; it ranges from Venice to Ecuador.

Any of these three crabs may be found under rocks at Laguna, La Jolla, or Ensenada—sometimes all of them under a single large rock, possibly with *Lophopanopeus* and a small octopus thrown in for good measure.

§134. *Hippolysmata californica* (Fig. 117) is a conspicuous, transparent shrimp, up to 1½ inches long, with broken red stripes running fore and aft. A dozen may occur on a single undersurface, but the collector who can capture more than one of them is doing well; for when the rock is moved, they hop around at a great rate and disappear very quickly. *Hippolysmata* has not been recorded north of Santa Barbara. We have taken specimens at Ensenada, and it no doubt ranges farther south.

§135. The much maligned octopus—which, in the vernacular, shares the name of devilfish with the giant manta ray of tropical waters—is not often found in the intertidal regions of the north, but from Santa Barbara south a small species, *Octopus* (formerly *Polypus*) *bimaculoides* (Fig. 118), is a common under-rock inhabitant of the outer tidelands. The octopus has eyes as highly developed as ours, and a larger and better functioning brain than any other invertebrate animal. Its relatively high degree of intelligence is quite likely a factor in its survival in the face of persistent collecting for food. Italians and Chinese justly relish the animals as food, and so does an occasional American. Delicious as octopus is, however, it is decidedly tough and rubbery, and the meat grinder treatment is recommended.

In the metropolitan areas around Los Angeles and San Diego, octopuses are no longer to be had in their former abundance, for, besides the many people who hunt them specifically, hunters of abalones and spiny lobsters capture them incidentally. Nevertheless, a good many are

'4. *Cycloxanthops novemdentatus* (§133).

115. *Paraxanthias taylori* (§133).

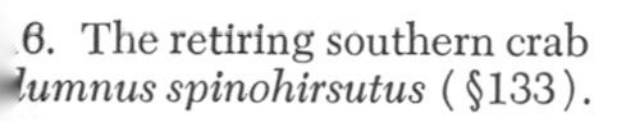

.6. The retiring southern crab *'lumnus spinohirsutus* (§133).

7. Below: the southern insparent shrimp *Hippolysmata lifornica* (§134).

still found, even in the areas mentioned, where it is probably safe to say that several thousand are taken for food every year. This observation is based on personal experience. For several years we have collected, observed, and photographed along the Corona del Mar shore at Newport Bay and in the region north of Laguna. On one visit to Corona del Mar we questioned one of several crews who we had supposed were collecting bait. It developed that they were capturing octopuses, and their rather pernicious method, which we afterward watched, was as follows: Two men, armed with gaff hooks and carrying gunnysacks, stationed themselves quietly by a large, isolated rock in a pool. Finding a hollow that indicated the probable entrance to an octopus's under-rock lair, one of the men poured into the water a bit of what he called "lye" solution, probably chloride of lime (bleach is now popular and works all too well). Usually the mollusk was soon forced to come out and seek less irritating water, whereupon he was promptly hooked and deposited in the gunnysack. These two men had captured 13 octopuses, totaling possibly 30 pounds, and several other crews appeared to have had equally good "luck." The whole procedure has been approximately duplicated on each of the dozen or more times we have visited the spot. So devastating a method of collecting as poisoning pools would be regrettable anywhere, unless the collectors were in genuine need of food. Along the coast of Baja California the octopuses are more fortunate, for the Mexicans we have talked with along the shore are interested in abalones and spiny lobsters only and regard the eating of octopuses with the same horror that most Americans do.

Between Tijuana and Ensenada we have found octopuses very numerous in April, May, and December, in several different years. In December 1930 they were so abundant that one could count on finding a specimen under at least every fourth rock overturned. Their clever hiding and escaping strategies make them difficult to see, difficult to capture after they are seen, and difficult to hold after they are captured. Two animals may be found under the same rock, but to take them both with bare hands is almost an impossibility. A captured specimen will sometimes cling to one's hand, and, if the animal is large enough, to one's forearm; but unless the collector has acquired some skill in handling the wily animals, the specimen will at once let go, shoot out of his hand like a bar of wet soap, and disappear. A little observation will convince the collector that in a given area probably half of the specimens escape notice despite the most careful searching—a highly desirable situation from the point of view of the octopus.

118. A young octopus, *Octopus* sp. (§135), in an aquarium.

This mud-flat octopus is known as *Octopus bimaculoides* because of its two large, round, eyelike spots. These are wider apart than the animal's real eyes, which are raised above the body surface in knobs reminiscent of the light towers on sailing ships. Each of the eight arms has two rows of suction disks, so powerful that when even a small specimen decides to fight it out by main strength it is very difficult to detach him from his rock. In an aquarium, very small specimens survive well, withstanding inclemencies of temperature and stale water most amazingly for animals so delicate; but large specimens (an arm spread of 2 feet is large) must be kept in cool and well-oxygenated water to survive.

O. bimaculoides is inferior to the rest of his tribe in the ability to change color, but even so he can produce startling and often beautiful effects and harmonize his color so perfectly with his surroundings as to be almost invisible until he moves. The usually larger *Octopus* that is occasionally found in Monterey Bay is a more versatile color-change artist, as is *O. dofleini,* of waters from Alaska to Oregon.

Like all of the octopods (and the related decapods, such as the squid) the octopus has an ink sac, opening near the anus, from which he can discharge a dense, sepia-colored fluid, creating a "smoke screen" that should be the envy of the navy. The junior writer once spent hours changing water in a 5-gallon jar so that he might better observe the half-dozen octopuses therein, but his arms were exhausted before the animals' ink sacs.

The octopus, together with the squid and *Nautilus,* belongs to the molluscan class Cephalopoda. Though the octopus has no shell, an internal vestige is to be found in most squids—in some forming the "cuttlebone" —and *Nautilus* has a well-developed external shell. The beautiful "shell" of the paper nautilus (*Argonauta*) is actually an egg case, secreted not by the mantle but by a specialized pair of arms in the female. The male argonaut has no shell.

The octopus can move rapidly over sand or rocks by the use of its arms and suckers; but in open water its arms trail away from the direction of motion with an efficient-looking, streamlined effect as it propels itself swiftly backward with powerful jets of water from its siphon tube. The animal's usual method of hunting is to lie quietly under rocks, dart out to capture passing fish or crustaceans (crabs seem to be a favorite food), and then kill them, presumably with the strong, beaklike jaws that are normally concealed inside the mouth opening.

The octopus's method of reproduction is decidedly unique. At breeding time one arm of the male enlarges and is modified (hectocotylized) as a copulatory organ. From the generative orifice he charges this arm with a packet of spermatozoa, which he deposits under the mantle skirt of the female. In some squids a portion of the arm is detached and carried in the mantle cavity of the female until fertilization takes place—often a matter of several days. This detached arm was formerly thought to be a separate animal that was parasitic in the female cephalopod. In dissecting male squid, the senior writer has seen the packet of spermatozoa explode when the air reached it, shooting out a long arrow-like streamer that was attached with a cord to the spermatophore.

The act of fertilization accomplished, the male octopus has but to retire to repair his damaged arm. The female, however, has before her a long vigil, for a description of which we quote from a paper by Dr. Fisher. A small octopus had been captured, late in June 1922, near the Hopkins Marine Station at Pacific Grove. It was kept alive in an aquarium for 3½ months, where it fed irregularly, and usually at night, on pieces of fish and abalone. Dr. Fisher writes:

"For reference in conversation, the octopus was named 'Mephisto,' later changed for obvious reasons to 'Mephista.' She had a little shelter of stones in a front corner of the aquarium. A movable board was so arranged as to exclude all but dim light, since in the language of animal behaviorists an octopus is negatively phototactic (unless, perchance, it is too hungry to care).

"During the night of July 4 two or three festoons of eggs were deposited on the sloping underside of the granite roof of Mephista's retreat. These eggs were nearly transparent, and resembled a long bunch of sultana grapes in miniature. The central axis, consisting of the twisted peduncles of the eggs, was brown. During the following week about forty other clusters were deposited, always at night. The process was never observed.

"In the meantime she had assumed a brooding position. The masses of eggs rested against the dorsal surface of the body and the arms were curled backwards to form a sort of basket or receptacle. [During one period of observation] the ends of the arms were in frequent gentle movement over the clusters and among them—almost a combing motion.

"Mephista was never observed away from the eggs during the day, unless badly irritated. If a morsel of fish or abalone was dropped close to her, she endeavored to blow it away with water from the funnel. If this were not sufficient, she caught the food in one or two arms and dropped it as far as she could reach. If then the fish was returned she flushed reddish-brown, seized the offending morsel, crawled irritably out of her little cell, and, with a curious straightening movement of all the tentacles, fairly hurled the food from her. She then quickly returned to the eggs. In the morning the rejected food would be found partly eaten.

"Although the eggs were laid in early July they showed no signs of development for nearly a month. By August 20 the embryos were well advanced and their chromatophores were contracting and expanding. From time to time bunches of eggs had been removed for preservation, without causing unusual reactions on Mephista's part. About ten bunches were left. They hatched during my absence from the laboratory near the middle of September. On my return, October 1, Mephista was still covering the remains of her brood—the dilapidated clusters of empty egg capsules. She refused to eat either fish or abalone, and had become noticeably smaller. On October 11 she was found dead at her post.... Possibly during the post-brooding period the food reaction is specialized —that is, limited to relatively few things. I find it hard to believe that in *Polypus* death is a normal sequel to egg-laying."

Two years later another specimen, named Mephista II, repeated the performance, with some variation in details, and provided further opportunities for observing the embryos and newly hatched young. The latter, a shade less than 1/16 inch long, are so translucent that some of the internal organs are visible, and "the beating of the systemic and branchial hearts may be observed." The body form is more like that of a squid than that of an adult octopus. The young of Mephista II swam as soon as they were hatched; but they would not eat any food that was offered, even bits of yolk from their own eggs, and none lived more than three or four days.

The process of hatching usually involves a violent struggle. First the embryo squirms until it splits the capsule at one end; then "by vigorous expansions and contractions the body is forced out through the opening.... The escape from the shell proceeds rather rapidly until the visceral sac and funnel are free. The somewhat elastic egg shell seems to close around the head, behind the eyes, and the animal has to struggle violently to extricate itself. This final stage may occupy from ten to forty minutes." Once, in what might be interpreted as a flurry of temper at the difficulty of being born, one of the tiny animals discharged ink into its capsule.

§136. Conchologists have celebrated the richness of the southern California shores, and even the collector interested in animals rather than empty shells will be impressed by it. A smooth brown cowry, *Cypraea spadicea* (Fig. 119), and a cone shell, *Conus californicus* (Fig. 120), are likely to be found, both representative of families well developed in the tropics and featured in shell collectors' cabinets. The California cone, like its numerous tropical relatives, is an active predator that immobilizes its prey with venom. Its radula is carried on a long, prehensile proboscis that is cautiously sneaked near enough to the prey for a final quick jab from the poison-bearing tooth. Some tropical cones paralyze small fish, and some are dangerous to man. The California species feeds on polychaetes and snails like *Nassarius*. Its feeding habits are less specific than those of cones found on tropical reefs.

The top shells and purple snails are amply represented. The smooth

9. A smooth brown cowry, *Cypraea spadicea* (§136).

120. *Conus californicus* (§136).

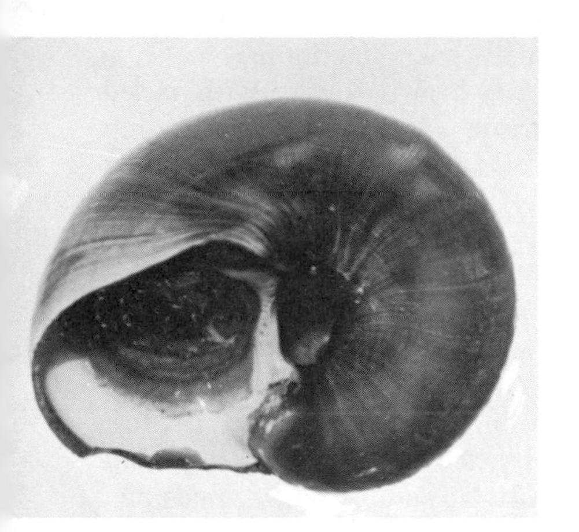

. The smooth turban snail
rrisia norrisi (§136).

122. *Septifer bifurcatus* (§136).

123. *Cardita carpenteri* (§136).

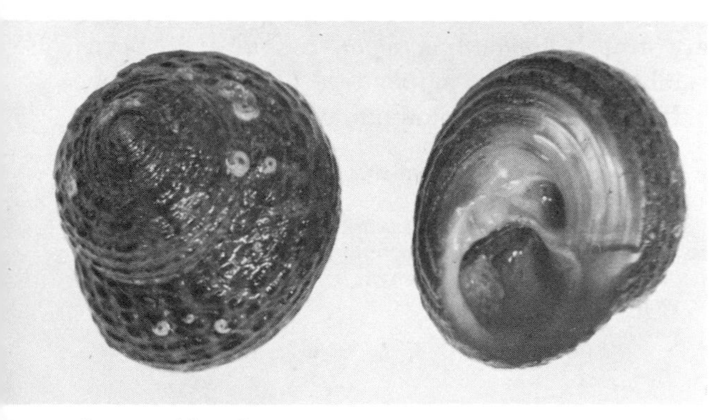

. *Tegula eiseni* (§137).

125. *Amphissa versicolor* (§137).

turban *Norrisia norrisi* (Fig. 121) will be noticed because of its size and numbers, and enjoyed because of the pleasing contrast that the vivid red flesh of the living animal forms with the lustrous brown of the shell. It is a satisfying experience to find crawling about in tide pools living specimens of a type that one has known from empty shells only—especially since, for some of these mollusks, there are no known methods of anesthetization that will preserve the relaxed animal. More often than not, individuals of *Norrisia* will have a pinkish slipper limpet attached to them. As its name implies, *Crepidula norrisiarum* is characteristically associated with *Norrisia*; but it may be found on other snails, or occasionally on a crab.

Septifer bifurcatus (Fig. 122), a ribbed and slightly hairy mussel not unlike *Mytilus,* is found in these parts, but only under rocks and stones, often well up in the intertidal zone. Specimens may be more than 1½ inches long, but the average is smaller. Another common bivalve is *Cardita carpenteri* (Fig. 123), a very small cockle-like form.

§137. *Tegula eiseni* (Fig. 124), a rusty-brown turban with raised beaded bands, is gaily colored by comparison with *T. funebralis,* its drab relative of the more shoreward rocks; and it is more retiring, hiding under rocks during the day. South of the Los Angeles area the two may occur on the same beach, but north of that region *T. eiseni* is rarely found. It is recorded, however, as far north as Monterey.*

No northern snail fills the place held by *T. eiseni* at, for instance, La Jolla. *Amphissa versicolor* (Fig. 125) possibly comes nearest. Restricted to the underside of rocks loosely buried in a gravel substratum, this excessively active form sports a proboscis longer than the average. Most snails retract into and extend from their shells very deliberately. *Amphissa* emulates his enemy the hermit crab. No sudden danger finds him napping, and he can retreat and advance a dozen times while *Tegula* is doing so once. This animal and expanded specimens of the giant moon snails (§252) are good for exhibition to whoever still believes that snails are mostly shell.

§138. A polyclad flatworm, *Notoplana acticola,* has already been mentioned (§17) as an inhabitant of the upper tide pools. Along the central California coast, the polyclads are not abundant in the lower tide

* This seems as appropriate a place as any to point out that many records for Monterey may indicate shells found perhaps but once or twice. Since the days of dear old Josiah Keep, Monterey and vicinity have been a favorite collecting ground for conchologists, and they have assiduously recorded their findings and preserved their rarities in cabinets. This has resulted in a certain skewness for many distribution records, for species north as well as south.

pools; but in the south it is under rocks in this lowest zone that they will be found—two of them, both probably unidentified, very commonly, and others occasionally. One of them we think of, for lack of any other name, as the "pepper-and-salt" flatworm. The other is an orange-red form. Two others, somewhat less common but more striking, have papillated furry backs and affix themselves to the rocks with suckers on their undersides. The first of them, also unidentified, is a soft brown form up to 1½ inches long. The second, probably a new species of *Thysanozoon,* the "scarf dancer," is not an impressive sight when found attached by its suckers and huddled up on the underside of a rock; but when it stretches itself to get under way in the clear water of a tide pool, it presents a different picture. It is brilliantly colored and swims gracefully, undulating its margins delicately but effectively and justifying its name. Italians living near Naples call the *Thysanozoon* there the "skirt dancer." Collectors who procure any of these undetermined forms will do well to preserve them carefully in 6 per cent formalin for forwarding to a specialist currently working with Pacific coast polyclads.

§139. The most common habitat of nemerteans, or ribbon worms, is in the mud of bays and estuaries, so they are considered in some detail in that section (§312). However, a good many forms, some of them large and common, will be found at home under rocks on fairly open coast, and therefore require mention here. The most conspicuous of these, and indeed one of the most striking of all intertidal animals, is *Micrura verrilli* (see color section), whose body is vividly banded with sparkling lavender and white. It may be up to 12 inches long, and ranges from Alaska to Monterey Bay, but unfortunately it occurs infrequently. *Paranemertes peregrina,* purple or brown with a white underside, is commoner, ranging from Bering Sea to southern California; it is up to 6 inches in length. Near the Big Sur, and at Laguna, La Jolla, and Ensenada, we have found a long, stringy, pinkish nemertean we take to be the remarkable *Cephalothrix major.* Its slightly flattened body is only about 1/16 inch wide, but it reaches a length of 4 feet and is comparatively strong. Its mouth is set back some 2 inches from the tip of the head, and this 2-inch portion is cylindrical. Still another rocky-shore form is *Lineus vegetus,* known from Pacific Grove and La Jolla, a slender animal up to 6 inches long, brown in color, with numerous encircling lighter lines. In contraction the body coils in a close spiral. This nemertean has been the subject of some interesting regeneration experiments (§312). A bright reddish-orange, soft-bodied nemertean, capable of considerable attenuation, is often common in spring and summer on Oregon beaches and is found under rocks in northern California; this is *Tubulanus polymorphus* (see color section). A good many other forms may occur, most of which will be found treated in Coe's papers. *Carinella rubra* (§239), of quiet waters, will turn up here sometimes.

§140. Some of the bryozoans have been treated previously, so that the collector who has come with us this far will recognize several common encrustations under the rocks as belonging to this group. *Eurystomella bilabiata* (Fig. 126) is characteristically old-rose colored and always encrusts flatly on stones and discarded shells. A close examination, even with the naked eye, reveals much loveliness, but the aesthetic appeal is much enhanced, as with other bryozoans, by the use of a binocular microscope, or even a ten-power hand lens. *Hippodiplosia insculpta* (Fig. 127) is deep buff, and sometimes mildly erect in a leaflike formation on sticks. When encrusting under rocks, it may have an apparent thickness of several layers, rolled over at the margin. The living zooids of both these bryozoans can be seen, in fresh colonies taken during the winter or spring, with the lowest power of a compound microscope. One of the so-called corallines, *Phidolopora pacifica,* is a bryozoan colony. Small, lacy clusters of this latticed calcareous form are not uncommon under rocks in the south. The large colonies, up to a size that would make a double handful, can only be collected by dredging. *E. bilabiata* ranges from the Queen Charlotte Islands to Pacific Grove, *H. insculpta* from Sitka to Pacific Grove (and on to southern California as a subtidal form), and *P. pacifica* from Puget Sound to southern California. Attached forms also include the encrusting red sponges mentioned in §32, some of which are more at home in this deep littoral than further up. The brownish red *Esperiopsis originalis,* not even remotely resembling the

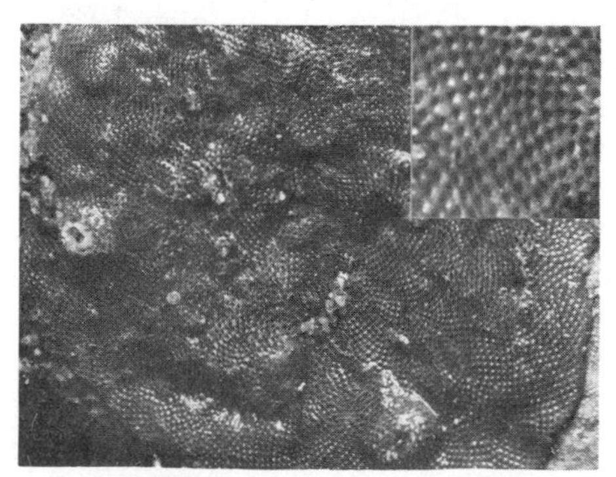

126. The bryozoan *Eurystomella bilabiata* (§140); enlarged portion inset.

127. The bryozoan *Hippodiplosia insculpta* (§140).

magnificent erect *Esperiopsis* of central British Columbia inner waters, rarely occurs higher up. It encrusts the undersides of stones, but also occurs in crevices. A hard, pearly white encrusting form, *Xestospongia vanilla,* characteristic of the entire Pacific coast, looks like thin, smooth cake frosting. The very lumpy, hemispherical, bright orange, woody-fibered *Tethya aurantia* may be taken at Carmel, far down in rocky crevices or on the undersides of rocks. *T. aurantia's* base, with its radiating core structure, is visible if the colony is detached; it is diagnostic, and there is nothing else even slightly like it in the tide pools. This is a cosmopolitan form, almost worldwide in distribution, and is more abundant than might be supposed at first glance, since many specimens are overgrown with green algae.

§141. Although the southern California cucumber, *Stichopus parvimensis,* has been treated (§77) as a protected-rock animal, it often occurs, especially when a low tide comes on in the heat of the day, in the under-rock habitat. North of central California the white *Eupentacta quinquesemita* occurs in this habitat, too, but it is much more common in sheltered waters like Puget Sound (§235).

Lissothuria nutriens, a small, red, flattened cucumber that is remarkable for carrying its young on its back, is occasionally found between Monterey and Santa Barbara. Strangely, however, central California, a region otherwise very fecund, has no shore cucumbers comparable in abundance with *Stichopus* in the south or the orange *Cucumaria miniata* in the north, though there are occasional patches of *C. miniata* in the Bodega region. Presumably, more efficient animals just happen to have crowded them out. Another viviparous holothurian (cucumber), red with a white underside, is the small *Thyone rubra. L. nutriens* or *T. rubra* may be found on kelp holdfasts..

§142. A definite form of the seastar *Pisaster giganteus,* called *capitatus* (Fig. 128), is a not uncommon under-rock member of the southern California fauna. It is by no means gigantic, despite its name, being usually smaller than the common *P. ochraceus.* It is certain to be taken for a different species, even by collectors familiar with the typical *P. giganteus* of Monterey Bay and north (Fig. 128), for it differs in habitat, shape, and color, as well as in size. The few spines are large, stumpy, and vividly colored a slaty purple on a background of ocher. The northern limit of its range is about San Luis Obispo, and we have taken it as far south as Ensenada. With it may be taken an occasional *Astrometis* (§72) or small *Patiria* (§26), most of the southern representatives of the latter species being stunted and limited to the under-rock habitat.

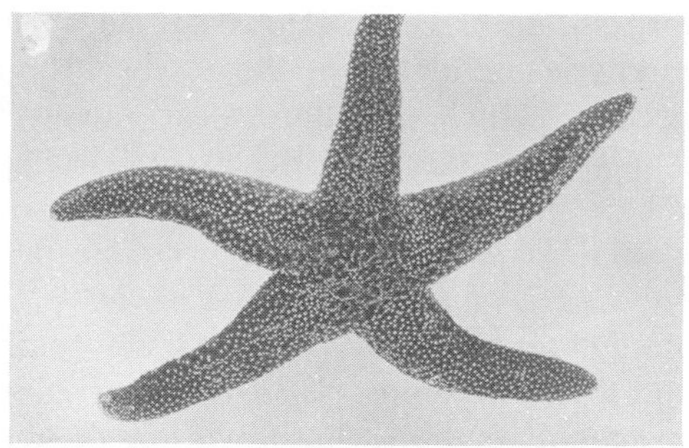

128. Two subspecies of the seastar *Pisaster giganteus*: left, *P. giganteus capitatus* (§142); right, *P. giganteus* (§§142, 157).

129. The pistol, or snapping, shrimp *Crangon dentipes* (§143); the large snapping claw is shown detached from the body.

§143. The noisy pistol shrimp *Crangon dentipes* (Fig. 129) may be taken from under rocks in the lower extremity of the rocky intertidal zone and from the sponges and bryozoans growing in kelp holdfasts. Unlike the good children of a past generation, it is more often heard than seen. Very often the collector will hear the metallic clicking of these shrimps all around him without being able to see a single specimen and, unless he is quick and industrious, without being able to take one. Captured and transferred to a jar, the animal is certain to cause excitement if his captor is inexperienced: it is hard to believe, on hearing the inevitable sound for the first few times, that the jar has not been violently cracked. This startling sound is made by the animal's snapping a thumb against the palm of its big claw; the suddenness of the motion is made possible by a trigger-like device at the joint. Any attempt at a close investigation of the operation is likely to result in the investigator's being left in possession of the detached claw while the denuded shrimp retires to grow himself a new snapper. What this snapping accomplishes, unless it frightens enemies, is difficult to guess. *Crangon* is a rather cosmopolitan species in northern tropic regions, and occurs also, for some unknown reason, as far north as San Francisco, extending southward on this coast to Cape San Lucas. A related species, *C. californiensis,* occurs in the south.

In similar environments the colorful shrimp *Spirontocaris* (§150), of several species, may be seen. Amphipods and isopods of several types occur, but none seems to be specifically characteristic here or of particular interest to anyone but a specialist. Sea spiders (pycnogonids) are seen occasionally. One, *Ammothella bi-unguiculata* var. *californica,* is common and fairly characteristic under stones at Laguna Beach and elsewhere in southern California, with specimens also being taken on the hydroids.

§144. Two "soft corals" (*Alcyonaria*) occur, most amazingly for deep-water forms, along Pacific coast rocky shores, even where there is some surf. From La Jolla south, one of them, probably a species of *Clavularia* (Fig. 130), is present in the very-low-tide zone along the sides of rocks just where they are buried in the substratum, or on the bottoms of rocks. Typically, there is a brown, ribbonlike stolon, which is not easy to detach, networked on the undersurface of the stone, with cartridge-like polyps ⅛ inch or so high in contraction beaded up from the stolon. Expanded (and they expand readily in aquaria), the polyps may be ⅜ inch high, spreading eight feathery tentacles. Because this typically alcyonarian polyp is abundant, easy to get, and hardy, it should be useful for experimental work. We have found it most abundantly in Todos Santos Bay, north of Ensenada. This is the animal that has been considered (at Laguna Beach) to be *Telesto ambigua,* a form known from several hundred fathoms in Monterey Bay. Miss Deichmann, however, refers the

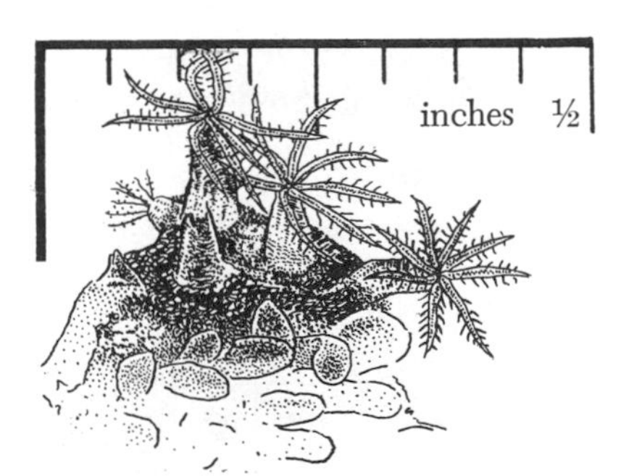

130. A "soft coral," or alcyonarian, *Clavularia* sp. (§144).

intertidal species to *Clavularia,* in spite of the resemblance of the polyps to those of our deep-water *Telesto.*

The second "soft coral" is found in Monterey Bay, and the late Dr. Hickson identified it as another species of *Clavularia.* It is white, clustered like a tunicate, and certainly different from the first-mentioned form. Both of these are related to the tropical pipe-organ corals seen in museums, but resemble them in neither appearance nor habitat.

Substratum Habitat

Limited observation has failed to indicate any outstanding difference between the substratum populations of the middle and lower tide pools. The ghost shrimps and blind gobies treated as substratum animals in the middle tide-pool zone belong more properly here, but they are considered in the higher association because the collector will find them there first.

§145. A larger and handsomer sipunculid, or peanut worm, than *Phascolosoma* (§52) is *Dendrostoma pyroides,* rich brown in color and having long plumed tentacles. It is this or a related species that is illustrated in Figure 131. The fully contracted specimens, as taken from their under-rock burrows, will usually flower out very beautifully if allowed to remain undisturbed in a dish of fresh seawater placed in subdued light. *D. pyroides* may be taken as far north as Monterey Bay, where it occurs in the lowest reaches of the intertidal zone; and a related species, *D. zostericolum* (described as D. *zostericola*), occurs at La Jolla and in northern Baja California under rocks and in the matted roots of the surf-swept eelgrass *Phyllospadix.* Quite possibly the deep substratum under-rock *Dendrostoma* of southern California and below is also *D. pyroides,* although it is ecologically different.

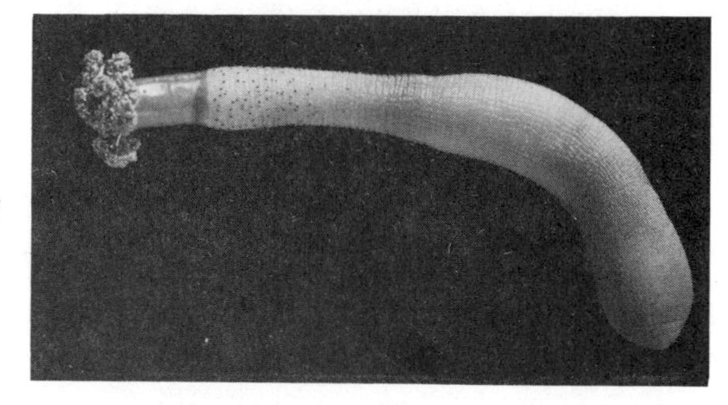

131. A small specimen of the flowering peanut worm, *Dendrostoma pyroides* (§145).

Sipunculids will be found living happily in foul black mud beneath stones and the rootstocks of *Phyllospadix*. They can apparently withstand living for a week or so in seawater with no detectable oxygen, and even survive several days in something like mineral oil. They are very difficult to narcotize, and it seems to be only an accident when one is preserved in a nicely expanded condition.

§146. The mantis shrimp *Pseudosquilla lessoni* (Fig. 132) is a shrimp-like crustacean up to 3 inches long, and is related to the mantis shrimp (*Squilla*) of Europe and the southeastern coast of the United States. It is a strictly southern form, not occurring north of Point Conception, and it lives so deep in the substratum beneath the lowest intertidal rocks that probably none but the most ardent collector will find it. When he does find the shrimp, he is likely to be rewarded for his labor by so severe a nip that he will be glad to let it go immediately. The nip will have been delivered by a unique instrument quite different from the plier-like claw of a crab. The last segment of the great claw has a row of sharp curved hooks that are forced, by muscular contraction, into sockets in the next segment. Thus, although the animal lacks the sheer strength of a crab, it compensates with greater mechanical efficiency.

Other mantis shrimps are reported from the southern region. We have failed to find them, but it is safe to assume that their habitat does not differ greatly from that of *Pseudosquilla*. The similar *Squilla polita* is known to occur even as far north as Monterey Bay (from one dredging record); although the National Museum reports no occurrence in less than 29 fathoms, *S. polita* is probably one of the forms dug occasionally by enthusiastic shore collectors in the Laguna–La Jolla region.

In these environments, along the San Mateo County shore, occasional gigantic annelids (up to several feet long) will be dug out. These are specimens of *Neanthes brandti* (§164), probably the largest polychaete on the coast; they are similar in appearance to the mussel worm illustrated in Figure 149.

§147. Where the substratum is formed of stiff clay, packed hard enough to provide attachment for sessile animals, any of several pholads, or boring clams, may occur, their sometimes highly colored siphons protruding from their burrows. The smallest of the lot is *Penitella gabbi* (Fig. 133), with a clean-cut, almost oval, shell and a creamy-lemon-colored siphon covered with tubercles. It ranges from Bering Sea to the Gulf of California. *Penitella penita* (§242), the form that also bores in concrete, is a bit larger—a little larger than a man's thumb—and has a longer shell, with leathery flaps at the siphon end. *Saxicava pholadis*

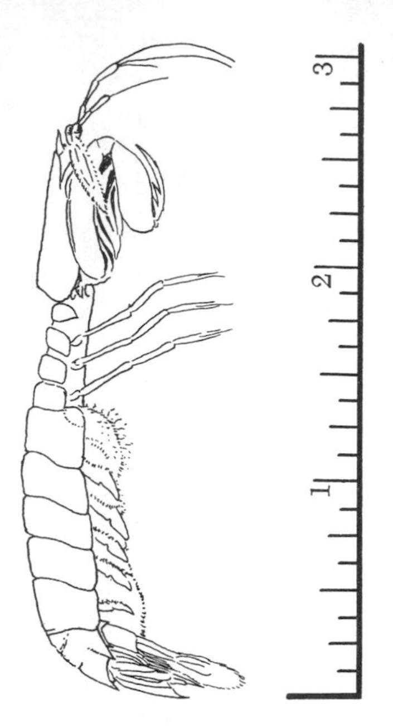

132. *Pseudosquilla lessoni* (§146), a mantis shrimp.

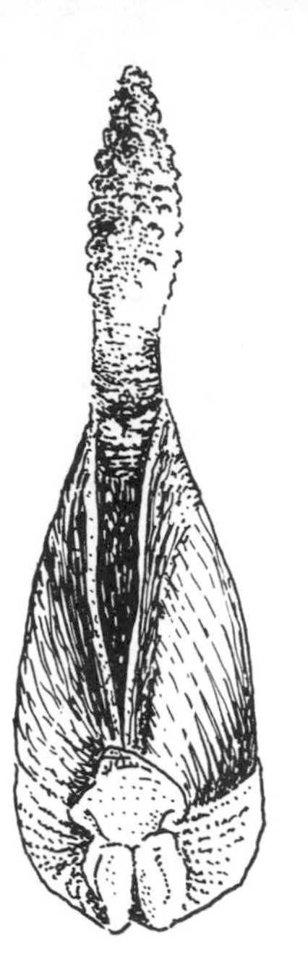

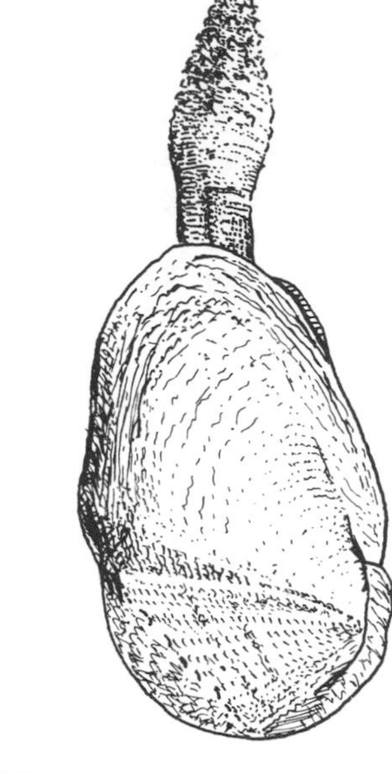

133. *Penitella gabbi* (§147), a clam that burrows in stiff clay, is found all along the Pacific coast.

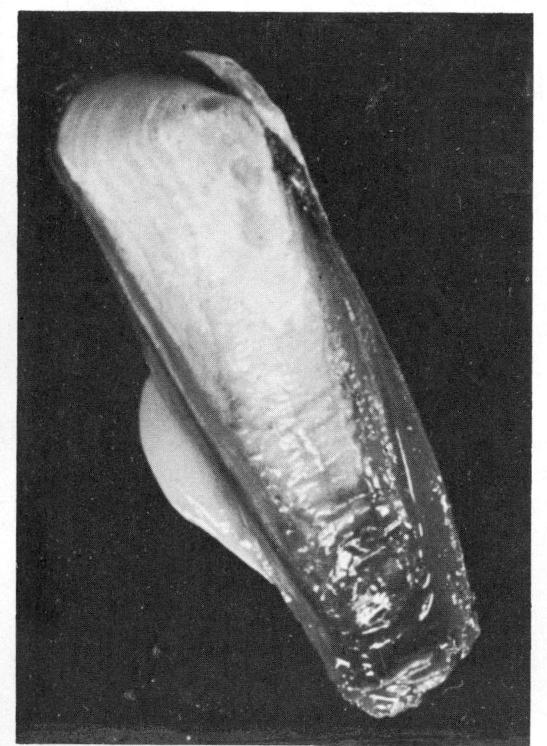

134. *Saxicava pholadis* (§147).

135. *Platyodon cancellatus* (§147).

(Fig. 134), a nestler, with red-tipped siphons, is chunky, often distorted, and ranges from the Arctic to Panama. *Platyodon cancellatus* (Fig. 135), a relative of the soft-shelled clam *Mya,* may honeycomb whole banks of stiff blue clay and is undoubtedly a factor in the disintegration of certain shores. On a trip to Santa Cruz we found a tremendous chunk of clay, weighing many tons, that had been broken off and upended by the waves, which were slowly pounding it to pieces. Hundreds of these boring clams were being exposed and killed by this disintegration—a process that they themselves had undoubtedly started by burrowing and making larger holes as they grew. A few specimens of *P. cancellatus,* whose recorded range is from Bolinas Bay to San Diego, have been found in low-grade concrete piles in Los Angeles Harbor. Another species, a giant, red-siphoned form (probably *Parapholas californica*), is abundant at Santa Cruz, but we have never succeeded in taking a specimen because of its occurrence on outer reefs, where the continual surf makes the prolonged labor of digging out so large an animal impossible. A smaller species that occurs with *Parapholas* in stiff blue clay is apparently *Barnea pacifica,* although that species is reputedly a mud-borer.

After these forms have completed their burrows their muscle feet degenerate and the shell closes over them, in contrast with *Zirfaea* (§305), which remains active throughout its life. The boring methods of the pholads are considered in §242.

The abandoned burrows of dead boring clams furnish safe and comfortable homes for various smaller animals. Among them are likely to be the crab *Pachycheles* (§318) and others—the flatworm *Stylochoplana* (§243) and the button shell *Trimusculus reticulata* (formerly *Gadinia*). The last looks like a small limpet, but has "lungs," is hermaphroditic, and is related to the common garden snail. It is often found under overhanging ledges.

5. *Low Intertidal: Pools, Roots, and Sand*

Two OTHER habitat areas in the low intertidal of protected rocky shores —tide pools and root or holdfast—have been treated separately here, since they provide similar protection no matter where they are found and may occur with any type of substratum. In addition, we have treated the protected sandy beaches here: their population is usually sparse and limited to a few species; and they are often mere pockets in the rock.

Low Intertidal Pool Habitat

§148. In the pools at the lowest tide level there is a wealth of life. Most of the individuals occur in other situations as well and are treated elsewhere; but some of the larger animals are entirely free-swimming and are essentially pool inhabitants. Only by the rarest accidents are they stranded high and dry. The famous spiny lobster, *Panulirus interruptus* (Fig. 136), which so often meets the fate of the innocent oysters that accompanied the Walrus and the Carpenter, belongs in this group. It is chiefly a subtidal form, but it so often allows itself to be trapped in pools that it may at least be said to have shore-loving tendencies.

Our spiny lobster lacks the pinching claw of its Atlantic relative, but it is nevertheless perfectly capable of making its own way in a world that is hostile, cruel, and uncooperative. Unless captured with a spear or a trap, it has a good chance to evade the path that leads to mayonnaise and seasoning. If taken with bare hands, it snaps its broad tail back powerfully, possibly driving home some of its spines.

To prevent extermination of the species, the females under 10½ inches in length are protected by law until they are mature enough to spawn. Former restrictions on the maximum size for females, based on the idea that larger females produce more eggs, have been rescinded. The "leaf-body" larvae of *Panulirus* are delicate, feathery creatures that seem to consist mainly of two large eyes, the rest of the body being highly transparent. After the larval and intermediate stages the lobster continues to grow, like the crabs, by a series of molts. The molted shell is left in such perfect condition that it may easily be mistaken for a live animal. Spiny lobsters are omnivorous feeders—almost anything will do, plant or animal, fresh or decayed—and they are themselves preyed upon by octopuses and large fish, particularly the jewfish. They move by walking —forward, sideways, or backward—or by propelling themselves stern first through the water by rapid flips of their powerful tails.

Panulirus is an entirely southern form, not authentically known north of Point Conception but occurring far down into Mexico. A similar lobster occurs off the Florida coast, where it is known as the Florida crayfish. Southern California divers call them "bugs."

§149. Another enlivener of southern tide pools is the vividly gold garibaldi, a fish known to science as *Hypsypops rubicunda*. Vertebrates have scant place in this account, since an adequate treatment of fish alone would require a separate book, but the garibaldi is too common and too obvious to omit entirely. Because of *H. rubicunda*'s abundance at La Jolla, one locality is known as Goldfish Point. The young garibaldi has vivid deep blue splotches that are lost as the fish grows older. Another fish that merits passing mention is variously known as the tide-pool sculpin, the rockpool johnny, and *Oligocottus maculosus*. It is a small fish with a sharply tapering body, a large ugly head, and large pectoral fins. It is red-brown and prettily marked.

§150. Several species of shrimps may sometimes be common in the lower pools. They are members of the common genus *Spirontocaris*, or "broken back" shrimps, which characteristically bend their tails suddenly under and forward and thus swim backward. The *Spirontocaris* found in the higher tide pools are small and transparent; those centering here (Fig. 137) are opaque and relatively massive. S. *prionota* ranges from Bering Sea to Monterey. S. *palpator* ranges from San Francisco to Magdalena Bay, and S. *cristata* (§278) from Sitka to San Diego, but almost invariably subtidally or in eelgrass in quiet water. Less commonly, any of half a dozen others may be found, including S. *taylori* in the south. A large and beautiful white shrimp, S. *brevirostris*, has been reported

136. *Panulirus interruptus* (§148), the California spiny lobster; reduced. These animals are uncommon in the north.

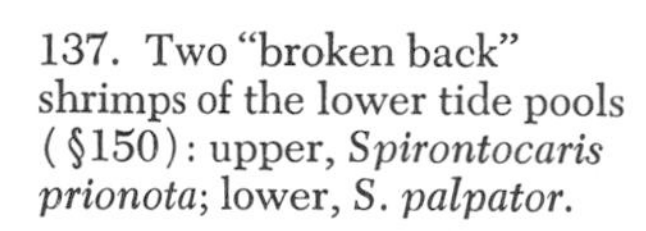

137. Two "broken back" shrimps of the lower tide pools (§150): upper, *Spirontocaris prionota*; lower, *S. palpator*.

from Alaska, and ranges at least as far south as the area between Carmel and Big Sur; we have found it there in some quantity, under rocks in pools where the water is very pure. The typical littoral specimens, however, are smaller than the large and robust "bastard" shrimps of the Alaska shrimp dredgers.

These shrimps show a great variety of brilliant coloring, which makes them conspicuous in a white tray but very difficult to see against the colorful background of a tide pool. There is no quick way of distinguishing the various species, reference to exact anatomical descriptions being the only method.

To the northward very much larger shrimps of the edible species of *Crago* and *Pandalus* (§283) may occasionally be trapped in the pools, but these must be regarded as rare visitors from another habitat.

§151. Some of the rhythms in nature are as completely baffling today as they have been since they were first observed. For instance, the palolo worm of the tropical seas near Samoa, living the year round in coral burrows at the sea bottom, is an entirely reliable calendar. To quote from an excellent English work, *The Seas,* by Russell and Yonge:

"But true to the very day, each year the worms come to the surface of the sea in vast swarms for their wedding dance. This occurs at dawn just for two days in each of the months October and November, the day before, and the day on which the moon is in its last quarter; the worms are most numerous on the second day, when the surface of the ocean appears covered with them. Actually it is not the whole worm that joins in the spawning swarm. The hinder portion of the worm becomes specially modified to carry the sexual products. On the morning of the great day each worm creeps backwards out of its burrow, and when the modified half is fully protruded it breaks off and wriggles to the surface, while the head end of the worm shrinks back into its hole. The worms are several inches in length, the males being light brown and ochre in colour and the females greyish indigo and green. At the time of spawning the sea becomes discoloured all around by the countless floating eggs.

"The natives are always ready for the spawning swarms, as they relish the worms as food. . . . The worms are eaten either cooked and wrapt up in bread-fruit leaves, or quite undressed. When cooked they are said to resemble spinach, and taste and smell not unlike fresh fish's roe."

All of which is far from home but not entirely disconnected, for we have at least two similar annelid worms on the Pacific coast. One is *Odontosyllis phosphorea,* whose swarming periods have been plotted at Departure Bay. Only a very persistent or a very lucky collector will see these worms, for except in their brief periods of sexual maturity, when they appear at night in countless hordes, not a single specimen is to be found.

In the tide pools of the California coast, another syllid, only a fraction of an inch in length, is more common than *O. phosphorea.* At dusk they may be seen swimming actively at the surface of pools, appearing as vivid pinpoints of phosphorescence.

The inevitable question with regard to these animals is: How do they know when the moon has reached a certain stage in a certain one of the 13 lunar months? The obvious answer would be that they do not know, but govern their swarming time according to tides, which, in turn, are governed by the moon. Unfortunately, this answer will not hold. The tides are governed not by the moon alone but by the sun and more than half a hundred additional factors. To suppose that the worms can take all of these factors into consideration passes the bounds of credulity. Furthermore, worms placed experimentally in floating tanks have spawned naturally at the usual times, although they had no means of sensing what the state of the tide might be. We are driven, then, back to the moon as the undoubted stimulus for their spawning, and confronted with a mystery beside which any of Sherlock Holmes's problems seem pale and insipid.

Nor are these worms the only sea animals that follow amazing cycles. The grunion (§192) is, if possible, an even more startling example.

§152. The occasional sandy pools—found rarely between rock outcroppings a few feet apart, as at Carmel and more frequently along the outer coast of Washington, as at Queets and Mora, and all along the Oregon coast—may be characterized by a colonial spionid worm, *Boccardia proboscidea. Boccardia* extends from Washington to southern California or beyond, penetrating clay or shale rocks with vertical, U-shaped burrows and protruding its two minute tentacles from one end of the burrow.

In these pools will also be found a few young edible crabs (*Cancer magister*), which stray up into the intertidal from deeper sandy bottoms. In the northern latitudes of Oregon and in Washington, where it is known as the Dungeness crab, *C. magister* often comes near shore or into bays in early summer to molt, and the resulting windrows of ecdysed "skeletons" inspire public worry that the crabs are dying off because of some catastrophe. The edible crab of the San Francisco region has become scarcer in recent years, and it appears that reproduction in the central California area has not been too successful. This seems to coincide with the increased use of pesticides, which are being drained into the San Francisco Bay system and consequently into the nearby ocean. The case is not yet "proved"; but it must be remembered that some agricultural pesticides are designed to kill arthropods like insects and mites, and are probably equally effective on crustaceans. Moreover, they are effective in extremely low concentrations, perhaps as low as parts per billion.

The exoskeleton of a crab presents a formidable barrier to love-making; and although it seems to be the general rule that mating of crabs requiring internal fertilization can take place only when the female is still soft from molting, the process has not often been observed. Dale Snow and John Neilsen of the Oregon Fish Commission were able to watch the mating behavior of the Dungeness crab in an aquarium (Fig. 138). They followed the process devotedly for 192 hours, and have provided us with a minute by minute description of the climactic moments.

A few days before the female is ready to molt (or does the male perhaps stimulate molting in the process?) the male, who does not molt at this time, clasps the female in a belly-to-belly premating embrace, and holds her this way for several days. If the female becomes too restless, the male holds her more firmly and strokes her carapace with his chelipeds "in an up and down motion that seemed to pacify her." When the female is ready to molt, she nibbles at her partner's eyestalks and is permitted to turn over. The female begins to molt while still confined in the basket formed by the male's appendages, and it appears that he may assist her in getting out of her old clothes and into something more comfortable. During actual shedding of the old exoskeleton the female is not held tightly; and when the ecdysis is complete, the male pushes the castoff "shell" away. Copulation does not take place immediately; possibly it is necessary for the new carapace to harden somewhat. Finally, at the right moment, the female allows herself to be turned over to the proper position, and extends her abdominal flap to receive the male, who inserts his gonopods into her spermathecae. As observed in the aquarium, this is not quite the end of the process; there is a postmating embrace that may last for two days. Perhaps this is associated with the close quarters of aquarium life. As a result of his affair, the male accumulates scars or marks on his chelipeds from the incessant stroking of the female before she molts; excessive wear has been attributed to polygamy, but it may be the result of a single, protracted mating.

138. Mating behavior in *Cancer magister* (§152), the Dungeness crab: top, premating embrace, male on top; center, the female molting her old carapace; bottom, position of the abdominal flaps in copulation.

Snow and Neilsen are not sure how the male knows when the female is about to molt; but since molting may be controlled by hormones, it does not seem improbable that some hormone may be released in the stages preliminary to molting, thus alerting the male. As Mae West is alleged to have said, "When you get over your cold, come up and see me sometime."

The average life span of this crab is about 8 years, the maximum, 10; crabs are not sexually mature until probably the fourth or fifth year.

Root and Holdfast Habitat

§153. A collector with sufficient strength and enthusiasm will find much of interest in pulling up, at extreme low tide, the holdfasts of various algae, or large clumps of surfgrass, roots and all. In a typical large holdfast or root he will find hundreds of individuals of dozens of different species. Nemerteans and occasional nestling clams live in the inner parts, and bryozoans, hydroids, sponges, tunicates, small anemones, holothurians (§141), and chitons cover the outside. Crawling over and through this living mass are hordes of snails, hermits, crabs, brittle stars, seastars, amphipods, and isopods.

Here is a perfect little self-sufficient universe, all within a volume of less than a cubic foot. A census of the holdfast inhabitants from various regions has been suggested as an index of richness for comparing different areas. Most of the forms occurring thus in the Pacific have been considered in connection with other habitats. The few that do not occur elsewhere are forms so small as to fall outside the scope of this work; but, if only for the sake of renewing a large number of old acquaintances, the collector will find holdfasts and roots worth investigating. Holdfasts cast up on the beach, however, are usually rather barren; because they have been torn off the rocks by wave action and tumbled about in the surf, most of the inhabitants have left for other shelter or have been washed out. However, holdfasts cast up from deeper water may occasionally contain such comparative rarities as the brachiopod *Terebratalia transversa* (Fig. 195).

The holdfasts shelter many small worms, particularly in the interstices. A *Lumbrineris*, possibly *L. erecta* (= *heteropoda*), green, and stouter than the attenuated *L. zonata* of §130, occurs in *Phyllospadix* roots from Monterey well into Baja California. A similar *Lumbrineris*-like form, very slim and breakable, green, and highly iridescent, is the cosmopolitan *Arabella iricolor*. *Anaitides medipapillata*, one of the "paddle-worms" (phyllodocids), so called because of its broad, leaf-shaped appendages, will be turned up in this assemblage—not frequently, but with admiration and enthusiasm on the part of the fortunate observer, who will admire its rich dark greens and brown or cream trimmings, with iridescent glints. Specimens of this sort may be several inches long. A similar form is illustrated on page 314, Volume 2 (1910), of the *Cambridge Natural History.* Also in *Phyllospadix,* sometimes in quantity sufficient to bind their roots together, will be found tubes of the small *Platynereis agassizi,* similar in appearance to *Nereis* but lacking the broadness of the posterior appendages. Specimens experimentally deprived of their white tubes will construct others very quickly in aquaria, utilizing and cementing together with mucus any bits of plant tissue available.

Sea spiders (pycnogonids) have also been taken in the holdfast. One is characteristic of *Phyllospadix* roots, but has been taken as well on hydroids, or even walking about on the pitted rocks; this is *Ammothella tuberculata,* which extends from northern California to Laguna Beach. Figure 7, Plate 12, of Cole (1904), illustrates this form, which has a spread of under ½ inch.

The amphipods occur here in profusion, and among the more than 70 different species (many of them undescribed) that we alone have turned up in Monterey Bay the following might be considered characteristic of kelp holdfasts and similar habitats: *Eurystheus tenuicornis,* from Puget Sound to San Diego, especially common in the south; *Aoroides columbiae,* known also from Puget Sound, and from many habitats, including coralline algae, clusters of *Leucosolenia,* and even *Pelvetia,* higher up; *Hyale,* probably of several species, possibly the commonest low-tide hoppers of the Monterey Bay region and north at least to Sonoma County.

Sandy Beaches

The sandy beaches of the protected outer coast are the most barren of the various intertidal regions discussed. They are distinct from the typical open Pacific beaches, which are exposed and subjected to strong surf but nevertheless support a highly specialized but limited fauna of a few abundant species, and from the sheltered sandy beaches of bays and estuaries, which support a rich fauna with many species. The protected sandy beaches are usually at most a few hundred yards long, often between rocky outcrops. Some of the animals of the surf-swept beaches do not tolerate these sheltered conditions, though these beaches are not quiet enough for the animals of the sheltered bay flats.

In more northern regions along the coasts of Oregon and Washington there are occasional small beaches of gravel and cobbles (there are also some cobble beaches in Baja California). These are the most barren of all shores because they can be sustained only in regions of heavy wave action (for example, a relatively "sheltered" situation where wave refraction may be heavy against the apparent lee of a headland), and there is no refuge for life among the wide spaces between gravel and cobbles. Sea lions and elephant seals may haul out on such beaches.

On some sandy protected beaches of this type in other parts of the world there is a varied microfauna living between the sand grains. This so-called interstitial fauna includes such oddities as a bryozoan reduced to a single polypide, marine tardigrades, and odd, copepod-like animals called mystacocarids. This fauna has not been found on our beaches, although there are representatives of the interstitial fauna in the sheltered sands of bay shores.

§154. Often the difference between a protected beach and an exposed beach is a matter of degree. This degree is in some ways subtly marked by the beach hoppers characteristic of most sandy beaches on our coast. The "short-horned" large beach hopper *Orchestoidea corniculata* (Fig. 140) frequents the smaller, short, steep beaches, and may occur at the sheltered ends of long beaches, where the "long-horned" beach hopper *O. californiana* (Fig. 139) is the most abundant hopper; but according to Darl Bowers, the reverse situation does not occur. *O. corniculata,* then, is a creature of steeper, coarser beaches, where sand sorting is poorer, but where the sand is perhaps damper and richer in oxygen. It does not retreat far up the beach, as does *O. californiana,* perhaps because its preferred beaches do not have much backshore. On some beaches the two species overlap, but their requirements and occurrences are complemental rather than competitive. Thus they are good examples of Gause's Principle—two species of the same requirements should not (or cannot) occupy the same ecological situation, or "niche."

During daytime the hoppers are usually in shallow burrows, and they are best observed at night or on cloudy days, when they come out to feed on seaweed or picnic leavings. Sometimes large hoppers may be found under piles of seaweed. During the night the hoppers leave their burrows open, and evidently are not too particular about finding a "home" burrow. Often they can be observed fighting for a burrow just before dawn, perhaps because it is easier to obtain a hole by fighting a weaker hopper than to dig a new one. The hoppers will also fight for bits of food and, presumably, for females.

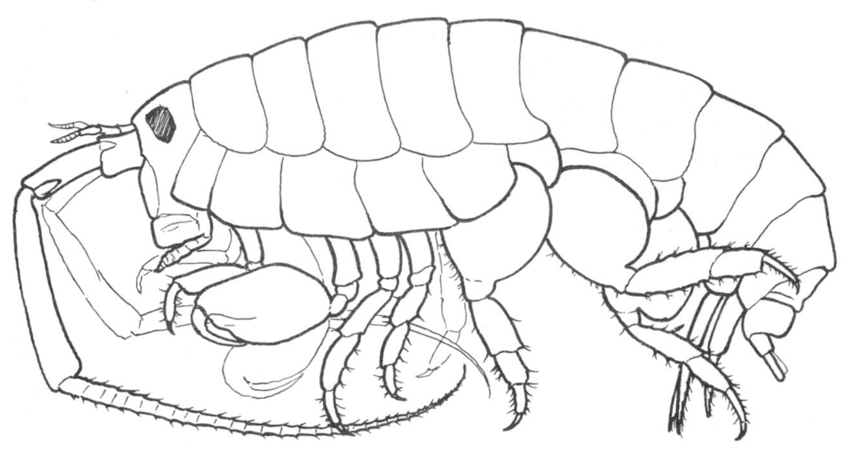

139. The large beach hopper *Orchestoidea californiana* (§154).

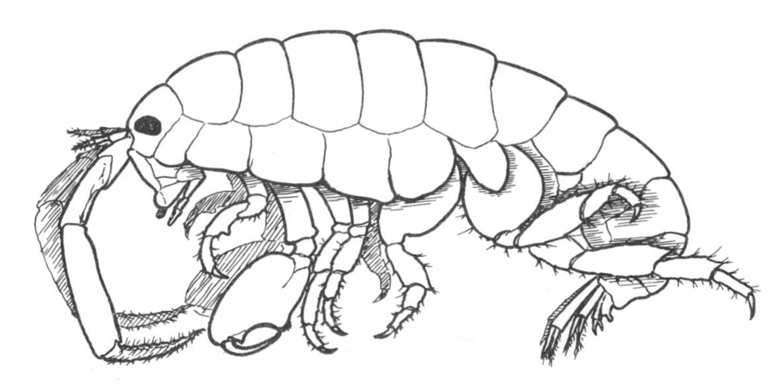

140. The beach hopper *Orchestoidea corniculata* (§154).

Another beach hopper, *Orchestoidea benedicti*, occurs on both open and protected beaches and is often seen during the daytime, apparently oblivious of its vulnerability to birds. It is smaller than the two big hoppers, and is perhaps protected from predation by its size.

There are five common species of beach hoppers on our various beaches (and a sixth in southern California). Anyone who consults the standard systematic literature will find the descriptions too intricate for an untutored mind, but Bowers has noted that the species have characteristic patterns of pigmented spots in life and may accordingly be distinguished on that basis (Fig. 141).

When beach hoppers come out at night, they move down the beach

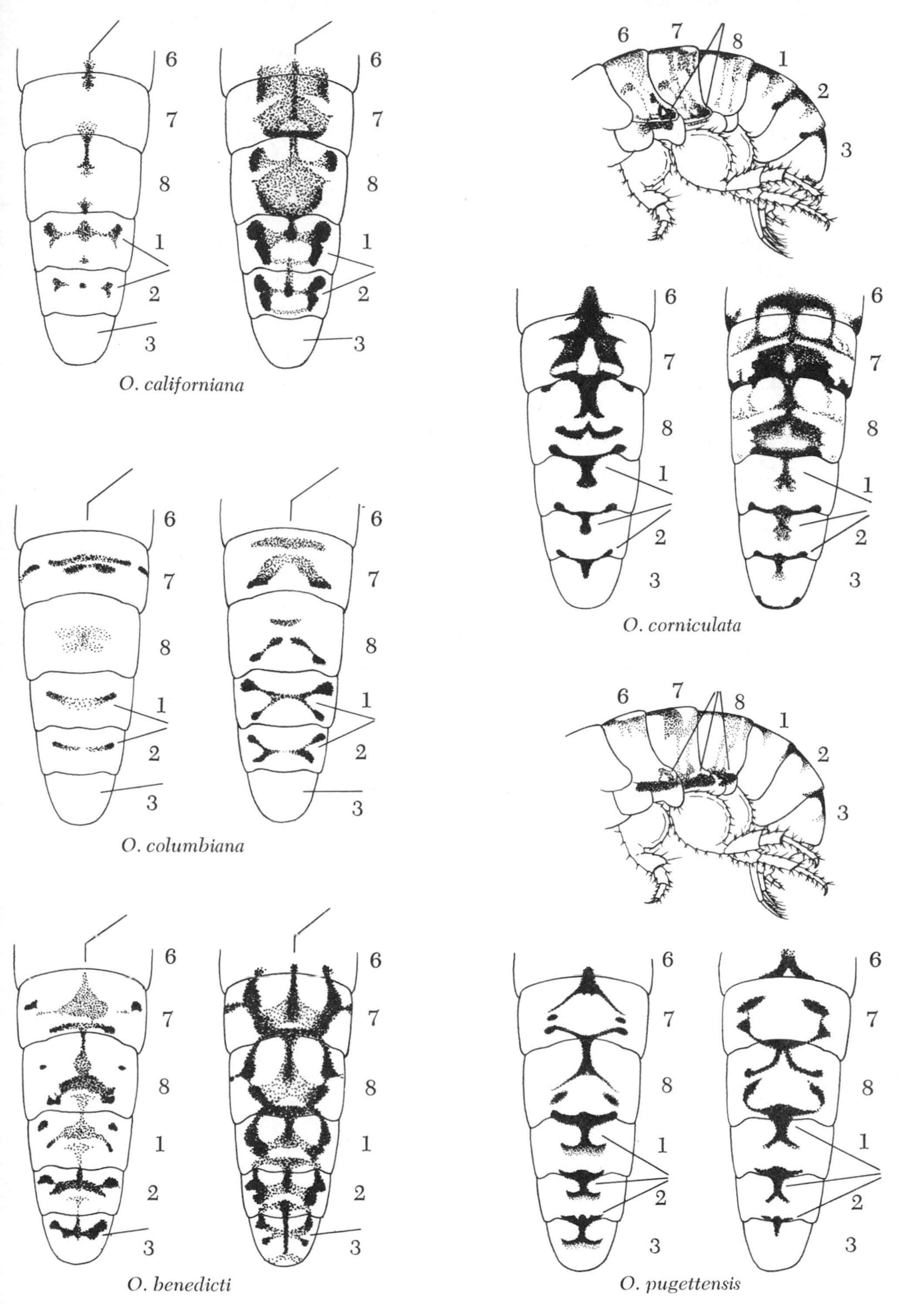

141. Identifying pigment patterns of the five species of *Orchestoidea* common from central California to Canada (§154). The patterns are present only in living animals.

to feed near the water's edge, and toward dawn they return to their own —or the other fellow's—burrows on the upper beach. How do they know which way to go? J. T. Enright, of Scripps Institution of Oceanography, has examined this question and finds that *Orchestoidea corniculata* can orient to the moon at an angle of about 120°, but that it cannot tell the difference between the real location of the moon and the moon's reflection in a mirror or compensate for changes in the moon's position. It appears that the animals may adjust themselves each night according to the positions of sunset and moonrise. In any event, it does not seem that *Orchestoidea* can automatically compensate by some internal lunar rhythm; if kept in the dark for several hours before the experiment and then exposed, it simply goes away from the moon's current position.

§155. The inhabitant *par excellence* of these protected beaches is the bloodworm, *Euzonus* (formerly *Thoracophelia*) *mucronata* (Fig. 142). This bright red opheliid worm, about 1½ to 2 inches long, lives in a narrow band at about mid-tide level (from near the surface at high tide to 10 or 12 inches down at low tide). Its range is from Vancouver Island to Punta Banda. The position of the most densely populated band of worms will change somewhat with the season and the tide cycle, but usually the worms are found where the sand is damp but not mushy. Sometimes, in particularly sheltered situations, the presence of the

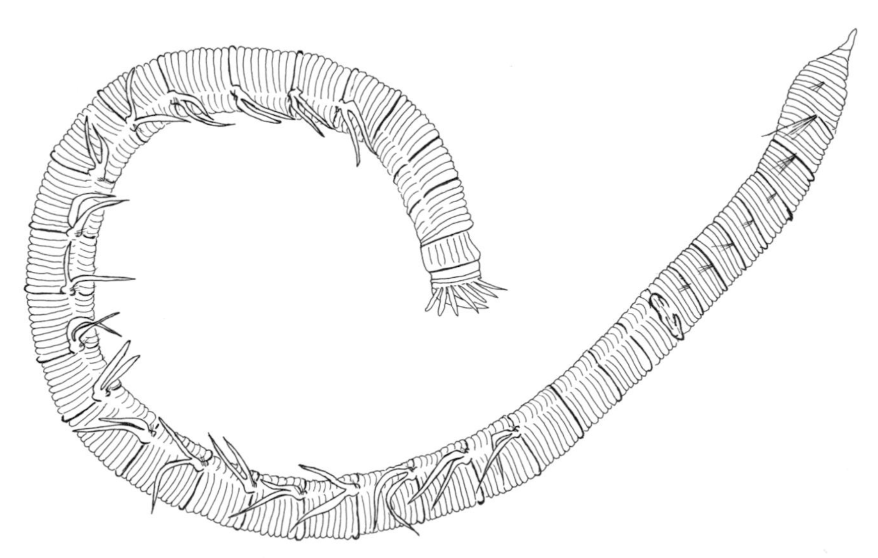

142. *Euzonus mucronata* (§155), the bloodworm.

worms will be revealed by small, close-set holes at the surface, as at the La Jolla beach studied by McConnaughey and Fox. In these places they may attain a density of 2,500 or 3,000 per square foot. The bright red color of the worm is due to the hemoglobin in its blood plasma. *Euzonus* evidently feeds on the nutrient material found on the surface of sand grains, since it passes sand through its alimentary tract. The worms do not swim or crawl, but burrow. When turned out of the sand, they curl up in a circle or coil like a watchspring.

The abundance of bloodworms on some beaches indicates the rich supply of nutrient material in sand that seems barren to us. McConnaughey and Fox estimated that a worm bed a mile long might contain 158 million worms, or 7 tons of them, which would cycle 14,000 tons of sand a year for a possible yield of 146 tons of organic matter.

A greenish, nereid-like polychaete, up to 5 or 6 inches long, is often found in the bloodworm zone. This is *Nephtys californiensis,* which is suspected of preying upon the bloodworm.

§156. So deep down as to be chiefly a subtidal form, especially where the sand is firm with an admixture of detritus, but sometimes to be seen from the shore, is a lugworm that belongs more properly in the mud flats of protected bays (§310). It is a burrowing animal that breathes by keeping two entrances open to the surface. About one of them characteristic castings of sand will be noticed. On these beaches it occurs, so far as we have observed, only where rocky outcrops divert the surf in such a way as to induce the deposition of some silt—the situation to be expected, since the worm feeds by eating the substratum and extracting the contained organic matter.

Weed-covered rocks surrounding a tide pool on the protected outer coast at Pacific Grove.

A semi-protected sandy shore at Dillon Beach.

A representative tide-pool scene. Specimens of the sea urchin *Strongylocentrotus purpuratus* and the green anemone *Anthopleura xanthogrammica* are visible.

The rock crab *Pachygrapsus crassipes* (§14), a common inhabitant of the high intertidal.

The purple shore crab *Hemigrapsus nudus* (§27), a dominant crab of the middle intertidal.

Collecting on mud flats in a sheltered bay.

Wave-swept rocks on the exposed outer coast. Mussels and goose barnacles cover the rocks, and common seastars (*Pisaster ochraceus*) of varying colors can be seen clinging near the mussel beds.

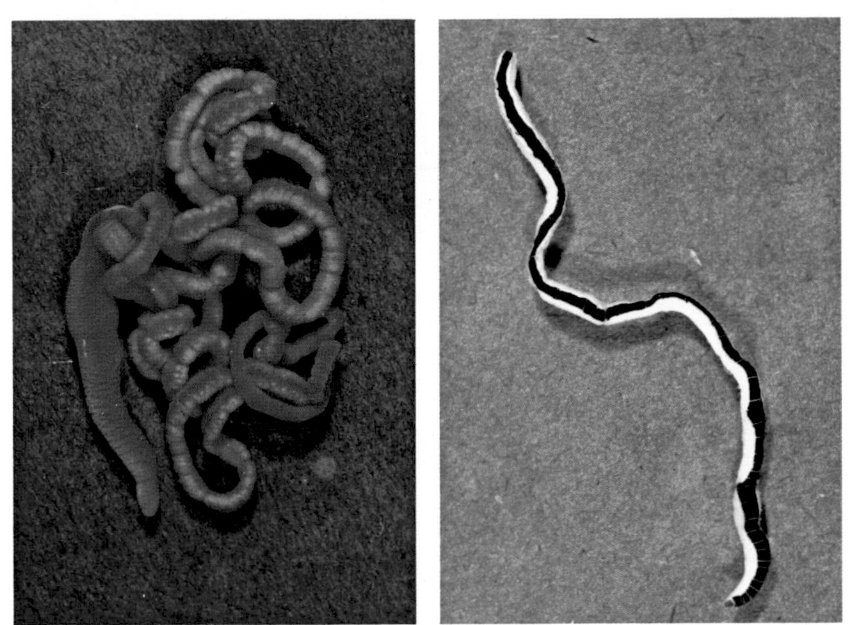

Two common nemerteans: left, *Tubulanus polymorphus* (§139); right, *Micrura verrilli* (§139).

Two sedentary polychaetes: left, *Thelepus crispus* (§53), with its tube; right, *Sabella crassicornis,* without tube.

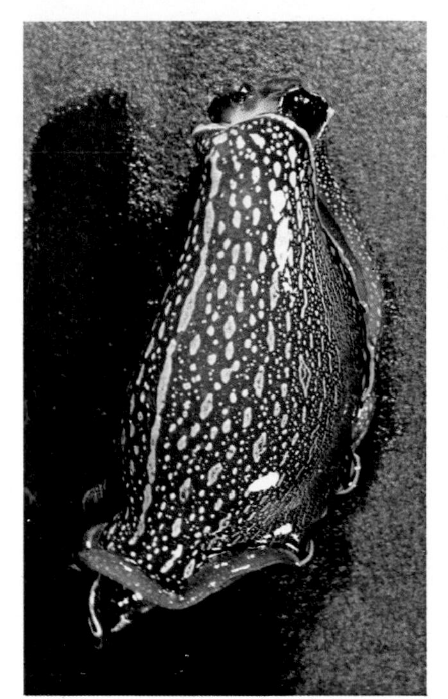

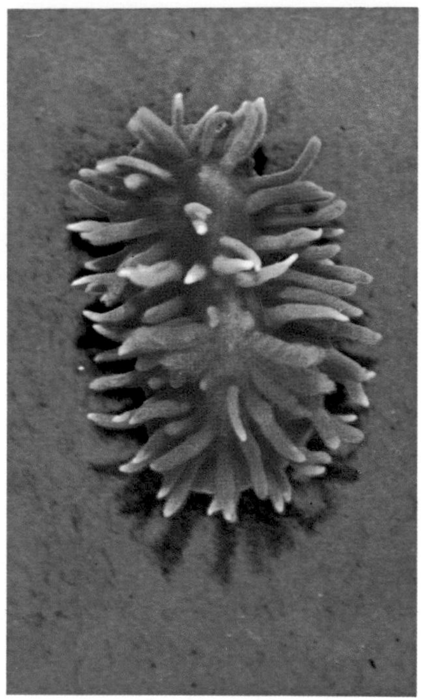

Left: the predatory tectibranch *Navanax inermis* (§294). Right: the cerise nudibranch *Hopkinsia rosacea* (§109).

The nudibranch *Hermissenda crassicornis* (§295) on a sand flat at Tomales Bay; the branchial cirri of the burrowing annelid *Cirriformia* may be seen protruding from the sand.

Left: the "ice-cream cone worm," *Pectinaria californica* (§281), in its tube. Right: the common hermit crab *Pagurus samuelis* (§12) in the shell of *Calliostoma annulatum.*

Left: the bat star, *Patiria miniata* (§26). Right: the proliferating anemone *Epiactis prolifera* (§37).

Two rock snails with their characteristic egg capsules: left, *Ceratostoma foliatum* (§117); right, *Thais lamellosa* (§200).

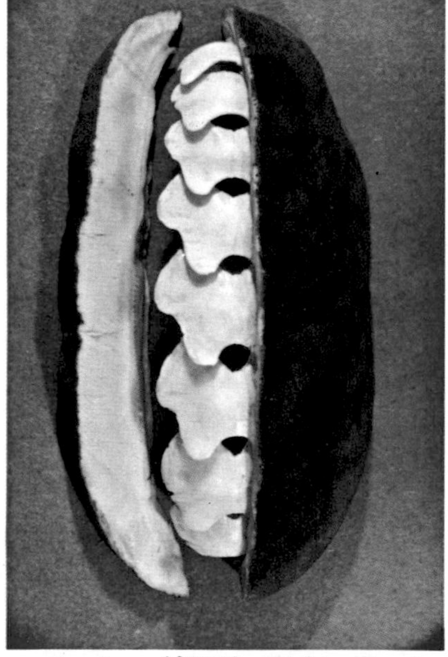

Left: shell of the rock cockle *Protothaca staminea* (§206). Right: *Cryptochiton stelleri* (§76), the gum boot chiton, cut away to show the valves, or "butterfly shells," commonly found on the beaches north of Point Conception.

PART II. OPEN COAST

It may seem amazing that in this surf-swept environment the same animals should occur almost unchanged all the way from Sitka to Point Conception (a few miles west and north of Santa Barbara), for the long intervening stretch of coast works an irregular traverse across 22½ degrees of latitude—across most of the North Temperate Zone. But this range seems less remarkable when we consider that conditions are actually very uniform. Along the California coast there is an upwelling of relatively cold water such that the resulting water temperature is lower than would be expected, considering the latitude; and in the north the Kuroshio, the warm Japan Current, sweeps inshore in direct proportion to the increase in latitude, with the result that the waters along the outside coast of British Columbia and southeastern Alaska are very nearly as warm as those along the California coast as far south as Point Conception. Also, the summer–winter variation in water temperature is only a few degrees. The water is almost constantly agitated, and throughout the Sitka to Point Conception range there are great depths close offshore and correspondingly uniform oceanic conditions. Finally, even the weather is relatively uniform along this immense coastline, fogs and cool weather during summer being the rule, with no great peaks of heat or cold.

Close inshore, however, at the shifting interface of the tidal level, seashore plants and animals can be exposed to a surprisingly wide range of temperatures. The ocean may be cold and relatively constant in temperature a few yards offshore, but the variation of water temperature at the sea's edge, especially in tide pools, may exceed within a day the annual range of offshore extremes. When this fluctuation is combined with the air temperatures (and sometimes there are sunny summer days, even at Pacific Grove!), the range may equal that experienced by terrestrial beings. Too often we have jumped to conclusions about the temperature relations of seashore animals from the data for seawater at the end of a dock, or that gathered at weather stations miles away. But the seashore is a rough, turbulent part of the world, and it is difficult to obtain good records of temperature and humidity in this active zone. Nevertheless the similarity of the common fauna suggests that these ranges must be about the same order of magnitude all along the coast, however they may differ seasonally and from place to place.

6. Open-Coast Rocky Shores

ALONG the surf-swept open coast, the rocky cliffs of the inaptly named Pacific have developed associations of animals with phenomenal staying power and resistance to wave shock. It is not the fact that these animals will not live elsewhere that makes them characteristic, but the fact that no other animals can tolerate the rigorous conditions of heavy surf. Obviously, the prime requisites here are the ability to hold on in the face of breaking waves and the possession of structures fit to resist the sudden impact of tons of water. These necessities have produced the tough skins and heavy shells generally present, the strong tube feet of seastars, and the horny threads of natural plastic by which mussels are attached.

Perhaps a word of warning is in order here. Especially from central California northward, completely exposed rocky points like those considered in the first part of this chapter are very dangerous places. It has been stated that in some regions any person within 20 feet of the water, vertically, is in constant danger of losing his life; and every year newspapers, with monotonous regularity, report the deaths of people who have been swept from the rocks by unexpected waves of great size. This loss of life is usually as unnecessary as it is regrettable. To lie down, cling to the rocks like a seastar, and let a great wave pour over one takes nerve and a cool head; but more often than not it is the only sane course of action. To run, unless the distance is very short, can be fatal. Like most mundane hazards, the danger from surf largely disappears when its force is recognized and respected; hence no one need be deterred by it from examining the animals on exposed points, always bearing in mind that a brief wetting is better than a permanent one. When collecting at the base of cliffs, we always keep handy a rope made fast at the top of the cliff. It is best to avoid dangerous-looking places, and in any event it is advisable to watch from a safe vantage point before venturing into an unfamiliar place, to observe the wave action and study out the lay of the sea. Above all, do not trust the old notion about every seventh or ninth wave being higher, with the implication that the high one may be followed by half a dozen "safe" ones. Sometimes there may be several high waves in succession. And, finally, never go collecting or observing alone in any place where there is the slightest danger.

Interzonal Animals

The force of the waves on stretches of completely open coast is so great that animals up near the high-tide line are wetted by spray on all but the lowest tides. Consequently, there is not the relatively sharp zonation that we find elsewhere, and some of the most obvious animals are fairly well distributed throughout the intertidal zone. Accordingly, we shall depart somewhat from our usual method and shall treat these wide-ranging animals first, taking up the more definitely zoned animals later.

§157. The most conspicuous of the open-coast animals is *Pisaster ochraceus* (Fig. 143), sometimes popularly called the purple or ocher seastar but more often the common seastar—an animal distinctly different from the common seastar of the Atlantic. Since these animals are not "fish" in any sense, "seastar" is perhaps a better name for them than starfish, and has been used throughout this book. Specimens varying from 6 to 14 inches in diameter and having three color phases—brown, purple, and yellow—are commonly seen on exposed rocks from Sitka to Point Sal (near Point Conception). In the warmer waters below that point a subspecies, *P. ochraceus* ssp. *segnis,* extends at least to Ensenada.

Up to 1 per cent of the *Pisaster* seen between northern California and Monterey Bay will be the more delicate and symmetrical *P. giganteus* (Fig. 128), with beautifully contrasting colors (which are lacking in *P. ochraceus*) on and surrounding the spines. This seastar is improperly named, since the average specimen is smaller than the average *P. ochraceus.* The explanation is that the species was originally named from a particularly large specimen, which was assumed to be typical.

Pisaster neither has nor seems to need protective coloration. Anything that can damage this thoroughly tough animal, short of the "acts of God" referred to in insurance policies, deserves respectful mention. To

. The common seastar, *Pisaster ochraceus* (§157).

detach a specimen from the rocks one heaves more or less mightily on a small crowbar, necessarily sacrificing a good many of the animal's tube feet but doing it no permanent injury thereby, since it will soon grow others to replace the loss. The detached tube feet continue to cling to the rock for an indefinite period. An eastern investigator finds that the tube feet of the common Atlantic seastar will live for several days in clean sea-water and will respond to stimuli for two days or more. Histological examination shows deep sensory cells and a net of material that is probably nerves.

The action of the pedicellariae, minute, pincerlike appendages that keep the skin of seastars and urchins free of parasitic growths, can be demonstrated particularly easily with *Pisaster*: allow the upper surface of a husky specimen to rest against the skin of your forearm for a few seconds; then jerk it away. The almost microscopic pincers will have attached themselves to the epidermis, and you will feel a distinct series of sharp nips. A classic experiment to demonstrate the function of pedicellariae is to drop crushed chalk on the exposed surface of a living seastar. The chalk is immediately ground to a powder so fine that it can be washed away by the action of the waving, microscopic cilia in the skin. Considering that sessile animals must struggle for even sufficient space for a foothold, and that seastars live in a region where the water may be filled with the minute larvae of barnacles seeking an attachment site, the value of these cleansing pedicellariae is very apparent. It is even possible that they have some value in obtaining food; for, despite their small size, they have been observed to catch and hold very small crabs, which may eventually be transferred to within reach of the tube feet.

Just how a seastar's breathing is effected is not definitely known, but it is assumed that the tube feet may play a part. There are, in addition, delicate, fingerlike extensions of the body wall scattered over the animal's upper surface. These extensions are lined, both inside and out, with cilia, and they probably assist in respiration. Many small shore animals have not even this degree of specialization of breathing apparatus—sponges, for example, get what oxygen they need through their skins. In connection with the general subject of respiration, it is interesting to note the great number of marine animals, including some fishes, that are able to get oxygen from the air as well as from the water. This is necessarily the case with a large proportion of the animals herein considered. Some, in fact, like the periwinkles, some of the pill bugs, and a fish that lives on the Great Barrier Reef, will drown if they cannot get to the air. Among the shore crabs the gill structure shows considerable progress toward air-breathing—a process that reaches its climax in some of the tropical land crabs, which have developed a true lung and return to the sea only to breed.

Pisaster will eat almost anything it can get its stomach onto, or around; the stomach is everted and thrust into the shell of a clam or mussel, or around a chiton or snail. Digestion, therefore, takes place outside the body of the diner, but the stomach must be in contact with the meal. Seastars are often observed humped up over mussels, apparently pulling the shells apart in order to intrude their stomach. But this is not entirely necessary, since the stomach can squeeze through a very small space, and since the mussel may not realize its danger until it is too late; also, many active clams cannot remain tightly closed very long. A seastar may well be strong enough to pull a mussel or oyster apart, but it may not often need to do so. The food of *Pisaster* is governed somewhat by circumstance: where there are mostly barnacles, it eats barnacles; and where there are mostly mussels, it eats mussels.

Some years ago E. C. Haderlie observed that limpets would move away from a seastar on a rock as rapidly as possible; they were thrown into a sort of hysteria, as he put it, and for a limpet, the speed was virtually a gallop. Many snails (including abalone, it turns out) will also dissociate themselves from the presence of *Pisaster,* usually with as much alacrity as they can muster. It is not necessary for them to be touched by the seastar; its nearby presence is enough to set them on the move. One common shore limpet (*Acmaea scabra*) does not evade *Pisaster,* however, and it is a frequent item on the seastar's diet. Since the evasion response may be elicited by crushed essence of *Pisaster,* it is evident that the seastar must release some substance that warns the snail to begone. The nature of this substance is still undetermined.

Little is known about the life history or rate of growth of this Pacific form. Atlantic seastars become adult in one season, but there is reason to suspect that *Pisaster* grows more slowly. There is no season of the year, so far as is known, when the representative specimens are larger or smaller than at any other season, and we have found only one or two regions where there was a high percentage of small specimens. If growth were rapid, one would expect to find a high proportion of young specimens in the community, whereas if growth were slow, the expectation would be just what we find—only a few small specimens in a representative area. Some indication of growth rate was observed by D. B. Quayle in British Columbia, who recorded that the rays increased in length from 22 to 64 mm. in about a year's time; if this is related to a steady rate of growth, it would suggest that at this rate of about 1½ inches a year (3 inches in total "diameter") a seastar about 9 inches across might be about 3 years old. Nobody knows how old a seastar may be; perhaps after a while, like many other animals, they cease growing and live for years.

The sexual products are extruded from pores between the rays and on the upper surface, as with *Patiria* (§26), but probably not so frequently. Observations extending over 5 years, on several thousand *Pisaster* that were spread out prior to preservation, showed only five cases of extrusion—one in March, one in June, two in October, and one in December. Then, in the warm summer of 1931, we found sexual products extruding from more than a score of specimens in one lot of a few hundred.

As far as we can tell, the principal enemies of seastars are people: hordes of schoolchildren on field trips who must each bring back one, tourists and casual visitors from inland who hope the star will dry nicely in its natural colors, collectors for curio stores, and divers who think that all seastars are some sort of marine vermin to be removed for the betterment of whatever it is they are after. Though it will probably be impossible to exterminate such a robust and abundant animal as *Pisaster ochraceus,* there are indications that it will cease to frequent heavily utilized beaches. One can almost tell where people have been by the scarcity of *Pisaster.*

It is this general disrespect for the humble *Pisaster,* as abundant and inexhaustible as it may seem to be, that has, as much as anything else, inspired the "wanton waste" laws of Oregon and Washington, as expressed on the poster to be seen on many Oregon beaches:

It is unlawful to wantonly waste or destroy any intertidal animal.

Fish Commission of Oregon

§158. The common seastar is the most obvious member of the triumvirate characteristic of this long stretch of surf-swept rocky coast. The others are the California mussel and the Pacific goose, or leaf, barnacle. The mussel, *Mytilus californianus* (Fig. 144), ranging from Alaska to Mexico, forms great beds that extend, in favorable localities, from above the half-tide line to well below extreme low water. Here is another animal that is distinctly at home in crashing waves. Indeed, it occurs only where there is surf, whereas some *Pisaster* may be found elsewhere.

Each animal is anchored to the rock by byssal hairs, tough threads extruded by a gland in the foot. This process of producing a fluid or viscous plastic that hardens in air to form a tough fiber has since been rediscovered and perfected by man, who now makes his fishing lines and some other fibers in this manner. Contrary to the natural expectation, however, *Mytilus* is actually capable of limited locomotion, provided it is not too tightly wedged into position by other mussels. It can achieve a slight downward motion by relaxing the muscle that controls the byssal hairs, but it can go still further than this by bringing its foot into play. By extending its foot and gripping the rock with it, it can exert a pull sufficient to break the byssal hairs a few at a time, or even pull the byssus out by the roots. Once the hairs are all broken, it can move ahead the length of its foot, attach new hairs, and repeat the process.

The mussel's chief enemies are man and *Pisaster,* but we have caught the snail *Thais* in the act of drilling neat holes through the shells of small specimens preparatory to extracting the body bit by bit. It is possible that some fish feed on them also. The mussels themselves feed, like other bivalves, on minute organisms or suspended material in the water.

The California mussel is fine eating, but cannot be recommended during the summer months, when it may accumulate the minute organisms whose toxin causes paralytic poisoning. This extremely powerful toxin is produced by dinoflagellates of the genus *Gonyaulax* (Fig. 145). A great abundance of these organisms gives a red cast to the sea, and excessive numbers of them may cause the death of fishes and invertebrates. There are several kinds of dinoflagellates associated with catastrophic red tides in various parts of the world, but *Gonyaulax* is responsible for the fatalities associated with mussel poisoning on this coast. If, after a hearty meal of mussels, your lips begin to feel numb, as though the dentist had just injected novocaine to work on your front teeth, and your fingertips begin to tingle, you had better take measures to rid yourself of your meal as soon as possible, and, of course, summon a physician. It is said that cases are less severe when mussels have been only part of the meal, but this hearsay should not be accepted as truth any more than the old notion that when a silver spoon does not turn black in a batch of mushrooms, they are safe to eat.

Gonyaulax becomes dangerously abundant only during the summer months, along the open coast. It apparently does not invade bays, or at least bay mussels and clams do not concentrate it. Periodically, public health officials test the toxicity of mussels by injecting mice with extract from the mussels; the amount needed to kill a laboratory mouse is now a standard unit. Other open-coast bivalves besides mussels may become

144. California mussels, *Mytilus californianus* (§158), with the goose barnacle *Pollicipes polymerus* (§16

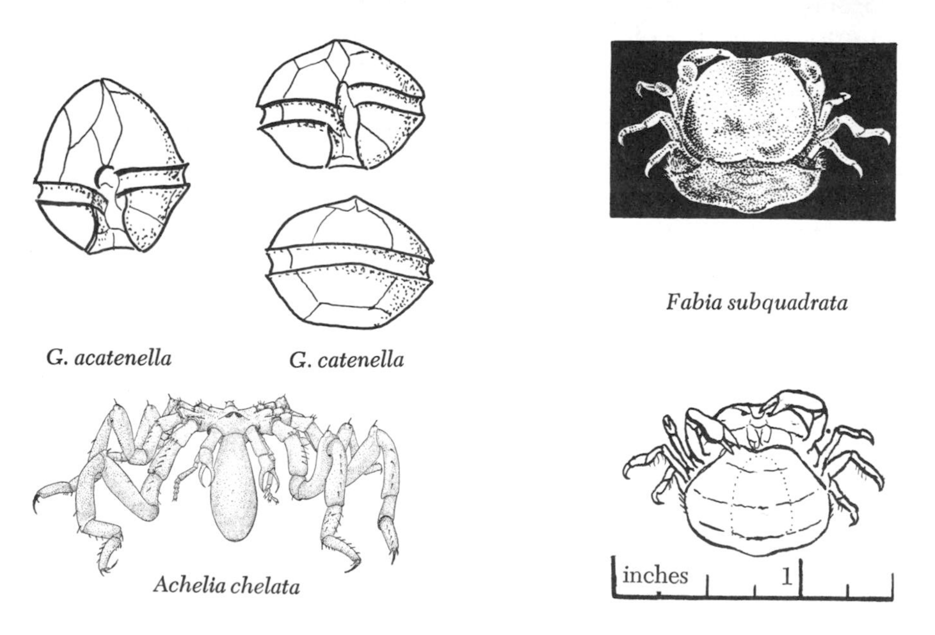

145. Associates of the California mussel. Left, above: dinoflagellates of the genus *Gonyaulax* (§158), the cause of mussel poisoning. Right: dorsal and ventral views of *Fabia subquadrata* (§159), the mussel crab. Left, below: the pycnogonid *Achelia chelata* (§159).

toxic, especially razor clams (§189). Other clams, such as the Washington and horse-neck clams (§§300, 301) in inlets near the open sea, may be suspect and should be avoided in summer. The sand crab *Emerita* (§186), which is a plankton feeder, also accumulates dangerous amounts of *Gonyaulax*. Oysters, *Protothaca* (rock cockles), *Mya*, and most of the Washington and horse-neck clams are perfectly safe, being restricted to quiet waters.

One of us (Calvin), working up Alaskan historical data at Sitka, found frequent references to mussel poisoning in translating the accounts of Father Veniaminoff and other early Russian explorers. Baranoff lost a hundred or more Aleuts all at once in Peril Strait; the stench arising from the stricken made even the others ill. One of Vancouver's men is said to have died thus, and others nearly lost their lives similarly.

The breeding season of *Mytilus californianus* has been investigated by Stohler (1930), who found July and December peaks in the proliferation of eggs and sperm. The sexes are separate.

§159. In perhaps 1 to 3 per cent of the California mussels at various places along the central California coast will be found a rather large pea crab, *Fabia subquadrata* (Fig. 145). In the waters of Puget Sound this crab infests by preference (or necessity?) the horse mussel, *Volsella modiolus* (§175). Jack Pearce, who studied the life cycle of this pea crab in Puget Sound, found evidence that *Fabia* is not a harmless or benign commensal, but is a true parasite that feeds upon its host, although its original intentions to dine upon the food accumulated by the host may have been innocent enough. The mature female crab is almost as large as the end of a thumb, and this is the one usually noticed in the mussel. Often the male is with her; but being much smaller, he may be overlooked. Many consider *Fabia* as tasty as the oyster crab that infests the eastern oyster. The eastern oyster crab is considered such a delicacy that an exception is made in the public health codes, which usually prohibit the marketing or serving of parasitized animals as food, to permit the sale of infested oysters.

In the mature stages of *Fabia* the female is so much larger than and so different from the male that they were originally described as separate species. Both sexes, in the first crab stages (just after the megalops), are rather similar in size and appearance, and both infest mussels. Some of these young crabs infest various other bivalves, but apparently do not return at a later stage. At the end of the first crab stage the males and females leave their hosts for a sort of mating swarm in the open water. The female then returns to a mussel and goes through five more stages in about 21–26 weeks before producing eggs. In Puget Sound, mating takes place in late May, and the eggs appear in November. The males usually return to a host also.

Pea crabs have not been found in either *Mytilus edulis* or *M. californianus* at San Juan Island, but on Vancouver Island 18 per cent of an intertidal population of *Volsella modiolus* was infested.

Another, and probably equally unwelcome, guest of the California mussel is a pycnogonid, *Achelia chelata* (Fig. 145). This pycnogonid is recorded from various places along the central California coast, but has only been found parasitizing mussels just south of the Golden Gate in San Francisco, and at Duxbury Reef in Marin County. Individual infestation may be as high as 20 or 25 pycnogonids in a large mussel. The damage to the host in these cases may be so extensive that the gills are almost completely destroyed, and even the gonads may be attacked. Another case of pycnogonid infestation of a bivalve has been reported from Japan; both host and guest are different species. This parasitism has not been observed a few miles north of Duxbury Reef at Tomales Point, although *Fabia* is frequently encountered in mussels there. So far, no case of parasitism by both pycnogonids and pea crabs in the same mussel has been found, and it may be that the two parasites are mutually exclusive; but this may not be the case in the Japanese situation.

§160. The Pacific goose barnacle, *Pollicipes polymerus* (Figs. 144, 146), the third member of an assemblage so common that we may call it the *Mytilus-Pollicipes-Pisaster* association,* is fairly well restricted to the upper two-thirds of the intertidal zone. These three animals, "horizon markers" in marine ecology if ever there were any, are almost certain to be associated wherever there is a stretch of rocky cliff exposed to the open Pacific. The goose barnacle extends the general range of the association, being recorded from Bering Strait to the middle of Baja California; specimens on the open coast of Alaska, however, are comparatively few and scattered.

Pollicipes is commonly chalky colored when dry. A vividly colored subspecies that is restricted to caves or rocks sheltered from direct sunlight is identified for us by the late Dr. Pilsbry as *P. polymerus* ssp. *echinata*. In the Santa Cruz region, clusters of this subspecies increase in brilliancy of markings as they range deeper into the darkness of the cave. Although the adult goose barnacles resemble but vaguely the squat and heavy-shelled acorn barnacles, the two are closely related, and both have substantially the same life history.

* Formerly, before the name-changers meddled around, the *Pisaster-Mytilus-Mitella* association.

46. Clusters of goose barnacles, *Pollicipes polymerus* (§160), ıowing the uniform alignment of individuals.

At Monterey, according to the studies of Galen Howard Hilgard, this barnacle is engaged in reproductive activity most of the year; embryos were brooded for 8 months in 1957, while the water temperature varied from 12.3° to 17°C. Both sexes in each hermaphroditic individual mature at the same time, but cross fertilization is necessary. A mature goose barnacle produces 3–4 broods a year, or perhaps 100 to 240 thousand larvae. Though there are a great many *Pollicipes* on our shores, it would seem that there would be no room for anything else if all this reproduction resulted in adult barnacles.

The goose barnacle often feeds on amphipods and other creatures up to the size of house flies, rather than on the finer fare of the acorn barnacles. The tendency of groups of this barnacle to line up with the direction of the wave or current movement (Fig. 146) may be related to this mode of feeding. Although it is apparently immobile at low tide, *Pollicipes* is capable of a slow, stately sort of motion, which may be elicited in sunlight by passing a hand over a cluster of barnacles. As the shadow moves across them, they will be observed (if they have not been tired out by some previous shadow) to bend or twist their stalks so that their orientation is shifted. The individuals of a cluster will all move in the same direction. The significance of this shadow reaction is not clear, at least at low tide; perhaps it has some value in capturing food when *Pollicipes* is covered with water.

The *California Fish and Game Commission Publication* for 1916 gives a recipe for preparing goose barnacles, the "neck" or fleshy stalk of which the Spanish and Italians consider a choice food. We have heard that these and other barnacles are canned in Chile, but also that they may be carcinogenic as a steady diet.

The name "goose barnacle" comes to us from the sixteenth-century writings of that amiable liar John Gerard, who ended his large volume on plants with ". . . this woonder of England, the Goosetree, Barnakle tree, or the tree bearing Geese." After declining to vouch for the authenticity of another man's report of a similar marvel, Gerard continued: "Moreover, it should seem that there is another sort heerof; the Historie of which is true, and of mine owne knowledge: for travelling upon the shores of our English coast between Dover and Rumney, I founde the trunke of an olde rotten tree, which (with some helpe that I procured by fishermens wives that were there attending on their husbandes returne from the sea) we drewe out of the water upon dry lande: on this rotten tree I founde growing many thousands of long crimson bladders, in shape like unto puddings newly filled before they be sodden, which were verie cleere and shining, at the neather end whereof did grow a shell fish, fashioned somewhat like a small Muskle, but much whiter, resembling a shell fish that groweth upon the rocks about Garnsey and Garsey, called a Lympit: many of these shels I brought with me to London, which after I had opened, I founde in them living things without forme or shape; in others which were neerer come to ripeness, I founde living things that were very naked, in shape like a Birde; in others, the Birds covered with soft downe, the shell halfe open, and the Birde readie to fall out, which no doubt were the foules called Barnakles. I dare not absolutely avouch every circumstance of the first part of this Historie concerning the tree that beareth those buds aforesaide, but will leave it to a further consideration: howbeit that which I have seen with mine eies, and handled with mine handes, I dare confidently avouch, and boldly put downe for veritie.”

Britannicæ Conchæ anatiferæ.
The breede of Barnakles.

147. The barnacle tree (§160); from Gerard's *Herball*, 1597.

Gerard was quite in accord with the modern tendency to popularize science. However, he perceived that pure scientists would desire certain definite information, so he added: “They spawne as it were in March and April; the Geese are formed in Maie and Iune, and come to fulnesse of feathers in the moneth after.” Also, he catered to the modern feeling that pictures cannot lie: his graphic representation of the birth of barnacle-geese is reproduced in Figure 147.

The remainder of the wave-swept rocky-shore animals lend themselves better than the foregoing to treatment according to tidal zones that correspond to the upper, middle, and lower intertidal regions of the protected outer coast.

Zone 1. Uppermost Horizon

§161. As in other regions, the first animal to be encountered will be the dingy little snail *Littorina*; but whereas it occurs a little above the high-tide line on protected outer shores, it will be found here 20 or 25 feet above the water. *Littorina* is famous for its independence of the ocean. Tropical species have taken to grass and herbage at the top of low oceanic cliffs, and even to trees bordering the water.

Next in the downward progression, but still many feet above the water, come very small specimens of the barnacle *Balanus glandula* (§4), whose sharp, encrusting shells are likely to exact a toll of flesh and blood from the careless visitor. In regions of excessively high surf, however, they may be absent, not because they lack the ability to hold on once they have colonized the region but because the delicate, free-swimming larvae may never be able to get a foothold if there is high surf during the season when they should attach. A tremendous and epic struggle for existence comprises the daily life of the animals of this region, where the only uncontested advantage is an abundance of oxygen. Barnacles and mussels begin to feed within 60 seconds after immersion, and their great haste is understandable when one considers that they are covered, in this highest zone, only when a wave dashes up over them, and that there is competition for every passing food particle. They must fight for a foothold, fight to keep it, and fight for their food.

Limpets occur here also, but not in the great numbers that characterize the highest zone of more protected shores. The commonest form is *Acmaea digitalis* (Figs. 13, 16), but others will be found, including *A. pelta* and the giant *Lottia,* which, on these open shores, ranges high up. *Lottia* was once common at Moss Beach.

§162. On vertical faces of rock, often well above tide level and wave action but in situations where the surface is damp from spray, especially in broad fissures and caves, the rock louse *Ligia* is often abundant. The animals are a general, dark, rock color, and are not easily seen until they move, especially in dark shadowed places. A visitor to such places becomes aware that the rock surface seems to be crawling as the *Ligia* scuttle away, usually upward, in fits and starts. Often the animals will evade capture by dropping from the rock surface when reached for, then running for safety among the jumble of boulders and cobbles at the base of the rock face. There are two species, the slender, spotted *Ligia occidentalis,* common from Sonoma County southward, and the much

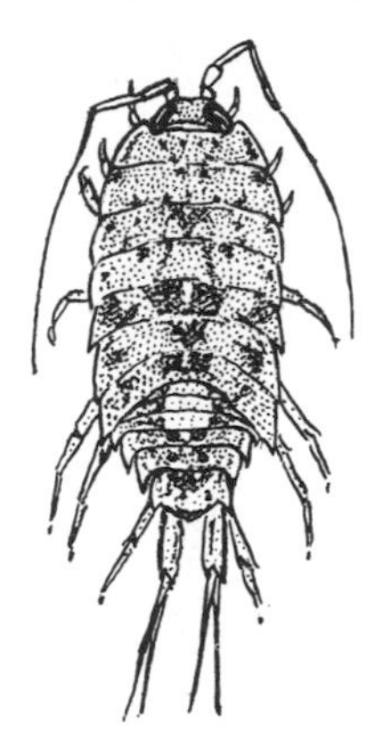

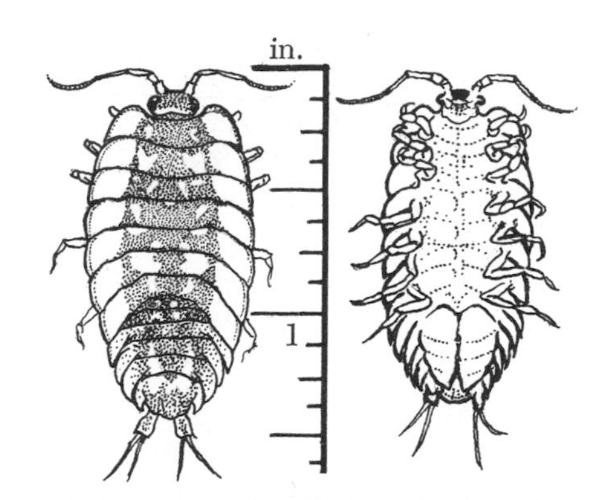

148. Two rock lice of the genus *Ligia* (§162): left, *L. occidentalis* (see also §2 and Fig. 9); right, *L. pallasi.* Both drawings are natural size, but *L. pallasi* is usually much larger north of Bodega Bay.

broader, more uniformly colored *Ligia pallasi* that occurs from Sonoma County northward to Alaska (Fig. 148). The larger northern species occurs sporadically as far south as Santa Cruz, but *L. occidentalis* is the common one at Monterey and in Marin County. *Ligia pallasi* has been observed carrying embryos in July on the Washington coast.

The beginnings of social order have been observed in a Japanese species. The animals move back and forth, in a more or less orderly procession, along apparently established routes from shelter among upper boulders at high tide down to the lower beach to feed. The procession is led by older members of the tribe. The casual impression gained from watching our own *Ligia* is that of an aimless, disorganized rabble, dispersing in various directions.

Zones 2 and 3. High and Middle Intertidal

§163. Here mussels and goose barnacles begin to occur in large beds, which afford shelter to a number of animals not themselves adapted to withstanding surf. Probably the most obvious of these is the predacious worm *Nereis vexillosa* (Fig. 149), known variously as the mussel worm, the clam worm, and the pile worm. Certainly it is the most important from the standpoint of sportsmen, who seek it along thousands of miles of coastline to use for bait. Possibly because of its proclivity for letting other animals provide its shelter from surf, *Nereis* is one of the few animals that may be found in nearly all types of regions, from the violently surf-swept shores to the shores of bays and completely protected inlets, and in such varying types of environment as rocky shores, gravel beaches, and wharf piling. Not only does it occur from Alaska to San Diego, but it seems to be abundant throughout its range, whereas other animals occupying so great a stretch of coast usually have optimum areas. Thus *Nereis* might be considered in almost any section of this handbook. It is treated in this particular place because the largest specimens are commonly found in the mussel beds of the open coast, where they vary from 2 to 12 inches in length. They are usually colored an iridescent green-brown.

The animals are very active, and they squirm violently when captured, protruding and withdrawing their chitinous jaws, which terminate a wicked-looking, protrusible pharynx and make carnivorous *Nereis* a formidable antagonist. These powerful jaws can deliver a businesslike bite to tender wrists and arms, but in collecting many hundreds of the worms barehanded we have rarely been bitten, always taking the precaution of not holding them too long.

149. The mussel worm *Nereis vexillosa* (§163).

In its breeding habits *Nereis* is one of the most spectacular of all shore animals. In common with other segmented worms, its sexual maturity is accompanied by such changes in appearance that early naturalists considered *Nereis* in this condition to be a distinct animal, *Heteronereis*—a name that has been retained to denote this phase of the animal's life cycle. The posterior segments (red in the female) containing the gonads swell up with eggs or sperm, and the appendages normally used for creeping become modified into paddles for swimming. When moon and tide are favorable, the male heteronereis, smaller than the female, leaves his protective shelter, seemingly flinging caution to the winds, and swims rapidly and violently through the water, shedding sperm as he goes. Probably a large number of the males become food for fishes during the process, and no doubt a similar fate awaits many of the females, who follow the males within a short time, releasing their eggs.

It has been determined experimentally in connection with related species that females liberate their eggs only in the presence of the male, or in water in which sperm has been introduced. If a specimen of each sex can be obtained before the ripe sexual products have been shed, the experiment can be performed easily. The female, in a glass dish of clean seawater, will writhe and contort herself for hours with frenzied and seemingly inexhaustible energy. Introduce the male—or some of the water from his dish, if he has shed his sperm in the meantime—and almost immediately the female will shed her eggs. When the process is over, the worms, in their natural environment, become collapsed, empty husks and die, or are devoured by fishes, birds, or other predacious animals.

We have many times found the large heteronereis of *N. vexillosa* in the mussel clusters on surf-swept points—abundantly in late January, less frequently in February or early March. Dr. Martin W. Johnson found spawning heteronereid adults at Friday Harbor in the summer of 1941 and identified gelatinous egg masses of the species. According to Dr. Johnson, *N. vexillosa* spawns only at night, usually an hour or two before midnight. Shedding specimens of a smaller wharf *Nereis* (*N. mediator*) are common in Monterey Harbor (§320).

§164. A second nereid, not as common as *N. vexillosa* but sufficiently startling to put the observer of its heteronereis stage in a pledge-signing state of mind, is the very large *Nereis brandti,* which differs from the similar *N. virens* in having many, instead of few, paragnaths on the proboscis. Specimens may be nearly 3 feet long, and are broad in proportion—a likely source for sea-serpent yarns. To the night collector, already a bit jumpy because of weird noises, phosphorescent animals, and the ominous swish of surf, the appearance of one of these heteronereids, threshing madly about at the surface of the water, must seem like the final attack of delirium tremens. We have found this great worm, not in the free-swimming stage, under the mussel beds at Moss Beach, San Mateo County. It has been reported from Puget Sound to San Pedro and has been collected from southeastern Alaska.

§165. Associated with *Nereis* in the mussel beds are the scale worm *Halosydna brevisetosa* (§49), the porcelain crab *Petrolisthes* (§18), and, at times, the pill bug *Cirolana harfordi* (§21). In some regions the nemertean worm *Emplectonema gracile* is so common that it forms tangled skeins among the mussels. This rubber-band-like animal is tinted pale yellow-green on its upper surface and white on its lower. It is slightly flattened, but is rounder than most ribbon worms, and is from 1 to 4 inches long when contracted. We have found great masses of these at Mora on the outer coast of Washington, and at Santa Cruz, California. Other nemerteans have the unpleasant habit of breaking up into bits when disturbed; but this one—a highly respectable animal from the collector's point of view—has that trait poorly developed and may therefore be taken with comparative ease. A fuller account of nemerteans is given in §312.

Another haunter of the mussel beds is a snail superficially resembling the rock snail *Thais*, but it often occurs higher up than *Thais*. This is *Acanthina spirata* (Fig. 150), the thorn shell; the name is derived from the short spine on the outer margin of its lip. It is recorded from Puget Sound to Socorro Island, approximately in the position occupied by *Tegula* on more sheltered rocks. Like some more southern relatives, *Acanthina* is probably a "barnacle specialist," as Robert Paine puts it. It had been thought that the spine was used to pry open the barnacle's shell but Paine concluded that it may be used to give the snail a better purchase while drilling into its prey, although in most cases he could observe no function for the spine.

Two small crustaceans will be found at this level. The minute amphipod *Elasmopus rapax* (Fig. 151) is exceedingly abundant in the middle-zone mussel beds. This is the commonest small crustacean at Santa Cruz, in numbers more than equivalent to the mussels that make up the clusters in which it characteristically occurs. The much larger isopod *Idothea wosnesenskii* (Fig. 152), up to 1½ inches long, ranges from Alaska to central California.

§166. Great honeycombed colonies, often dome-shaped, of the tube worm *Phragmatopoma californica* (Fig. 153) may share available areas

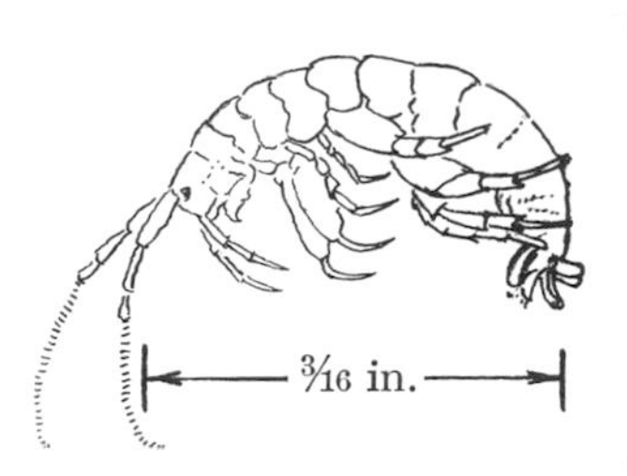

151. *Elasmopus rapax* (§165).

50. *Acanthina spirata* (§165), the thorn shell.

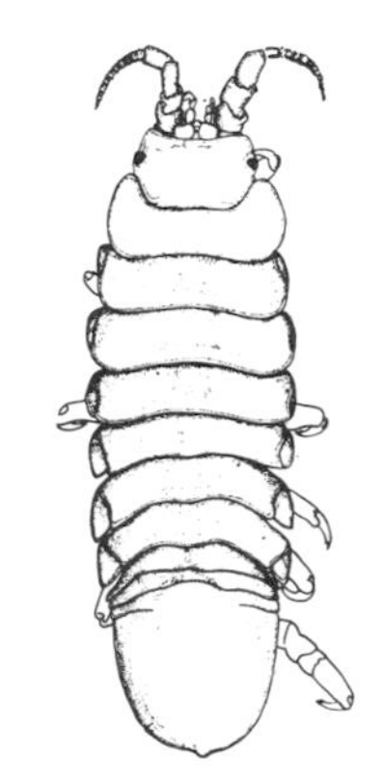

152. *Idothea wosnesenski* (§165).

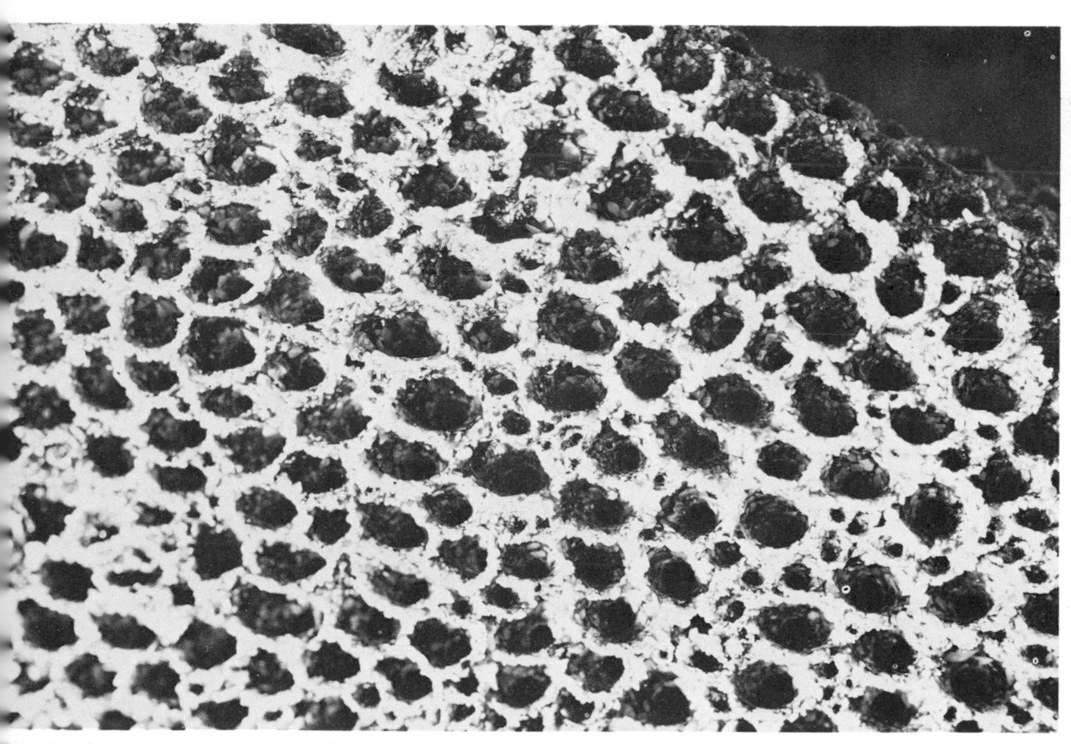

3. A colony of the tubed worm *Phragmatopoma californica* (§166).

with the mussel beds. They are very likely to be found taking advantage of the slightest bits of shelter, such as overhanging ledges and concave shorelines. The thin-walled tubes that form the colonies are made, apparently, of sand cemented together. The worms themselves, rarely seen unless one chops into a colony, are firm, chunky, and dark, with a black operculum that stoppers up the tubes when the animals are retracted. The similar, more northern *Sabellaria cementarium* has a rough, amber-colored dome. Like other tube worms, these are dependent for food on what chance brings their way, although they can assist the process by setting up currents with the cilia on their protruded gill-filaments.

Phragmatopoma occurs from Sonoma County (Shell Beach) to northern Baja California. The colony shown was photographed at Shell Beach.

§167. In situations similar to the above, but where the shore formation is tipped more horizontally, beds of the small aggregated anemone *Anthopleura elegantissima* often occur. They are not so characteristic here, however, as in the protected outer coast environment (§24).

§168. A fairly large barnacle, dull brick-red in color, conical in shape, and with the semiporous consistency of volcanic slag, is *Tetraclita squamosa* ssp. *rubescens*. Average specimens are easily twice the size of the *Balanus* (§4) found in the same association, and they are more solitary, rarely growing bunched together or one on top of another, as is the case with the white barnacle.

§169. *Nuttallina californica* (Fig. 154) is a small sea cradle, or chiton,

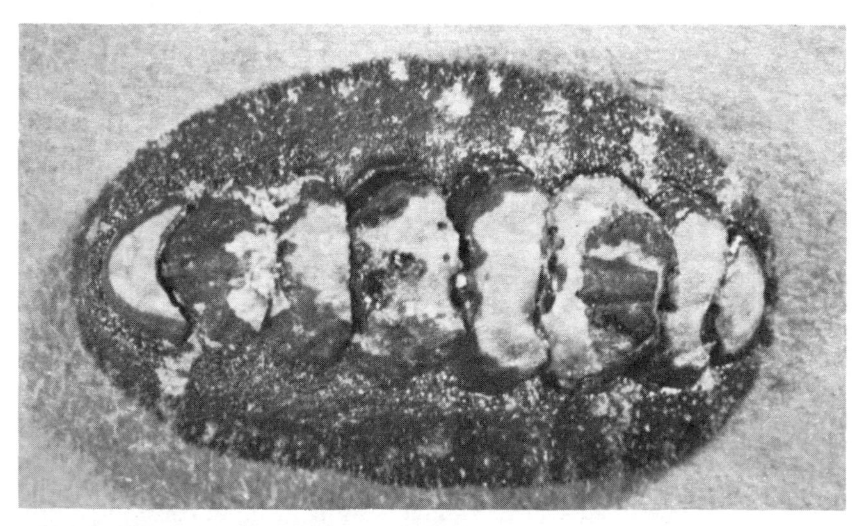

154. The chiton *Nuttallina californica* (§169).

rarely more than 1½ inches long, that is pretty well restricted to the middle intertidal of fairly exposed shores. It can be distinguished by its rough uncouth appearance, spiny girdle, and color — dull brown streaked with white. It may occasionally be found well up toward the high-tide line, but never much below the middle zone. At Laguna Beach and other places in the south this chiton lives in sculptured furrows in soft sandstone. If these furrows are of the chiton's own making, as they appear to be, they are comparable to the excavations made in rock by the owl limpet and the purple urchin. In all three cases the object is apparently to gain security of footing against the surf. *Nuttallina*'s recorded range is Puget Sound to the Coronado Islands.

Below the middle zone, *Nuttallina*'s place is taken by the larger *Katharina tunicata* (§177). Both of these chitons are perfectly able to take care of themselves at times of high surf, for they are very tough and attach to the rocks so tightly that a flat, sharp instrument must be used to pry them away. Detached from their supports, chitons promptly curl up like pill bugs.

§170. Where there are rock pools at this level, congregations of purple urchins may be found, with occasional hermit crabs (*Pagurus samuelis*), rock crabs (§14), and purple shore crabs (§27); but these are all treated elsewhere, in the zones and habitats where they find optimum conditions. Theoretically, any animals common in the middle tide-pool region of protected outer shores may be found also in these more or less protected rock pools, but they are scarce or local, and not characteristic of the region in general, as is the *Mytilus-Pollicipes-Pisaster* association, which finds its climax in this zone.

§171. Plants are outside the province of this book, but in connection with methods of resisting wave shock it is worthwhile to mention the brown alga called the sea palm, *Postelsia palmaeformis* (Fig. 155). It is restricted to the temperate Pacific coast of North America, and occurs only where the surf is continuous and high. Along the central California coast, great forests of these beautiful plants are found on rocky benches and flats. Instead of sustaining the shock of the towering breakers by rigid strength, as do most of the animals in the region, the sea palms give to it. Under the force of a powerful breaker a row of them will bend over until all fairly touch the rock, only to right themselves immediately and uniformly as the wave passes. This particular form of resistance is made possible by the sea palm's tremendously tough and flexible stalk, which is attached to the rock by a bunchy holdfast of the same material. It will be noted that specialized individuals of a limpet considered previously, *Acmaea pelta* (§11), may be found on the swaying stalks of

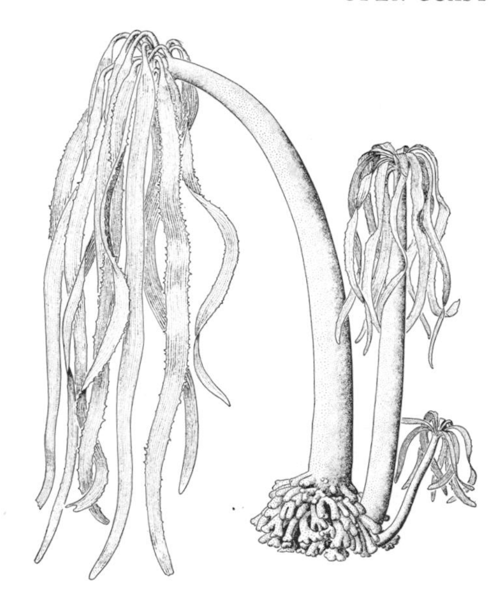

155. The brown alga *Postelsia palmaeformis* (§171), the sea palm; about one-fourth natural size.

Postelsia, thus extending their range into an environment otherwise too rigorous on account of surf—were it not for the protection and amelioration offered by this plant's flexibility.

Zone 4. Low Intertidal

§172. On violently surf-swept cliffs and tablelands where even the sea palm cannot gain a foothold, bare rock constitutes the typical tidal landscape. A few scrawny laminarians may occur, but never on the boldest headlands; and not until the collector descends (the surf permitting) to the extreme low-tide level will he find any common plant. Even then he is likely to overlook it unless he accidentally knocks a bit of inconspicuous red crust from a rock. This very flat encrusting material is the alga *Lithophyton,* under the surface of which most of the small animals of the region are to be found. The animals so occurring, however, are on the completely exposed shore only by virtue of the protection that this encrusting plant gives them, for all are protected coast forms (and hence treated elsewhere). Many of them also occur in kelp holdfasts.

The great green anemone *Anthopleura xanthogrammica* is very much at home here, growing to a size that makes it a close second to the ane-

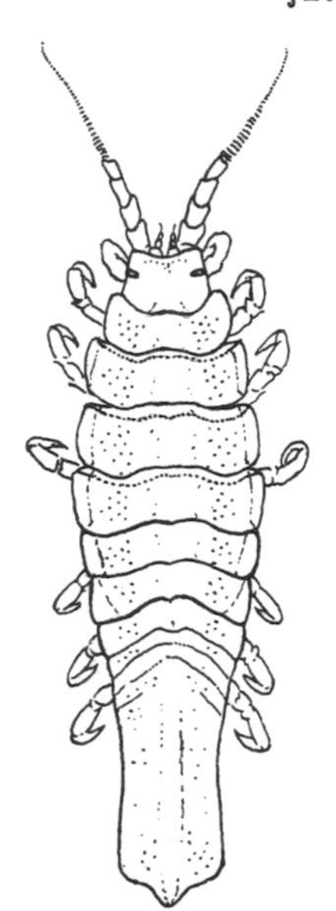

156. One of the largest of the isopods, *Idothea stenops* (§172); about life size.

mones of the Australian Barrier Reef. This giant beauty will always be found, however, in situations where it is reasonably sheltered from the full force of the surf. For this reason, and because it is quite likely to have been first seen in the tide-pool regions, it has been treated as a member of the protected outer coast fauna (§64).

A more characteristic animal, occurring also under flat rocks and embedded boulders, is *Idothea stenops* (Fig. 156), one of the largest of the isopods, recorded from Coos Bay to Monterey Bay. The smaller *I. schmitti* will be found under similar circumstances, and at San Remo, below Carmel, we have taken both at the same place. *I. schmitti* ranges from Bering Sea to Monterey Bay.

§173. A great ostrich-plume hydroid, *Aglaophenia struthionides,* occurs in rock crevices, especially in large vertical crevices in the line of surf. This form is considerably larger than members of the genus found on relatively protected shores, and the largest clusters are so dark as to be nearly black. Otherwise there is little difference, and the reader is referred to Figure 67.

This is one of the few hydroids that occur in bona fide surf-swept environments, except on the open coast at Sitka. There tufted colonies of two great, coarse sertularians will be found. The ranker of the two, *Abietinaria turgida* (reported as common in the Bering Sea–Aleutian region), occurs in coarse growths up to 6 or 8 inches high, with short branches (which may be branched again) almost as coarse as the main stem, and with the cuplike hydrothecae crowded close on stem and branch. The more delicate and elongate *Thuiaria dalli,* ranging from Bering Sea to Puget Sound, has a straight, heavy main stem; but the branches, which take off at right angles, are slender and without secondary branches, and the cuplike receptacles into which the tentacles are withdrawn are small and comparatively few.

However, where the crevices are deep enough, even along the California coast, or where overhanging ledges simulate the semisheltered conditions of reef and tide pool, such other hydroids as *Eudendrium* (§79), *Abietinaria anguina* (§83), and even *Tubularia marina* (§80) may occur, with the sponge *Rhabdodermella* (§94), the hydrocoral *Allopora porphyra* (§87), and the bryozoan *Bugula californica* (§113), all of which are treated in Part I.

Although it is mostly a subtidal species, especially in the southern parts of its range, the sea strawberry *Gersemia rubiformis* is occasionally found intertidally on almost barren rock in Alaska. This circumpolar, boreal-arctic species, a beautiful bright pink with eight-tentacled polyps, is now recorded as far south as Trinidad Head in Humboldt County, California. It is a fairly common subtidal species in the Cape Arago region, where it may sometimes be found harboring an interesting pycnogonid, *Tanystylum anthomasti.* This pycnogonid was described from soft corals in Japan, but Oregon, Point Barrow, and its type locality are the only places from which it has been recorded, although its potential hosts are widely distributed (Fig. 157). Possibly it has been overlooked as it retreats into the crevices between the lobes of the *Gersemia* colony.

§174. Here the purple urchin, *Strongylocentrotus purpuratus* (Fig. 158), is distinctly at home, having made one of the most interesting adaptations of all to the pounding surf. The animal will commonly be found with at least half of its bristling bulk sunk into an excavation in the rock. For more than a hundred years the method of producing these excavations has been a subject of controversy, but it is generally agreed now that Fewkes stated the situation correctly in 1890. He believed that the teeth and spines of the animals, aided by motions produced by waves and tides, were sufficient to account for the pits, however hard the rock and however frayable the urchins' spines, for the spines, of course, would be continually renewed by growth. Sometimes an urchin will be found that has imprisoned itself for life, having, as it were, gouged out a cavity

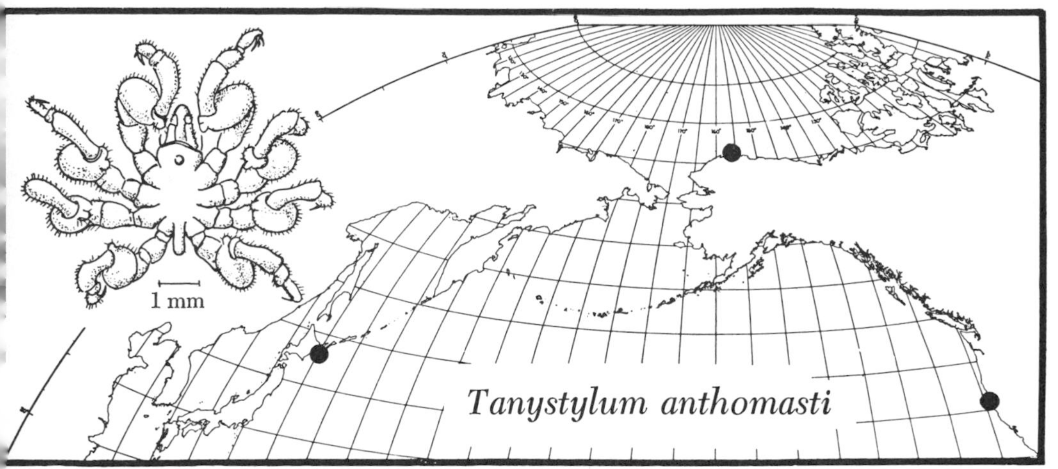

57. The pycnogonid *Tanystylum anthomasti* (§173), with a chart showing its distribution.

58. *Strongylocentrotus purpuratus* (§174), the purple urchin; 2-inch specimens, in pits.

larger than the entrance hole made when it was young (or, more probably, made by another urchin, since it seems probable that these holes may be formed by successive generations of urchins). Holes abandoned by urchins may provide refuge for all sorts of nestlers, such as dunce-cap limpets, small chitons, and the like.

This burrowing habit probably has some protective value for the urchins against their chief enemy, the sunflower star (§66), although it would appear that an animal able to manage such a prickly meal as a sea urchin would not be deterred by a shallow burrow.

Urchins have been reported as capable of burrowing through steel girders used to reinforce piling; but more probably they set up conditions that accelerate corrosion beneath them, and the steel is attacked by the environment caused by the urchins, rather than by the urchins directly.

The purple urchin ranges from Alaska to Cedros Island, probably entirely on the outer coast, unlike the green urchin of Puget Sound and northern inside waters. The reproductive season at Monterey is during March and April, at least. The food-getting, reproductive, and fighting habits of urchins have been discussed in connection with the giant red urchin (§73). A good many red urchins occur here at a level below the pits of the purple urchins. In fact, they are possibly as common here as in the outer tide pools of the more protected areas.

The purple urchin is not as sedentary as its habit of living in holes might suggest. Many urchins do not live in holes, especially where the rock is too hard; and when sand is piled up in their usual habitat, they may move upward. They have been seen perched on top of rocks when unusual wave conditions cover the lower levels with sand. Sea urchin eggs are important research material, and some developmental biologists have claimed that removal of large numbers of urchins is followed by a replacement from lower levels; but others are not so sure that the recruitment is adequate and prefer to take a more conservative view of the source of supply.

All three of the urchins, red, purple, and green, have a number of hangers-on, both inside and out. An intestinal flatworm has already been mentioned in connection with the giant red form. In addition to the flatworm, Lynch finds practically 100 per cent infestation of the intestine by one or more of 12 distinct species of protozoans, some of them comparatively large (¼ mm.). There are also isopods (*Colidotea rostrāta*) that occur nowhere except clinging to the spines of urchins.

§175. The gigantic horse mussel *Volsella modiolus* (formerly *Modio-*

lus), up to 9 inches in length, is no longer common, its depletion being the result, probably, of too many chowders, too many conchologists, and the animal's presumably slow rate of growth. Although it occurs a bit below the low-tide line, the horse mussel's brown shell, naked except for a fine beard, is very noticeable, especially by contrast with the California mussels alongside it. Such of the California mussels (§158) as occur this far down are giants of the tribe but are never clean-shelled, as are the colonies higher up in the intertidal; many plants, from coralline algae to small laminarians, grow on their shells, providing almost perfect concealment. Some common seastars (§157) range down this far also, undeceived by the mussels' disguise; but their apparent base of distribution is higher up. The smaller and more southern *V. recta* may also be seen, particularly along less strongly exposed shores, as at La Jolla.

§176. In the Monterey Bay region, beds of the foliose coralline *Cheilosporum* (the coralline being stiffened by the same encrusting red alga that grows on rocks in this zone) are certain to harbor very tiny cucumbers, *Cucumaria curata,* that look like bits of tar. The animals are so small that they can be mounted whole on slides, with the internal anatomy diagrammatically visible under the microscope. We do not know of their occurrence anywhere except in the region mentioned, where, in addition to their coralline habitat, they are occasionally found under mussels in this lowest zone. We have found ovigerous specimens in December, the eggs being brooded on the underside, as in the related species discussed below.

On the same coralline there occurs a very similar but much larger species, *Cucumaria lubrica* (Fig. 159), which is often more than an inch long. Size alone will distinguish the two, but an additional difference is

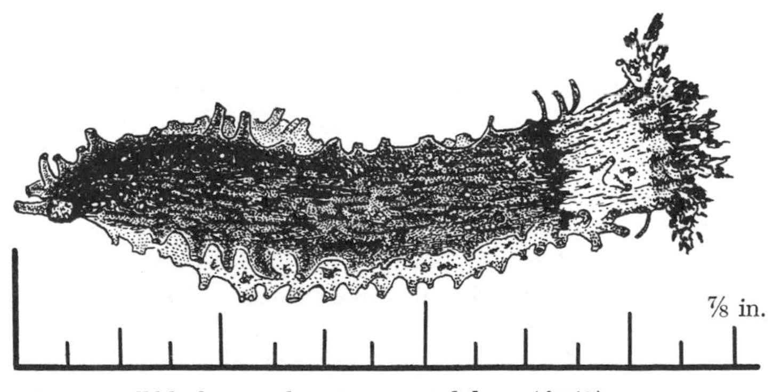

159. A small black cucumber, *Cucumaria lubrica* (§176).

that *C. lubrica* has 10 large tentacles, whereas *C. curata* has 8 large and 2 small ones. We have specimens of *C. lubrica* from the vicinity of Juneau and from Puget Sound (the northern specimens being larger and lighter in color), as well as from Monterey Bay; it also occurs on the Sonoma County coast in masses on sheltered rock faces.

§177. *Katharina tunicata* (Fig. 160), whose dead black tunic has almost overgrown the plates of its shell, is one of the few chitons that do not retreat before daylight, or even sunlight; and next to the "gum boot" (*Cryptochiton*) it is the largest of the family. In the low-tide area *Katharina* assumes the position held in the middle zone of the same surf-swept regions by *Nuttallina. Katharina,* however, shows more definite zonation, occurring, in suitable locations, in well-defined belts a little above the zero of the tide tables. In Alaska, *Katharina* is the most abundant intertidal chiton; it ranges plentifully as far south as Point Conception, where there is a great colony, and less commonly below there to the Coronado Islands.

On the most exposed parts of the open coast *Katharina*'s eggs are shed in July, at least in the central California region, but in Puget Sound sexual maturity comes a month or two earlier. Strangely enough, on at least one occasion *Katharina* was found to be not only the commonest chiton but one of the most prevalent of all littoral forms, in a British Columbia locality almost completely protected from wave shock (but directly fronting a channel to the open coast)—an anomalous situation recalling the remarks of the ecologist W. C. Allee (1923) to the effect that if the search is long enough, one can turn out very nearly any animal in any environment, however far-fetched.

§178. Where laminarians or other algae provide the least bit of shelter, the sea lemon, *Anisodoris nobilis,* one of the largest of all nudibranchs, may be found. Its average length is around 4 inches, but 8-inch specimens have been taken. Their usual vivid yellow color sometimes tends toward orange; all have background splotches of dark brown or black, while the knoblike tubercles that cover the back are yellow. In common with other dorid nudibranchs, especially the yellow ones, *Anisodoris* has a fruity, penetrating, and persistent odor. This species ranges from Vancouver Island to Laguna Beach. MacFarland wrote, in 1906, that it occurred on piling in Monterey Harbor; but it disappeared from there, along with many other animals, after the oil fire of 1925. We have seen these hermaphroditic animals depositing strings of eggs in November.

§179. The black abalone, *Haliotis cracherodii* (Fig. 161), prefers more surf than its red cousin or, more likely, tolerates more surf because

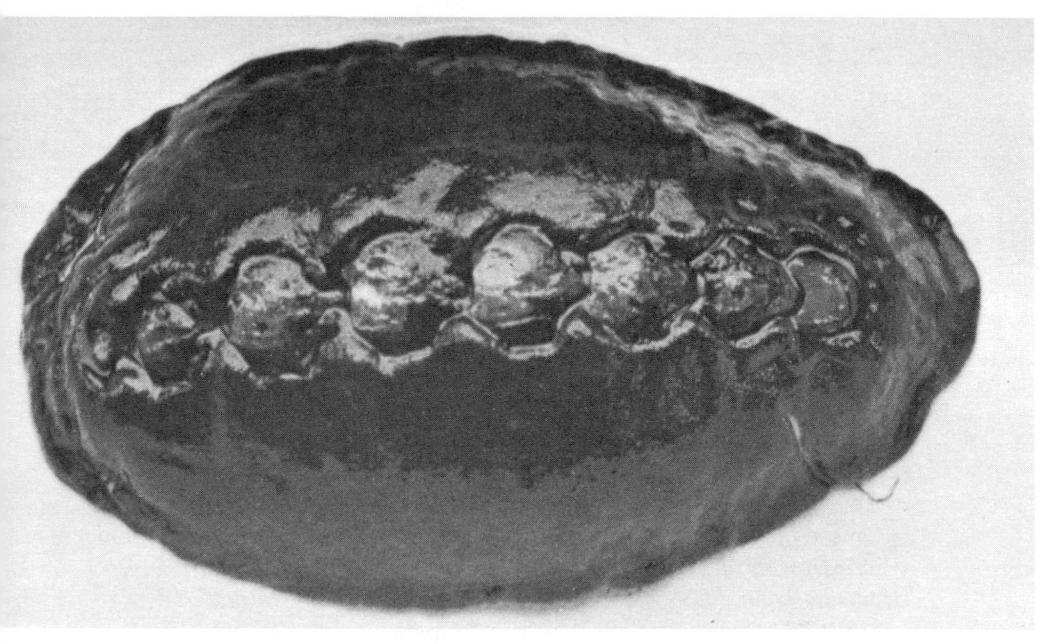

▪. *Katharina tunicata* (§177), the black chiton.

. The black abalone, *Haliotis cracherodii* (§179); this specimen,
wn life size, is under the legal limit of 5½ inches.

it finds food-getting conditions more to its liking on surf-swept shores. Some abalones, this black form in particular, seem to derive much of their nourishment from plankton—which explains why the black abalone is so often found in the crevices of barren rocks, where there is no visible food in the way of fixed vegetation. Again unlike the red form, which invariably carries a small forest on its back, the black abalone keeps its shell clean and shining. It would be interesting to know how this is managed; perhaps it has something to do with the animal's location.

The black abalone is smaller than the red, with 5 to 8 open holes in its shell. The legal size is 5½ inches. The range is from Coos Bay (rarely seen) to Cape San Lucas; it is most abundant from central California south.

On the outer coast of British Columbia and Alaska is another abalone, whose thin, pink, and wavy shell is the prettiest of the lot; this is *H. kamtschatkana* (§74), commoner in semiprotected waters.

§180. The largest of the keyhole limpets, which are related to the abalones and no more than a family or so removed from them in the phylogenetic scale, is *Megathura crenulata* (Fig. 162), also known as *Lucapina.* A good-sized specimen is 7 inches long and massive in proportion. The flesh of the underside of the foot is yellow, and the oval shell, which has many fine radiating ridges, is nearly covered by the black mantle. Many animals lose their color after death, but this giant keyhole limpet is unique in that its black will come off in life if the mantle is rubbed. The range is from Monterey to Baja California, and we have found specimens particularly common in Todos Santos Bay.

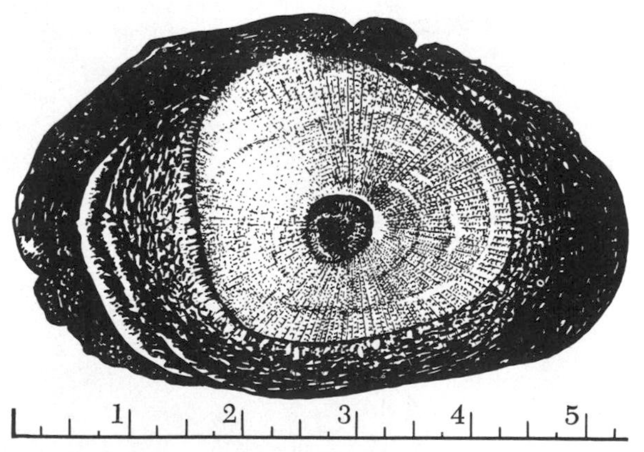

162. The giant keyhole limpet *Megathura crenulata* (§180), whose mantle almost covers its shell.

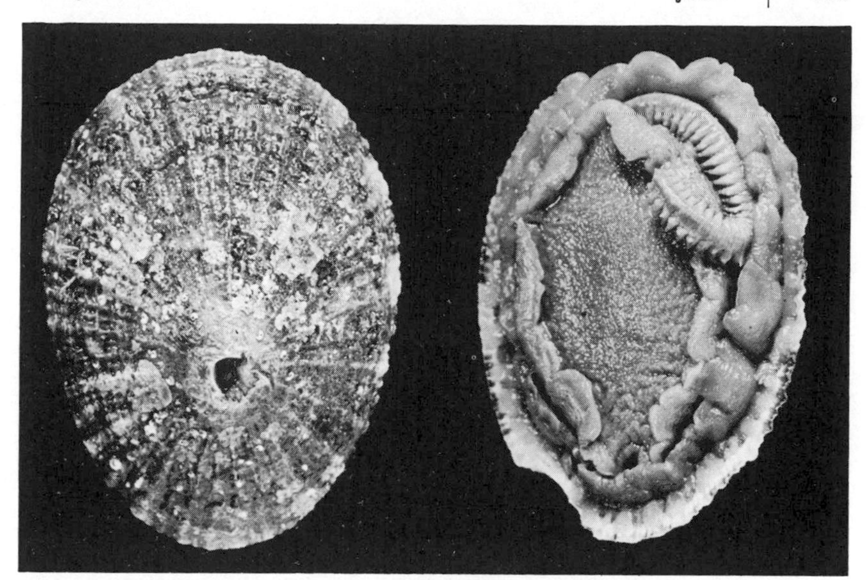

163. The keyhole limpet *Diodora aspera* (§181): left, dorsal view, showing the "keyhole" at the top of the shell; right, view of underside, showing the commensal scale worm *Arctonoë vittata* (§76). Both pictures are half natural size.

§181. *Diodora aspera* (Fig. 163), another of the keyhole limpets, is small by comparison with the *Megathura,* but still large for a limpet. Large specimens will measure 2 to 3 inches long, and the recorded range is Alaska to Baja California. Any specimens found should be examined for the commensal scale worm *Arctonoë vittata,* which has been mentioned as occurring with *Cryptochiton* (§76). In 1866, Lord, in British Columbia, remarked that the position of the worm—like a ribbon on edge—prevents its being crushed when the host clamps down to the rock. The length of the worm, considering the size of the host, is remarkable, and the worm often has to curl around the mantle so completely that its ends almost touch.

A third member of the group, the volcano-shell limpet, is the smallest of the lot, but is handsome nevertheless. It has been treated (§15) as a protected-outer-coast form, but it is practically as common in this environment.

§182. The leafy horn-mouth snail, *Ceratostoma foliatum* (§117), is occasionally found in surf-swept areas where there is a bit of shelter just at the low-tide line. Occasional also, but sufficiently spectacular to cause comment, is the prettily marked ring top shell, *Calliostoma an-*

nulatum (see color section), up to an inch in diameter and ranging from Alaska to San Diego. The body is salmon pink, and the yellow shell has a purple band along the beaded spiral. This handsome species may be found on the giant kelps just offshore. *C. canaliculatum* is characteristically an inhabitant of the rocky shore, from the lowest zone far down into the sublittoral. *C. ligatum* (Fig. 164) occurs sparsely in Zone 4, but it is most characteristically a Puget Sound quiet-water form. *C. ligatum* may be differentiated from *C. canaliculatum* by its more rounded shape and the blue color of the lower layer of the shell that is apparent in worn specimens. The shell of *C. canaliculatum* is sharply conical in outline and uniformly light brown in color.

Most characteristic of all is the short-spired purple, or rock, snail *Thais emarginata* (Fig. 165), which is not only abundant on rocky shores where the surf is fairly heavy but is almost entirely restricted to these regions. Writing in 1871, Verrill and Smith said that a whelk similar to our *Thais* (*Purpura lapillus*) "is seldom found living much below the low water mark, and prefers the exposed rocky headlands on the ocean shores, where it flourishes in defiance to the breakers. It lays its eggs in smooth, vase-shaped capsules, attached to the sides or under surfaces of stones by a short stalk, and usually arranged in groups." If they had been describing our purple snail, they could not have hit it more closely. We would add only that the egg cases, called "sea oats," are yellow, and are likely to be mistaken at first sight for purse sponges.

Purple snails are so named not for the color of their shells (usually black and white) but because they are reputed to be the snails from which the Tyrian dye of ancient times was made. Irreverent modern investigators find the dye to be a rather dull purple that fades badly. It is suspected that the color sense of ancient peoples was based on standards that would seem strange to us. No doubt the colors that thrilled the ancient Romans would arouse contempt in a schoolboy artist of today, since the civilization of Tiberius knew nothing of our lacquers and brilliant coal-tar dyes.* Our *Thais* is usually gray-brown or greenish in color, or white, with darker bands; young specimens are occasionally a bright orange. The rock snail ranges from Alaska to Mexico and usually feeds on barnacles and small mussels.

In connection with *Thais*, a mild dissertation on birth control, prenatal care, and race suicide seems to be almost mandatory, for the tribe

* Unfortunately, Ed Ricketts did not see Pompeii; he would have revised his opinion of Roman talent for pigments as well as enjoyed the subject matter on the walls. *J. W. H.*

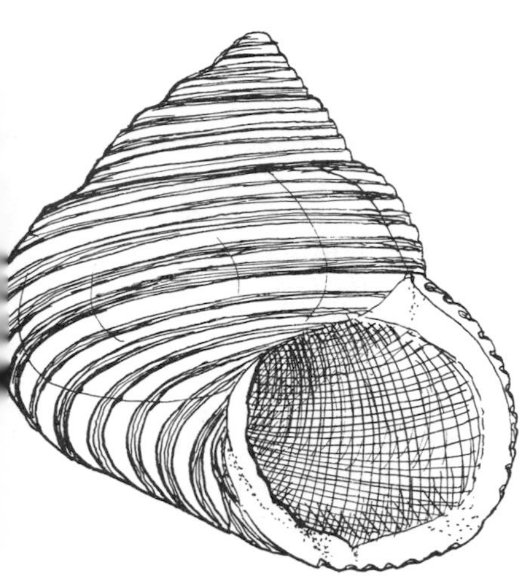

164. *Calliostoma ligatum* (§182), a top shell.

165. *Thais emarginata* (§182), the rock snail.

An expanded cluster of tubed worms, *Salmacina tribranchiata* (§183).

of *Thais* persists in spite of the most arrant cannibalism. The adults eat snail eggs indiscriminately, including their own, and the larval *Thais* eat each other while still in the egg capsule. Instead of there being a single egg, with a sufficient supply of nutritive yolk, there are many eggs in each capsule, with not enough yolk to carry the embryos through to hatching time. Consequently they eat each other until only one—and that one presumably cock of the walk—is left to emerge from the capsule into a world where its chances of survival are still poor.

§183. Two encrusting forms may puzzle the amateur collector. A roughly circular growth that looks like a tightly adherent mass of corallines is composed of tiny-tubed annelid worms, the serpulid *Salmacina* (Fig. 166). They are not unlike intertwining tubes of *Serpula* or *Sabellaria,* but on a pinpoint scale. At first sight, the mass may appear dingy, but on closer examination it will be seen to have an encrusted, dainty, filigree pattern. En masse the gills are a rusty red, individually a brighter red and rather attractive.

The other form, occurring in similar situations on vertical cliff faces almost at the lowest-water mark, is the sponge *Hymeniacidon sinapium,* which may thinly encrust areas of several square feet. This is the yellow-green, slightly slimy form, with nipple-like papillae, recognizable by the European observer as similar to *Axinella,* and originally so called in this area. De Laubenfels notes that the open-coast habitat is rare, specimens occurring more characteristically in quiet water (§225), on oyster shells and the like.

7. *Open-Coast Sandy Beaches*

WHEREAS the animals living on surf-swept rocky shores have solved the problem of wave shock by developing powerful attachment devices, the inhabitants of surf-swept sandy beaches achieve the same end by burying themselves in the sand. Some, like the mole crab and the razor clam, are able to burrow with extraordinary rapidity. Others, like the Pismo clam, burrow more slowly, depending on the pressure-distributing strength of their hard, rounded shells. These animals have achieved the necessary great strength and resistance to crushing not by the development of such obvious structural reinforcements as ribbing, with the consequent economy of material, but by means of shells that are thick and heavy throughout. Ribbing would provide footholds for surf-created currents that could whisk the animal out of its securely buried position in a hurry. Natural selection has presumably produced a race of clams with thick, smooth shells, over which the streaming and crushing surf can pour without effect. In addition, the actively burrowing forms, such as *Emerita* (§186), would seem to be provided with a sense of orientation not dependent on sight.

That the obstacles faced by sand dwellers on an exposed coast are all but insuperable is indicated by the fact that few animals are able to hold their own there. These beaches are sparsely populated in comparison with similar rocky shores. Actually, we know of only six or seven common forms that occur in any abundance on heavily surf-swept sand beaches, and two of them are already well along toward extinction because of human activities. This reflects a situation quite different from that assumed by most amateur collectors, who would have one but turn over a spadeful of beach sand to reveal a wealth of hidden life.

Most members of the sandy beach community are uncertain in their occurrence: they have good years and bad years, and a beach that was teeming with sand crabs or hoppers one year may be unoccupied a year later. The population may have been wiped out by a change in the shape of the beach—the waves may have shifted direction, thus cutting the beach back to the dunes, or building it up faster than it can be populated. Most of the animals living on sandy beaches have pelagic larval stages, so the young must be set adrift and may therefore settle in another part of the world than their parents. Food is also uncertain—little is produced in the sand itself except minute algae among the sand grains, for the most part available only to animals adapted to ingesting sand grains and eating whatever may adhere to the surface. The major source of food in this region is either the plankton washed ashore by the waves or the seaweeds and occasional corpses of fishes, birds, and sea mammals cast ashore. In late years many beaches have probably been significantly enriched by the leavings of man.

Nevertheless, the species that have adapted to living in the sand are, as species, successful in terms of numbers. In good years they may be there by the millions. They have learned to shift up and down the beach with the tide and the waves and to subsist on an uncertain source of supply. Most of them, except for the clams, are short-lived, living perhaps a year or two. But the Pismo clam and the razor clam, with comparatively long life spans, are more vulnerable to the culinary ravages of man, for it is more difficult for these populations to become reestablished once the breeding stock has been reduced.

As for the beach itself, it is a phenomenon of the meeting forces of sea and land, and an experienced observer can tell much about the condition of the sea from the slope of the beach, the sizes of the sand grains, and the kinds of minerals the sand is made of. Since this is beyond our scope, the reader is referred to the excellent small book by Willard Bascom, an ardent and lifelong student of beaches.

§184. We have already cited several examples (notably the periwinkles) of animals with marked landward tendencies. On the sandy beaches of the open coast there is another, the little pill bug *Alloniscus perconvexus,* about ⅝ inch long. This isopod is an air-breathing form that will drown in seawater. It will be found, therefore, in the highest zone, above the high-tide line; and because of the obvious nature of its burrows it is often one of the first animals to be noticed in this environment. The molelike burrows are just beneath the surface, and in making them the animal humps up the surface sand into ridges. Another air-breather, the isopod *Tylos punctatus,* a ¼- to ½-inch oval form resembling *Gnorimosphaeroma* (§198), is restricted to the southern California beaches.

§185. During the night, or most noticeably at dusk or at dawn, the foreshore seems to be alive with jumping hordes of the great beach hopper *Orchestoidea californiana* (§154). They are pleasant and handsome animals, with white or old-ivory-colored bodies and head region and long antennae of bright orange. The bodies of large specimens are considerably more than an inch long; adding the antennae, an over-all length of 2½ inches is not uncommon. Like the other beach hoppers, this form avoids being wetted by the waves, always retreating up the beach a little ahead of the tide. These hoppers seem always to keep their bodies damp, however, and to that end spend their daylight hours buried deep in the moist sand, where they are very difficult to find. Night is the time to see them. Observers with a trace of sympathy for bohemian life should walk with a flashlight along a familiar surfy beach at half-tide on a quiet evening. The huge hoppers will be holding high carnival—leaping about with vast enthusiasm and pausing to wiggle their antennae over likely looking bits of flotsam seaweed. They will rise up before the intruder in great windrows, for all the world like grasshoppers in a summer meadow. Too closely pursued, they dig rapidly into the sand, head first, and disappear very quickly. Ovigerous females have been taken in March in Monterey Bay.

§186. *Emerita analoga* (Fig. 167), the mole crab or sand crab, is the "sand bug" of the beach-frequenting small boy. Its shell is almost egg-shaped—a contour that is efficient for dwellers in shifting sands where the surf is high, since the pressure is distributed too evenly to throw the animal out of balance. Most of the crabs characteristically move in any direction—forward, backward, or sideways—but whether the mole crab is swimming, crawling, or burrowing, it always moves backward. Crawling is apparently its least efficient mode of locomotion and swift burrowing its most developed, but it is also a fairly good swimmer—an action achieved by beating its hindmost, paddle-like appendages above the after margin of the shell. The same appendages assist in burrowing, but most of the work of burrowing is done by the other legs.

When in the sand, the mole crab always stands on end, head end up and facing down the beach toward the surf. Characteristically, the entire body is buried, while the eyes (tiny knobs on the end of long stalks)

167. The sand crab *Emerita analoga* (§186), feeding: left, the feathery antennae are extended into the current; right, the accumulated food particles are scraped from one antenna by specialized appendages.

and the small first pair of antennae (which form a short tube for respiration) project above the sand. When a wave starts to recede down the beach the sand crab uncoils its large second pair of antennae (like small feathers) and projects them in a V against the flowing water to gather minute organisms, mostly dinoflagellates but also small plant particles and some sand grains. The feeding antennae work rapidly and are pulled through an elaborate arrangement of bottlebrush-like appendages to be scraped clean of food, perhaps several times during a receding wave. *Emerita* may gather food particles as small as 4 or 5 microns in this manner (a micron is 1/1000 mm.), and can handle objects up to 2 mm. in diameter. Ian Efford, who has made the sand crab his life's study, has not been able to confirm opinions that the sand crab may feed on bacteria or on food adhering to sand grains, or that the animal can sift the sand out of its food.

Sand crabs often occur in dense patches or aggregations on the beach, the largest individuals at the lowest part of the beach and the smallest higher up the beach. Since the males are smaller than the females, they will also be found somewhat higher up the beach than the females (large females are about 1¼ inches long). The reason for this aggregation is not understood, but the crabs may be packed so tightly that there is virtually no room between them; perhaps such a dense stand has some value in reducing the amount of sand taken in during feeding. Aggregation does not seem to be of much value for protection from predators; in any event it does not appear that predation by birds is very serious, since birds tend to space out when feeding and thus pass over a colony of *Emerita* somewhat lightly.

The aggregations tend to remain in discrete patches on the beach. Efford has demonstrated that if enough individuals are moved, forming a "critical mass," a new aggregation may be started; but that if only a few are moved, they disperse.

Sand crabs move up and down the beach with the tide, staying more or less in the zone of breaking waves, where feeding is evidently best (although animals will live and feed submerged in an aquarium if the water is flowing). This movement requires some adjustment to tidal rhythms, but it may also be to some extent a simple physical reaction to the digging properties of the sand: the crabs cannot dig in dry sand and evidently function best when the sand has a certain "index of penetration." This property of sand is extremely difficult to measure objectively, however. Certainly if *Emerita*'s instinct were to move downhill, as one investigator (Mead) thought, it would be difficult to explain its movement uphill with the tide, except that the slope of a real beach is usually less than the experimental slopes.

Emerita analoga has a larval life of about 4 months, during which time the zoea and megalops stages may be carried for long distances. In some years, when the northward set of currents is favorable, larval swarms may populate beaches in Oregon and even Vancouver Island in British Columbia; but *Emerita* is most abundant, within our range, from central California southward to Ensenada. The species has been recorded from Alaska to Patagonia! In southern California the megalops larvae arrive on the beach in the greatest numbers from May to July. *Emerita* reproduces during its first year of life, and may not live more than 2 years. Further south there is a tropical species, *Emerita rathbunae,* in which the males are minute and attached to the legs of the female, thus carrying the tendency for small males in this genus almost to the verge of parasitism by the males.

The spiny sand crab, *Blepharipoda occidentalis* (Fig. 168), is an inhabitant of subtidal regions of sandy beaches, and may occasionally be found at very low tide. It is larger, with a flattened carapace 2 or 3 inches long. It evidently occurs as far north as Drake's Beach, since molted carapaces may be found along the Limantour Spit. Unlike its nicely

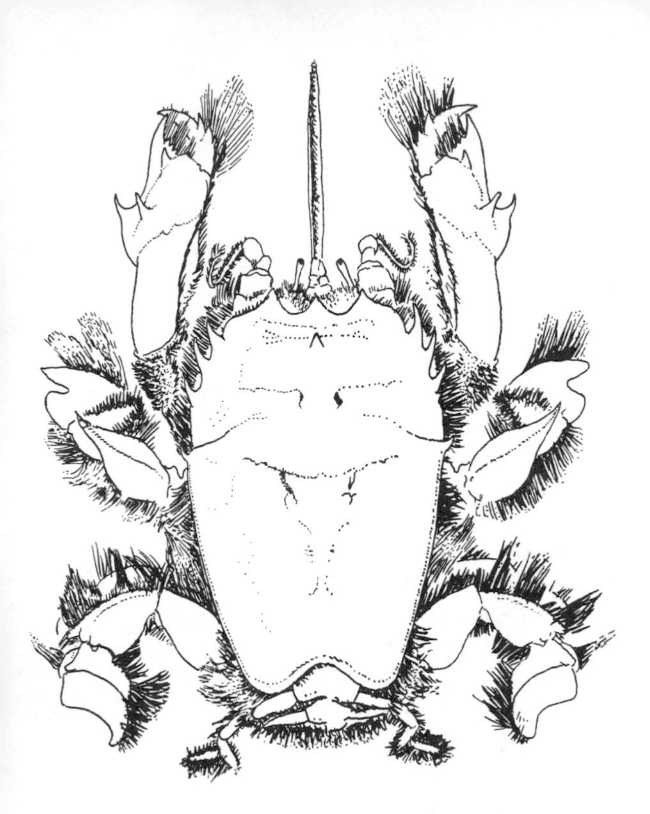

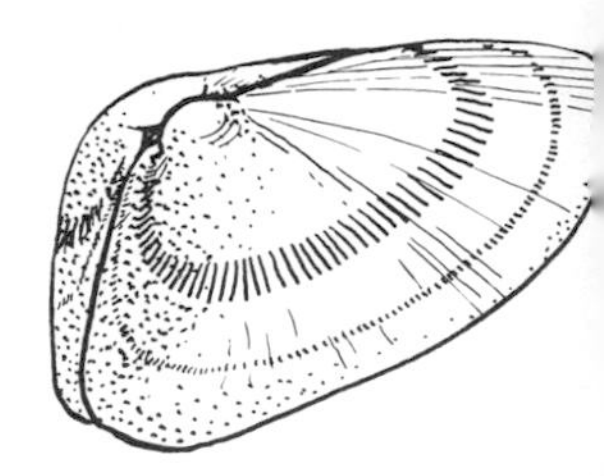

169. *Donax gouldi* (§188), the bean clam.

168. Left: the spiny sand crab *Blepharipoda occidentalis* (§186).

170. The Pacific razor clam, *Siliqua patula* (§189); foot and siphons extended.

adapted relative *Emerita, Blepharipoda* is a generalized feeder, a scavenger and predator like most of the more ordinary crabs.

§187. At about the sand crab's level are small, shrimplike crustaceans, *Archaeomysis maculata,* called opossum shrimps because, like the other mysids, they retain the young in a marsupial pouch under the thorax. Often these animals hide just below the surface of the moist sand, but they may be brought into the open and collected when water filling a hole dug in the sand causes them to swim about. They may also be netted from the surf with a fine net. This form is related to the more visible mysids of the tide pools, although it lacks gills; some mysids have small gills.

§188. The bean clam, *Donax gouldi* (Fig. 169), is common from the San Luis Obispo region to Mexico. This small, wedge-shaped clam, averaging an inch in length, is said to have been so common at one time that it was canned commercially at Long Beach. For many years it has not been available in commercial quantities, but the individual collector can still find enough for a delicious chowder by combing the sand just beneath the surface. The bean clam's hiding place is commonly revealed by tufts of a hydroid that grows on the shell and protrudes above the surface of the sand. This elongate hydroid, *Clytia bakeri,* related to *Obelia* and occurring also on the Pismo clam, is the only hydroid found on exposed sandy shores.

Like the sand crab, *Donax* forms dense aggregations and is sporadic from year to year in some places. When it is abundant, it may give the beach a pebbled appearance, since it is not completely buried in the sand. Our *Donax* is sedentary; species of *Donax* in some other parts of the world, like Japan and Texas, migrate up and down the beach, as *Emerita* does on Pacific shores.

§189. A razor clam, *Siliqua patula* (Fig. 170), corresponds ecologically, on the open sandy beaches of Washington, to the different-looking Pismo clam of similar stretches in California and Mexico. *Siliqua* is long (shell up to 6 inches) and thin, with fragile, shiny valves—just the opposite of what one would expect in a surf-loving animal. Apparently it depends on speed of digging for protection from wave shock. A clam that has been displaced by a particularly vicious wave can certainly be reburied under several inches of sand before the next comber strikes, for specimens laid on top of the sand have buried themselves completely in less than 7 seconds. The foot, projected half the length of the shell and pointed, is pushed into the sand. Below the surface, the tip expands greatly to form an anchor, and the muscle, contracting, pulls the clam downward. The movement is repeated several times in rapid succession before the clam disappears. A digger must work quickly to capture the animal before it attains depths impossible to reach.

Along the Washington coast, from the mouth of the Columbia northward, there is a stretch of 38 miles of beach once worked by commercial diggers for local canneries, and the average annual pack, from 1915 to 1927, was 37,000 cases.* In Alaska, too, the canning of razor clams is an industry of some importance, the fisheries and canneries being at Cordova, Chisik Island, Cook Inlet, and Kukak Bay. It is estimated that tourists take from the Washington coast at least a third as many clams per year as the canneries use. Many are also taken for crab bait and for consumption by local residents. The drain on the beds is altogether too great for an animal that takes (in Washington) 3½ years to attain its legal size of 4½ inches, and the supply is decreasing steadily. At Cordova the clams require 6 years or more to reach the same size.

A Bureau of Fisheries Bulletin records some further interesting differences between the northern and southern clams. In the south (Washington, etc.) the spawning is simultaneous, all the clams along several miles of beach spawning on the same day when there is a sudden rise in water temperature. This usually occurs at the end of May or early in June. Sometimes the set of young is enormous, but in other years the animals almost fail to spawn at all. In Alaska they are likely to spawn without fail, but not suddenly and not simultaneously, and the set of young is more uniform. Spawning is usually in July and early August. The larval stage for both regions lasts for 8 weeks. The average maximum age in Washington is 12 years; in Alaska, specimens up to 17 and 18 years old are known.

Fraser (1930) found that 86 per cent of the 3-year specimens on Queen Charlotte Island beaches were mature or maturing. Spawning took place from late July on, depending on the minimum water temperature of 13° C. The following is from Weymouth and McMillan's (1930–31) summary of a statistical study of 14,000 individuals of the four species of razor clams recognized: "Over the wide range of the species a high correlation is found between the relative growth rate, age, length, and geographic position. . . . Under the conditions of the southern beds there is

* Fisheries statistics have a way of being several years behind, and the latest we now have available are for the year 1965, when only 1,047 cases of razor clams were canned; most of the catch is now going to restaurants, but the clams are indeed becoming as scarce as predicted in early editions of this book, even if we allow for the possibly declining popularity of canned clams, because there are many more restaurants now than there had been.

initially a more rapid relative growth rate which, declining more rapidly, reaches a low final level, and the clams show a smaller final size and a shorter life span than in the northern beds." This observation was found (by the same authors in another paper) to apply to Washington and Alaska cockles (*Clinocardium nuttalli,* §259).

The species illustrated, ranging down from Bering Sea, is said to extend as far south as San Luis Obispo, but it is rarely seen along the central California coast. In this connection it should be noted that specimens may be lethal as food at times during the summer, owing to mussel poisoning (§158).

Up to 80 per cent of these clams, according to canners, carry an internal commensal, the nemertean worm *Malacobdella grossa.* The percentage of clams so accompanied increases toward the north. MacGinitie, in a verbal communication, reported this or a related form from clams at Humboldt Bay. We have taken it also, but sparsely (only one was found in a lot of several dozen *Siliqua*), along the open sandy beaches near Queets, central Washington. These flat nemerteans, up to 1½ inches in length, attach themselves to the clams' gills by a sucker and feed on minute plankton in the water passing through the gills. Apparently they have no harmful effect on the host.

§190. The Pismo clam, *Tivela stultorum* (Fig. 171) does not merely tolerate surf; it requires it. Clams removed from their surf-swept habitat to lagoons and sheltered bays to await shipment live but a few days, even though tidal exposure, temperature, and salinity are the same. Apparently *Tivela* has accustomed itself, through generations of living in an environment so rigorous that it would be fatal to most animals, to the high oxygen content of constantly agitated water and has carried the adaptation to such an extreme that it can no longer survive under less violent conditions. An interesting feature of the clam's adaptation is its inhalant siphon. Most clams live in relatively clear and undisturbed waters, but the Pismo clam, living in waters that are commonly filled with swirling sand, must provide against taking too much sand into its body with the water that it must inhale in order to get food and oxygen. It does this by means of a very fine net of delicately branched papillae across the opening of the siphon, forming a screen that excludes grains of sand and at the same time permits the passage of water and the microscopic food that it contains.

Many years ago, when the Pismo clam was as common on exposed beaches in southern California as are sand dollars on bay and estuary beaches, teams of horses drew plows through the sand, turning up the

171. *Tivela stultorum* (§190), the Pismo clam.

clams by the wagonload. Now, adults of the species are almost unobtainable in the intertidal zone. Experienced diggers with rakes or forks work at low tide, wading out waist deep or even shoulder deep, where the surf frequently breaks over them. It is hazardous work, for the diggers must feel their way along bars that are separated from the shore by deeper channels, and now and then the surf claims a victim.

Because of this clam's commercial importance, a good deal is known of its natural history, particularly through the work of Weymouth and Herrington. During stormy weather the clams go down to considerable depths to avoid being washed out, but normally they lie at a depth about equal to the length of their shells. They show a strikingly constant orientation in the direction of wave action, always being found with the hinged side toward the ocean. Though they are not as active diggers as the razor clams, they can, especially when young, move rather rapidly; and, given a reasonable chance, they can dig themselves back in from the surface. Weymouth observed that "the ordinary action of the foot in burrowing appears to be supplemented in the Pismo clam by the ejection of the water within the mantle cavity . . . recalling the method of 'jetting' a pile, . . . an important factor in moving such a bulky shell through the sand."

The Pismo clam is a very slow-growing species, requiring from 4 to 5 years to reach its present legal size of 4½ inches, and continuing to grow, but not so rapidly, until it is at least 15 years old. Many legal-sized clams are at least 24 years old; the oldest clam reported was estimated to be 53 years of age! The rate of growth varies according to a definite seasonal rhythm—rapid in summer and slow in winter. The winter growth of the shell is darker and is deposited in narrow bands. The animal is hermaphroditic; after it becomes sexually mature at the end of its second or third year, it spawns in late summer and produces eggs in quantities directly proportionate to its size, a large clam producing an estimated 75 million in one season.

The Fish and Game Commission takes an annual census of the animals, based on test counts in strips of beach running from the upper limit of the intertidal zone out to the waterline at extreme low water. The results indicate that despite the present restrictions (10 clams, no smaller than 4½ inches, per person per day; all undersized clams must be immediately returned to the hole from which they are dug) the species is in danger of becoming seriously reduced on the beaches. So far, recruitment from subtidal populations seems to keep the shore population at a level where limits may be safely dug.

With the exception of man, the Pismo clam's natural enemies are few. Seastars do not frequent sandy beaches, and boring snails are unable to penetrate the hard, thick shells of adults. Gulls will devour small specimens that are turned up by diggers and left lying on the surface, but they cannot get at the clams in their normal buried habitat. The nearest relatives of this clam are tropical forms, and it can be considered subtropical. It ranges from Half Moon Bay to Socorro Island, off Mexico, but is not common in the northern part of its range. In Monterey Bay it is well known, but it is still not abundant by comparison with its occurrence on the southern beaches. The most famous region for Pismos at present is south of San Luis Obispo, where several towns are chiefly supported by tourists lured by the possibility of capturing this large delicacy. There is reason to believe that the coast of Baja California may supply large quantities of Pismos when our own beaches are depleted. It is to be hoped that the Mexican Government, which has shown itself to be wide awake in the matter of conserving wildlife, will enact and properly enforce legislation that will help *Tivela* to hold its own. Like the bean clam, *Tivela* will be found occasionally harboring a cluster, several inches long, of the hydroid *Clytia bakeri* (§188), which, protruding above the surface, reveals the hidden presence of the clam.

§191. *Euzonus mucronata,* perhaps more abundant on protected beaches (§155), is not a stranger to some exposed beaches. On the exposed beach at Long Bay on the west coast of Vancouver Island the Berkeleys reported it "in vast numbers . . . , whole stretches of sand being tunnelled by countless millions. Judging by the complex system of furrows on the sand beds they inhabit, they seem to emerge from their burrows and crawl on the surface of the sand, but none were found exposed. Large flocks of sandpipers are frequently seen at low tide, extracting these worms and feeding on them."

In certain exposed sandy beaches at and below the lowest-tide level, bait-gatherers frequently dig out the polychaete worm *Nainereis dendritica,* which may be had sometimes by the shovelful. Except for its smaller size, this so nearly resembles *Nereis vexillosa* (§163) that bait collectors confuse them, swearing up and down that it is the height of folly to dig under mussels when the desired worms are so available in easily dug sand.

§192. A good many popular food fish, notably the striped bass, are found along sandy shores just outside, or even within, the line of breakers. The live-bearing surf perch occurs similarly, but it is usually too small for food. Although these can scarcely be considered intertidal forms, one interesting southern fish (which has been reported in Monterey Bay also) actually comes high into the intertidal zone for egg deposition. This is the famous grunion, *Leuresthes tenuis,* a smeltlike fish about 6 inches long (Fig. 172).

The egg-laying time of the grunion is holiday time for tremendous numbers of southern Californians. Along the coast highways cars are parked bumper to bumper for many miles, and the moon and thousands of beach fires light up the scene. The fish are caught with anything available, from hats to bare hands, and are roasted over the fires, making fine fare indeed. It is illegal to use nets, however.

The grunion's extraordinary spawning habits are as perfectly timed as those of the palolo worm of the South Seas, and the timing force is as mysterious. On the second, third, and fourth nights after the full moon—in other words on the highest spring tides—in the months of March, April, May, and June, and just after the tide has turned, the fish swim up the beach with the breaking waves to the highest point they can reach. They come in pairs, male and female. The female digs into the sand, tail foremost, and deposits her eggs some 3 inches below the sur-

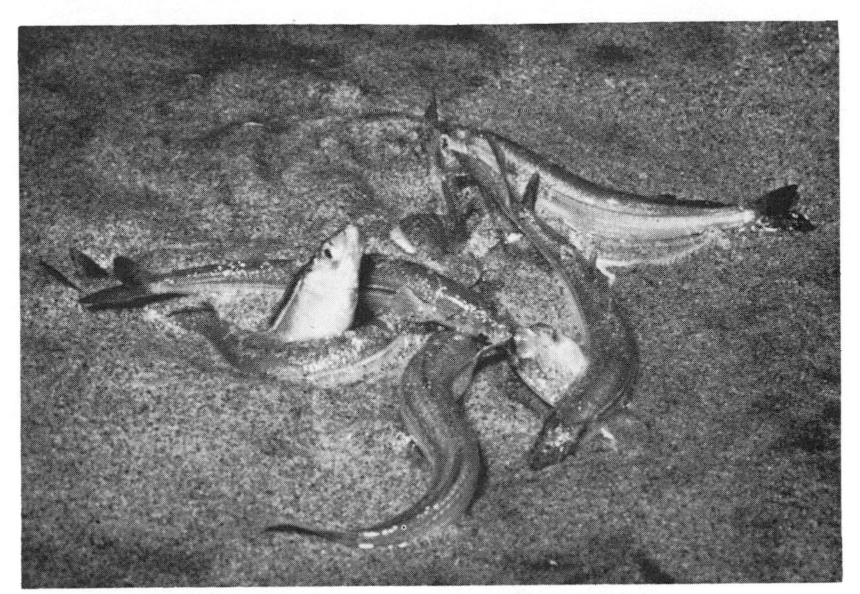

172. The grunion, *Leuresthes tenuis* (§192), spawning at La Jolla; the female is half-buried in the sand, depositing the eggs.

face. During the brief process the male lies arched around her and fertilizes the eggs. With the wash of the next wave the fish slip back into the sea. Normally the eggs remain there, high and dry, until the next high spring tides, some 10 days later, come to wash them out of the sand. Immediately on being immersed the eggs hatch, and the larvae swim down to the sea.

It is an astounding performance. If the eggs were laid on any other tide, or even an hour earlier on the same tide, they would probably be washed out and so destroyed. If they were laid at the dark of the moon, they would have to wait a month to be hatched, for the full-moon tides are never as high as those of the dark of the moon. But the interval between two sets of spring tides is the proper period for the gestation of the eggs, and the grunion contrives to utilize it. Incidentally, the succeeding lower high tides actually bury the eggs deeper by piling up sand.

The fish mature and spawn at the end of their first year, and they spawn on each set of tides during the season. During the spawning season their growth ceases, to be resumed afterward at a slower rate. Only 25 per cent of the fish spawn the second year, however, 7 per cent the third, and none the fourth.

§193. The storm wrack and flotsam cast up on the sandy beaches is sure to contain the usually incomplete remains of animals from other kingdoms—representatives of floating and drifting life and of bottom life below the range of the tides.

Shells of deep-water scallops, snails, piddocks, and other clams are very commonly washed up. Although perfect specimens of this sort are adequate for the conchologist, to the biologist they are merely evidence that the living animals probably occur offshore.

Some years in early spring vast swarms of the by-the-wind sailors, *Velella velella,* often mistakenly called "Portuguese man-of-war," are blown toward our coast, and great numbers of the little cellophane-like floats, with their erect triangular sails, may be cast ashore in windrows (Fig. 173). Often the fresh specimen is intact enough to place in an aquarium or jar of water to observe its details. The animal, beneath its transparent float, is bluish to purple; contrary to older zoological opinion, it is not a colony of specialized individuals, like the Portuguese man-of-war, but a highly modified individual hydroid polyp that has taken up life on the high seas (Fig. 174). This can be visualized as the upside-down hydranth of a hydroid like *Polymorpha,* which has been developed from a larva that did not descend to the bottom to settle and grow a stalk but has instead settled at the surface and grown a float, so to speak. *Velella* and two or three lesser-known relatives are properly called chondrophores, and the term siphonophore is used for such animals as the Portuguese man-of-war and a host of lovely, delicate colonial creatures of the high seas. *Velella* is one of the few examples of high seas life that a beachcomber may expect to find.

Examination of windrows of cast-up *Velella* will soon reveal something of interest to the casual beachcomber: the sail is situated on a diagonal to the long axis of the animal, and this diagonal is in the direction of northwest to southeast on the specimens cast up on northeast Pacific beaches (we should get used to the circumstance that the northeast Pacific is our side of the ocean; the northwest Pacific is the side of Japan and Kamchatka). Another form of *Velella,* with the sail running from northeast to southwest, occurs on the western side of the Pacific. In the southern hemisphere the distribution of the two forms of *Velella* is reversed. It is suspected, although no one yet has enough data to prove it, that the two kinds of *Velella* may be mixed together out in the middle of the Pacific Ocean, perhaps in about equal numbers; and that they are sorted by the action of the wind, so that the "right-handed" and "left-handed" kinds are concentrated on different sides of the ocean.

173. Above: drifts of the chondrophore *Velella velella* (§193) cast up at Dillon Beach in spring 1962.

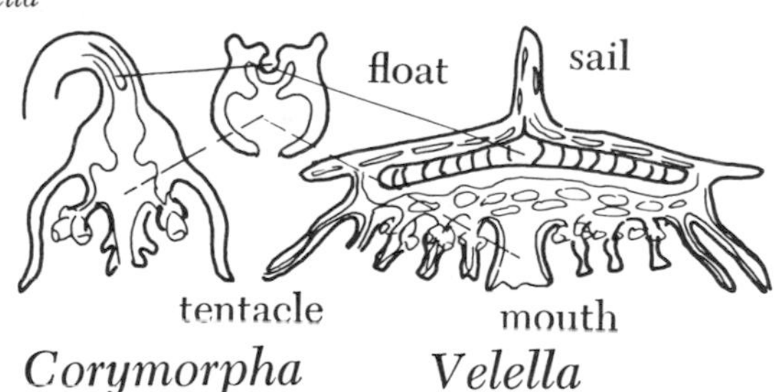

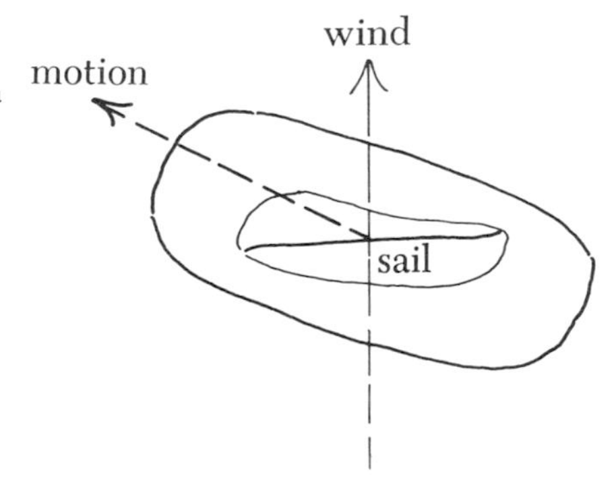

174. Structure and sailing ability of *Velella*: above, *Velella* compared with *Corymorpha*, a hydroid polyp, to show the corresponding structures; below, the relation of *Velella*'s movement to wind direction.

When blown before a moderate wind, *Velella* tacks at about 45° away from the following wind when viewed with its long axis at right angles to the wind direction (Fig. 174). The *Velella* that occurs on our beaches tacks to the left, so that light southerly winds tend to blow it away from the shore. However, when the wind is strong, *Velella* tends to spin around rapidly and follow the wind at a much closer angle. This explains why the animal may appear on our shores after the first strong southerly or westerly winds of the year.

The sailing ability of *Velella* may be demonstrated in a broad shallow pool on the beach. George Mackie, one summer at Dillon Beach, designed an ingeniously simple experiment, using plastic bottle tops for controls (they sailed straight across the pool as good controls should). Computing the angle of arrival of *Velella* on the opposite side, he found that the "best sailors" could tack as much as 63° to the left of the wind. Thus it would seem that the name "by-the-wind sailor" is not strictly accurate; anyhow, *Velella velella* Linnaeus is a more musical name for this stray from the high seas.

After a few days on the beach the animals die and disintegrate, leaving their skeletons to drift across the beach and into the dunes like bits of candy or cigar wrappers. Often the arrival of *Velella* on the shore is the omen that summer is not far behind.

"Gooseberries," the "cat's eyes" of the fishermen, are occasionally cast up on the beach, where succeeding waves roll them around until they are broken. These are comb jellies, or ctenophores, usually *Pleurobrachia bachei* (Fig. 175). The nearly transparent spheres, usually ½ to ¾ inch in diameter, carry two long tentacles. Of these also we have picked up fresh and living specimens so perfect that the iridescent paddles of the plate rows started to vibrate the moment the animals were placed in seawater.

Various kinds of jellyfish, or scyphozoans, are stranded on the beach from time to time. *Aurellia* (§330) is commonly seen, as a flattened transparent blob; if not too broken up, it can be identified by its four horseshoe-shaped gonads, which are yellowish in the female and lavender in the male. A smaller and firmer transparent jellyfish may be *Polyorchis* (§330), sometimes in good enough condition to be revived in a bucket of water. Two large jellyfish that are occasionally stranded in tide pools as well as on the sandy beach are potentially dangerous to anyone who may have allergic reactions. One of these is the dark brown-and-purple striped *Chrysaora melanaster*. The other may appear as a shapeless mass of yellowish or pale orange jelly with a great tangle of tentacles; this is *Cyanea capillata*, the lion's mane. Either of these may cause dangerous

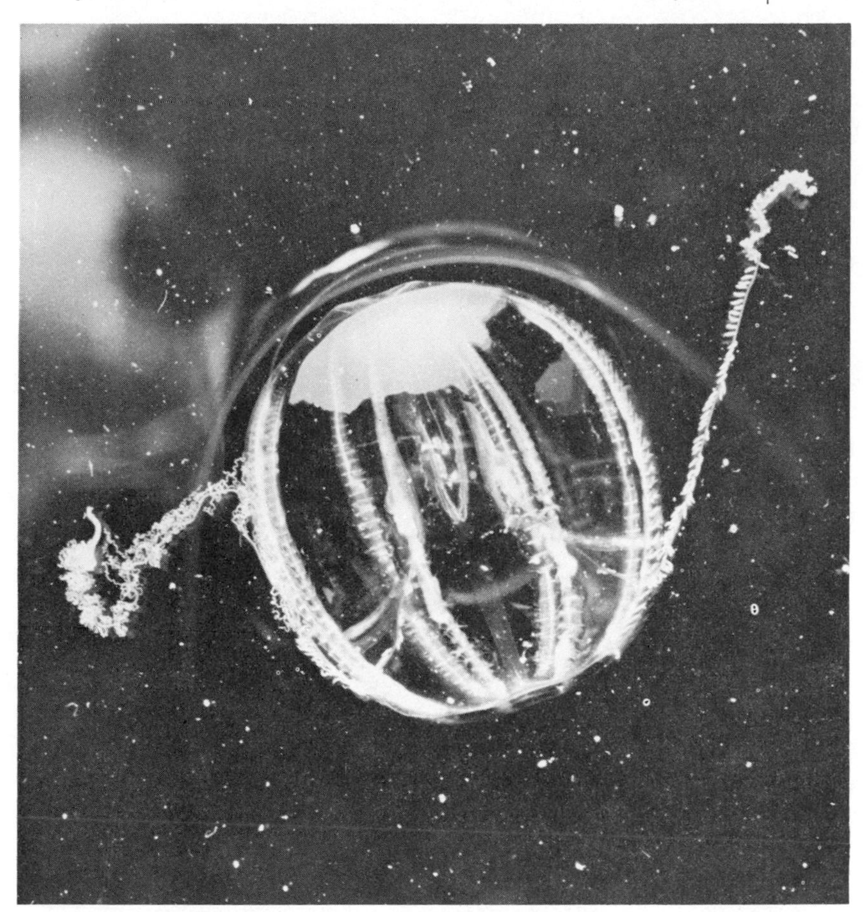

175. The sea gooseberry, or cat's eye, *Pleurobrachia bachei* (§193); this is one of the ctenophores, or comb jellies.

reactions when handled—from a severe rash to acute allergy shock, depending on individual susceptibility and past history of exposure. Sherlock Holmes once investigated, unsuccessfully, a mortality caused by the lion's mane.

One of the charms of strolling on a beach is the possibility that almost anything may turn up, sooner or later—a case of scotch, perhaps, or some relic from the wreckage of a ship offshore. But usually it will turn out to be a block of wood or piece of timber bored by *Teredo* and festooned with a dense growth of a goose barnacle, *Lepas anatifera*, and perhaps with other exotic creatures crawling among the crevices.

PART III. BAY AND ESTUARY

The chief environmental factor that distinguishes the animals of bay shores, sounds, sloughs, and estuaries from all the others so far considered is their complete, or almost complete, protection from surf. This single factor, it is true, will segregate only a few animals, for of the many already mentioned a scant half dozen actually require surf, however many may tolerate it. The absence of surf, however, will alter the habits of many animals, and from several incidental standpoints a coherent treatment of the animals living on completely sheltered shores is desirable. For one thing, the tides in these locations are invariably later—often, as in parts of Puget Sound and the inside waters of British Columbia and southeastern Alaska, many hours later. The result is that the intertidal areas are bared during the heat of the day or the chill of the night, and the shore animals must withstand temperature changes far greater than animals in corresponding positions on the outer coast ever have to meet. The "completely protected" shore animals must also tolerate more variable salinity, which is decreased by freshwater streams and sometimes increased to a considerable degree by evaporation from quiet, shallow areas. Altogether, the animal communities in this environment are characteristic and very different, however much of a potpourri they may be from a geographical and ecological standpoint.

Such previously encountered environments as rocky shores and sandy beaches will be found in these sheltered waters; in addition, there are gravel beaches and mud flats (see color section). Each type of shore has an assemblage of specific forms not found elsewhere, mingled with ani-

mals that occur almost anywhere and a few strays from neighboring environments.

An estuary is a mixing zone between the sea and the fresh waters of the land, and the term "estuarine" connotes conditions between the two extremes. The mixing is effected primarily by the tidal action from the sea, and most of the inhabitants of the estuary, both plant and animal, are derived from the sea. A true estuary, receiving a steady supply of nutrient materials washed down by the rivers and flushed out of the bordering marshlands, is a rich and thriving place—or was, before man came on the scene. The estuaries of the Atlantic seaboard and the Gulf coast once supported teeming populations of fishes, crabs, and oysters in their waters and on the bottom, and beds of clams in the bottom mud and sand; but only a few isolated estuarine areas are still rich in life.

Estuaries are transient phenomena in the geological sense, and the greater abundance of organisms adapted to the changing salinity conditions of eastern and Gulf coast estuaries may be a reflection of the greater period of time that estuaries have had to develop there. On the geologically younger Pacific coast there are no old estuaries, so there are few estuarine fish, and no native estuarine oyster or large crab; the large prawns of warm temperate waters are also missing. These unoccupied ecological niches are easily taken over by exotic species. One of the most astounding examples of fish transplantation is that of the striped bass from the Atlantic seaboard to the estuarine regions of the Sacramento–San Joaquin river system. This was accomplished in 1879 by Livingston Stone, who released 150 striped bass at Martinez in upper San Francisco Bay. Within 3 years the fish had established themselves. Obviously this could not have happened if the environment needed by the striped bass were already occupied by some other enterprising species; but California waters were poor in corresponding types of fish. Quite a few estuarine animals have needed no encouragement (indeed, they have invited themselves in to settle); but others, especially the desirable oysters, seem most reluctant to oblige man and become naturalized in Pacific coast estuaries. This may be due in part to different temperature requirements and in part to different food requirements at early stages.

Recent studies of estuaries indicate that a well-developed, mature estuary may be a closely interrelated system, where changes in the primary microscopic plants of the plankton with different seasons support different species of small herbivorous copepods (perhaps two species of the same genus), successive schools of young fish at the different periods, and so on. In short, every ecological possibility is being utilized.

The inhabitants of estuaries are characteristically euryhaline, that is, they can adapt themselves to changes in the salinity of the water; but most of them can survive, if not thrive, in oceanic water as well. This may account for the wide distribution of some estuarine species, and for the tendency of some of them to appear in scattered parts of the world, including the Pacific coast (see below). Interestingly enough, some organisms from the sea are more capable of adjusting to reduced salinities than freshwater ones are of accepting increased salinities; very few representatives from fresh water occur in salinities higher than 2 or 3 parts per thousand.

There are only a few estuaries on the Pacific coast: San Francisco and Humboldt bays in California, Coos Bay and the smaller bays of Oregon, Willapa Bay in Washington, and the branches of Puget Sound. Most of the bays of the Pacific coast are essentially marine bays, not estuaries (Fig. 176). A few, especially in southern California, may be closed off from the sea and form hypersaline lagoons. However, it is now rare to find a bay left in southern California. The bays are being chopped up into sterile marinas, and the marshlands have become parking lots. From the naturalist's point of view, the transformation of Mission Bay, once an "unimproved" environment of sloughs and marshlands, into a complex of artificial islands, swank restaurants, and speedboat courses is hardly an improvement, Roger Revelle to the contrary notwithstanding. Bays and estuaries, alas, are particularly vulnerable to progress, since so many of them lie near cities and are treated either as sewers or as undesirable wastelands to be filled up and converted to subdivisions of ticky-tacky boxes. It is not much consolation to reflect that in due time the sea may rise again and do away with all this, and that life of some sort should prevail as long as the sun lasts; for we are concerned for our own times, which are brief enough on earth.

8. *Rocky Shores of Bays and Estuaries*

Here we shall find many animals that will be recognized as having occurred outside, plus many that creep up from deeper water on finding no rigorous wave shock. The fauna of this environment is particularly well developed on the reefs and cliffs at the south end of Newport Bay, along the railway embankment in Elkhorn Slough, on the rocky shores of Tomales Bay, in the San Juan Islands and other parts of the Puget Sound region, and in hundreds of places in British Columbia and southeastern Alaska.

Zones 1 and 2. Uppermost Horizon and High Intertidal

Although, as on the outer coast, the highest fixed or lethargic animals are barnacles, limpets, and periwinkles, it follows from the lack of surf that their absolute level (allowing for tidal differences) is lower. That is, since spray and waves will not ordinarily come up to them, they must go down to the water; and instead of ranging several feet above high-tide line they will be found at or a little below it (see Fig. 4).

§194. An extremely common small barnacle of fully protected waters throughout our area is *Balanus glandula,* which has already been discussed (§4) as the dominant barnacle of the protected outer coast. As was pointed out in the previous account, this great variability of habitat bespeaks a tolerance and generalization that are uncommon among marine invertebrates. In the outer-coast environment there are wave shock, low temperature, high salinity, and plenty of oxygen. In bays and estu-

176. Bodega harbor and Bodega Head. Bodega harbor is a small marine bay, without permanent stream drainage; beyond the harbor is Bodega Bay, actually a broad bight. At the top of the picture are Dillon Beach and the entrance to Tomales Bay. The photograph was taken before the nuclear reactor project on the Head was begun.

aries the animals must put up with variable and often high temperatures, variable and often low salinities (because of the influx of fresh water), and relatively low oxygen content in the water, gaining no apparent advantage except escape from wave shock. To make the situation even more puzzling, *B. glandula* actually thrives best, according to Shelford *et al.* (1930), in such regions as Puget Sound, living in enclosed bays where these apparent disadvantages are intensified and avoiding the more "oceanic" conditions of rocky channels that have swift currents and more direct communication with the sea. The term "thrive" translated into figures means that although *B. glandula* is never compressed by crowding in Puget Sound, there may be as many as 70,000 individuals per square meter.

Perhaps the ecological catholicity of *Balanus glandula* will enable it to resist invasion by *Elminius modestus* (the New Zealand barnacle that now appears to be taking over the barnacle lebensraum on many European shores), should *Elminius* arrive on our shores. It should be noted that colonization of open ocean shores by invaders seems to be rare.

Another barnacle, the much larger *Balanus cariosus,* comes very near to reversing *B. glandula's* strange predilections. In Puget Sound this form prefers steep shores with strong currents and considerable wave action, that is, the nearest approach to oceanic conditions that these waters afford. When this same barnacle moves to the protected outer coast, however (and it is fairly common as far south as Monterey Bay), it avoids oceanic conditions most assiduously, occurring only in deep crevices and under overhanging ledges in the low zone, where it finds the maximum protection available. The net result, of course, is that it maintains itself under nearly identical conditions everywhere—a much more logical procedure than that of *B. glandula.*

The young of *B. cariosus* have strong radiating ridges in a starry pattern and are beautifully symmetrical. An adult specimen that has not been distorted by crowding is conical, sometimes has shingle-like thatches of downward-pointing spines, and is up to 2 inches in basal diameter and slightly less in height. The largest specimens are found to the northward of Puget Sound; but under ideal Sound conditions the animals sometimes grow in such profusion and are so closely packed together that "lead-pencil" specimens develop, which reach a height of 4 inches with a diameter of less than ½ inch. Test counts have shown 15,000 of these per square meter. *B. cariosus* occurs, as a rule, at a lower level than *B. glandula,* and where the two are intermingled *B. glandula* will often be found attached to the larger form rather than to the rock.

Still a third barnacle, *Chthamalus dalli,* occurs abundantly from Alaska to Puget Sound, filling in that region the position held by *Chthamalus fissus* in the south and even extending, though not in great numbers, as far south as Monterey. It is the smallest of the lot, being about ¼ inch in basal diameter and half as high. It is a definite and clean-cut form, never crowded and piled together like *B. cariosus,* although one test count in the San Juan Islands showed 72,000 to the square meter. In the Puget Sound region it will be found chiefly interspersed with *B. glandula,* rarely or never competing with the lower *B. cariosus. C. fissus* is very hardy, however: it can tolerate both high salinity and small enclosed bays where there is so much fresh water or decaying vegetation that no other barnacles can survive. Only in the latter environment will it ever be found in pure stands. This small form shows particularly well the six equal plates or compartments that form the shell. The various species of *Balanus* have six also, but they are unequal in size and overlap irregularly, so that it is often difficult to make them out.

Many species of barnacles resemble each other so closely that it is usually hopeless for the beginner to try to identify them surely. However, the collector who is sufficiently interested to dry the specimens and separate the valves can probably determine their species by consulting Pilsbry (1916).

§195. Several familiar limpets occur in these quiet waters. *Acmaea pelta olympica* is foremost; a variety of the shield limpet mentioned in §11, it is by far the commonest limpet in southern Puget Sound and is often the only one. Great colonies of small specimens occur high up with the small barnacles, on smooth rocks, and on the shells of mussels. Farther down—and these animals occur at least to the zero of tide tables—they are solitary and larger. The dingy little *A. digitalis* and/or *A. paradigitalis* (§6) also occur, but only in the more exposed parts of the Sound, and both the plate limpet (§25) and the low-tide dunce-cap limpet (§117) exceed it in abundance, according to the published lists.

A common Puget Sound periwinkle, especially in the quiet waters where the barnacle *B. glandula* and the quiet-water mussel thrive, is *Littorina sitkana* (Fig. 7), usually under ½ inch in diameter and having a gaping aperture. In the less protected parts of the Sound, with the *Balanus cariosus*–California mussel association, the small, cross-stitched snail *L. scutulata* (§10) is the most common form. Farther north a few of the variety of *L. sitkana* formerly called *L. rudis* will be seen. Recent tagging experiments indicate that the littorines are great stay-at-homes, never migrating from the immediate neighborhood of the pool in which they have been born. *Littorina sitkana* is also the larger periwinkle of somewhat protected ocean shores in Oregon.

The purple shore crab *Hemigrapsus nudus* (§27) cavorts around rocky beaches in Puget Sound, and on gravel shores it mixes with its pale and hairy brother *H. oregonensis* (§285). The latter, however, distinctly prefers regions where there is some mud, whereas the former is equally partial to rocks. In Oregon bays and in San Francisco Bay *H. oregonensis* is the common form.

§196. A small hermit crab with granulated hands. *Pagurus granosimanus,* appropriates the shells of the northern periwinkles. These hermits, which we have previously noted as occurring with the dominant *P. samuelis* (§12) in Monterey Bay, differ from their very similar relatives in a few slight anatomical characters. Their external characteristics, however, will identify them in a reasonable percentage of cases. In the first place, *P. granosimanus* lacks the more southern form's slight hairiness; and in the second, it is typically the smallest intertidal hermit to be found in Puget Sound. Beyond that lies color—and confusion. Not only is there likely to be great color variation in a large number of the animals herein listed, but it is apparently impossible to find any two persons who will use the same color terms in describing an animal, even if they are looking at the same specimen. As an experiment, we submitted one specimen to three people and got three sets of color terms with almost no overlapping. One writer describes this particular hermit as "buffy olive to olive" with "porcelain blue" granules on the claws. The harassed amateur collector can do no better than find an undoubted specimen and make his own color chart—a chart that will very likely have to be altered to fit the next specimen he finds.

P. granosimanus ranges from Unalaska to Todos Santos Bay, but it is relatively uncommon below Puget Sound.

§197. The animals mentioned so far differ only in species from comparable forms familiar to the reader. The same thing is true of the mussel *Mytilus edulis,* but it is a predominantly quiet-water animal and notable for its wide distribution—literally around the world in northern temperate regions. The rich beds along the coasts of England and France provide great quantities for food and bait for commercial fisheries, and on the west coast of France the mussels are cultivated very much as oysters are cultivated in other places. When "farmed" they grow larger than in the typically overcrowded natural beds.

M. edulis is wedge-shaped and rarely more than 2 inches long, and its shell is smooth, whereas that of *M. californianus* (§158) is ridged. Sometimes the battle for living space drives it to establish itself on gravel beaches. We found a particularly fine example of this in Hood Canal, a long arm of Puget Sound, where the mussel beds formed a continuous belt along the beach. The byssal threads were so closely intertwined that one could tear the bed away from the gravel in a solid blanket-like mass, exposing the hosts of small animals that lived beneath it.

In Puget Sound and northward to Alaska this mussel occurs in tremendous quantities. It may also be expected in nearly every quiet-water bay southward of Puget Sound as far as Baja California, but never as commonly as in the north. A few anemic-looking specimens can be found on gravel banks and rocky points in San Francisco Bay, but most of those that have survived the Bay's contamination occur on wharf piling.

Mytilus edulis occasionally (or regularly?) settles on rocks exposed to surf near the mouths of bays, and it may form bands just below those of *M. californianus.* This has been observed on the south side of Cascadia Head at the mouth of the Salmon River in Oregon, and at Dillon Beach. These settlements may be seasonal, or at best temporary; at Dillon Beach an incipient colony disappeared after a few months, and *M. californianus* was the only mussel present. *M. edulis* has finer byssal threads than *M. californianus* and may not be able to withstand heavy winter surf.

§**198.** Under the crust formed by mussels on gravelly shores there is an important and characteristic assemblage of animals. There are few species but tremendous numbers of individuals; and the commonest of the larger forms, to the eternal joy of bait gatherers, is a small, quiet-water phase of the segmented worm *Nereis vexillosa* (§163), rarely exceeding 4 or 5 inches.

The little pill bug *Gnorimosphaeroma oregonenis* (Fig. 177), less than ¼ inch long, flaunts its belief in large and frequent families. The average under-crust population of this squat isopod with widely separated eyes will run many dozen to the square foot in Puget Sound, and it is the commonest pill bug in San Francisco Bay.

There are two subspecies or varieties of *Gnorimosphaeroma oregonensis*: *G. o. oregonensis* frequents the more saline parts of bays from Alaska to San Francisco Bay, and *G. o. lutea* is found in dilute estuarine water and in fresh water near the sea from Alaska to the Salinas River in Monterey County. Another subspecies, *G. o. noblei,* occurs in Tomales Bay from Marshall south.

There are several ribbon worms here, notably our acquaintance *Emplectonema gracile,* the pale yellow-green rubber-band nemertean

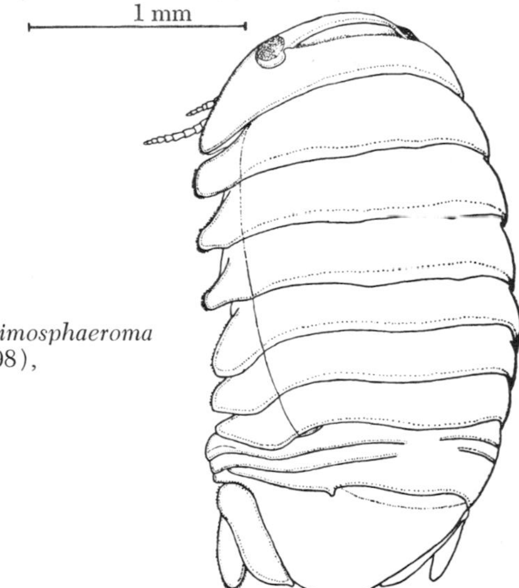

177. The isopod *Gnorimosphaeroma oregonensis lutea* (§198), greatly enlarged.

(§165) from under open-coast mussel clusters, and *Paranemertes peregrina* (§139), the purple or brown form that occurs under rocks. Small individuals of the northern blenny (*Anoplarchus purpurescens,* §240) will also be seen.

The infrequent bare spaces on the sides of mussel-covered stones, especially in crevices, may be occupied by an encrusting sponge, *Haliclona rufescens* (similar to the outer-coast form treated in §31), described originally as a *Reneira.*

Zone 3. Middle Intertidal

§199. The quiet-water form of the seastar *Pisaster ochraceus* (§157), *P. o. confertus,* haunts the mussel beds in completely sheltered waters as does its rough-water relative the beds of larger mussels on the open coast. *P. o. confertus* may be found in the lower zone as well as in the middle zone, but we have never seen it below the low-tide line—a fact entirely consistent with its unfriendly relationship with mussels. On certain quiet-water gravelly and rocky shores in Puget Sound and British Columbia, to which region the animal is restricted, most of the specimens are colored a vivid violet.

§200. The common quiet-water snails are the purples—various species of *Thais,* especially *T. canaliculata* (Fig. 178), the channeled purple, and *T. lamellosa* (Fig. 178), the highly variable wrinkled purple. Both range from Alaska to Monterey Bay and are common from Puget Sound to San Francisco Bay. They also occur on sheltered outer-coast rocky shores and on jetties.

T. lamellosa is famous for its great variation in shell structure. Specimens living on rough rock surfaces where there are currents or where there is some wave action have their projecting plates coarse, rugged, and relatively abbreviated, whereas specimens from rocks in quiet backwaters develop exquisite ornamentations. Animals living in swift currents would be at a disadvantage if they possessed projecting flat surfaces; thus utilitarianism appears to be, in this case, the determining factor in limiting what is presumably the animal's natural tendency toward ornamentation. Professor Trevor Kincaid has published (on the press in his garage) an interesting book of illustrations showing the variations of *T. lamellosa* in Puget Sound.

Locally, *Searlesia dira,* the elongate "dire whelk," may be abundant, especially to the north, where it becomes the commonest littoral snail in the quiet pools and channels of northern British Columbia. A murex,

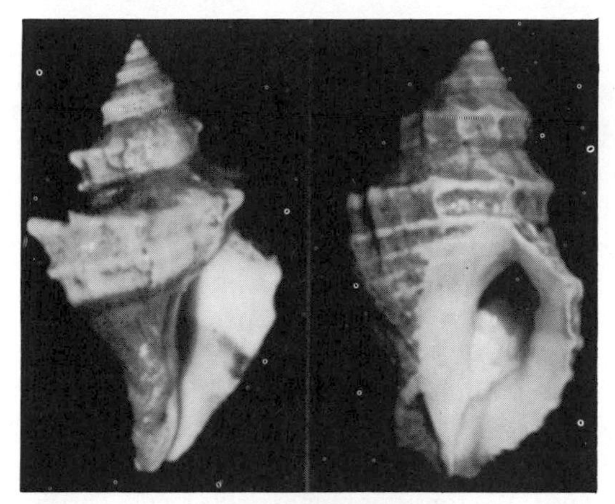

178. Snails of the genus *Thais*. Left: two shells of *T. lamellosa* (§200), the wrinkled purple snail; the light and fragile shell on the extreme left was taken from very quiet waters, and the heavier shell next to it came from a relatively exposed position. Right: the channeled purple snail, *T. canaliculata* (§200).

Ceratostoma foliatum of §117, also occurs in these northern waters—surprisingly for a relative of these tropical shells—as one of the three very common snails in certain locally rich pockets, like Fisherman's Cove, south of Prince Rupert.

§201. The shells of *Thais* provide portable homes for the hairy hermit, *Pagurus hirsutiusculus*, which, in quiet northern waters, attains great size. At Friday Harbor, specimens have been taken with 2-inch bodies and an overall length of more than 3 inches. The animal's extreme range is Alaska to Baja California, but south of Puget Sound the specimens are smaller and less hairy. They are fairly plentiful at Elkhorn Slough, however, and are the common hermits of San Francisco Bay.

To one whose acquaintance with hermits has begun with outer-coast forms, those found in Puget Sound and northern inside waters will exhibit a strange trait—a trait that we deduce to be a direct result of the great difference in environment. The outer-coast hermits will never desert their shells except when changing to another or when dying. Usually they cannot be removed by force unless they are caught unawares (a difficult thing to do) or are literally pulled apart. In Puget Sound, however, the animals will abandon their shells very readily—so readily, indeed, that on more than one occasion a collector has found himself holding an empty shell, the hermit having deserted in midair. Our unproved inference is that the presence or absence of surf is the factor determining how stubbornly a soft-bodied hermit will cling to its protecting shell.

§202. The ubiquitous warty anemone *Anthopleura elegantissima* (§24) is found in the quiet waters of Puget Sound as an attractive red and green form. This anemone, with its many variants, is certainly an efficient animal, for it is universally characteristic of the northern Pacific. A recent visitor from Japan, Professor Uchida, reports it as common there, but he considers the center of distribution to be on the American coast, since it is more variable and more highly developed here.

On erect rocks of the inner waters at Sitka an anemone that may possibly be *Charisea saxicola* is very abundant. It grows fairly high up, colonizing rock crevices and vertical and sharply sloping surfaces; the average specimen is longer (to 2 inches) than it is thick, and is generally dingy white to buff in color. We found it nowhere else, and Torrey (1902) records it only from "the shore rocks at Sitka," where it was taken abundantly by the Harriman expedition.

§203. Everywhere on the Puget Sound rocky shores the purple shore crab, *Hemigrapsus nudus* (§27), replaces *Pachygrapsus,* the rock crab of the California coast. Conditions in Puget Sound would lead one to expect *Pachygrapsus* there also; but for some reason, possibly the somewhat lower winter temperatures, it does not extend that far north. The purple crab modifies its habitat somewhat with the change in regions, occurring on the California coast chiefly in the middle zone, below *Pachygrapsus,* but reaching considerably higher in Puget Sound. Northern specimens at their best, however, are smaller than those on the California coast, and the highest specimens are very small. Hart (1935) records females in "berry" (ovigerous) in April and May at Departure Bay, and has studied the development, finding five zoeal stages and one megalops.

Numerous other species already treated are frequent in Puget Sound, such as the lined chiton (§36) and the keyhole limpet *Diodora* (§181). Small and undetermined polyclad worms similar to *Notoplana* (§17) are very common. The estuary fauna in the south derives largely from open-coast species. In Newport Bay great twisted masses of the tubed snail *Aletes* (§47) may almost cover the rocks, and the rock oyster *Chama* (§126) grows so thickly that it appears to form reefs.

§204. Along the shores of Hood Canal and generally in the Puget Sound region we have found great hordes of the small crustacean *Epinebalia pugettensis,* mostly in pockets of silt among the rocks and in the organic mud under half-decayed masses of seaweed (*Fucus* and *Enteromorpha*). It also occurs in abundance among the beds of green algae and eelgrass in upper Tomales Bay, and has been recognized from Pacific Grove. This or a closely related species occurs in similar sheltered situations in Morro Bay, Los Angeles Harbor, San Diego, and southward to San Quintín. Though it thrives on organic detritus, it seems to require situations where the tidal action is enough to prevent completely foul conditions. Cannon, who studied the feeding habits of the North Atlantic *Nebalia bipes,* a closely related species (Fig. 179), describes the manner in which this animal feeds. Like many other small "filter feeding" arthropods, the nebaliid sets up a food current by oscillating its abdominal appendages. The food is then strained by the bristles, or setae, on the anterior appendages. *Nebalia* can also feed on larger particles.

These curious little crustaceans are of particular interest to students of classification because they are representatives of the subclass Leptostraca, which possesses characters transitional between the smaller, less specialized crustacea (the Entomostraca of the older textbooks) and the more specialized crustacea (the Malacostraca). They have a small bivalve shell, somewhat like that of the ostracods, but their loss of the nauplius larva characteristic of the entomostracans and their arrangement of appendages and sex openings have earned them a place as the lowest division of the Malacostraca.

§205. Under the rocks there are the usual pill bugs and worms, occurring in such variety and abundance as to distract the specialist. As in other environments, the algae provide homes for numerous isopods, the predominant species being *Idothea resecata* (Fig. 180) (also common on eelgrass) and *Idothea urotoma.* The first is a yellow-brown form, just the color of the kelp to which it may often be found clinging with great

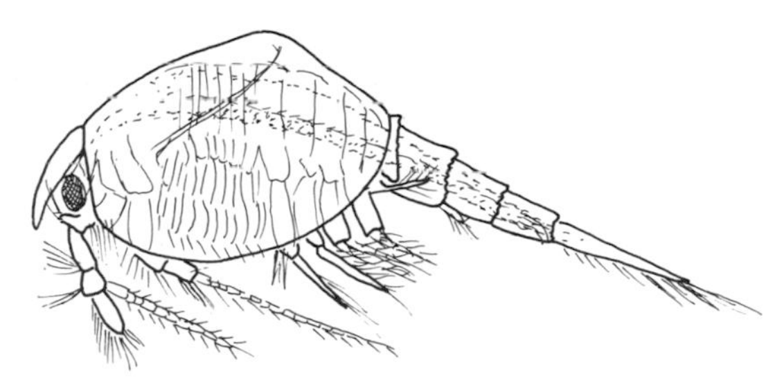

179. Above: *Nebalia bipes* of the Atlantic, which resembles *Epinebalia pugettensis* (§204), the Pacific coast leptostracan.

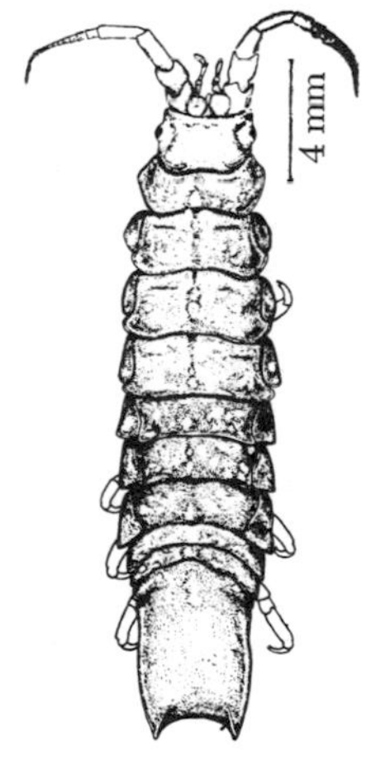

180. *Idothea resecata* (§205).

tenacity. Forcibly removed from its kelp, it seems lost until it can regain a similar position. *I. resecata*'s body is long (nearly 2 inches) and narrow, and terminates in two jawlike points. We found a particularly rich culture near Port Townsend, and it occurs from British Columbia to southern California. The second species mentioned is usually less than an inch long and is relatively wider. It has about the same range.

Another abundant isopod of this genus is *Idothea montereyensis,* found from Puget Sound to Estero Bay, San Luis Obispo County. This isopod really seems to prefer the surf zone where *Phyllospadix* flourishes, but the young that are hatched there cannot hang on, and they move inshore. As they settle on the red algae, their color changes from green to red. After they grow older in this more sheltered situation, they move out toward the *Phyllospadix,* where they can now hang on without being washed away. In the process their color changes from red to green. There are transitional brown stages, and an occasional piebald one. Thus it appears that we have a neat adaptation to circumstances by the populations of old and young animals; Welton Lee, who studied this animal at Pacific Grove and Dillon Beach, refers to the process as "remarkably precise and beautiful."

§206. Occasional geyser-like spurtings that will be noticed wherever a bit of substratum has no covering rock can usually be traced to clams, often to the chalky-shelled *Protothaca* (formerly *Paphia*) *staminea,* the rock cockle (see color section). This well-rounded clam (called a cockle because its strong radiating ridges suggest the true cockle) is also known popularly as the littleneck clam, the hardshell clam, the Tomales Bay cockle, and the rock clam. As the number of popular names would suggest, it is well known and is widely used for food. In suitable localities it occurs so thickly that the valves often touch, and on the level gravel beaches of Hood Canal and such southern Puget Sound places as Whollochet Bay we have found it in such superabundance that two or three shovelfuls of substratum would contain enough clams to provide a meal for several hungry people.

The rock cockle is a poor digger, and for that reason it never lives in shifting sand, where rapid digging is essential. It will be found in packed mud or in gravel mixed with sand but seems to prefer clayey gravel, where it usually lies less than 8 inches below the surface. Where these favorable conditions exist it grows almost equally well in isolated bits of the protected outer coast. At several points, especially Tomales Bay, the clam assumes some commercial importance, and at many other places it is an attraction to tourists. Open-coast specimens should be used warily during the summer, on suspicion of mussel poisoning (§158).

Typical specimens seldom exceed 3 inches in length, although larger ones may be taken from below the low-tide level.

In the vicinity of Sydney, British Columbia, according to work done by Fraser and Smith, spawning takes place in February and March, with no evidence of summer spawning. The rate of growth, being determined by the extent and constancy of the food supply, seems to depend on the animal's position in relation to tidal currents and on its degree of protection from storms, rather than on the character of the beach. Growth is relatively slow: at the end of the second year, when about half of the animals spawn, the average length is only 25 mm.; when the balance spawn, a year later, it is only 35 mm.

The Japanese littleneck, *Protothaca semidecussata*, is now the abundant small cockle in many parts of Puget Sound, and is common in Tomales Bay. Perhaps it is crowding the native species out.

Psephidia lordi is a small clam of some importance in Puget Sound and to the northward, having a total range of Unalaska to Baja California. It has a polished and gleaming white or light olive shell that is somewhat triangular in shape, flatter than that of the rock cockle, and usually only a fraction of its size.

Zone 4. Low Intertidal

Here, as on the protected outer coast, the low intertidal region is the most prolific. Hence it seems desirable to divide the animals further, according to their most characteristic habitats.

Rock and Rockweed Habitat

In the Puget Sound area the most obvious and abundant animals of the low-tide rocky channels are a large seastar, a green urchin, a tubed serpulid worm of previous acquaintance, a great snail, and a vividly red tunicate.

§207. The seastar *Evasterias troschelii* can, it is suspected, be considered a mainly subtidal animal, but in quiet waters that are too stagnant for *Pisaster* it is very much a feature of the low intertidal zone and therefore deserves consideration here even though its proper habitat is below our scope. Elsewhere it takes up the seastar's ecological niche where the common *Pisaster* drops it. Although the average *Evasterias* has a smaller disk, more tapering arms, and a slimmer and more symmetrical appearance than the common seastar, the two are much alike, and they are sure to be confused by the amateur collector. Since both occur so frequently in the same region, the interested observer will want to differentiate them. This can be done readily with a pocket lens, or even with a keen unaided eye. On the underside of the rays, and among the spines just bordering the groove through which the tube feet are protruded, *Evasterias* has clusters of pedicellariae (§157). These are lacking in the common *Pisaster*.

At Departure Bay, and elsewhere on the inside coast of Vancouver Island, this seastar occurs in great numbers; but although it is recorded from Unalaska to Carmel Bay, the shore collector will seldom see it south of Puget Sound. The late Dr. Walter K. Fisher distinguished several forms, but considered the animal to be one of the most variable of seastars. Specimens with a diameter of 2 feet or more are not rare. Sexual maturity comes in June, July, and August.

Evasterias apparently has somewhat different table manners than its relatives. Aage Møller Christiansen found that it prefers force to stealth when feeding on a bivalve, because digestion is slowed down unless all lobes of the stomach can encompass the meal; therefore, the clam must be opened as wide as possible.

A good many *Evasterias* have polynoid worms living commensally in the grooves under the rays. Johnson noted more than 30 years ago that several of these, *Arctonoë* (formerly *Halosydna*) *fragilis*, might be found in a single specimen.

§208. The very fragile seastar *Orthasterias koehleri*, also chiefly subtidal and occurring probably less than 1 per cent as frequently as *Evasterias*, may occasionally be found in the lowest intertidal zone at Departure Bay and Friday Harbor. It is a striking and brilliant animal. The noticeable spines, often surrounded by wreaths of pedicellariae, are white or light purple and are set against a background of red marked with yellow. Strangely, this apparently quiet-water animal is fairly common in the subtidal zone in Monterey Bay, straying, on rare occasions, just above the lowest of low tides. Even more strangely, it has been found there in surf-swept regions, but probably creeps up into the influence of waves only when the surf is light. Occasionally it is found on the seaward side of Tomales Point at very low tide. Dr. Fisher noted the range as Yakutat Bay, Alaska, to Santa Rosa Island, off southern California, and remarked: "Northern specimens shed their rays readily and are difficult to preserve well. The Monterey form is not at all difficult to preserve and does not detach its rays readily. I have handled both sorts, alive, and have kept the Monterey form in aquaria where it is fairly active and eats a variety of food including dead squid, crabs, and fish."

§209. In Puget Sound *Pisaster brevispinus* (Fig. 181), a pink-skinned, short-spined edition of the common seastar, attains gigantic proportions.

1. Left and right: upper and lower surfaces of *Pisaster brevispinus* (§209) and the common seastar *ochraceus* (§157). *P. brevispinus* is the smaller, paler specimen.

. The small northern seastar *tasterias hexactis* (§210), slightly ırged.

183. The leather star, *Dermasterias imbricata* (§211).

We have seen many specimens, notably in Hood Canal, that were more than 2 feet in diameter and massive in proportion. They occurred there on soft bottoms below rocky or gravelly foreshores, usually just below the line of low spring tides. *P. brevispinus,* normally a deep-water form (drag-boat hauls from 60-fathom mud bottom in Monterey Bay bring them up by the thousand), ranges into the intertidal on sand, mud, or wharf pilings only in the most quiet waters. Its softness and collapsibility mark it as an animal that is obviously not built for long exposure to the air. This pink form ranges as far south as Monterey Bay and at least as far north as British Columbia.

§210. Species of the small six-rayed seastar *Leptasterias* occur in these quiet waters, the common form in Puget Sound being *L. hexactis* (Fig. 182), an Alaskan species that ranges as far south as Cape Flattery. *L. hexactis* is rarely more than 3 inches in diameter, usually less. As with the more southern species (§68), the females incubate their eggs, carrying the young until they are fully formed miniature seastars. In Alaska, ovigerous females have been taken in April, and in Puget Sound the breeding season is known to be finished before summer.

§211. A number of other seastars may be seen. The slim-armed red seastar *Henricia leviuscula* (Fig. 55) is common in these quiet waters of the north as well as on the protected outer coast. We have taken one specimen that harbored a commensal scale worm, *Arctonoë vittata,* much as *Evasterias* entertains its similar guest.

Dermasterias imbricata, the leather star (Fig. 183), is a "web-footed" form resembling *Patiria.* This is another of the very beautiful animals for which the tide pools are justly famous. The smooth covering of skin is commonly delicate purple with red markings, and the tips of the rays are often turned up. Ranging from Sitka to Monterey, the leather star, nowhere very common, is most numerous in the northern half of its range, and the specimens found in Puget Sound are gigantic (diameter up to 10 inches) by comparison with those in the tide pools of the protected outer coast.

The sunflower star, *Pycnopodia* (§66), also occurs here, as well as the sun star, *Solaster* (§67).

§212. The common urchin in the Sound and northward into southeastern Alaska is the mildly green *Strongylocentrotus dröbachiensis,* of circumpolar range. Except for its color, its pointed spines, and its usually smaller size it resembles the southern purple urchin (Fig. 158). Many specimens occur in the rocky San Juan Islands, and they appear to be crowding out the purple urchins that were once common there. The latter still retain unquestioned dominance on surf-swept shores along this coast, although on the east coast and in Europe the green urchin populates the open shore as well as protected regions. This cosmopolitan green urchin, in common with others, passes through several free-swimming stages not even remotely resembling the adult. It is an omnivorous feeder but lives mainly on fixed algae. Juvenile specimens of *S. purpuratus* are often greenish white, and the beginner may confuse these with *S. dröbachiensis.*

The giant red urchin (§73), so common along the California coast, occurs also in northern inside waters but is found mainly below the low-tide line. It does range into the intertidal, however, at least as far north as Juneau.

§213. Great twisted masses of the limy-tubed worm *Serpula vermicularis* (Fig. 184), formerly called *S. columbiana,* cover the rocky reefs in the Puget Sound–Strait of Georgia region. Except that these worms have red "stoppers" to their tubes and exhibit vividly red gills that snap back into shelter on slight provocation, they occur just as do the tubed snails on the rocks in Newport Bay. Even as far south as Elkhorn Slough near Monterey this serpulid may be found on the sides of rocks in quiet circumstances.

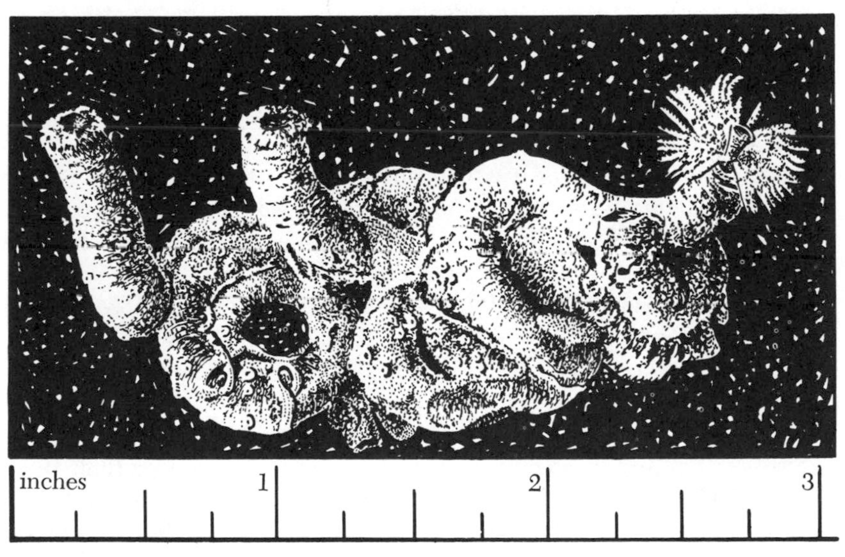

184. The limy-tubed worm *Serpula vermicularis* (§213), one specimen with gills exposed. Many of these clusters are much larger than the one shown.

Serpula, like other tube worms, is dependent for its food on such microscopic particules of organic matter as chance may bring it. It assists the process, however, by setting up currents with the hairlike cilia on its delicate gill filaments. Puget Sound specimens are sexually mature in July and early August; and if the removed animals there extrude sexual products as readily as do the related Pacific Grove forms (§129), they should provide very accessible embryological material.

§214. The hairy Oregon triton, *Fusitriton oregonensis* (Fig. 185), is the largest of the rocky-intertidal snails in our entire territory, for large specimens reach a length of almost 6 inches. The animal, with its handsome coiled shell, brown and hairy, is a feature of the extreme low-tide rocks and reefs near Friday Harbor and has a recorded range of Alaska to La Jolla. During the summer, egg capsules are found on the same rocks. A murex, *Ceratostoma foliatum* (§117), may be found this far down also, associated with the Oregon triton, but its center of distribution is thought to be higher in the intertidal.

Only the largest intertidal hermit crab on the coast would have any use for so huge a shell as the Oregon triton's, and we find this hermit in *Pagurus beringanus.* It has granules and short scarlet spines on the hands, but no true hair. The general color is brown with scarlet and light green markings. The only other large hermit common on shore in the Puget Sound region and lower British Columbia is the smaller, slimmer, hairy-handed hermit, *P. hirsutiusculus* (§201).

§215. Two other common shelled snails, *Trichotropis cancellata,* the checkered hairy snail, and *Crepidula nummaria,* the northern slipper shell, are illustrated in Figures 186 and 187. The blue top shell *Calliostoma ligatum* (Fig. 164), which ranges from Alaska to San Diego on the protected outer coast, occurs in these quiet waters also and is particularly abundant in Puget Sound, under overhangs and in rocky crevices fairly well up, and from there clear into the subtidal.

§216. The tunicate *Cnemidocarpa finmarkiensis* (Fig. 188) has an extraordinarily bright red hidelike tunic, or test, which makes it one of the most conspicuous animals in the Puget Sound region. The craterlike projections—the excurrent and incurrent siphons that provide the animal with water for respiration and feeding—are rather prominent. In collecting and preserving specimens we have noted that these orifices have considerable power of contraction, closing almost completely under unfavorable conditions. We have found eggs in August.

The stalked *Styela gibbsi,* as found under rocks at Pysht (near the entrance to the Strait of Juan de Fuca) and probably in many other

5. *Fusitriton oregonensis* (§214), the hairy Oregon triton; this is
 largest intertidal rock snail of the Pacific coast.

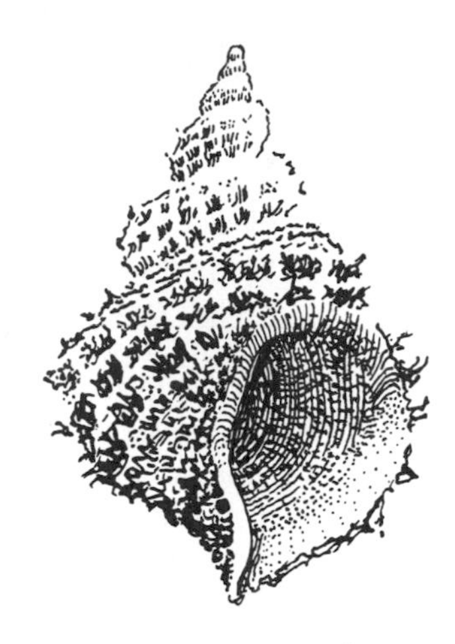

186. The checkered *Trichotropis cancellata* (§215).

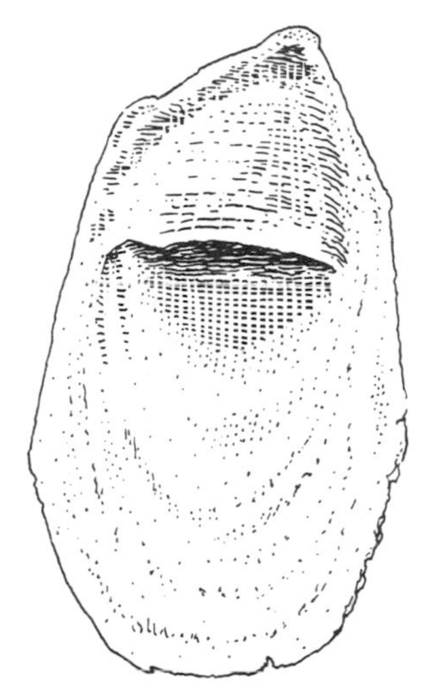

187. The slipper shell *Crepidula nummaria* (§215).

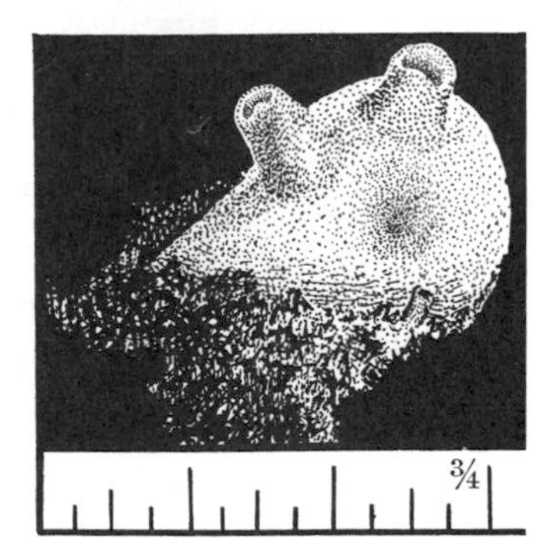

188. The tunicate *Cnemidocarpa finmarkiensis* (§216).

places, is a small and short-stalked edition of the tunicate so well developed on wharf piling. S. *gibbsi* is interesting in that it plays host to several other animals: a commensal pea crab; the parasitic and degenerate cirripede *Peltogaster*, one species of which is sometimes found on hermit crabs; and a vermiform copepod, *Scolecimorpha huntsmani*, which more nearly resembles a degenerate isopod.

Pyura haustor, another large Puget Sound tunicate (often several inches long), has a bumpy test covered with all sorts of foreign material —a veritable decorator crab among tunicates. A subspecies, *P. haustor johnsoni*, is common under and alongside rocks in Newport Bay, and more rarely in Elkhorn Slough.

Several other tunicates occur, some obviously and many commonly, but for their sometimes difficult determination the reader must be referred to the scattered literature.

§217. The rock oyster or jingle shell, *Pododesmus macroschisma* (Fig. 189), occurs fastened to Puget Sound reefs, just as a similar but smaller jingle, *Anomia peruviana*, occurs in the quiet bays of southern California. Curiously, both of these are associated with similar-looking but widely differing tube masses, *Serpula* in Puget Sound and *Aletes* in southern California. *Pododesmus* is one of various animals that are called jingles, presumably because of the sounds made by dead shells on shingly beaches. The valves are thin, and the adherent valve has a notched foramen, or opening, for the byssus that attaches the animal to its support. Large specimens may be more than 3 inches in diameter. The range is from Bering Sea to Baja California. Below Puget Sound, specimens will be found almost exclusively on wharf piling. The flesh of *Pododesmus* is bright orange, and it has an excellent flavor.

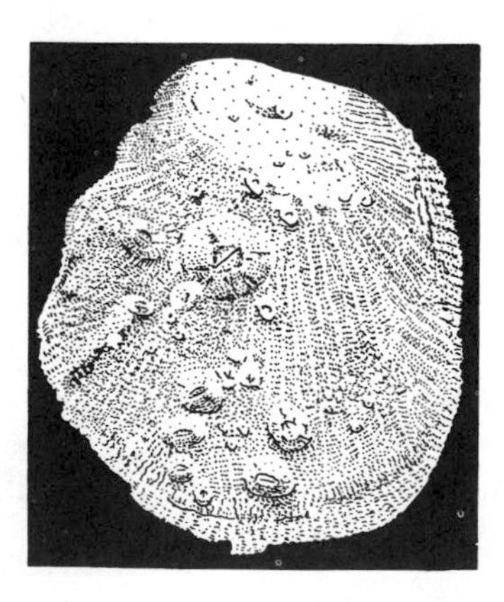

189. *Pododesmus macroschisma* (§217), the rock oyster or jingle shell; about natural size.

Old shells of this and other mollusks, north of Puget Sound and clear into Alaska, are likely to be honeycombed to the point of disintegration by small, yellow-lined pores, evidence of the work of *Cliona celata,* the boring sponge. Break apart one of the shells so eaten—they are fragile and crumbly if the process is well advanced—and note how the interior is composed almost entirely of the yellow, spongy mass. In the Monterey Bay region *Cliona* attacks the shells of abalones, but usually only those growing in deep water, and the sponge is not in any case the important littoral feature that it becomes up north. Oyster shells in southern California estuary regions may be overgrown with a sponge, *Hymeniacidon* (§183), encountered also in wave-swept crevices.

§218. Enclosed in Newport Bay is a stretch of fully protected reef that is the counterpart of the Puget Sound reefs. In this small area at least two Panamanian animals, neither of which commonly occurs as an intertidal animal elsewhere along the California coast, maintain an extreme northerly outpost. The first is a rather beautiful gorgonian, *Muricea californica,* which is related to the "sea fans" and "sea whips" and hangs in graceful, tree-shaped, brown clusters from the vertical cliff face. The minute zooids that extend from the branches when the animal is undisturbed are a pleasantly contrasting white. Clusters may be more than 6 inches in length and as large around as one's fist. This gorgonian is common in the kelp beds.

The second animal is the pale urchin *Lytechinus pictus,* about 1½ inches in diameter, a light gray form that occurs abundantly in the Gulf of California. This Newport Bay region appears to be more plentifully supplied with urchins (in number of species) than any other similar area on the Pacific coast of the U.S. Both the purple urchin and the giant red urchin may be taken with this pale-gray form on the same rocky reef; on the neighboring sand flats the ubiquitous sand dollar occurs, with an occasional heart urchin or sea porcupine (§247).

§219. The small native oyster, with its irregular white shell, is familiar to most coast dwellers from Sitka to Cape San Lucas. *Ostrea lurida* (Fig. 190) occurs characteristically in masses that nearly hide the rock to which the animals are fastened. For many years it was scorned as food because of its small size, and even now this is probably the chief factor in preventing the animal from being used for food as extensively as it deserves. People acquainted with its delectable flavor, however, often pronounce it more delicious than the larger Eastern oyster, and probably its virtues will come to be more generally appreciated. The cultivation of *O. lurida* is a considerable industry in Puget Sound, where it is known as the Olympia oyster. Thirty years ago, Olympia oyster production had a total value of about $750,000 a year; in recent years the value has declined, and in 1965 (admittedly a low year) the total oyster production was 40,600 pounds, with a value of $121,800. In part this decline represents the determination of unreconstructed individualists who persist in cultivating oysters in the face of continually rising costs, in part it may be a reflection of such factors as pulp mill pollution. It may be difficult to prove legally that the pulp mills of Puget Sound have something to do with poor oyster production; but it does seem that after a pulp mill shuts down, the oysters thrive again on nearby beds. This circumstance is called to the attention of the traveler by billboards reading: "Once again . . . Relish Olympia oysters from beds cleared of pulp-mill pollution." Conversely, in a blighted area, the billboard may read: "Pollution laws are not enforced against the B—— Pulp Mill."

In Oyster Bay, near Olympia, the method of cultivation is as follows: Low concrete dikes are built to partition off many acres of mud flats, which are carefully leveled; the object is to retain water when the tide goes out, so that the growing oysters are never exposed. Since oysters would smother in mud, quantities of broken shell are scattered over the surface, and to these bits of shell the "spat," or free-swimming young, attach. Nelson studied the method of attachment of the Eastern oyster and found that a larva moves about over the selected area, testing it out with its foot (it will not settle on a shell that is badly riddled). After choosing an attachment site it circles about, wiping its large and turgid foot over the area and extruding enough cement to affix the left valve. The cement, which hardens in ten minutes, is secreted by a gland similar to the one that produces a mussel's byssal hairs.* After 2 years the spat are transferred to more extensive beds, and after 5 or 6 years they reach maturity and a diameter of about 2 inches; they are then ready to market. It takes from 1,600 to 2,000 "shucked" oysters to make a gallon, solid pack.

The introduced Eastern oyster, *Crassostrea virginica* (Fig. 190), has never established itself on this coast, although considerable quantities are grown in Tomales Bay and Puget Sound from imported spat, usually in subtidal beds (Fig. 191). A third oyster is being grown commercially from imported spat of the Japanese oyster, *C. gigas*. Both of these impor-

* The attachment of the Olympia oyster also has been investigated. Hopkins (1935) found spat more commonly on the undersurface of shells—probably not because of their greater survival there, but because the normal swimming position of the larva, with foot upward, enables it to affix more readily under overhangs.

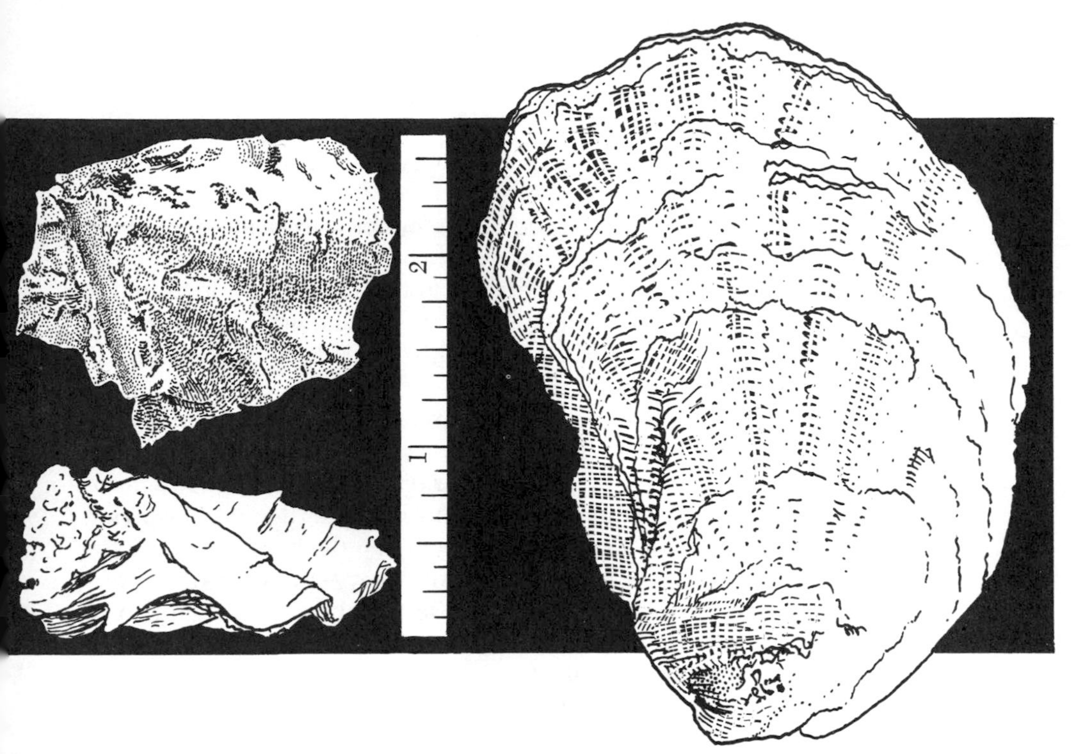

90. Left: the native oyster of the Pacific coast, *Ostrea lurida* (§219), generally known as the Olympia yster. Right: the larger eastern oyster, *Crassostrea virginica* (§219).

91. A cultivated bed of *Crassostrea virginica*, exposed by low tide in Tomales Bay; the stake fence keeps t stingrays and angel sharks.

tations have interesting sidelights in that unwanted settlers came with them.

The Japanese oyster (called the Pacific oyster since some unpleasant recent history) does reproduce naturally to some extent in Puget Sound and British Columbia, and considerable efforts are being made to collect the young oysters (spat) as they settle by suspending shells as cultch from floating racks; but the industry still depends to a considerable extent on spat from Japan.

Oyster culture in Pacific coast waters seems to be at the crossroads: there are opportunities for a growing market that offset the uncertainties of raising bivalves in the presence of various enemies, foreign and domestic (notably, in northern waters, the great hordes of burrowing shrimp, *Upogebia* and *Callianassa,* which dig up the beds and cover the oysters with mud). At least seastars are not the oyster pests in our waters that they are in the east, perhaps because they have never learned to eat oysters or because they do not care for the reduced salinity in some regions. The oysterman's chief enemy is man himself, who builds roads across choice grounds, dumps his wastes into streams and bays, and digs up bottoms for boat basins. Along with these troubles is the one of ever-increasing expense: oyster-growing takes a lot of tedious labor, and labor is becoming more and more expensive, even in Japan.

What is hoped for is a tasty oyster that grows rapidly and without much care. We have tried the eastern oyster, which neither grows nor reproduces well, and the Japanese oyster, which grows well but does not reproduce satisfactorily. For that matter, the standard Japanese or Pacific oyster is about as appetizing to a gourmet as a blob of mucus. Recently a smaller, tastier variety, the Kumamoto, has been introduced. Now we are going to try the European oyster. Perhaps this will be the answer. In the most recent year for which statistics are available (1965), Pacific oyster production in the three Pacific states amounted to 9,114,600 pounds, with a value of $2,085,749; landings are estimated to be worth $3,000,000 for 1967.

The sexual vagaries of the oyster are worth detailed mention. It had been known for some time that the European oyster changed its sex, and it was later discovered that our eastern oyster is also what is known as a protandrous hermaphrodite. The late Wesley R. Coe, working in retirement at La Jolla, showed that *Ostrea lurida* belongs in the same category. It is never completely male or completely female, but goes through alternate male and female phases. The gonads first appear, showing no sexual differentiation, when the animal is about 8 weeks old. At 12 to 16 weeks the gonads show primitive characteristics of both sexes, but thereafter the male aspect develops more rapidly until, at the age of about 5 months, the first spermatozoa are ready to be discharged. Before the clusters of sperm are discharged, however, the gonads have begun to change to female, and immediately after the release of the sperm the first female phase begins—a phase that reaches its climax when the animal is about 6 months old. Coe reported: "Ovulation then occurs, the eggs being retained in the mantle cavity of the parent during fertilization and cleavage and through development until the embryos have become provided with a bivalved, straight-hinged shell—a period of approximately ten to twelve days, perhaps. It is not improbable that ovulation takes place only when the animal is stimulated by the presence in the water of spermatozoa of other individuals." As before, the phases overlap, so that the second male phase has begun while the embryos are still developing within the mantle cavity. The second male phase produces vastly greater numbers of sperm than the first. When they are discharged (except a few clusters that are retained, as always) the body of the oyster is "soft, flabby and translucent, presenting a marked contrast to its plump whitish condition preceding the discharge of the gametes." A period of recuperation follows before the assumption of the next female role. Apparently this cycle of female, male, and recuperative phases is continued throughout the animal's life. A lower temperature will arrest or prolong a particular phase. Coe remarks: "It is not unlikely that in the colder portion of the range of the species a single annual rhythm or even a biennial rhythm may be found to occur, as is the case with the European oyster in some localities." In Puget Sound, sexual maturity comes in early summer, but in southern California, where Coe worked, spawning takes place during at least 7 months of the year.

The period from the fertilization of the egg to the time when the shelled larva leaves the parent is 16 or 17 days. The free-swimming period lasts about 30 to 40 days, after which the larva settles down as previously described.

Recent studies of the sexuality of oysters suggest a complicated interplay of genetic and environmental circumstances. In *Crassostrea virginica,* for example, a small percentage of the individuals in a population are permanently males or females with no capacity to shift to the other sex. The rest of them are potentially hermaphroditic, with varying degrees of expression. Some of these function only as males, others only as females, while about 30 per cent may have equal male and female phases. Usually the male phase is first (protandry); examples of complete hermaphroditism, with both sexes functional simultaneously, are rare. Larviparous oysters such as the Olympia oyster are more regularly alternating male and female. Though the different kinds of sexuality may possibly be determined by genetic factors, the actual expression of sexuality may be a response to environmental circumstances. Somehow the matter is arranged in nature so that there is an adequate release of the gametes of both sexes simultaneously—or nearly so—to ensure continuation of the species.

§220. In the Puget Sound area the scallops, *Pecten hericius* and other species, are the basis of an important industry. Motor and steam trawlers sweep the bottom with nets, scooping up hundreds in a single haul. Although scallops in commercially valuable quantities occur only in deep water, the collector along rocky shores will be certain to see them in the intertidal zone and below the low-tide line. All scallops are characterized by projecting "ears" on either side of the hinge and by prominent radiating ridges on their handsome shells. The species mentioned, which ranges from Bering Sea to San Diego, is essentially similar in outline to *P. circularis* (§255 and Fig. 192) but has finer and more numerous radiating ridges; also, the left valve is a darker pink than the right. *P. hericius*, like most other species, remains symmetrical throughout its life. Scallops have only one large muscle for closing the shell, instead of a large one and a small one, as in the mussels and clams. It is usually this single muscle that is marketed, the rest of the body being thrown away, although the entire animal is edible—delicious, in fact.

Next to the octopus, the scallop is the "cleverest" of all the mollusks. It is quick to take alarm, being warned of impending danger by a row of shining functional eyes along the mantle fringe, and darts away by clapping the two valves of its shell together. The motive power for this peculiar swimming is provided by two jets of water that the sudden closing of the shell drives out through openings in the mantle. Necessarily, then, the motion is hinge-first, or backwards, a reversal of the animal's usual movement, which is slower and in the direction of the shell opening. This emergency reversal of the normal direction of movement is characteristic of several other animals, including the octopus.

Scallops may be stimulated into galloping away in the aquarium by placing a seastar in the tank. Almost immediately they will flap precipitously away from the source of danger. It may not be necessary for the scallop to see or touch the seastar; some essence of seastars will stimulate this avoidance reaction in several kinds of potential prey.

De Laubenfels remarks that very few *Pecten* in the Puget Sound area

192. *Pecten circularis* (§220), the thick scallop.

lack the symbiotic sponge *Ectyodoryx parasitica* (formerly *Myxilla*), or the less frequent *Mycale adherens.* Both of these occur only on scallops, and no other sponges occur thus. The *Mycale* "may be distinguished by comparatively coarser structure, and best by this: when torn, it reveals prominent fibers thicker than thread, absent in *Myxilla.*"

§221. Of chitons, the black *Katharina* (§177) can be expected on relatively exposed shores. The giant, brick-red *Cryptochiton* (§76) occurs also. One or more species of *Mopalia* (§128) will be found. *Mopalia hindsii* is known to occur in such sheltered localities as Elkhorn Slough and San Francisco Bay. The closely related *M. muscosa* ranges from the Shumagin Islands to Baja California, and Puget Sound specimens are sexually mature in July and August.

§222. Not many crabs will be found walking about on top of the rocks in this low zone—a situation quite the opposite of that existing higher up, where the purple shore crabs scamper over every square foot of rock surface. The small Oregon *Cancer* may be taken occasionally on the rocks, but is commoner in the under-rock habitat (§231). A lanky, "all legs" spider crab (*Oregonia gracilis,* §334), which occurs more characteristically on wharf piling, also strays into this rocky-shore region. Another small spider crab, *Pugettia gracilis,* the graceful kelp crab, seems to be characteristic here, but it occurs also in eelgrass and in kelp. It looks like a small and slim *P. producta* (Fig. 85), which also is occasional in this association. Like that larger kelp crab, *P. gracilis* keeps its brown, yellow, or red carapace naked and smooth, not deigning to disguise itself with bits of sponge, bryozoans, or algae, as the more sluggish spider crabs do. These kelp crabs, very active and having hooked and clawed legs that can reach almost backward, are amply able to defend themselves without recourse to such measures as masking. Most crabs can be grasped safely by the middle of the back, but the safe belt in the kelp crabs is narrow, and he who misses it on the first try will pay the penalty. *P. gracilis* ranges southward from Alaska as far as southern California.

§223. The only other noticeable exposed crustaceans are barnacles. Except in Puget Sound and comparable waters to the northward, barnacles are rare in the very low intertidal zone, although they are everywhere tremendously common in the higher zones. One might expect that these low-zone barnacles would be of species different from those previously treated, since the relative periods of exposure and immersion are so different. Until recently this has been assumed to be the case; but it seems to be well established now that the commonest barnacle of this low zone is *Balanus cariosus* (§194), the same form that occurs in such tremendous numbers in the upper zone of quiet-water regions. Field identifications, however, of barnacles are so uncertain that some of these may turn out to be *Balanus nubilis* (§317), the second commonest form, which seems to achieve its greatest development on fairly exposed wharf pilings. On Puget Sound rocks the largest examples reach the considerable size of 3 inches in basal diameter by 2½ inches tall.

§224. Anemones, abundant on all the quiet-water rocky shores from Alaska south, are especially noticeable in Puget Sound. Elsewhere in the Pacific the white-plumed *Metridium senile* (§321) occurs only on piling and in deep water, but in the Sound it is often found on the lowest intertidal rocks and ledges and in the pools, just as it occurs on the coast of Maine. North of Nanaimo, British Columbia, we have seen large numbers of *M. senile* hanging from the underside of strongly overhanging ledges.

The large, red-tuberculated *Tealia* (§65) is occasional, and the small, semitransparent *Epiactis prolifera* (§37), with its eggs and young in brood pits around the base, certainly attains its maximum in size, though probably not in abundance, in these quiet waters.

One of the common Japanese and cosmopolitan anemones, *Haliplanella luciae,* has been introduced on this coast—probably with oysters, for it has appeared at or near places where Japanese oysters are known to have been introduced. We have seen it above the Strait of Georgia, on Puget Sound floats with *Metridium,* and in Tomales Bay and Elkhorn Slough, in central California. It is a small form, usually under ½ inch in diameter, but it is attractively colored—dark green, with 4 to 48 orange stripes. If the form in question is identical with the Atlantic *H. luciae,* a good deal of information is available in Hausman (1919). A circumboreal form, it appeared suddenly at Woods Hole, Massachusetts; and owing to its competitive ability and rapid asexual reproduction it colonized very quickly in areas where its colors and small size conceal it effectively. Hausman found it to be a voracious feeder on small crustacea and annelids.

§225. Reaching its apparent southern limit at Pender Harbor, British Columbia, a large erect sponge occurs (Fig. 193) that is similar in appearance to the *Neoesperiopsis rigida* reported by Lambe from deep water north of Vancouver Island. Specimens from Refuge Cove are up

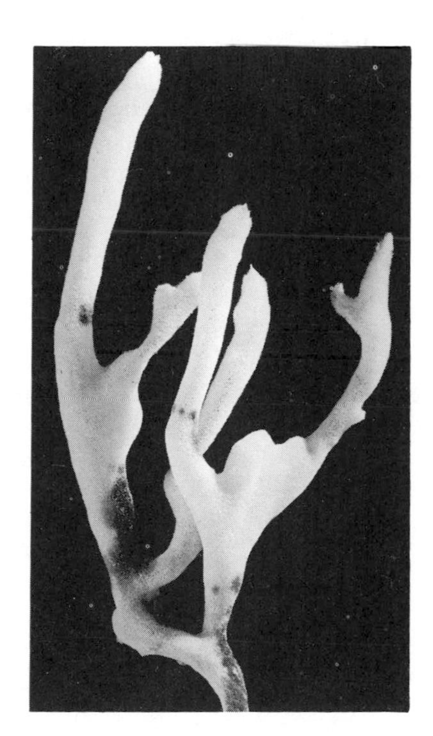

193. An erect sponge similar to *Neoesperiopsis rigida* (§225); the actual height of this specimen was 8 inches.

to 8 inches long. They are brightly colored with reds or lavenders, which fade immediately in preservative. Forests of these upright sponges, growing at or below the lowest tide line, add something of a coral-reef atmosphere to the northern fauna.

Although we have not been successful in finding it along the shore, a most interesting sponge, *Choanites suberea* var. *lata*, the home of the hermit crab *Pagurus dalli*, is recorded from Puget Sound and northward as occurring in the low-tide zone as well as in deep water. Clusters of this yellow "cheesey" or carrot-like sponge may be 3 inches in diameter and are almost invariably pierced with an opening that harbors the large hermit. The mutual advantages are obvious: the crab gets a protective covering, and the sponge gets plenty of food and oxygen by being so much on the move.

De Laubenfels remarks that the slimy, papillated sponge *Hymeniacidon* (§183) occurs abundantly in these quiet waters; we have noted it in connection with oyster shells.

§226. In this low zone of the quiet-water rocky environment, sea cucumbers are common in the south, entirely absent in Elkhorn Slough, and occasional in Puget Sound. Under rocks, however, they are common in the northern regions (§235). One of the strange, sluglike, creeping cucumbers, *Psolus chitonoides* (Fig. 194), is a fairly frequent migrant into the intertidal zone from deep water. If seen below the surface, and unmolested, *Psolus* will be noted and recognized at once. There will be a crown of gorgeously expanded tentacles at the head end, often in rosy reds or purples contrasting with the yellow or white of the scales that cover the back; on the soft belly are the tube feet that provide the not overly rapid locomotion.

§227. *Terebratalia transversa* (Fig. 195) represents a group of animals not previously mentioned, which will be met only once again in this account. The brachiopods, bivalved animals resembling clams but in no way related to them, comprise a line tremendously important in past geological ages but now restricted to a few hundred species, most of which occur in deep water. The valves of clams lie on either side of the body and are connected by a hinge; those of brachiopods (sometimes called "lamp shells") are upper and lower, and are unconnected except by muscles. Brachiopod shells are dissimilar in size and shape. The upper is the larger, and through an opening in its overhanging margin protrudes the fleshy peduncle by which the animal is attached to the rock. Although brachiopods are usually (and correctly) assumed to be sessile animals, we observed that specimens taken in British Columbia

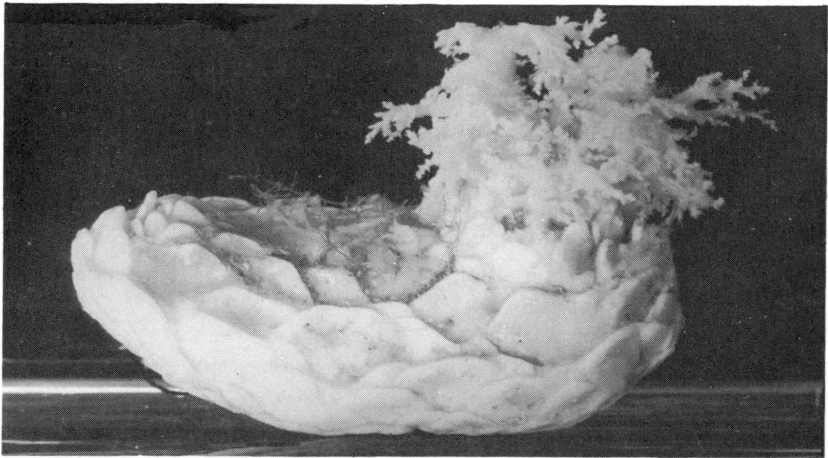

194. The creeping cucumber *Psolus chitinoides* (§226).

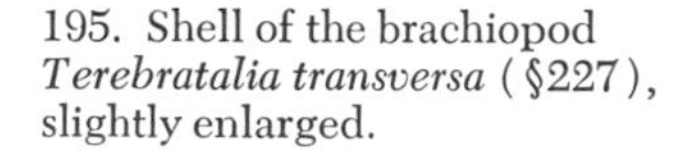

195. Shell of the brachiopod *Terebratalia transversa* (§227), slightly enlarged.

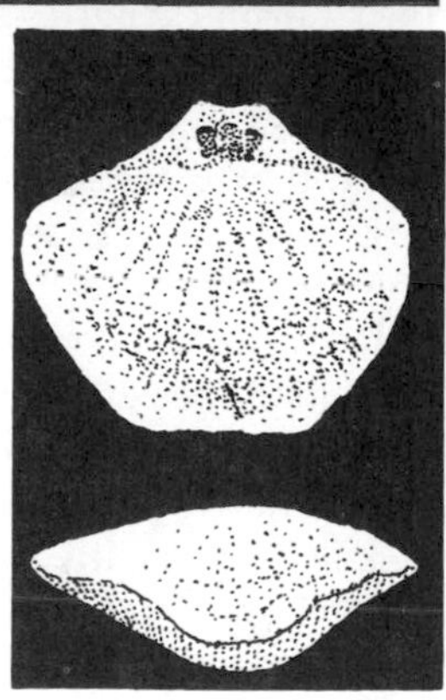

were able to rotate the shell considerably on the contractile peduncle. A cluster of the living animals will show all the shells slightly agape but ready to snap shut immediately at the least sign of danger. They feed on minute organic particles brought to them by the tidal currents or the slight currents created by waving the ciliated tentacles on the lophophore—a "coil spring" organ that transmits food into the mouth and stomach. Thus, like other plankton-feeding forms, they must keep constantly on the job, opening their shells the instant they are unmolested. *Terebratalia* is restricted to the lowest part of the low intertidal zone, where it is a feature of the rocky fauna of British Columbia. It is taken occasionally as far south as Friday Harbor, and in deep water it extends to southern California. In the summer of 1932 we found a great colony of *T. transversa* in a land-locked marine lake, Squirrel Cove, in south-central British Columbia. This unusual environmental condition was reflected in the great size of the individual brachiopods, in their large number, and in the dark color of their shells. Interestingly, the very abundant *Stichopus californicus* (§77) taken in this rocky lagoon were also more blackly purple than usual, higher in the littoral (this species is more common subtidally), and small, climbing all about the tops and sides of the rocks. *Dermasterias* (§211) taken in this assemblage were also darker than usual.

§228. At Elkhorn Slough (and probably elsewhere along the coast, although there are no other records) the observing amateur will sooner or later find a group of ivory-white polyps, each about ½ inch long, attached to a rock on the side sheltered from the current. If disturbed, they immediately contract into stalked translucent lumps less than half of their normal length. These tiny, unspectacular polyps are the parents of highly spectacular pelagic animals—the great jellyfish (the scyphomedusae) that are sometimes stranded on the beach. It will be remembered that there is an alternation of generation between the relatively large and plantlike hydroids and the small, free-swimming jellyfish. The giant jellyfish also have an alternation of generation, but the attached polyp is small and rarely seen. It is this polyp, the scyphistoma of *Aurellia* or similar species, that will be found at or below the lowest tide line at Elkhorn Slough. We have transferred a good many examples to aquaria, where they live well, being extraordinarily tough and resistant, and have watched them with interest. They are accomplished contortionists. Fully relaxed specimens are pendulous, with dainty tentacles extending the equivalent of the length of the polyp; the mouth forms a raised crater in the center of the tentacles. At other times the internal anatomy will, for no apparent reason, be everted through the enormously stretched mouth. When the animal contracts, the tentacles almost disappear, and the scyphistoma becomes merely a ball at the end of a stalk. When producing jellyfish, the polyp is modified into a strobila by constriction, so that it is topped by a "stack of saucers," which, when budded off, become the free-swimming young, the ephyrae, of the giant jellyfish. The jellyfish, normally to be seen only offshore (except in northern inside waters, §330), produce eggs that are liberated as ciliated larvae, the planulae. These larvae, frequently drifting inshore, settle down to become polyps, and the cycle is complete.

§229. Along these sheltered rocky shores there are, finally, many hydroids and bryozoans, and occasional sponges, mostly of types already mentioned. Some of the ostrich-plume hydroids (*Aglaophenia,* §84), but more frequently the "sea firs" (*Abietinaria,* §83) and the delicate, almost invisible *Plumularia* (§78) will be seen. Collectors from the south will recognize a sponge cluster, *Leucosolenia nautilia,* similar to the California species considered in §93.

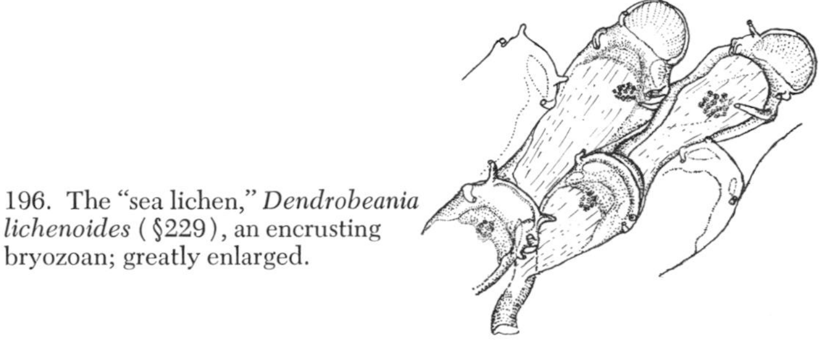

196. The "sea lichen," *Dendrobeania lichenoides* (§229), an encrusting bryozoan; greatly enlarged.

In the Puget Sound area we have found the erect, treelike bryozoan *Bugula pacifica* (§114), with some of its exposed clusters colored a vivid salmon pink; but the more flabby European form *Dendrobeania murrayana,* restricted to this region in the Pacific, may be seen more commonly. The white, crinkly bryozoan *Tricellaria* (§112) occurs in these northern sounds on the roots and holdfasts of marine plants just as it does in less sheltered areas along the California coast.

The "sea lichen" *Dendrobeania lichenoides* (Fig. 196) occurs on shells, rocks, and worm tubes as a leafy encrustation. It ranges from Alaska to central California and is common in Puget Sound. The white *Membranipora* (§110) encrusts on kelp, as do the small and dainty-tubed colonies of *Tubulipora flabellaris.* Colonies of the latter are ¼ to ½ inch across, and the tubes are as hard as those of serpulid worms. This form ranges from the Sound to San Diego. A southern form, known from San Diego Bay only, is the soft *Zoobotryon pellucidum,* which forms great, flexible, treelike masses. The listing of bryozoans might go on indefinitely, for more than 150 species are known from Puget Sound alone. In the Strait of Georgia Dr. O'Donoghue took some 76 species in a single dredge haul, and almost half of this total may be taken inshore. Most of us can scarcely hope to recognize more than a few.

Exceedingly common in the Nanaimo region, but restricted apparently to such areas (unless this is the animal taken rarely on Pacific Grove rocks), will be found great gelatinous masses of a rare and curious worm-like form, *Phoronis vancouverensis,* related to but not resembling the solitary phoronids (§297).

Under-Rock Habitat

It would be expected that bay and estuary regions in the south would have their rocky-shore fauna pretty well concentrated under rocks as protection from sun and desiccation, and that is what we find. In Puget Sound, however, the under-rock habitat is also well populated—a fact probably attributable to the late-afternoon low tides, which bare the inter-tidal areas during the heat of summer days.

§230. The energetic upturner of rocks will find that in sheltered waters the most obvious under-rock animals are crabs, worms, and sea cucumbers, and, in the north, nemerteans. Other previously encountered animals are the eye-shaded shrimp *Betaeus* (§46) and the small and active broken-back shrimp *Spirontocaris* (§§62, 150), which occurs throughout our territory. Another shrimp, *Hippolyte californiensis,* is fairly well restricted to quiet waters and will be found, in the daytime, hiding under rocks and in crevices. At night it may be seen in great congregations swimming about among the blades of eelgrass. This graceful, green (quickly turning white after preservation), and excessively slender shrimp has an absurd rostrum, or forward projection of the carapace. It is known from Bodega Bay to the Gulf of California, but is common only as far south as Elkhorn Slough. The sometimes bulging carapace is a pathological condition due to infection by a parasitic isopod, *Bopyrina striata.*

§231. The Oregon cancer crab *C. oregonensis* (Fig. 197), the roundest of the family, is the commonest *Cancer* of the Puget Sound region and is almost entirely restricted to the under-rock habitat. Its extreme range is from the Aleutians to Baja California, but it is rare south of Washington.

The black-clawed *Lophopanopeus bellus* (similar to *L. heathi,* Fig. 110, but with slimmer big claws) cannot be mistaken, since it is the only member of its genus in the Puget Sound region and is uncommon below there, although ranging south to Monterey Bay. Eggs are carried in April and hatch from May to August, as determined for British Columbia specimens by Hart (1935). Northern specimens of both of these crabs are often parasitized by the obvious *Sacculina* (§327), and at Sitka possibly 25 per cent of the tide-pool specimens will be so afflicted.

197. *Cancer oregonensis* (§231).

A quiet-water anomuran crab, *Petrolisthes eriomerus,* may be taken in Puget Sound and in Newport Bay. It differs only slightly in appearance from the outer-coast form shown in Figure 21. A reddish or brown hairy crab, *Hapalogaster mertensii* (resembling *H. cavicauda* except that it has tufts of bristles instead of fine hairs), also occurs in the under-rock environment to the north, ranging from the Aleutians to Puget Sound.

§232. Tube worms of several types will be found here, some with pebble-encrusted, black tubes up to 10 inches long. Dark-colored, almost purple, specimens may be turgid with eggs, or, if light-colored, with sperm. *Amphitrite robusta* (similar to *Thelepus,* in color section) is one of the largest and commonest. Field directions for distinguishing several genera will be found in §53; but only a specialist, interested also in ecology, could straighten out this situation. The varying environments have undoubtedly exerted a selective influence on the various species. We ourselves have noted differences in tube construction and appearance between the worms predominating in wave-swept, cold-water, California habitats and those in quiet-water sounds but have not been able so far to correlate the data.

198. An unidentified pea crab similar to *Pinnixa tubicola* (§232).

Any of the several scale worms may take to living in the tubes of these larger and more self-reliant worms. Our friend *Halosydna* (§49), of the profitable degeneracy, is frequent; *Hololepidella tuta* (§234) also occurs. Meticulous search will also reveal, in practically every tube, one or more of several pea crabs—*Pinnixa tubicola* (§53, Fig. 198) or *P. franciscana* juveniles. Juveniles of *P. schmitti,* the echiurid commensal (§306), have also been taken, but *P. tubicola* is by far the commonest pea crab found thus, everywhere on the coast. The combination of tube builder, polynoid, and pea crab is particularly characteristic of quiet waters. The smaller commensals apparently have adapted themselves to gathering the proverbial crumbs. There is a nice economy about the situation, for what one animal rejects as too large or too small exactly suits one of its partners.

§233. For a number of years we have been taking a gray, sluggish "bristle worm" from the muddy interstices of rocks dredged up from below 40 fathoms in Monterey Bay. During the summer of 1930 we were surprised to find this *Stylarioides papillata,* apparently perfectly at home, at several points along the shore in Puget Sound. Here is another case of an animal quite competent to live in the intertidal zone once the menace of surf is done away with. It has often been said of Pacific-coast invertebrates that northern-shore forms may be found in the south, but in deep water; and it has been assumed that the determining factor is temperature. In Puget Sound, however, the summer midday temperature that these animals must tolerate at low tide is greater than the maximum along the Monterey shore at any time, and the northern winter temperatures are correspondingly lower. It is easy to believe, therefore, that the absence of wave shock permits a normally deep-water animal to migrate upward until it has established itself successfully in the lowest intertidal zone.

§234. Many worms will be turned up here. *Neanthes virens* (§311) makes semipermanent burrows where the substratum is readily penetrable. *Phascolosoma* (§52) occurs; and under Puget Sound mussel beds on a gravel substratum one may find *Nereis vexillosa* (§163), a variety of the glycerid worm (§280), and the slaty blue *Hemipodus borealis.*

But there could be almost no end to the enumerating of worms. Certainly more than a hundred could be taken, and the unfortunate aspect of the worm situation is that any of them might prove to be common at some of the good collecting places that we have necessarily missed. Furthermore, worms abundant one year may be rare or absent the next.

In any event, one more dainty little polynoid must be mentioned: this is *Hololepidella tuta.* It has a serpentine gait, and will literally wiggle itself to pieces when disturbed too persistently. When first molested, it will protestingly shed a few of its beautifully colored scales; lifted from its favorite under-rock surface, it immediately breaks in two, and before the transfer to jar of water or vial of alcohol can be completed the disappointed collector is left with a number of short fragments that continue the characteristic wiggle. We have never been able to keep a specimen intact long enough to photograph it. It occurs frequently as a commensal with various tube worms (terebellids, §§53, 232).

§235. The large reddish cucumber seen so abundantly under rocks in Puget Sound, or between the layers of friable shale ledges, is *Cucumaria miniata* (Fig. 199), which reaches a length of 10 inches. The tube feet, which are arranged in rows paralleling the long axis of the body, hold fast to the rock very efficiently but rarely seem to be used for locomotion. Surrounding the mouth are long and much-branched tentacles that are drawn in when the animal is disturbed or when the ebbing tide leaves it high and dry. In an aquarium they will flower out beautifully. The animal's color varies, in different specimens, from yellow and red to deep purple. *C. miniata* has been variously considered as synonymous with an Oriental cucumber and with a cucumber occurring on both sides of the Atlantic; in any case, it is the North Pacific representative of a circumpolar and almost cosmopolitan form. The pentamerous symmetry that marks cucumbers as relatives of the seastars, brittle stars, and urchins is rather apparent in *C. miniata.*

C. miniata occurs commonly at least as far north as Sitka, south to Sonoma County, and sparsely at Carmel. In British Columbian channels such as Canoe Pass it is one of the most spectacular of animals; great beds are zoned at 1–5 feet below mean lower low water. The delicate beauty of these expanded colonies, with diagrammatically extended, translucent coral-red tentacles literally paving the rocks, surprises the zoologist who has known them only from preserved specimens or has seen them only as contracted animals exposed at low tide.

A smaller white cucumber, *Eupentacta quinquesemita* (Fig. 200), may be considered as living primarily under rocks, since it is rarely found elsewhere in the intertidal, although it occurs in all sorts of situations below the low-tide line. The two double rows of long tube feet give the animal a bristly appearance. This form and *C. miniata* differ from the *Stichopus* we have considered earlier in having their tentacles around the mouth fairly long, branched, and yellow or orange. *E. quinquesemita* is very temperamental: specimens we took in the summer of 1930 at Pysht, in Juan de Fuca Strait, achieved a record of 100 per cent evisceration, in a few hours and before we could get them into anesthetizing trays. This is the more surprising when it is realized that the temperatures in the seaweed-filled buckets could not have been appreciably higher than those customarily tolerated on the tide flats during the late midsummer lows, when the sun and wind dry the algae to the point of crispness. This white cucumber ranges from Sitka to Pacific Grove, but it occurs infrequently, and then so far down as to be practically a subtidal form. In Puget Sound it is often found associated with the reddish *Cucumaria miniata.* In southeastern Alaska a very similar form occurs, described by Deichmann as *Eupentacta pseudoquinquesemita,* which has heretofore been known from subtidal depths only. It is scarcely distinguishable from the more southerly-ranging white cucumber except that it is thinner-skinned, so that in preservation it partially constricts itself, producing many ringed wrinkles in the body wall.

Cucumaria vegae is a small black holothurian taken from low tide rock crevices at Auke Bay, north of Juneau. Another occasionally intertidal form in the Puget Sound area (but pulled up at Pacific Grove in subtidal holdfasts of kelp only) is the white, pepper-and-salt *Cucu-*

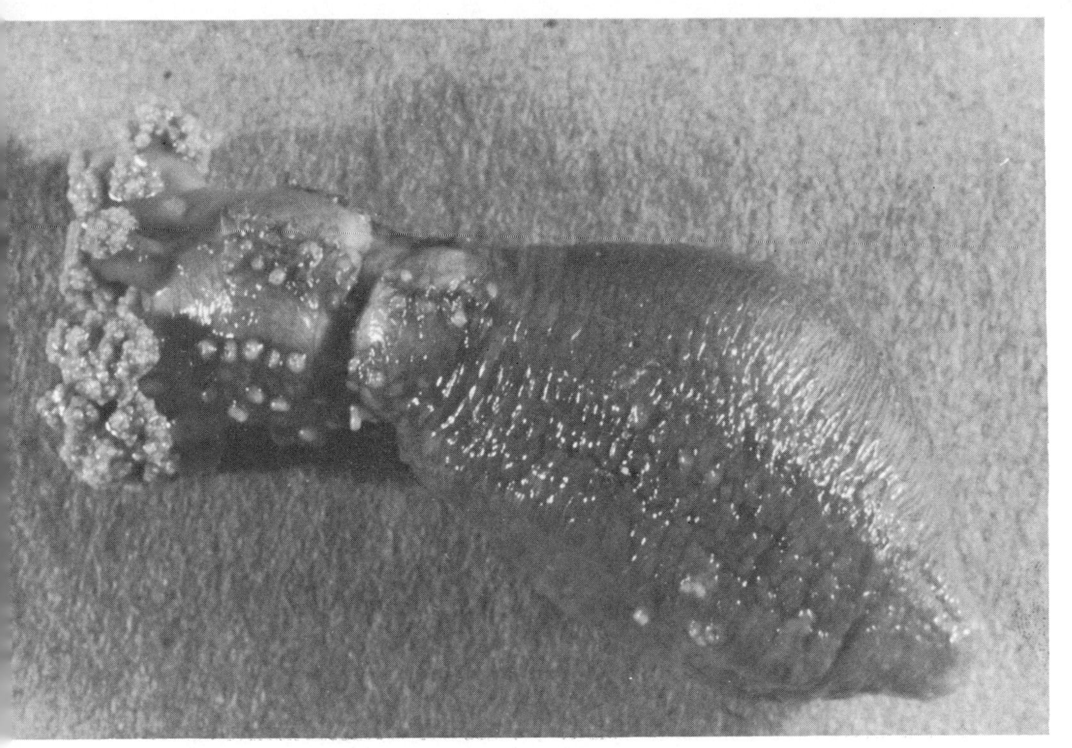

). The reddish *Cucumaria miniata* (§235), very common in Puget Sound; a circumpolar species.
is specimen is contracted, and the characteristic rows of tube feet appear as slight bumps.

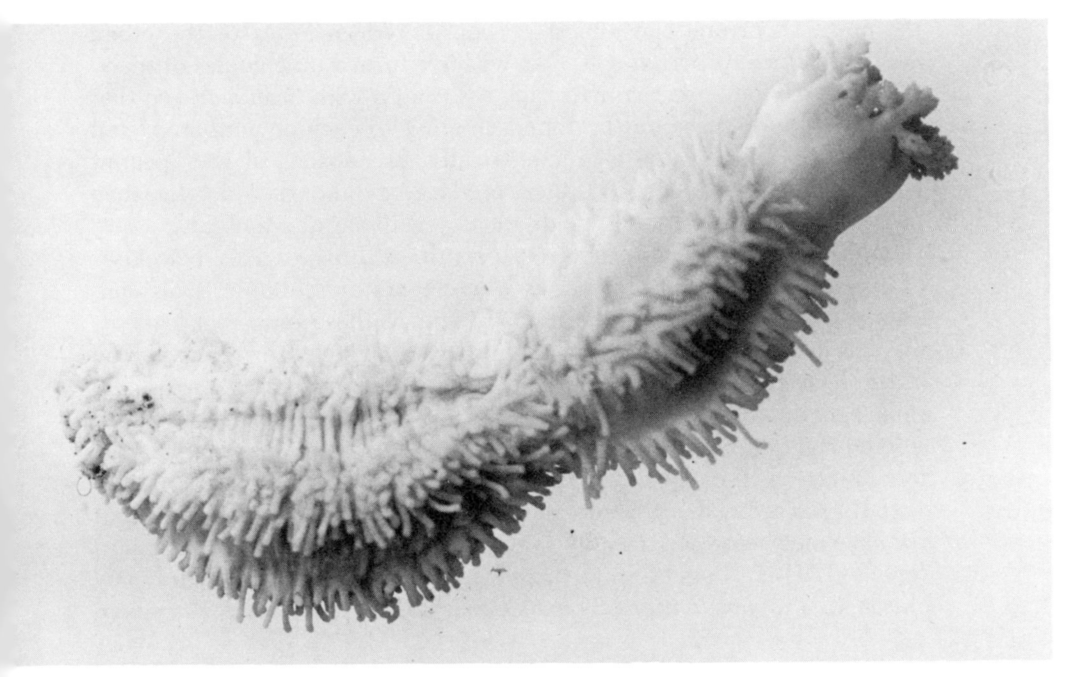

). A small white cucumber, *Eupentacta quinquesemita* (§235).

maria piperata. The large and light-colored *C. lubrica* (§176) may also occur here.

With the northern cucumbers, and elsewhere under rocks, a really gigantic flatworm, tough and firm, will sometimes be turned out, large enough to be the paternal ancestor of all the polyclads the southern collector will have seen. This is *Kaburakia excelsa,* which we have taken at Sitka, in British Columbia at several points, and on the rocks near Port Orchard; it is reported to be common in the San Juan region. Individuals are reported up to 10 by 7 cm. Most of the polyclads are photonegative—light-shunning—and *K. excelsa,* taken on the undersides of rocks or in deep crevices, is no exception, although a diver working in these rocks at high tide or at night would probably find them crawling about on the upper surfaces actively enough. A specimen taken during the summer of 1935 seems to have attempted to engulf the soft parts of a limpet when, en route from tide pool to laboratory, it and others were placed in a jar of miscellaneous animals.

Another smaller (to 4 by 1 cm.), thinner, and more delicate polyclad, often secured with the above, is probably the *Freemania litoricola* reported as *Notoplana segnis* by Freeman, and said by Freeman to be abundant around Friday Harbor.

The brittle star *Ophiopholis aculeata* (§122) occurs frequently enough in beds of *Cucumaria miniata* (especially north of Puget Sound) that the presence of one can almost be considered an index of the other. South of Prince Rupert and at Sitka we have turned up colonies of specimens up to 5 inches in diameter, and associated with *Cucumaria* so that several individuals would be found clinging to each cucumber. Mixed up into the assemblage, characteristically, are dozens of the "peanut worms," *Phascolosoma* (§52); these are larger, and darker in color, than the possibly more numerous individuals available at Monterey.

§236. Along the California coast south of Los Angeles octopuses (§135) have moved into estuaries where there are suitable pools containing rocks arched over the mud. Living in these mud-rock caverns in quiet waters, they attain, when unmolested (which they rarely can be in this heavily populated country), sizes that are large for their usually small species. Wherever they are numerous enough, they might be considered the dominant animals, for their size, strength, and cunning put them pretty well in control of their surroundings.

§237. In the northern waters there are several under-rock chitons that we have met before, noticeably the small, red-marked *Lepidozona mertensi* (§128). In Puget Sound there is a still smaller, greenish to tawny chiton, *Cyanoplax dentiens.* This species occurs as far south as Monterey, and in waters south of Monterey will be found the very similar but slightly larger *Lepidochitona keepiana.*

§238. Under rocks in such varied quiet waters as those of Departure Bay and Pender Harbor, British Columbia, and Elkhorn Slough, California, we have found a flat encrusting sponge that was described originally from the central California estuary as *Mycale macginitiei;* it is yellow, smooth, and slightly slippery to the touch, although it has stiffening spicules and is marked by furrows. *Mycale* adheres closely to the rock, probably by slightly penetrating the rock pores. A species of *Halisarca, H. sacra,* is known to occur at Elkhorn Slough along with *Mycale,* but always below low water. There are various additional species, which, on the basis of field characters alone, could be confused with the above and with previously treated species, and the value of careful laboratory work should be kept in mind.

§239. The rubbery nemertean worms will often be seen under rocks in the north. Two especially common forms are sure to be seen sooner or later and will occasion excited comment. Both are a brilliant scarlet that is brighter than most reds even in the brilliant tide pools, both are large (to 3 meters, although the average will be no more than a couple of feet), and both occur under loosely embedded stones, especially on gravel substratum, from Dutch Harbor clear into Puget Sound. But the two may easily be differentiated. *Carinella rubra* is soft and rounded, very distensible, sluggish, and apt to be coiled in angular folds. The head is wide, rounded, and pinched off from the rest of the body. We have found this to be one of the characteristic under-rock inhabitants of quiet waters. In lower Saanich Inlet, about Deep Cove, for instance, in the summer of 1935, almost every suitable stone upturned revealed one of these brilliant worms in an apparently fabricated mild depression. *Cerebratulus montgomeryi,* by contrast, is ribbonlike and flattened, with longitudinal slits on the sides of its white-tipped head, and with lateral thickenings that extend to thin edges (probably correlated with its swimming ability) and run the entire length of the body. *Cerebratulus* is famous for fragmenting at even the slightest provocation; but this species, Coe remarks, lacks that trait to the extent that specimens may be killed whole by immersion in hot formalin.

§240. At least one vertebrate belongs with the under-rock fauna from Puget Sound clear into Elkhorn Slough, the grunting fish *Porichthys notatus*—possibly better known as the midshipman, since its pattern of luminescent organs resembles the buttons on a naval uniform. The eggs, each the size of a small pea, are deposited in handful clusters, which may be seen under and around rocks during the summer. On one Puget

Sound beach, in the summer of 1929, we found a small boy capturing the fish for food, using a method that must have seemed, to the fish, to violate Queensbury rules. He poked a slender stick under each likely rock and listened. If a midshipman grunted, he reached under with a heavier stick made into a gaff by the attachment of a large fishhook. A quick jerk, and another betrayed grunter was added to his string. One of them came off the string, however, after certain negotiations, and was added to our chowder, with satisfactory but not phenomenal results.

Blennies will also be found commonly in this zone under rocks in Puget Sound—usually *Anoplarchus purpurescens* and similar forms (see *Epigeichthys,* §198) that will have been seen higher up.

Burrowing Habitat

§241. The animals of the under-rock substratum in quiet waters sometimes differ from those outside. For instance, the rocks in Newport Bay are often completely honeycombed by the burrowing date mussel, *Lithophaga plumula* (Fig. 201), which we have never seen on the exposed coast. Quite possibly the answer in this case is temperature. The date mussel is a tropical form and might reasonably be expected to find the warm, sun-steeped climate of Newport Bay more to its liking than the colder, wave-swept shore outside.

The date mussel is a close relative of the common mussel (*Mytilus,* §158), but it has two important differences directly connected with its mode of life. First, it produces an acid that attacks calcareous rocks, such as limestone. Using this acid, which is secreted by a special gland, the animal excavates a safe home; it maintains communication with the outside world through its siphons, obtaining its oxygen and microscopic food, like all of its relatives, from the stream of water that flows in and out.

Obviously, this chemical boring involves another problem, which might be stated in the form of a syllogism. Major premise: the date mussel's acid secretion attacks calcareous matter. Minor premise: the date mussel's own shell is calcareous. Conclusion: the date mussel's acid secretion attacks its own shell. The premise overlooked is the animal's second difference from related forms. Covering the date mussel's shell is a thick brown layer of horny material, which is resistant to the acid. It is the color of this protective layer, as well as the shape of the shell, that has given the animal its popular name.

§242. Another bivalve, *Penitella penita,* the common piddock or rock clam (Fig. 202), drills into rock so hard that nothing short of a sledge hammer powerfully swung will break into the burrows; and it appar-

201. The date mussel, *Lithophaga plumula* (§241), a rock-boring form; about life size.

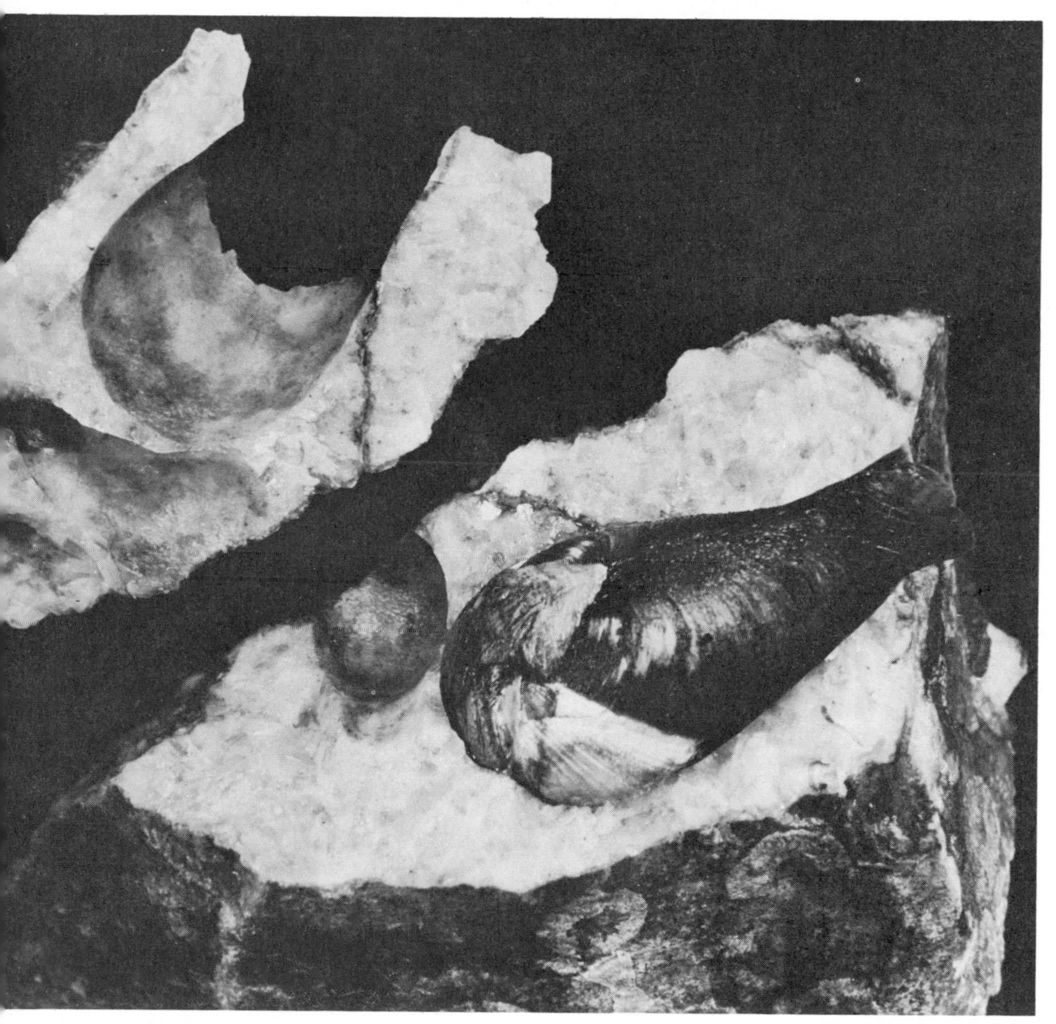

. *Penitella penita* (§242), the common piddock or rock clam; the rock has been split open to show animal in place in its burrow; slightly enlarged.

ently drills without the aid of chemicals, using mechanical means only. The animal twists and rocks itself on its round foot (which takes the form of a suction disk that grips the rock), pressing its rough valves outward at the same time so as to bring them into contact with the walls of the burrow. This method of drilling is necessarily slow in direct ratio to the hardness of the rock, partly because the continual wearing away of the shell by the grinding process must be compensated for by growth. In their well-known book *The Seas,* Russell and Yonge state that although no acid-forming gland can be found in the piddocks, there must nevertheless be some chemical action, since, although the head of the burrow is unquestionably excavated by mechanical action, the outer part of the burrow, through which the soft siphons protrude, is also enlarged. Piddocks, however, sometimes bore in rock that is immune to chemical action (e.g., sandstone).

This rock clam occurs, in deep water at least, throughout its range from Alaska to Baja California, but it seems to colonize the intertidal zone in quiet water only. It has caused considerable damage to concrete harbor works on this coast. One investigation revealed that more than half of the concrete-jacketed piles at four different places in Los Angeles Harbor were being attacked, sometimes with seven or eight borers to the square foot. In some cases the animals had turned aside after penetrating the concrete jacket in order to avoid entering the wood.

The rock-boring clams are sought after as food, mostly in areas where they burrow into relatively soft sandstones or mudstones that are easily broken up with a pick. Unfortunately, when a size limit is set, the digger will reject the undersized clams and break away more rock. In places like Duxbury Reef and parts of the Oregon coast (where the rock is not eternal basalt) the clammer has become a potent geological force.

§243. The burrows of dead rock-borers are likely to be appropriated by any of several homeseekers. The most numerous of these is the porcelain crab of wharf pilings (§318), but it has not adapted itself as completely to the borrowed home as has a large suckered flatworm (probably *Notoplana inquieta*) that resembles *Alloioplana* (Fig. 10) in outline. This worm has the gray-brown color so common to retiring animals, and is very sluggish. It is firm in texture and outline, but when routed out of its quiet nest, it curls up very slowly. Most polyclad worms crawl about and eat detritus that works into rock crevices. It would be interesting to determine if this one depends for its food on what the slight currents may bring it, or if it actually leaves the burrow and goes hustling out for something to eat.

§**244.** In the substratum between rocks, and ranging in its circumpolar distribution at least as far south as Puget Sound, is the burrowing anemone *Edwardsiella sipunculoides,* which is similar to the *Edwardsiella* of Figure 235. It is a dull-brown form, with long transparent tentacles, and shows but faintly the eight longitudinal bands that are reputed to distinguish it from other anemones. A more readily apparent difference is that it has no attachment disk, but instead lives loosely buried in the soil.

§**245.** Nearly all of the animals that occur in the low-zone substratum in protected outer-coast areas may be taken here also—after all, the degree of protection in the substratum does not vary greatly between the two regions. Because of their abundance, mention should be made of the transparent, wormlike cucumber *Leptosynapta* (§55), the soft blue burrowing shrimp *Upogebia* (§313), and the clam *Protothaca* (*Paphia*) (§206), which spouts from between tide-pool rocks in Puget Sound. In the south the mantis shrimp *Pseudosquilla* (§146) is taken infrequently far under loosely embedded rocks. In the fully protected waters of Alaska (as at Jamestown Bay, Sitka, and Auke Bay, Juneau) there are such characteristic forms as the echiuroid worm *Echiuris echiuris alaskensis* (treated in §306), the ubiquitous peanut worm (§52), and a slimy, apodous cucumber, *Chiridota albatrossi.* This last slim and sluglike form, covered with peppery black dots, may be 5 or 6 inches long, the otherwise smooth skin bearing calcareous white particles—the "anchors" already mentioned in connection with *Leptosynapta* (§55).

9. *Sand Flats*

In quiet-water beaches of fairly pure sand, where there is no eelgrass, we find a region that must seem utopian to many animals. The usual environmental problems are almost entirely lacking—which undoubtedly accounts, in part, for the extraordinary richness of the sand bars in Newport Bay in its former days, in Tomales Bay, and in parts of Puget Sound. There is no wave shock to be withstood; the substratum is soft enough that feeble burrowing powers will suffice, and yet not so soft that it requires special adaptations to avoid suffocation; there is no attachment problem, for a very little burrowing is enough to secure the animals against being washed out by the innocuous currents; and a little more burrowing protects them from sunlight and drying winds. These pure sands are likely to contain little organic food by comparison with the rich stores of mud flats and eelgrass beds, but every tide brings in quantities of plankton, and predacious animals that require firmer food can prey on each other.

There is usually, however, something wrong with apparent utopias. In this case, the elimination of the struggle against natural conditions results in proportionately keener competition among the animals, and overcrowding is the rule. There is no obvious zonation in areas of pure sand, but the region corresponding to the lowest zone of other shores is, for most of the animals, the only one that is habitable. The crowding is therefore accentuated, and the upper beach is sparsely inhabited.

Estuarine sand flats offer an added obstacle to prolific colonization by marine animals—the presence of fresh water. Where this occurs abundantly or sporadically, the faunal aspect is correspondingly changed. But a good many of the crabs and some of the shellfish tolerate greatly lowered salinities, and Reid (1932) has shown that fresh water, even if actively flowing over the flats, has little effect on the seawater retained in sand below the surface; thus even minimal burrowing ability would protect a sessile form until the return of the tide.

A startling example of this retention of salinity is found in a small Alaskan stream that is used as a spawning ground by pink salmon. The lower reaches of spawning gravel are covered only by high tides; even so, the salinity around the salmon eggs, which normally develop in fresh water a few miles from the sea, is about 8 parts per thousand. Nevertheless, spawning here seems to be reasonably successful.

§246. The sand dollars definitely belong in this environment, and they occur in tremendous numbers. Their round tests are 3 to 4 inches in diameter, very flat, and colored so deep a purple as to be almost black. At first glance, they seem to have little resemblance to other urchins, but urchins they are. The spines, instead of being 3 or 4 inches long, as in the giant red urchin of the outer coast, are about 1/16 inch long and so closely packed that the animal looks and feels as though it were covered with velvet. The tube feet are there, too, and are visible to the unaided eye if one looks closely enough. In motion, the spines and tube feet give an impression of changing light rather than actual motion. The five-pointed design on the back, another urchin characteristic, is visible in the living animal, but it is more obvious in the familiar white skeletons that are commonly cast up on exposed beaches. Many people believe that the living animals must occur near these beaches. They do—in deep water. The sand dollar avoids surf at all times, and will seldom be found alive by the shore collector except on completely sheltered flats.

When a low tide leaves the animals exposed, they will ordinarily be found lying flat, partly or completely buried. When still water covers them, they stand vertically, with two-thirds of the disk buried in the sand; but when there is a current, they lean away from it at an angle that is uniform throughout the entire bed and apparently in direct proportion to the strength of the current.

Sand dollars scour the sand for minute edible particles, selecting particularly tasty grains covered with diatoms and using the tube feet to pass them into the mouth—a food-getting process in contrast with that of many other sand-living forms, which pass great quantities of sand through their intestines in order to extract whatever nourishment it may contain. Reproduction is similar to that of other urchins, eggs and sperm

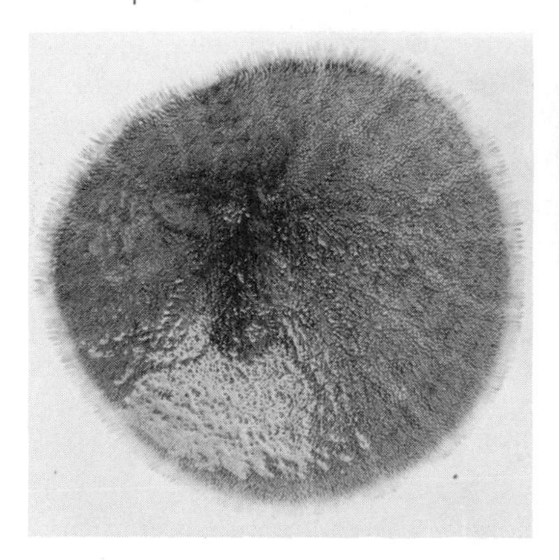

203. *Dendraster excentricus* (§246), the common sand dollar: left, a living animal; right, the dead test as found on the beach.

being extruded from different individuals for chance union and the development of free-swimming larvae that bear no resemblance to the adults. In Puget Sound sexual maturity comes in late spring, and in Elkhorn Slough specimens with ripe eggs and sperm have been taken in August.

Two species of sand dollars occur on this coast. The common form from Alaska to Baja California is *Dendraster excentricus* (Fig. 203), which has the star design off center and consequently a bit lopsided. We have seen great beds of them in southern Puget Sound, Elkhorn Slough, Newport Bay, San Diego Bay, and in El Estero de Punta Banda, just south of Ensenada.* The other species, *Echinarachnius parma,* is even flatter and is more symmetrical, the pentamerous design centering at the apex of the shell. It is a circumpolar form, known on this coast from Alaska to Puget Sound. Working with *E. parma,* Parker and Van Alstyne (1932) found that burrowing and most locomotion are a function not of the tube feet, as in the case of seastars, but of "waves of coordinated spine movements," the cilia being "concerned only in feeding, cleansing and respiration."

§247. A red to rose-lavender heart urchin, or sea porcupine, *Lovenia cordiformis* (Fig. 204), will be found occasionally on the southern sand flats, where it lies half buried near the extreme low-tide line. This animal, by no means common in the intertidal, still occurs often enough to pro-

* In several of these places (e.g., Newport Bay), "great beds" have not been seen recently (1960's).

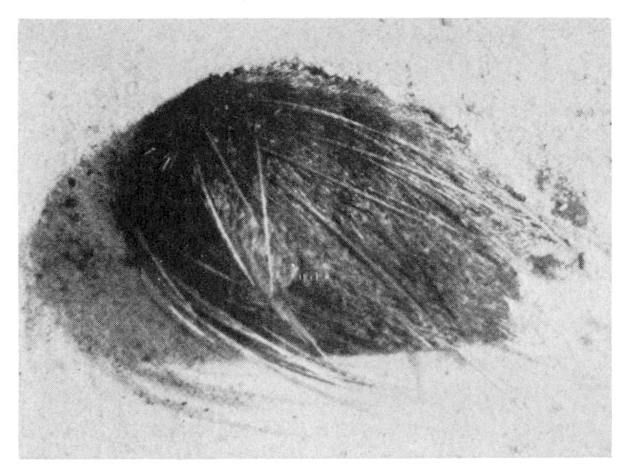

204. The heart urchin *Lovenia cordiformis* (§247).

vide an excellent argument against barefoot collecting; and unless it is handled carefully, it can inflict painful wounds on bare hands. Ordinarily the long sharp spines point backward, and their thrusting effect helps the animal walk through the sand; but they bristle up in a formidable manner when their owner is disturbed. We have seen specimens during the winter at Newport Bay.

§248. *Renilla köllikeri* (formerly called *R. amethystina*), the sea pansy, is one of the most obvious animals of the southern mud flats. When seen in their natural habitat, sea pansies seem misnamed, for they lie with their heart-shaped disks almost covered with sand and their stalks buried. A specimen should be transferred to a porcelain or glass tray containing clean seawater and left undisturbed for a time. The disk will there enlarge to three times its size when taken from the sand, and scores of tiny, hand-shaped polyps, beautifully transparent, will expand over the purple disk. Not the least of the sea pansy's charms is the blue phosphorescent light that it will almost invariably exhibit if mildly stimulated with a blunt instrument after being kept in the dark for an hour or so.

In Newport Bay, sea pansies occur by the hundred, in El Estero de Punta Banda by the thousand. They are said to have been plentiful in San Diego and Mission Bays before the days of commerce and industry. Southern California is the northernmost part of the range. The similar, broad, kidney-shaped, true *R. amethystina* extends at least to Panama.

The eight-tentacled polyps of *Renilla* are nearly always infested with minute parasitic copepods, *Lamippe* sp. The same genus is said to infest all alcyonarians all over the world.

§249. The three true crabs that occur on sand flats are found nowhere else in the intertidal zone. *Heterocrypta occidentalis* (Fig. 205), since it lacks any other name, might be called the elbow crab, for its large forelegs, which terminate in absurdly small claws, suggest an energetic customer at a bargain counter. This weird creature is the only northern representative of a group of crabs well represented in tropical and equatorial regions. We have taken it, with the other two, at Newport Bay, and it is known from deep water elsewhere along the coast.

Randallia ornata (Fig. 206) has the round, bulbous body and gangly legs of a spider, but it is not closely related to the "spider" crabs. The carapace is mottled purple, brown, or dull orange, and the legs are usually curled underneath it as the animal lies half buried in the sand.

Portunus xantusi (Fig. 207) is a swimming crab, one of the liveliest of its kind. Its last two legs are paddle-shaped, and with them the crab swims sideways rather rapidly, with its claws folded. Two long spines, one at each side of the shell, are effective weapons of defense, and in addition the crab is unduly hasty about using its sharp and powerful pincers, so that a collector who captures one without shedding any of his blood is either very lucky or very skillful—probably both. *Portunus* is a small relative of the prized blue crab of the Atlantic. It occurs as far north as Santa Barbara, but the other two crabs are not likely to be found in the intertidal above Balboa.

§250. A seastar, *Astropecten armatus* (Fig. 208), occurs on sand flats from Newport Bay to Panama, usually below the waterline but occasionally exposed. Now and then a specimen will be found having an arm spread of more than 10 inches. They are beautifully symmetrical animals, sandy gray in color with a beaded margin and a fringe of bristling spines. Normally they lie half buried in the sand.

§251. Several snails and at least one nudibranch are members of the sand-flat assemblages. In El Estero de Punta Banda we once saw tremendous hosts of the nudibranch *Armina californica* (Fig. 209), which were living, peculiarly for nudibranchs, half buried in the sand—another of the interesting adaptations to this fruitful environment. Thin, white, longitudinal stripes, alternating with broader stripes of dark brown or black, make *Armina* a conspicuous animal. In aquaria it has been observed to feed on the sea pansy *Renilla,* and it undoubtedly does the same thing in its natural habitat, for the two always occur together. A similar association involving other species of *Armina* and *Renilla* occurs in the Gulf of Mexico.

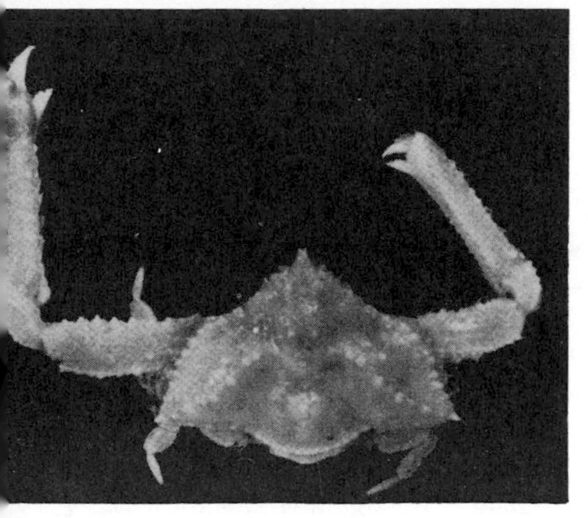

)5. *Heterocrypta occidentalis* (§249).

206. *Randallia ornata* (§249).

207. *Portunus xantusi* (§249).

8. *Astropecten armatus* (§250).

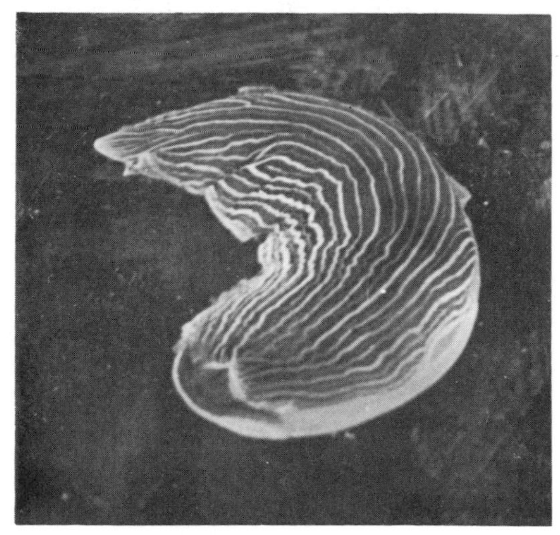

209. *Armina californica* (§251).

The range of this opisthobranch is from Vancouver Island to Panama; in the northern part of its range it must feed on something else besides sea pansies; in any event it is characteristic of clean sand bottoms. Occasional specimens turn up in Tomales Bay, but sea pansies have not been found there—yet.

§252. The moon snails, or sand-collar snails, are common inhabitants of the sand flats throughout the whole stretch of coast from Alaska to Mexico. There are two similar species. The larger (4-inch shell length), heavier, somewhat high-spired species is *Polinices lewisi* (Fig. 210), which may occur in somewhat muddy sand; *P. draconis* is somewhat smaller (3–4 inches) and shorter in height. When expanded, the body of the moon snail is much larger than its shell.

The strictly southern moon snail *P. recluzianus* is much smaller—usually less than 2 inches in shell length—and, like *P. lewisi,* shows a preference for muddy sand. It can be distinguished from the other moon snails by its occluded umbilicus; that is, the "navel" is grown over so that that part of the shell is smooth. Frequently the shells of *P. recluzianus,* which ranges from Crescent City to Chile, are occupied by a giant hermit crab (§262).

Any of the three will commonly be found partially buried in the substratum, through which they plow with apparent ease. At low tide a lump in the sand may indicate the presence of a moon snail an inch or so below the surface. All are capable, also, of contracting the immense fleshy foot enough to stow it away completely within the shell, which is then closed with the horny door common to other snails. This process involves the ejection of considerable quantities of water from perforations around the edge of the foot, and the animal cannot live long when contracted, since it cannot breathe.

The moon snail's method of food-getting is varied, but generally it clamps its foot around a clam, mussel, or other mollusk and drills a neat, countersunk hole through the shell with its radula. Shells perforated by the moon snail or some of its relatives are often found along the beach. A less spectacular method consists of suffocating the victim by holding it inside the foot until dead or by sucking the foot over the siphon of a clam. In this case, the proboscis is merely inserted between the shells of the dead animal and the contents are cut out. Apparently the great snail will also feed on any dead flesh that comes its way.

Cannibalistic members of its own tribe and the multi-rayed sunflower star *Pycnopodia* seem to be the moon snail's only natural enemies. In an aquarium a sunflower star has devoured two moon snails in three days.

210. The giant moon snail *Polinices lewisi* (§252), about ½ natural size. Right: the collar-like egg case of the moon snail.

The egg cases of these giant snails are, to most laymen, one of the puzzles of the intertidal world, for they look like nothing so much as discarded rubber plungers of the type plumbers use to open clogged drains (Fig. 210). Many years ago the junior writer of this book squandered a great deal of nervous energy on a fruitless search through unfamiliar literature in an attempt to find out what manner of animal they might be. It was some consolation to him to learn eventually that for a long while naturalists had made the same mistake. Certainly there is no obvious reason for connecting the rubbery, collar-shaped egg cases with the snails that make them. The eggs are extruded from the mantle cavity in a continuous gelatinous sheet, which, as fast as it emerges, is covered with sand cemented together with a mucous secretion. The growing case travels around the snail, taking its shape from the snail's foot as it is formed. In time the egg case crumbles, releasing a half million or so free-swimming larvae. There is some poetic justice in the assumption that many of the minute larvae will provide food for mussels, which will later on be eaten by adult moon snails.

§253. The beautifully marked and polished *Olivella biplicata* (Fig. 211), the purple olive snail, leaves a betraying trail on the sand (although it is itself out of sight) by plowing along just under the surface. Sometimes the shell is above the surface and the foot below. This species ranges from Vancouver to Baja California; *O. baetica,* a smaller slimmer species, will be found most commonly south of Point Conception, although its total range is about the same as the plumper olives.

The fragile bubble shell may be first noticed in pools on bare sand flats, but it is much more at home in beds of eelgrass (§287).

§254. Two edible animals, a shrimp and a scallop, are found here, usually in pools in the sand or buried in the sand, but occasionally exposed. *Crago nigricauda* (Fig. 212), the black-tailed shrimp, positively cannot be captured without a dip net unless it is accidentally stranded on the sand. It has, even on its legs and antennae, pepper-and-salt markings that make it extremely difficult to see against a sand background. This is one of the common market shrimps in California, along with several others found only in deep water. Shrimp-dredging boats net these animals from depths of from a few fathoms in San Francisco Bay to more than a hundred in Puget Sound, and it is in such depths that *Crago* presumably finds its optimum conditions, since only occasional specimens are found in the intertidal. The black-tailed shrimp ranges from Alaska to Baja California. Market specimens will occasionally be found with the carapace swollen out on one side into a blister, the result of infection with an asymmetrical isopod (bopyrid), *Argeia pugettensis,* which is especially common on commercially netted *Crago communis* in Puget Sound, where one out of possibly every 20 or 30 specimens will be noticeably afflicted.

§255. The thick scallop *Pecten circularis* (Fig. 192), frequently found buried in the sand, is fairly common. The peculiar swimming habits of the relatively "intelligent" scallops have been mentioned before (§220), as have their toothsome qualities. In addition to this sand-flat form, which ranges from Monterey to Peru, and the rocky-shore form (*P. hindsi*), another scallop will be found in eelgrass (§276). For reasons not understood, there are no scallops in Tomales Bay.

11. *Olivella biplicata* (§253), the purple olive snail.

12. *Crago nigricauda* (§254), the black-tailed shrimp.

§256. Although all the sand-flat animals so far considered burrow somewhat, the remainder are strictly burrowing forms. Perhaps the most obvious of these, when it occurs at all (it does not occur intertidally north of southern California), is a great burrowing anemone (similar to *Cerianthus aestuari*), which lives in a black, papery tube that is covered with muck and lined with slime. More energetic diggers than the writers have taken specimens with tubes 6 feet long. The muscular lower end of the animal is pointed and adapted to digging, and at the upper end there are two concentric sets of tentacles—a short stubby series about the mouth and a long waving series around the border of the lividly chocolate-colored disk. Unfortunately, specialists have not had an opportunity to study this animal and ascertain its identity.

We have found a much smaller *Cerianthus,* probably a different species, in the sand flats at Ensenada.

§257. Another burrowing anemone, common in southern California, is *Harenactis attenuata* (Fig. 213), which may occur north to Humboldt Bay. It is a tubeless form, sandy gray in color. The long, wrinkled, wormlike body is largest in diameter at the disk, tapering downward until it swells into an anchoring bulb. When annoyed, *Harenactis* pulls in its tentacles in short order, retracts into the sand, and apparently sends all its reserve body fluids into this bulb, thus swelling it so that it is impossible for a hungry bird or an avid human collector to pull the animal out. The bulb of a large specimen may be 18 inches below the surface, and digging out a specimen with a spade is very difficult, since the shift-

213. *Harenactis attenuata* (§257), the southern burrowing anemone.

ing sand fills up the hole almost as rapidly as it is made. While the excavation is in progress the anemone retracts still more and makes its bulb even more turgid. Having done that, it can do nothing but sit tight and rely on Providence—a fatalistic philosophy that is usually justified in this case. Only if the sand can be shoveled away from the side of the bulb can a specimen be secured uninjured.

§258. To the extreme north of our range is a third anemone, the burrowing *Anthopleura artemisia,* which is related to the great green anemone of the lower rocky tide pools. This species occurs in flats of rather pure sand, but only where small cobblestones or cockleshells are available underneath the surface for basal attachment. At low tide *A. artemesia* retracts completely beneath the surface and is to be found only by raking or shoveling. The contracted specimens are white or gray-green, and may easily be mistaken for *Metridium* (§321). Placed in a jar of seawater, they soon flower out beautifully, their long, wavy tentacles colored with lovely greens and pinks and their disks semitransparent.

§259. According to Weymouth there are nearly 500 species of bivalves on this coast. Many, of course, are too small or too rare to be considered in this account; and so that the volume will not grow out of all proportion, we must further limit ourselves to the large, the obvious, or the interesting forms. After the scallop, with its peculiar flapping-swimming proclivities, the basket cockle *Clinocardium nuttalli* (Fig. 214) is the bivalve most likely to be noticed by the casual collector. Reaching its maximum development in British Columbia and Puget Sound, the basket cockle extends southward in decreasing numbers as far as Baja California and northward to the Bering Sea. In the north, specimens are found very frequently in flats of "corn meal" sand that would be very shifty if surf could ever get to it. Basket cockles are occasionally found entirely exposed, but they are active animals and can dig in quickly with their large feet. They burrow shallowly, however, for they have no siphon tubes, and their siphon holes are merely openings in the margin of the mantle. This cockle makes excellent (if rather tough) food, but it is too scarce to be used extensively. Specimens will frequently be captured with tiny commensal crabs, *Pinnixa littoralis* (§300), in the mantle cavity.

Dr. Fraser (1931) has reported on the natural history of this form in British Columbia. Most of the 760 specimens examined were 3 or 4 years old, and none were more than 7 years. This genus is hermaphroditic, with ova and sperm usually shed simultaneously, during a long

214. *Clinocardium nuttalli* (§259), the basket cockle, with foot extended.

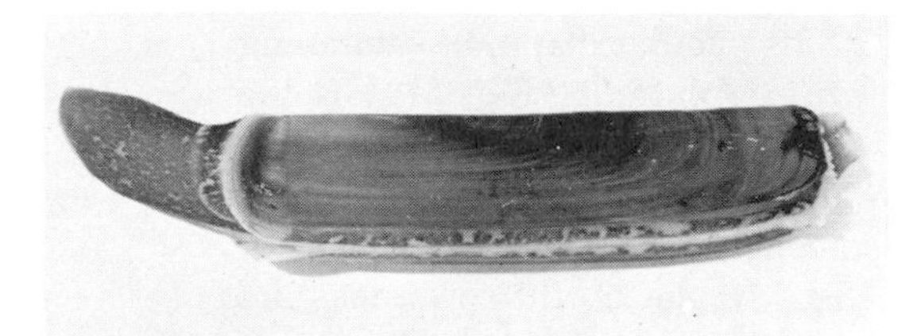

215. The jackknife clam, *Solen rosaceus* (§260).

216. *Chione undatella* (§260), a hardshell cockle.

217. *Sanguinolaria nuttalli* (§260).

spring spawning period that affects all animals over 2 years old. Fraser suspects that the free-swimming larvae may settle only in the sublittoral, moving inshore with age, which possibly explains why so few small individuals are found alongshore. The §189 data on the growth rates of northern and southern razor clams are known to apply also to the races of *Clinocardium*.

§260. One would scarcely expect to find a clam throwing off a part of itself as a crab throws off a leg, but *Solen rosaceus* (Fig. 215), one of the jackknife clams, actually autotomizes rather readily. *Solen*'s siphon, too large to be entirely retracted within its pink-tinged, nearly cylindrical shell, has prominent annulations. If the animal is roughly handled, or if its water gets somewhat stale, it drops off its siphon bit by bit, the divisions taking place along the lines of annulation. Presumably this autotomy was developed as a defense against the early bird in search of a worm and not averse to substituting a juicy clam siphon. This clam, which averages around 2 inches in shell length, ranges from Santa Barbara to the Gulf of California, and a similar but somewhat larger species, *Solen sicarius* (§279), occurs occasionally in mud flats and commonly in eelgrass roots as far north as Vancouver Island. S. *sicarius* has a slightly bent, glossy yellow shell. Both species dig rapidly and well, but not as efficiently as their cousin *Tagelus* of the mud flats.

Other common sand-flat clams are: *Chione* (Fig. 216), of three species, all known as the hardshell cockle and all ranging from San Pedro into Mexico; the purple clam *Sanguinolaria nuttalli* (Fig. 217), another southerner; and any of several small tellens, *Tellina bodegensis,* etc., most of which extend throughout the range considered in this book. A tellen, *T. tenuis,* is one of the characteristic components of the Scottish fjord fauna and its natural history is well known. Stephens (1928, 1929), investigating the Scottish species' distribution, found small individuals (but in great numbers, up to 1,000 per ¼ square meter) at the low level of spring tides. The optimum zone for larger specimens, however, was higher up, toward high-water mark, where there were fewer individuals—an interesting zonation that reverses the usual situation. He also found that growth was more rapid higher up (probably a function of the lessened competition), that it decreased in regular progression with the age of the animal, and that it almost stopped in winter. There were four age groups, representing four annual spawnings; the males became sexually mature in May, the females in June. Specimens were found to feed on plant detritus, and in the spring on diatoms. These data on growth rates coincide with what is known of razor clams (§189).

Still another sand-flat clam is the white sand clam, *Macoma secta,* which reaches a length of 4 inches. It rather closely resembles its mud-flat relative, the bent-nosed clam (§303 and Fig. 247), but is typically larger and lacks the "bent-nose" twist in the shell at the siphon end. The shell is thin, and the left valve is flatter than the right. This species makes very good eating; but the intestine is invariably full of sand, and the clams should be kept in a pan of clean seawater for a day or so before eating. The range is British Columbia to the Gulf of California.

§261. A mole crab occurs in this environment—*Lepidopa myops* (Fig. 218)—that never under any circumstances subjects itself to the pounding waves that delight its surf-loving cousin *Emerita* (§186). Visitors to the sheltered sand beaches at Newport Bay and southward will find it submerged in the soft substratum, with its long, hairy antennae extending to the surface to form a breathing tube. This form burrows rapidly in the shifting sand, but not with the skill of *Emerita,* whose very life depends on its ability to dig in quickly enough to escape an oncoming breaker.

§262. One December afternoon we found, in the unpacked sand along the channel in El Estero de Punta Banda, thousands of hermit crabs, *Holopagurus pilosus,* all living in the shells of moon snails. The largest were giants of their tribe—larger than any intertidal hermits we have seen south of Puget Sound—but the really remarkable thing about them was their habitat. Many of them, probably most of them, were completely buried in the sand. Above the waterline they could be detected only by their breathing holes, below the water by occasional bubbles. These very hairy and rather beautifully colored hermits have heretofore been known only as an offshore form, in water up to 30 fathoms deep between San Francisco and San Diego. Here hosts of them could be turned out by merely raking one's fingers through the sand.

§263. At the same time and place, and in the same manner, a considerable number of brittle stars were taken. Occasionally a waving arm showed above the surface, but the vast majority were completely buried. These were *Amphiodia barbarae,* also formerly thought to be exclusively deep-water animals, having been reported as ranging from San Pedro to San Diego in depths up to 120 fathoms. Except that its disk is thicker, this animal resembles *A. occidentalis* (Fig. 41).

§264. A large cucumber with a most un-holothurian appearance, *Molpadia arenicola* (Fig. 219), is taken now and then at Newport Bay, in El Estero de Punta Banda, and probably other places. Habitués of the Newport intertidal regions call it the "sweet potato," and the name is rather appropriate. A sweet potato as large and well polished as one of

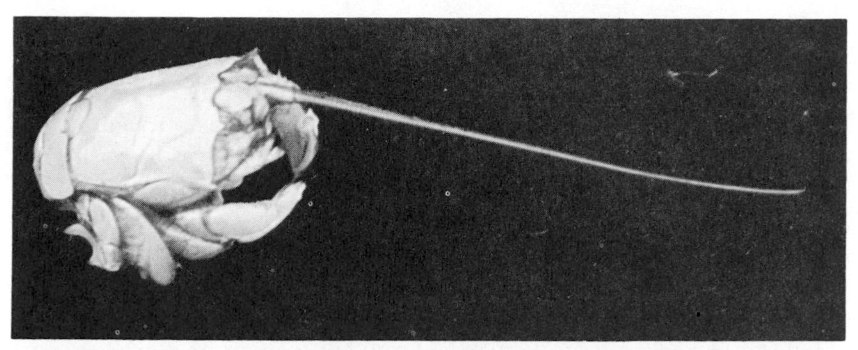

218. The quiet-water mole crab, *Lepidopa myops* (§261).

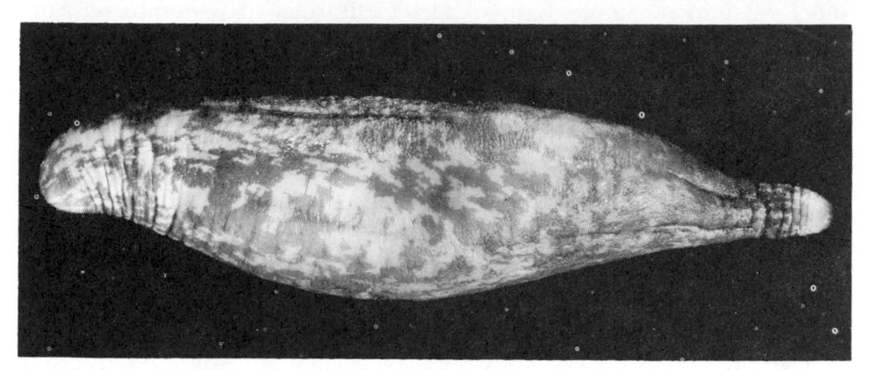

219. The "sweet potato," *Molpadia arenicola* (§264), a cucumber.

these animals, however, would be a sure prize winner at a county fair. The first specimens we saw were dug near Balboa and put in an aquarium for the edification of a collecting party. We took them to be giant echiurid worms, and certainly there is little about them to suggest their actual identity: the mottled, yellowish-brown skin is tough, smooth, and slippery; there are no tube feet, and no obvious tentacles. *Molpadia* feeds by passing masses of sand through its digestive tract for the sake of the contained detritus. Since it lives in sand that appears to be fairly clean and free from organic matter, it must be compelled to eat enormous quantities of inert matter to get a little food. We have no notes on the speed with which the sand mass moves through the animal, but the better part of the weight of a living specimen, and much of its bulk, is in the contained sand. Remove the sand, and the rotund "sweet potato" collapses.

Molpadia has one cucumber trait, however, in that it always has guests in its cloaca, the pea crab *Pinnixa barnharti* occurring so commonly as to be almost diagnostic. When specimens are being narcotized with Epsom salts for relaxed preservation, the pea crabs are likely to come out, just as the pea crab *Opisthopus* escapes from *Stichopus* (§77) under the same circumstances.

§265. Sand flats support the usual host of segmented worms. The tubed *Mesochaetopterus taylori* is a striking and usually solitary form. Unfortunately, it breaks so readily that it can rarely be taken whole; the hindmost segments are the parts first shed and the least often seen. For so delicate a creature this worm has an unusual range, occurring intertidally at Departure Bay, British Columbia; a similar or identical form may be found at Elkhorn Slough, Morro Bay, Newport Bay, California, and El Estero de Punta Banda, Baja California—it seems to be rare intertidally in the south, however. In some localities around Puget Sound it occurs in colonies. It is somewhat similar in appearance to *Chaetopterus* (Fig. 252).

The tubed and sociable "jointworm" *Axiothella* occurs in this environment, but it is more characteristic of areas of hard-packed muddy sand (§293).

Every shovelful of sand turned up in search of lancelets (§266) will rout out one or two specimens of a third segmented worm, a small and nereid-like species of *Nephtys.* (The species of sheltered waters is *N. caecoides*; *N. californiensis* lives on outer beaches.) The collector intent on snatching the swift lancelets the instant they are exposed will often grab a specimen of this smaller and slower animal before he can check the motion of his hand. A similar shorter form taken in similar environments of loose-packed sand is *Armandia binoculata.*

MacGinitie (1935) says of *Nephtys*: "It can burrow very rapidly, and in its activities reminds one somewhat of *Amphioxus.* . . . can swim through loose sand as rapidly as some worms are able to swim in the water."

§266. The lancelet, a famous animal that is seldom seen, comes into its own on this coast in southern California and (at least) northern Baja California.* Where there is a low-tide sand bar opposite the mouth of a sheltered but pure-water bay and far enough in to be protected from wave shock, these primitive vertebrates are likely to be found. Our species, *Branchiostoma californiense* (Fig. 220), is reputed to be the

* Hubbs (1922) reports it from Monterey Bay, California, to San Luis Gonzales Bay, Gulf of California. It certainly never occurs littorally in the north, however, and we have never been successful in even dredging it at Monterey. In the late summer of 1967 a single female with eggs was collected in Tomales Bay.

largest lancelet in the world, adult or sexually mature individuals being up to 3 inches long.* Specimens of that size will very rarely be taken, however, and the average size is less than half of the maximum. Violent stamping on the packed sand in just the right area will cause some of the animals to pop out. They will writhe about for a moment and, unless captured quickly, dive back into the sand. The usual method of capture is to turn up a spadeful of sand, trusting to one's quickness and skill to snatch some of the animals before all of them disappear. Until one has actually seen them in the act, it is hard to believe that any animal can burrow as rapidly as amphioxus, for the tiny, eel-like creatures are as quick as the proverbial greased lightning. Seemingly, they can burrow through packed sand as rapidly as most fish can swim. They will lie quietly on the bottom of a dish of seawater, but if disturbed, they will swim rapidly with a spiral motion.

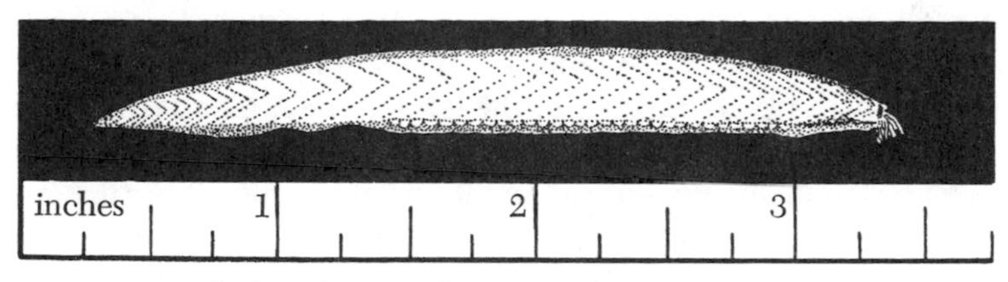

220. The lancelet *Branchiostoma californiense* (§266).

Though Figure 220 shows the shape of the animal, it is utterly impossible for any illustration, with the possible exception of a motion picture, to convey a clear notion of the appearance of a living lancelet in its natural habitat.

Amphioxus has achieved unique fame in the last half century through being interpreted as a form ancestral to the other vertebrates. An important theory of development makes this lancelet a vital step in the evolution of all the higher animals. We have taken it at Newport Bay, at San Diego Bay, and in El Estero de Punta Banda. It is known from the Gulf of California also, and is no doubt common in deeper water below the effect of waves, for it has been dredged off San Diego and in the Gulf.

* *Amphioxus* was the original name of the genus *Branchiostoma.* By the time the name was changed, biologists (the only people who discuss the animal very much) had got into the habit of referring to the lancelet as "*Amphioxus*," and the name had become almost colloquial. It is now customary to print it as "amphioxus," without italics or capital letter. The word, however, is still used grammatically as if it were a proper generic name, to the occasional bewilderment of editors and proofreaders.

Off the coast of China a similar species is dredged by the ton for food, and the Chinese, with their epicurean palates, consider it a great delicacy. Probably amphioxus is not as rare as has been supposed, even on our warmer shores, for its small size, retiring habits, and great speed in escape make it a form easily overlooked.

Living submerged in the sand with amphioxus is a small fish, *Gillichthys mirabilis,* the sand goby. We have no data concerning its habits.

§267. The acorn-tongue worm *Saccoglossus pusillus,* shares with amphioxus the honor of being considered an ancestral vertebrate, or at least of coming from similar primitive stock. Thirty years ago it was abundant in San Diego Bay and in what is now Los Angeles harbor, but suction dredging and industrial wastes have very materially changed the shorelines where it was once found. Ritter, when working up the embryology of the animal, wrote of taking more than a hundred on one tide; but now even the most proficient collector must be locally knowledgeable if he is to procure any in the southern region. It is reported to occur in Puget Sound, but we have never found any there. There was once a fine bed of them in front of the laboratory at Charleston, Oregon; but that sand flat is now a large parking lot, and the laboratory no longer has a waterfront. An occasional small specimen, possibly this species, has been dredged in Tomales Bay.

The males are pale orange, the females brighter. Various observers differ on whether the animals do or do not have castings at the mouths of their burrows and whether they do or do not protrude their orange or red proboscises at low tide so that these lie flaccid on the surface. Probably the conflicting statements arise from the fact that there are several species, all of them undescribed, that have different habits and habitats. We have never taken a whole specimen, but we have seen fragments of a size to indicate an animal at least 24 inches long—much larger than has been reported for S. *pusillus.* The excessively soft bodies are long, round, and very likely to be broken, and the animals secrete more than their share of slime. The various species of acorn-tongue worm have a not unpleasant but extremely pervading and persistent "estuary" odor that will scent the hands of the successful digger for several days.

10. Eelgrass Flats

SINCE the eelgrass, *Zostera,* occurs on flats of many types, from almost pure sand to almost pure muck, it seems desirable to list the animals associated with this seed plant before going on to the mud-flat forms.

§268. The start of a bed of eelgrass is an important step in the conversion of a former ocean region into wet meadowland and, ultimately, into dry land. The matted roots prevent the sand from being readily carried away by wind and tide, provide permanent homes for less nomadic groups of animals than inhabit the flats of shifting sand, and enrich the substratum with decaying organic matter until it can be classed as mud. Eelgrass supports a rather characteristic group of animals, which live on its blades, about its bases, and among its roots in the substratum.

§269. In El Estero de Punta Banda we have found small and delicate sea urchins crawling about on the blades of eelgrass, possibly scraping off a precarious sustenance of minute encrusting animals but more likely feeding on the grass itself. They are so pale as to be almost white, but with a faint coloring of brown or pink. This urchin, *Lytechinus anamesus,* is known from shallow water in the Gulf of California and from the west coast of Baja California. The largest recorded specimens are just over an inch in diameter. Ours are less than ½ inch; they resemble bleached, miniature, purple urchins.

§270. In an adjoining eelgrass pond in the same region were many thousands of small snails, the high-spired *Nassarius tegula,* also known as *Nassa* and *Alectrion.* Apparently these dainty little snails lead a precarious life, for about half of their shells were no longer the property of the original tenant, having been preempted by a small hermit crab.

These snails and hermits, and the urchins mentioned in the previous section, were all seen in the winter. The summer sunlight and temperature conditions in this region being decidedly unfavorable, it is likely that every animal able to do so would move downward beyond the chance of exposure. It is quite possible that El Estero, which we have not visited in hot weather, might then be a relatively barren place.

Many of the little *Nassarius* shells, whatever their tenants, are covered with white, lacy encrustations of the coralline bryozoan *Idmonea californica*. On superficial examination, the erect, calcareous branches of this form cannot be distinguished from tiny staghorn coral. It is fairly common in southern California; and it will be quickly noticed wherever found, and almost certainly mistaken for a coral. When encrusted on *Nassarius*, it makes the shells at least half again as heavy.

§271. Any of several hydroids—*Aglaophenia, Plumularia,* or *Sertularia*—will be found attached to eelgrass, as previously mentioned, or in other forms specific to this enviroment. The cosmopolitan *Obelia dichotoma* is common in situations of this sort all up and down the coast, and, in fact, in temperate regions throughout the world. Compared with other delicate and temperamental species of *Obelia, O. dichotoma* is very hardy and obliging. It withstands transportation well even when its water becomes somewhat stale, lives for several days under adverse conditions, and expands beautifully when its proper conditions are approximated. The delicate white clusters of this and of *O. longissima* (§316) are conspicuous on the tips of eelgrass in such places as Elkhorn Slough. Judging from specimens determined by Dr. Fraser, it is this form and *O. dubia* (§332) that have frequently been called *O. gracilis* locally; in any case, the differences are slight. *O. dichotoma* ranges on this coast from Alaska to San Diego; there is a good description, with an illustration of a living colony, in Johnson and Snook, pp. 59–60. We have not so far taken this form on the outer coast, but it occurs on Puget Sound wharf pilings with *O. longissima,* etc. At least some of the eelgrass hydroids taken at Jamestown Bay, Sitka, turn out to be the ubiquitous *O. longissima,* and it may be that the two are mingled there in the same habitat as they seem to be at Elkhorn Slough.

A minute nudibranch, *Galvina olivacea,* lives among these hydroids, presumably feeds on them, and tangles them up at certain times of the year with its white, jelly-covered egg strings. Some of the *Obelia* nudibranchs are among the smallest occurring within our geographical range

—a small fraction of an inch in length and identifiable only with a binocular microscope.

§272. Associated with the eelgrass of Elkhorn Slough, Tomales Bay, and similar places where the bottom is compact, sandy mud, there is a skeleton shrimp, *Caprella californica.* It is large, even by comparison with *Metacaprella kennerlyi* (§89) of the rock-pool hydroid colonies. Sometimes great numbers of them may be seen at night in the beam of a flashlight, under the surface of the water. Occasionally they will even be found on shores of pure mud. The animal may be bright green, red, or dull-colored, depending on whether it occurs on eelgrass, red algae, or muddy bottoms.

§273. Of the many eelgrass beds in Puget Sound, we know of one west of Port Townsend (and there are undoubtedly others) that supports a tremendous population of the fixed jellyfish *Haliclystus* (probably *H. stejnegeri,* Fig. 221). This species is actually one of the scyphozoan jellyfish, like those seen offshore in summer, pulsating lazily at the surface of the sea; yet it is small and has lost almost all power of motion. The small, stalked attachment disk has some power of contraction, and the tentacle-studded mouth disk is able to fold somewhat, but the animal swims not at all. We have never seen any of them so much as change their position on the grass blades, but it is presumed that they can glide about slowly on a smooth surface, like an anemone. Specimens detached from their grass will get a new hold with their tentacles and bend the stalk down until it can be reattached. An interesting feature of the digestive system is the presence of filaments that mix with the food in the body cavity, thus greatly increasing the area that the enzymes can act upon. The animal's colors are subtle rather than striking, but it is large enough (up to an inch in diameter) to be seen without difficulty.

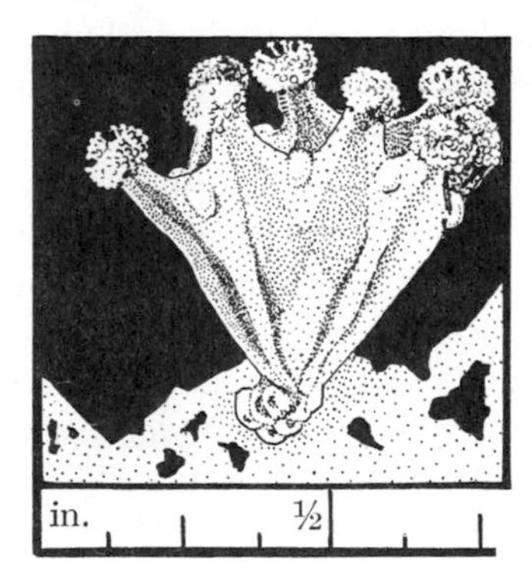

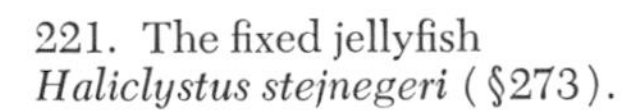
221. The fixed jellyfish *Haliclystus stejnegeri* (§273).

Originally described from the Commander Islands, in Bering Sea, *Haliclystus* is known to occur around Friday Harbor and may be expected in many other places in Puget Sound and to the northward. Some years ago Dr. Fisher took a single specimen at Point Joe, near Monterey, but its occurrence there was presumed to be accidental. No others were reported until, in May 1931, we discovered to our amazement that the animals had suddenly colonized at Monterey almost as abundantly as we found them in Puget Sound. They have continued to appear at Monterey each spring, have already been seen in Carmel Bay, and very likely have established themselves in other unpolluted spots where it is to be hoped they will become a permanent part of the fauna. Recent experience with this form, and the fact that it is not reported from the Nanaimo region, indicates that, although abundant in quiet spots, it requires fairly oceanic water. A few specimens have been found on coralline algae in deep tide pools on the Sonoma County coast.

§274. Necessarily small snails of several kinds frequent the slender leaves of eelgrass. *Lacuna porrecta,* shaped like *Littorina* but only one-fourth to one-half its size and having an aperture that takes up most of the front of the shell, may be found either on the grass or by screening the soil about the roots. MacGinitie remarks that this form waddles like a duck. It ranges from the Bering Sea to San Diego. The related *L. variegata,* of zigzag markings, may be found similarly on *Zostera* in the Puget Sound–Georgia Strait region.

Haminaea vesicula is a small, white, bubble-shell snail, up to almost 1 inch in shell length but usually about half of that, ranging from Vancouver Island to the Gulf of California. Like the great *Bulla* (§287) with which it sometimes occurs, it can scarcely stow away its comparatively large body in its fragile shell. In some specimens the beating heart can be seen through the shell. The eggs of this species are laid in deep yellow ribbons about ⅜ inch wide and up to nearly 8 inches long.

These two snails, and also a minute black *Philine* found at Elkhorn Slough, are tectibranchs, related to the large and shell-less sea hares. *Philine* always moves through a slime envelope of its own secreting. The larger *Navanax* (§294), also shell-less, is a feature of the southern eelgrass beds and the adjacent mud flats, where it feeds on *Haminaea.*

Another tectibranch, which looks much more like a nudibranch than like its relatives already mentioned, is a pretty green form with India-ink lines of stippling. It is *Phyllaplysia taylori,* reported from Nanaimo to San Diego and very common in Tomales Bay. Ovipositing individuals of *Nassarius fossatus* (§286) will be found among blades of eelgrass during the late summer, depositing encapsulated eggs. Even an annelid worm, the small *Nereis procera,* known more commonly from dredging in Monterey Bay, will be found at home here on the *Zostera* (and on *Enteromorpha*), where it constructs permanent membranous tubes on the blades and about the roots.

A very narrow limpet, the painted *Acmaea depicta,* is restricted to this habitat from Santa Barbara to Baja California. Though not at all obvious, it is readily discoverable if one runs a quantity of eelgrass through one's hands; where common, it may occur on every tenth blade of eelgrass. A similar limpet, but without the well-marked pattern of brown lines on a white surface, is *Acmaea paleacea,* which ranges from Vancouver Island to somewhat north of *A. depicta*'s southern limit, so that the two overlap in southern California. A third limpet found associated with plants, one somewhat broader and with a more striking pattern of markings than *A. depicta,* occurs as far north as Sitka; this is *Acmaea triangularis,* whose southern limit is about the same, but whose habitat is coralline algae rather than eelgrass.

§275. *Chioraera* (known also as *Melibe*) *leonina,* a large nudibranch of eelgrass areas, seems to be in the process of becoming a pelagic animal—that is, a drifting or mildly swimming animal. The head end of *Chioraera* is expanded into a broad, elliptical hood bordered by tentacle-like processes, and the cerata along the back are flattened, leaflike structures. It is pale yellow, with the grayish branches of the liver showing clearly in the cerata. When the water is smooth, and especially during cloudy weather, this queer creatures puffs out its hood with air, thus floating to the surface, where its tentacles rake in small crustaceans for food. It is common in some years in the Friday Harbor region, particularly on Brown Island. In 1928 approximately 1,000 appeared at one time, and of these an estimated 150 pairs were copulating, mostly in mutual coitus. Guberlet wrote: "During the act of copulation the animals lie side by side, or with the bottom of the foot of each together, with their heads in opposite directions, and the penis of each inserted in the vaginal pore of the other . . . and the animals so firmly attached that they can be handled rather roughly without being separated." After they have spawned the animals take on a shrunken appearance, and many, if not all of them, die, the indications being that they live only one season and spawn but once. The eggs are laid during the summer in broad spiral ribbons 3 or 4 inches long.

The recorded range is from Nanaimo to San Diego, but south of Puget Sound it must be rare or sporadic. In southern Puget Sound we have found occasional solitary forms only, but in central British Columbia we once came upon a great host of them about one of the rare eelgrass beds in this ordinarily precipitous region.

§276. The southern scallop, *Pecten latiauratus,* is commonly seen at Newport Bay and southward, attached to eelgrass clusters or swimming about their bases in the peculiar manner of the scallop tribe. It differs from *P. circularis* in being flatter and in having wider "ears" at the hinge line.

§277. A massive and most amazing sponge, *Tetilla* (§298), is sometimes attached to eelgrass clusters, but it is more common on the mud flats, where it leads an apparently normal life under abnormal conditions.

§278. In and about the roots of eelgrass lives a pistol (snapping) shrimp, *Crangon californiensis,* that differs from its relative of the outside pools only in minor characters, such as having a slightly smaller snapping claw of somewhat different shape.

Another quite different shrimp to be found here, and also in mudflat pools, is the lovely, transparent *Spirontocaris paludicola.* It is scarcely distinguishable from the *S. picta* described in §62. In Puget Sound a grass shrimp, *Hippolyte clarki,* occurs commonly in this association even in the daytime, its green color matching that of the eelgrass to which it clings so well that the observer may be working in a colony of them without realizing it until one of the graceful animals swims away. *Idothea resecata* (§205) will also be found clinging lengthwise to the blades, often abundantly, without its presence being suspected. Further north other crustacea may be found; at Sitka, in addition to the forms mentioned above, *Spirontocaris cristata* and *S. camtschatica* have been taken, among others.

§279. The jackknife clam *Solen sicarius* is, for its kind, an extremely active animal. It digs so rapidly that it can bury itself in 30 seconds with only a few thrusts and draws of its foot. But it has even more striking accomplishments: according to MacGinitie, it can jump a few inches vertically and can swim by either of two methods. Both involve the familiar "rocket" principle. In the first method, water is forcibly expelled from the siphons, thus propelling the clam forward; in the second, water is expelled from the mantle cavity through an opening around the foot, and the resulting darting motion is accelerated by a flip of the extended foot, the combined forces sending the animal some two feet through the water in a reverse direction.

S. sicarius has a glossy yellow, bent shell, reaches a length of a little more than 3 inches, and ranges from Vancouver Island to Baja California. Like its somewhat smaller, straight-shelled relative of the sand flats, S. *rosaceus* (§260), it autotomizes its siphons at the annular constrictions. The two may at times be found together in the sand flats, but S. *sicarius,* although it is occasional in mud flats also, is far more common among the roots of eelgrass.

§280. Of many of the worms that writhe about in the roots we freely confess our ignorance. Of others, which we can recognize, so little is known that a full list would "little profite the Reader" and of them all only three will be named. The first of these is the slender and iridescent *Glycera americana,* which looks not unlike a *Nereis* with a pointed head. The difference becomes obvious when *Glycera* unrolls a fearful introvert almost a third as long as itself—an instrument armed with four black terminal jaws that are obviously made for biting. When the introvert is extended the head loses its characteristic pointedness. Specimens occur singly and may be 8 inches or even more in length.

§281. The second worm to be mentioned is *Cistenides* (formerly *Pectinaria*) *brevicoma,* whose body is tapered to fit the slender, cone-shaped tube that it builds of sand grains cemented together. This worm digs by shoveling with its mouth bristles, which may be of brilliant gold, as in the case of one specimen that we dredged from the mud bottom of Pender Harbor, British Columbia. A model woven from cloth of gold could not be more striking. We have specimens from the vicinity of Juneau also, and they occur in Tomales Bay and Elkhorn Slough. The minute and highly specialized pea crab *Pinnixa longipes* (§293), widest of the lot, is a common commensal. The common southern California species, found from Redondo Beach to Todos Santos Bay, is *Pectinaria californiensis* (see color section); it is very similar, perhaps with a somewhat straighter shell than *Cistenides.*

Watson worked with Fabre-like patience on these animals, and his observations, hidden away from popular reach in an English biological periodical, seem to us as popularly fascinating as those of that French entomologist. "Each tube is the life work of the tenant," he writes, "and is most beautifully built with grains of sand, each grain placed in position with all the skill and accuracy of a human builder." Each grain is fitted as a mason fits stones, so that projecting angles will fill hollows, and one appreciates the nicety of the work when one perceives that the finished wall is only one grain thick. Furthermore, no superfluous cement is used. Suitable grains for building are selected by the mouth and applied to the two lobes of the cement organ, "after which the worm applies its ventral shields to the newly formed wallings, and rubs them up and down four or five times, apparently to make all smooth inside the tube. . . . The moment when an exact fit has been obtained is evidently ascertained by an exquisite sense of touch. On one occasion I saw the worm slightly alter (before cementing) the position of a sand grain which it had just deposited." Only very fine grains are used, but one specimen in captivity built ¾ inch of tube in two months.

Cistenides lives head downward, usually with the tube buried vertically in the substratum, although sometimes, and especially at night, an inch or so of the small end of the tube may project above the surface. Digging is accomplished by means of the golden bristle-combs that project from the head, "tossing up the sand by left and right strokes alternately." The worm eats selected parts of the sand that it digs: that is, it passes sand through its body for the contained food. The digging therefore creates a hollow about the digger's head, and frequent cave-ins result. Thus the worm is not only able to dig its way to new feeding grounds, but able to remain in the same place for a time and make its food fall within reach of its mouth. There are strong hooks at the hind end of the body that enable the animal to keep a firm grip on its tube while digging. For respiration and the elimination of waste matter, two currents of water are pumped through the tube in opposite directions. *Cistenides* is able to repair breaks in its tube, appears to be light-sensitive, and probably has a tasting organ. Like some other sand-burrowing forms, it may be here one year and somewhere else the next; a whole colony may simply disappear over the winter.

§282. The third worm, perhaps the most striking of all but the least common, is a magnificent species, *Dendrostoma perimeces*. Large specimens of this slim sipunculid are almost 10 inches long when expanded. Except for its slimmer body and usually larger size, this worm does not differ greatly from *D. pyroides* (the species illustrated in Figure 131); but closer examination will reveal that the tentacles of *D. perimeces* branch clear to the base, whereas the lower part of the smaller form's tentacles is bare. The delicately branched and beautifully flowering tendrils presumably gather food particles, which are transferred to the intestine by rolling in the introvert. This form has been taken only at Elkhorn Slough.

§283. In a good many of the quiet-water channels of Puget Sound the subtidal gravel shores support a variety of eelgrass that is always at least partially submerged on even the lowest tides. The characteristic fauna that belongs to these deep plants can scarcely be considered intertidal,

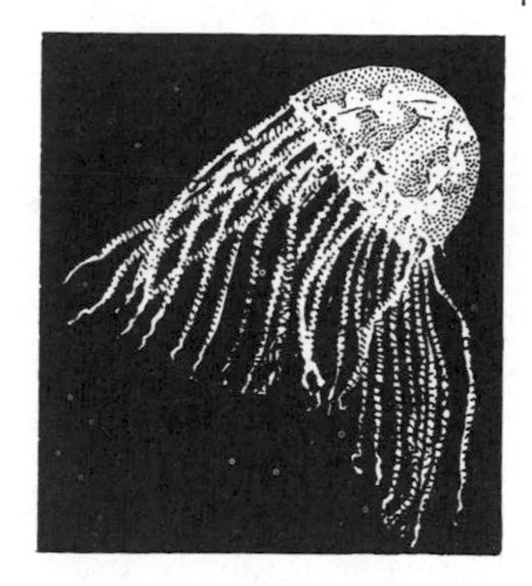

222. A hydroid jellyfish, *Gonionemus vertens* (§283); natural size.

but the shore collector will fringe this zone and will surely see, among other things, the hydromedusa *Gonionemus vertens* (Fig. 222). This graceful little bell-shaped medusa (½ to ¾ inch in diameter), almost invisible except for an apparent cross inside the bell formed by its reddish-brown gonads, has some 60 to 70 long, slender tentacles, with which it stings and captures small fish and crustaceans. It seems to be restricted to these particular growths of deep eelgrass, but every such location that we have examined in summer has, without exception, turned out a few specimens. When they are not visible, thrusting an oar into the eelgrass, or merely wading slowly through it, will send some of them pulsing toward the surface. On overcast days, however, they will be found near the surface, often in considerable numbers. In the winter there may not be a single specimen to be found.

Watching and timing the life cycle of this form through the hydroid stage is one of the many tasks that await some enthusiast who has the requisite time and patience. In this case the knowledge would be particularly welcome, for *Gonionemus* has for many years been the classic type of its group for teaching purposes. Some work has been done in this connection. It is known, for instance, that the fertilized egg develops into a ciliated, free-swimming larva that is oval in shape and somewhat drawn out at one end. The larva settles down on its broad end and develops at its drawn-out upper end first a mouth and, a few days later, four tentacles, two at a time. In this hydroid stage the animal is scarcely more than 1/16 inch tall with its tentacles expanded. Additional hydranths are budded off, but how the medusae are produced is not known.

Rugh has shown, after extensive experimentation with an east-coast species, that under normal conditions *Gonionemus* never lays its eggs in the light, although light is probably necessary for the maturation of the egg cells. During the breeding season the animals prepare to discharge their eggs within one minute after being placed in artificial darkness. Apparently the eggs are laid in several installments, the dates and duration of which the experimenter does not mention. Rugh remarks that the individual egg is large for so small an animal.

Occasional visitants from deep water may sometimes be seen in this environment. The cucumber *Stichopus californicus* (§77) is common underfoot. Shrimps occur, especially at night, including such commercially important forms as the "coon stripe," *Pandalus danae*; but they are rare and solitary. Although considerations of this nature are beyond the scope of this work, it may be interesting to note the answer (worked out by Berkeley, 1930) to a question frequently asked by shrimp fishermen. They have wondered why it is that the only males taken of a given species are small, while all the females are large. The rather spectacular answer is that the five species investigated by Berkeley invariably undergo a sexual metamorphosis. Young specimens are always male. At the age of from 18 to 40 months, depending on the species, they become females and presumably spend the rest of their lives as functioning members of that sex. This type of sexual alternation is known as protandry.

11. Mud Flats

THE PROBLEMS of respiration and food-getting in mud flats and the lack of attachment sites make for a rather specialized fauna. The skin breathing of seastars and urchins makes them unsuited to this environment, and, with the exception of an occasional *Pisaster brevispinus,* none occur. There are only one sponge and one hydroid, and there are no chitons or bryozoans; but the many worms, clams, and snails make up for the scarcity of other animals.

Even in the south the animals extend farther up into the intertidal on the mud flats proper than they do on the sand flats anywhere on the coast. Some of them, like the fiddler crab, seem to be changing into land dwellers—a tendency that reaches an extreme in Japan, where one of the land crabs has worked up to almost 2,000 feet above sea level.

Zones 1 and 2. Uppermost Horizon and High Intertidal

§284. The little fiddler crab *Uca crenulata* (Fig. 223) is characterized, in the males, by the possession of one relatively enormous claw, which is normally carried close to the body. It looks like a formidable weapon, but is used to signal for females and to fend off other males and as a secondary aid to digging. In battle the two contestants lock claws, each apparently endeavoring to tear the other's claw from its body. Sometimes, according to one capable observer, a crab loses its hold on the substratum and is "thrown back over his opponent for a distance of a foot or more"—a performance that suggests a judo wrestling match. At breeding time, and particularly when a female is nearby, the males make a peculiar gesture, extending the big claw to its full length and then whipping it suddenly back toward the body. One might conclude, with some logic, that the big claw is a distinctly sexual attribute comparable to the horns of a stag.

The Japanese call the fiddler crab *siho maneki,* which translates as "beckoning for the return of the tide." It is too picturesque a name to quibble over, but one might reasonably ask why Mahomet does not go to the mountain, for the presumably free-willed fiddler digs its burrow as far away from the tide as it can get without abandoning the sea entirely. There it feeds, like so many other animals occurring on this type of shore, on whatever minute plants and animals are contained in the substratum. Instead of passing quantities of inert matter through its body, however, the fiddler crab daintily selects the morsels that appeal to it. The selecting process begins with the little claws (the female using both claws and the male his one small one) and is completed at the mouth, the rejected mud collecting below the mouth in little balls, which either drop off of their own weight or are removed by the crab.

Dembowski, working with the sand fiddler of the east coast, made careful observations of the animal's behavior and came to the interesting conclusion that its behavior is no more stereotyped than that of man. He found that in their natural habitat the fiddlers construct oblique burrows up to 3 feet long, which never branch and usually end in a horizontal chamber. In sand-filled glass jars in the laboratory they actually sought the light, digging their burrows along the sides of the jars. Apparently it was an advantage to be able to see what they were doing; and although this condition had never existed in nature, they made use of the opportunity when it did occur. They dig by packing the wet sand between their legs and carapaces and pressing it into pellets, which they then carefully remove from the burrow. The smallest leg prevents loose grains from falling back while the pellet is being carried. Sometimes hours are spent in making the end chamber: Dembowski thinks that its depth is determined by how far down the fiddler must go to reach sand that is very moist but not wet. In pure sand water would filter into the burrow at high tide, but the crab always takes care to locate in substratum that has so high a mud content that it is practically impervious to water. Furthermore, it lines its burrow and plugs the entrance before each high tide. Thus at high tide or during heavy rains the animal lives in an airtight compartment.

Under experimentally unchanging conditions the crabs showed no

3. *Uca crenulata* (§284), the fiddler crab: above, a male specimen about 1½ inches in carapace width; low, fiddlers on the sand in upper Newport Bay (photo taken July 1962).

signs of the periodicity that might be expected in an animal that stoppers up its burrow against each high tide. Evidently the door-closing process is due not to memory of the tidal rhythm but to the stimulus of an actually rising tide, which would moisten the air chamber before the surface of the ground was covered. If a little water was poured into their jars, the crabs started carrying pellets to build a door. When the water was poured in rapidly, the crabs rushed to the entrance and pulled pellets in with great haste, not pausing to construct a door. If the jar was then filled up carefully, so as not to damage the burrows, the crabs would stay in their air chambers and not stir for a week. Dumping water in, however, filled the burrows and caved them in, burying the animals. Their response this time was to dig out and wait on the surface of the sandy mud for the "tide" to ebb.

Each "high" high tide destroys part of the burrow, and the animal must dig itself out by somewhat reversing its former operations. By digging, it "causes the chamber to rise slowly in an oblique direction, still keeping its volume unaltered, as the sand is always carried from the roof of it to the bottom," again by means of pellets. The procedure varies widely with the "individuality" of the animal.

Dembowski insists that this and other reactions are variable, very little being automatic. Thus the fiddler crab has to adjust not merely a pellet of sand but "always *this single pellet* with all its individual particularities." The animal's nervous system (including its nearest approach to a brain) is too simple to account for this variety of reactions: "We are compelled to admit a plasticity of the nervous centers; they must possess a certain creative power that enables them to become adapted to entirely new situations. The number of possible nervous connections is limited, but the number of possible reactions is infinite. This discrepancy may be avoided only by admitting that each nervous center may perform an infinity of functions." Since the days of Dembowski, however, the field of behaviorism, of critical study of the actions of lesser creatures (having nothing to do with the almost extinct branch of psychology promoted by the late Dr. Watson—not of Baker Street), has obliged us to reconsider our notions of creative and plastic behavior on the part of animals like fiddler crabs. They are, in fact, rather stereotyped animals, and their courtship behavior in particular is so ritualistic, both in time between gestures and in sequence of gestures, that different species can be recognized in the field by their characteristic courtship motions. For a more modern treatment of the behavior of fiddler crabs the serious student should consult the work of Peters (1955).

Under natural conditions the fiddlers are probably never in water longer than is necessary to moisten their gills, and they can live in air for several weeks without changing the water in their gill chambers. Nevertheless, several animals lived in seawater for 6 weeks with no apparent ill effects. Dembowski concludes that they are true water-breathers but have interesting land-living adaptations; the gills are probably filled with seawater rather than with nearly fresh water, as has been assumed.

Uca crenulata is the only fiddler crab north of southern Baja California, and it is known only as far north as Anaheim Bay. Our *Uca* is a creature of higher marshlands—regions peculiarly susceptible to improvement by man. Where any of the right kind of tidal marsh is left, a few *Uca* will still be found, as in upper Newport Bay, but fiddlers are becoming rarer every day.

It has been remarked that specimens are common only in summer. We got an inkling of why that should be, in January several years back, when we persistently dug out a burrow in the estuary south of Ensenada. The living animal was at the 40-inch level in a tunnel that extended to a measured 48 inches from the surface. The water in the pool at the bottom was bitterly salt.

§285. Another burrowing crab, superficially resembling a fiddler crab and with somewhat the same mannerisms, but living at all tidal levels, was once abundant in Newport Bay. This animal, *Speocarcinus californiensis* (Fig. 224), is apparently now rare. This crab shares some mannerisms with the fiddler: both bristle up at the mouths of their burrows to intimidate an intruder, and both scurry down into the security of their burrows before danger comes near. The lively fiddler crab never goes much below the high-tide line, however, and the less active *Speocarcinus* roams through the whole intertidal region. *Speocarcinus* has fringes of hair on its legs and carapace edges, and short eye stalks; the fiddler is hairless and has long and prominent eye stalks.

North of *Speocarcinus*'s range, which is Los Angeles to San Diego, the common mud-flat crab of the higher tidal reaches is *Hemigrapsus oregonensis* (Fig. 225). This species looks like a small edition of the purple crab of rocky shores (*H. nudus*) that has been bleached to a uniform light gray or muddy yellow on top and white underneath, deprived of its underside spots, and given, in compensation, a slight hairiness about the walking legs. Night, half tide, and an estuary like Elkhorn Slough furnish the ideal combination for seeing an impressive panoply of crab armor. Hordes of these aggressive yellow shore crabs rear up in formi-

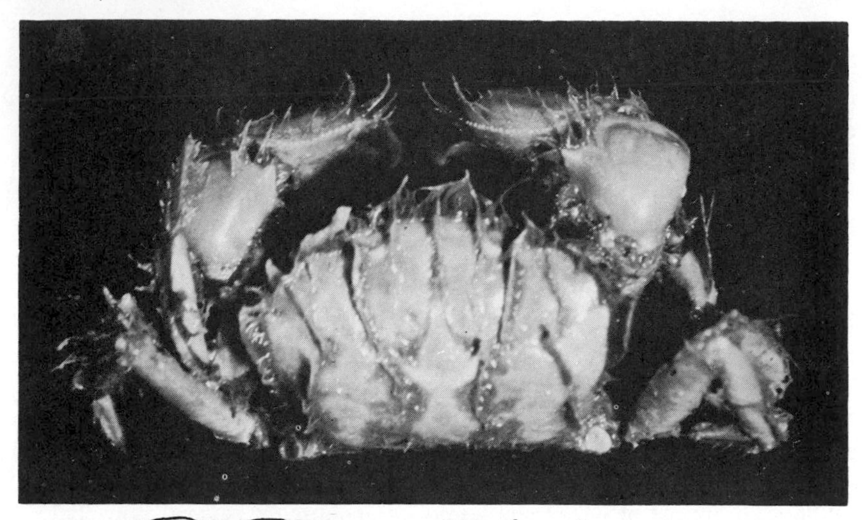

224. Above: *Speocarcinus californiensis* (§285), a burrowing crab of high intertidal mud flats.

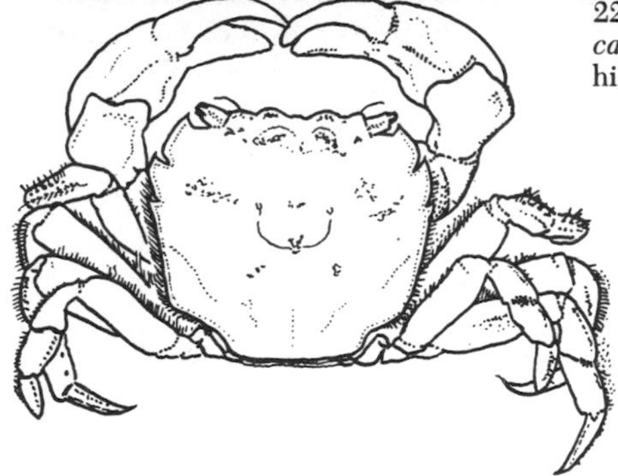

225. Left: *Hemigrapsus oregonensis* (§285), the shore crab of Puget Sound and San Francisco Bay; natural size.

dable attitudes, seeming to invite combat. The gesture is largely bluff: if hard pressed they resort to the comforting philosophy that he who runs away will live to fight some other time when he is more in the mood. *H. oregonensis* ranges southward from Alaska and is the common shore crab of San Francisco Bay. In Puget Sound and British Columbia it is extraordinarily abundant. Hart (1935), investigating the development, found that the young are hatched from May until August at Departure Bay.

The purple shore crab occurs to a limited extent along the clay banks in the upper parts of sloughs, but will generally be found pretty closely associated with rocks.

Zone 3. Middle Intertidal

Here, again, it is convenient to divide this prolific zone into different subhabitats—surface and substratum.

Surface Habitat

§286. The channeled basket shell, *Nassarius fossatus* (Fig. 226), also known as *Nassa* and as *Alectrion,* is the commonest of the large predacious snails on suitable mud flats between British Columbia and Baja California, although it is predominantly a northern species. Incidentally, it is one of the largest known species of this genus, large specimens being nearly 2 inches long, and the clean-cut beauty of its brown shell makes it a very noticeable animal. Ordinarily it will be seen plowing its way through flats of mud or muddy sand, probably in search of food. MacGinitie found that the animals are attracted by either fresh or decaying meat and will reach up for meat held in the air above them. When feeding, *N. fossatus* wraps its foot completely around the food, hiding it until it is consumed.

In August, at least, many individuals may be found depositing eggs in "shingled" capsules on the blades of eelgrass. MacGinitie also watched this process. The snail first explores the surface with its sensory siphon and then cleans a spot with its radula. Next it forms a fold in its foot to connect the genital pore with the pedal gland, and through this fold the naked egg mass is passed. The lip of the pedal gland is pressed against the cleaned surface for 9½ minutes during the process. When the eggs have entered the pedal gland, the infolded tube disappears, and the foot is used to form a water chamber, presumably for aerating the egg capsules already laid. During this 9½ minutes both the body and the shell oscillate, for some unknown reason, and after this period "the anterior part of the foot is lifted upward and backward, and, as the capsule is now cemented to the object to which it is being attached, it pulls out of the pedal gland as a completed egg capsule." The entire process has occupied 12½ minutes, and the animal is ready to move on and repeat the operation on a newly cleaned area just ahead. One typical string a little more than 2½ inches long contained 45 capsules and was 9 hours in the making. The flattened capsules, averaging just over ⅛ inch long, are laid with the base of one overlapping the base of the next, producing the "shingled" effect mentioned.

Another considerably smaller predacious snail is the oyster drill, *Uro-*

226. The channeled basket shell, *Nassarius fossatus* (§286).

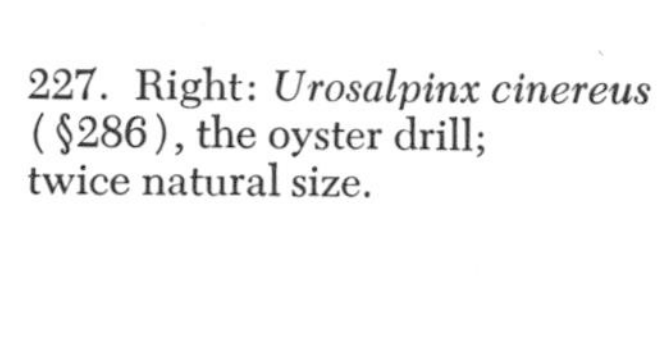

227. Right: *Urosalpinx cinereus* (§286), the oyster drill; twice natural size.

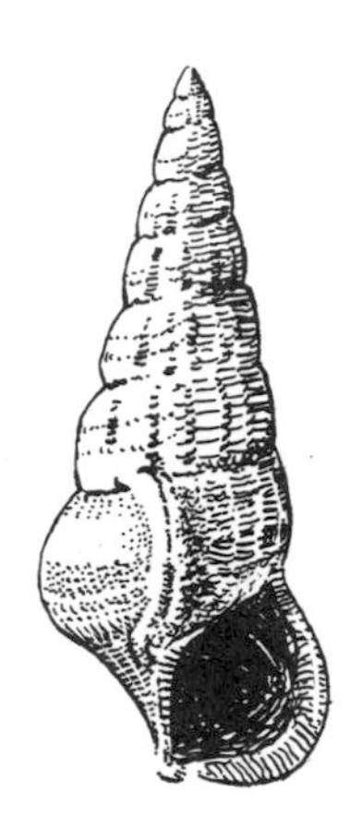

228. Left: the tall-spired horn shell, *Cerithidea californica* (§286); twice natural size.

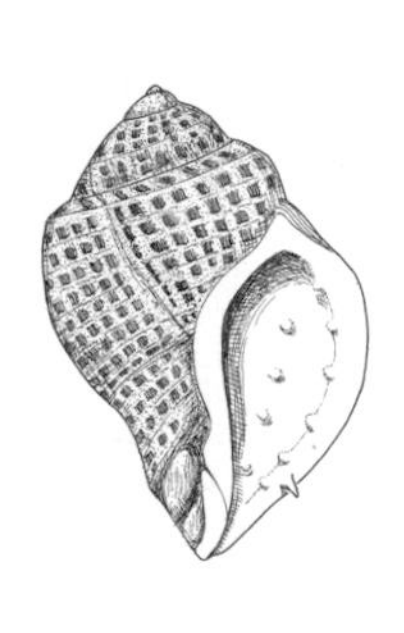

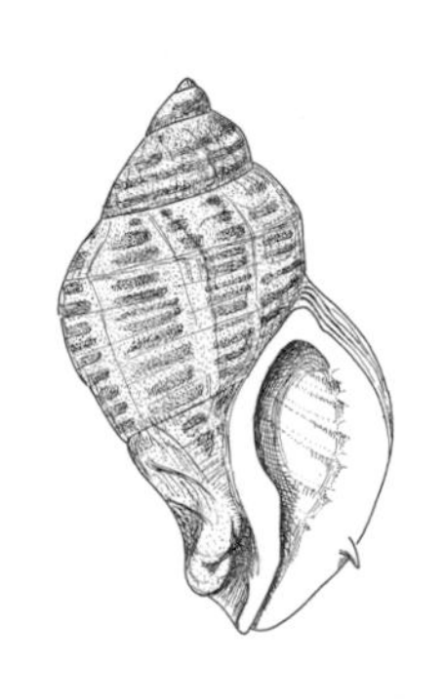

229. Two thorn snails (*Acanthina*). On the extreme left is *A. paucilirata* (§286), found in southern bays and estuaries; the larger shell, on the right, is *A. spirata* (§165), of open-coast mussel beds.

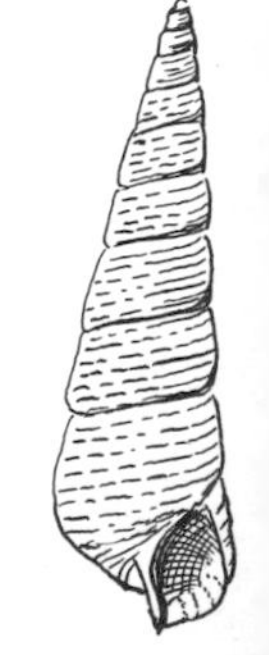

230. Right: *Batillaria zonalis* (§286).

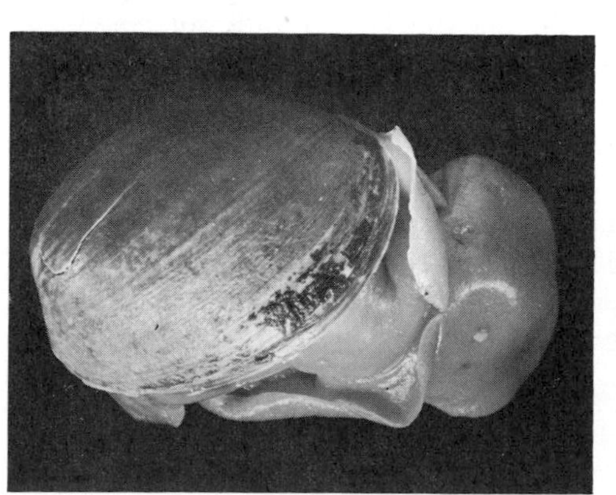

231. The bubble snail *Bulla gouldiana* (§287).

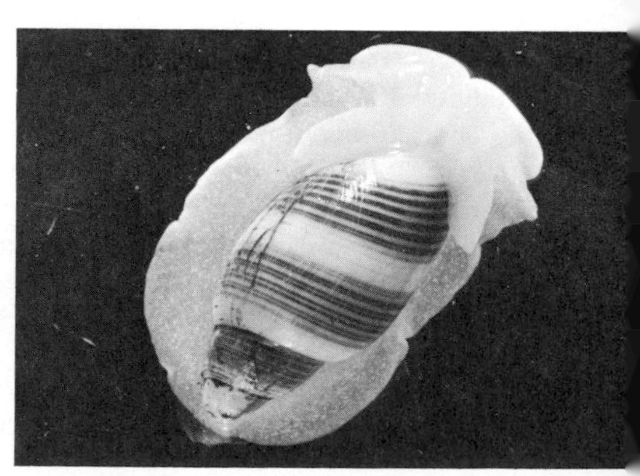

232. *Acteon punctocaelatus* (§287), a barrel shell.

salpinx cinereus (Fig. 227), a snail that was accidentally introduced from the east coast along with the Virginia oyster. It is an inoffensive-looking animal, like a small dingy gray *Thais,* but once it gets established it can work considerable havoc with the oyster beds. Its method of attack—the same as that employed by *Thais* and the great moon snail—is to drill a hole through the bivalve's shell and hack out the soft body.

Southern mud flats have several characteristic snails. The tall-spired horn shell *Cerithidea californica* (Fig. 228) extends to the upper reaches of such sloughs as Newport Bay, even where there is some admixture of fresh water; and *Acanthina paucilirata* (Fig. 229) occurs in the stretches more frequently washed by ocean water. The shell of *Cerithidea* is dark brown or black; that of *Acanthina* is *Thais*-like, with four to six whorls that are crossed by broad black longitudinal stripes and narrower white stripes, giving a somewhat checkered appearance.

From Moss Landing northwards, perhaps wherever Japanese oysters have been imported, will be found a tall spired snail superficially similar in appearance to *Cerithidea* and even more similar in its ecological requirements. This is *Batillaria zonalis* (Fig. 230); a Japanese native, it was first noticed about 1935, and is evidently doing very well.

§287. In the south a large tectibranch, the bubble-shell snail *Bulla gouldiana* (Fig. 231), with a shell some 2 inches long, is one of the most common of all gastropods in the Newport Bay and Mission Bay region. The paper-thin shell is mottled brown, and the body, which is too large for the shell, is yellow. When the animal is completely extended and crawling about in the mud, the mantle covers most of the shell. One observer says that the bubble-shell snails eat bivalves and smaller snails, swallowing them whole. The long strings of eggs, lying about on the mud or tangled in eelgrass, are familiar sights in summer.

The barrel shell *Acteon punctocaelatus* (Fig. 232) is sporadically common in central California bays. The animal is less than ¾ inch long but is so striking that it is sure to be picked out of its surroundings, for narrow black bands follow the whorls around its white shell.

§288. Mud flats would seem to be the most unlikely localities of all to harbor hydroids, but the tubularian *Corymorpha palma* (Fig. 233) is famous in the Newport–San Diego Bay area. Specimens attaining the very great size (for a shore hydroid) of 3 inches have been taken, and in some areas whole forests of these delicate fairy palms grow on the flats. The solitary individuals fasten themselves to the substratum by ramifying roots that come away as a chunk of mud. The animals are

233. The large mud-flat hydroid *Corymorpha palma* (§288); natural size.

watery and transparent, and therefore will often be overlooked if the light is poor. When the ebbing tide leaves them without support, they collapse on the mud, there to lie exposed to sun and drying winds until the tide floods again. It is remarkable that so delicate an animal can survive this exposure, but it apparently suffers no harm.

Around the single flowerlike "head" of the animal (the hydranth) the tiny jellyfish are developed, which, although they pulsate like free jellyfish in the act of swimming, apparently never succeed in freeing themselves from the parent hydroid. There they stay until they reach maturity, deposit their eggs, and wither away. The eggs may be transported some distance by currents, but wherever they touch they settle and begin to grow as hydroids, so that there is no free-swimming stage whatever.

Corymorpha is one of that minority of Pacific coast intertidal animals that have been closely observed and studied. In aquaria it remains erect with spread tentacles so long as a current of water is maintained, but begins a slow and regular bowing movement as soon as the current is stopped. During each bow it sweeps the mud with its tentacles; then, as it straightens up again, it rolls up its long tentacles until its short tentacles can scrape the food from them and pass it to the mouth. The animal is well known for its remarkable powers of regeneration. If the entire hydranth is cut off, a new one will grow within a few days, and sections cut out of the "stem" will grow complete new individuals.

Like other hydroids, *Corymorpha* has its associated sea spiders. Near Balboa we once found, in August, a bank of muddy sand that supported a *Corymorpha* bed about which were strewn large numbers of the straw-colored and quite visible gangly pycnogonid *Anoplodactylus erectus*, which occurs also with southern *Tubularia* (§333). Presumably it was a seasonal occurrence.* The sea spiders had probably just grown

* Today (1967) these things are rare in the area, mostly because of the changes associated with progress.

to an independent stage and released themselves from their hydroid nurses—quite likely by eating their way out.

Burrowing Habitat

§289. One of the noticeable substratum animals in the Newport Bay flats of sandy mud is the sea pen *Stylatula elongata,* but whether or not it should be considered a burrowing animal will depend on the observer's reaction—which, in turn, will depend on the state of the tide. The collector who appears on the scene between half tide and low tide will be willing to swear before all the courts in the land that he is digging out a thoroughly buried animal. But let him row over the spot at flood tide, and he will see in the shallow water beneath his boat a pleasant meadow of waving green sea pens, like a field of young wheat. The explanation comes when he reaches down with an oar in an attempt to unearth some of the "plants." They snap down into the ground instantly, leaving nothing visible but the short, spiky tips of their stalks. Like the anemone *Harenactis* (§257), they have a bulbous anchor that is permanently buried deep in the bottom. When molested, or when the tide leaves them exposed, they retract the polyps that cover the stalk and pull themselves completely beneath the surface. When cool water again flows over the region, they expand slowly until each pennatulid looks like a narrow green feather waving in the current.

Pennatulids as a group are notably phosphorescent, and our local species are no exception. Both *Stylatula elongata* and *Acanthoptilum gracile* (Fig. 234), the second common in Tomales Bay, exhibit startling flashes of light when adapted for about half an hour in a dark room and touched with a needle or stimulated with an electric current.

§290. Even in this middle intertidal zone there are many burrowing animals besides the spectacular *Stylatula.* Another relative of the hydroids, jellyfish, and corals is the burrowing anemone *Edwardsiella californica* (Fig. 235), which also occurs in great profusion. Early in March 1931 we found a sandy mudbank east of Corona del Mar that must have averaged more than 50 of these animals to the square foot. It goes without saying that they are not large, a good-sized, extended specimen being possibly 2 inches long. The wormlike body, which normally protrudes slightly above the surface, is almost covered with a brown and wrinkled tube. When first dug up, the animal looks very much like a small peanut worm (sipunculid). In an aquarium the disk flowers out, during the afternoon, with symmetrically arranged tentacles, and the attachment bulb expands and becomes transparent, so that

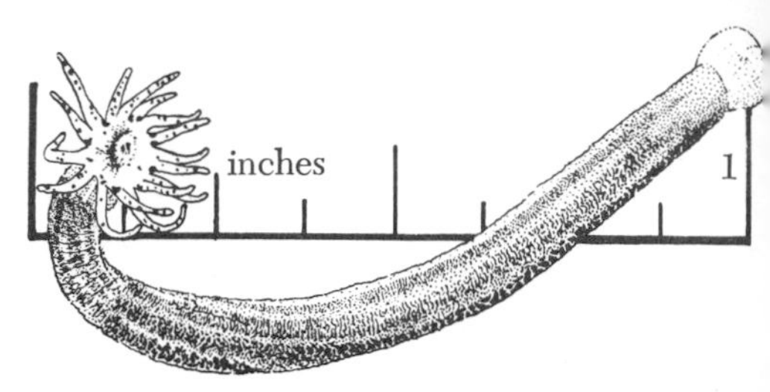

235. A burrowing anemone of the mud flats, *Edwardsiella californica* (§290).

234. The sea pen *Acanthoptilum gracile* (§289).

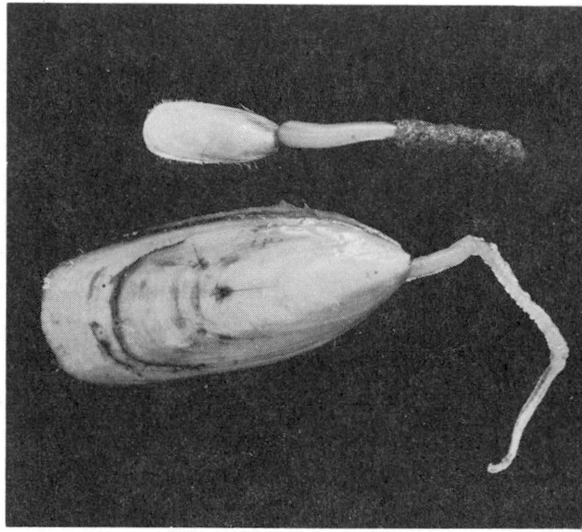

236. The brachiopod *Glottidia albida* (§291).

237. *Callianassa californiensis* (§292), the red ghost shrimp.

the partitions are visible as eight longitudinal bands. This species of *Edwardsiella* is known from southern California only, but a similar or identical form occurs (or occurred) rarely at Elkhorn Slough.

§291. During some years the southern California muddy sand contains a considerable population of the stalked brachiopod *Glottidia albida* (Fig. 236), the second and last of its group to concern us. This small bivalved animal (which, although it resembles a thin, stalked clam, is not even distantly related to the mollusks) is almost a facsimile of the large tropical *Lingula,* the oldest living genus in the world. Remains of *Lingula* date back to the Ordovician period, the second oldest geological period in which undoubted fossils occur. Millions of years after *Lingula* was established, myriads of other animals developed, most of which perished a few more millions of years before the first man learned to stand on his hind legs. It certainly speaks much for the tolerance and adaptability of an animal that it has lived from such a remote geological age to the present time and still persists in such numbers that races bordering the Indian Ocean use it for food, selling it in public markets as we sell clams.

Our Pacific coast specimens of *Glottidia* live in burrows, retracting at low tide so that only a slitted opening can be seen at the surface. The white, horny shells are rarely more than an inch long. The fleshy peduncle with which the animal burrows and anchors itself may be two or three times the length of the shell. The animal is not helpless when deprived of its peduncle, however: one specimen so handicapped was able to bury itself headfirst, presumably with the aid of the bristles that protrude from the mantle edge. When covered with water and undisturbed, the animal projects its body halfway out of its burrow, and at such times the bristles are used to direct the currents of water that carry food and oxygen into the body.

Glottidia is more abundant as a subtidal species in southern California waters, although there was once an intertidal bed of them near the laboratory at Corona del Mar. One bright morning years ago a group of conchologists raked them all up, every one, and they did not return; the memory of this infamous excursion used to turn George MacGinitie purple for years afterward. An occasional stray *Glottidia* has been collected in Tomales Bay.

§292. One of the burrowing ghost shrimps, the red *Callianassa californiensis* (Fig. 237), will be found in this zone. Next to the fiddler crab and the somewhat similar *Speocarcinus,* this pink-and-white, soft-bodied ghost shrimp seems to be the most shoreward of the burrowing forms, with an optimum zone of 0 to plus 1 foot. It occurs, too, over a wide geographical range, reaching southeastern Alaska to the northward. We have seen it in the Strait of Georgia, Puget Sound, Tomales Bay, Elkhorn Slough, Newport Bay, and El Estero de Punta Banda.

Callianassa is neatly adapted to its life in burrows, but it will survive in aquaria without the reassurance of walls around it. Earlier reports that it would not survive outside the burrow may have been based on injured specimens. According to MacGinitie, the animal digs its burrow with the claws of the first and second legs, which draw the sandy mud backward, collecting it in a receptacle formed by another pair of legs. When enough material has been collected to make a load, the shrimp backs out of its burrow and deposits the load outside. All of the legs are specialized, some being used for walking, others for bracing the animal against the sides of the burrow, and still others for personal cleansing operations.

The burrows are much branched and complicated, with many enlarged places for turning around, but the work of adding new tunnels or extending the old ones is unending. When the shrimps pause in their digging operations, they devote the time to cleaning themselves; seemingly they are obsessed with the Puritan philosophy of work. *Callianassa* feeds by extracting detritus, of which bacteria are thought to form an important part, from the continuous stream of mud that passes through its digestive tract.

This incessant digging overturns the sediment of the flats in a manner suggesting the labors of earthworms; but it also makes the ghost shrimps undesirable pests on oyster beds in Washington and Oregon, where studies are in progress to find some chemical that will kill them off without leaving undesirable residues in fish and oysters.

Judging from their age grouping, these shrimps are surprisingly long-lived, perhaps reaching an age of 15 or 16 years. In central California breeding is continuous throughout the year, with an optimum season during June and July.

As happens with many other animals that make permanent or semi-permanent burrows, the ghost shrimp takes in boarders, which are likely in this case to be a scale worm, *Hesperonoë complanata,* and the same pea crab, *Scleroplax granulata,* that occurs with the remarkable inn-keeper *Urechis* (§306). Five other commensals are known to occur, including the copepod *Hemicyclops callianassae* and, in Baja California, *Betaeus ensenadensis,* both on the gills. But the most common commensal is the red copepod *Clausidium vancouverense.*

§293. It is almost axiomatic to remark that there are the usual hosts of worms here. The slim and very long *Notomastus tenuis,* looking like a piece of frayed red wrapping twine, elastic, and up to 12 inches long, may be present in nearly every spadeful of mud, sometimes occurring so thickly as to bind the soil together; but it is so common and so ready to break that one soon loses interest in it. In the south will be found a long, white, wrinkled sipunculid worm with a threadlike "neck." This species, *Golfingia hespera,* sometimes occurs with the burrowing anemone *Edwardsiella* or the brachiopod *Glottidia.* The long and slim *Lumbrineris zonata* is exceedingly common at Elkhorn Slough. MacGinitie (1935) records a specimen with 604 segments.

Sandy tubes projecting a little above the surface in areas of hard-packed muddy sand are likely to be occupied by the fragile "jointworm" *Axiothella rubrocincta* (Fig. 238), whose segmented body is banded in dull ruby red. Large specimens extend 6 or 7 inches below the surface.

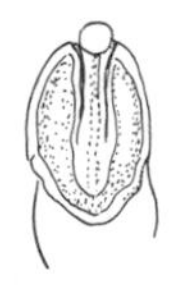

238. *Axiothella rubrocincta* (§293), the jointworm. Sketches, left to right: lateral view of head; dorsal view of cephalic plate; anal rosette and cirri.

The anal end of the worm bears a funnel-shaped rosette, and the head end, which is slightly enlarged, embodies a plug for stopping up the tube when the animal is retracted. *Axiothella* has been taken in Puget Sound and Tomales Bay, and at San Pedro.

Occurring in *Axiothella* tubes is the most specialized of all commensal crabs, the pea crab *Pinnixa longipes,* which, in adapting itself to moving sideways in the narrow tube, has become several times as wide as it is long and has achieved tremendous development of the third pair of walking legs. Seeing a specimen outside a tube, one can scarcely believe that it could possibly insinuate itself through the small opening. The passage takes some pains, but the tiny crab manages it, inserting one huge leg first and then carefully edging itself in sideways. Despite this high degree of specialization, *Pinnixa longipes* is an animal that could probably live almost anywhere. On the east coast this species occurs with the tubed worm *Chaetopterus,* but it has not been found with that worm on the west coast.

Two other burrowing worms will be taken frequently in this habitat. The first, *Cirriformia spirabrancha* (see §16 for similar form), with its coiled tentacles writhing about on the mud of such estuaries as Elkhorn Slough, will at first be taken for a cluster of round worms; but a little serious work with the shovel will turn out the pale, yellowish or greenish cirratulid, which lies hidden safely beneath the surface while its active tentacles keep contact with the outside world of aerated water. The muddy eastern shores of San Francisco Bay at Berkeley—for instance—are characterized by great beds, acres in extent, of the cosmopolitan tubed worm *Capitella capitata,* several inches long and lying head up in vertical, dirt-encrusted, black membranous tubes.

Zone 4. Low Intertidal

Here there are many channels and holes, which form pools when the tide is out. Therefore, besides surface and burrowing animals we will meet several that are characteristically free-swimming.

Surface Habitat

§294. Many snails crawl about on the mud flats in this low zone. The largest are slugs, not popularly recognized as true snails but actually belonging with this group of animals. *Navanax inermis* (see color section), an animal closely related to the sea hares, is large (up to 6 or 7 inches) and strikingly colored with many yellow dots and a few blue ones on a brown background. It is common in southern California and northern Mexico, and has been taken frequently at Morrow Bay and occasionally at Elkhorn Slough. We find it commonly on bare mud, but almost as often in association with eelgrass.

Navanax is a voracious feeder on other opisthobranchs, which it tracks down by following their mucus trails. In bays it feeds mostly on the shelled opisthobranchs *Bulla* and *Haminoea,* and on *Aglaja.* When on rocky shores, it will eat all sorts of nudibranchs at the rate of three to five a day, including *Hermissenda* and others presumably loaded with nematocysts from a coelenterate diet (which does not seem to say much for the idea that the nematocysts in the cerata of *Hermissenda* serve a protective function). *Navanax* will even eat small fish, and often consumes the young of its own species; it is so voracious that it may deplete the nudibranch population of an area.

A stranded *Navanax* has an interesting way of protecting itself by accumulating a covering of sand and mud when rolled about in shallow waves; it may have a layer of mud several millimeters thick.

Like all opisthobranchs, *Navanax* has never heard of birth control. In the La Jolla area it produces eggs the year round; large egg masses may contain 800,000 eggs. These are laid in light yellow, stringy coils, woven together in pleasing designs.

The sea hare *Aplysia* (§104) is a frequent visitor to the mud flats. At Elkhorn Slough huge specimens weighing 10 pounds or more (most of this is water) may frequently be seen when none are to be found on the rocky shore outside.

§295. *Hermissenda crassicornis* (see color section), a yellow-green nudibranch about an inch long with brown or orange processes along the back and a well-defined pattern of blue lines along the sides, is often common in bays (like all opisthobranchs, it is of uncertain occurrence in any given part of its range) but is also found in rocky tide pools. It ranges from Alaska to San Diego.

§296. The heavy-shelled bivalve *Chione* (§260) may be taken on flats of sandy mud, often, strangely, at the surface. There are also many kinds of shelled snails: a tropical brown cowry (*Cypraea spadicea,* §136), which ranges as far north as Newport, California, and a cone shell (*Conus californicus,* §136) are fairly common. The second packs about more than its allotment of the dark brown *Crepidula onyx* (Fig. 239) —a combination that extends into the upper reaches of the estuary at Newport. Ovipositing specimens of the olive snails (§253) have been noted in August at Elkhorn Slough. One of the "muricks," *Ceratostoma*

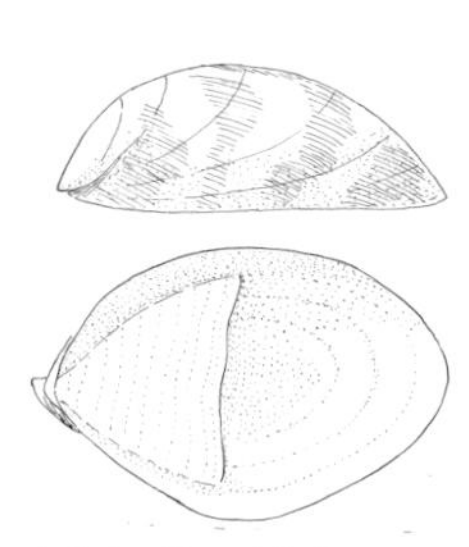
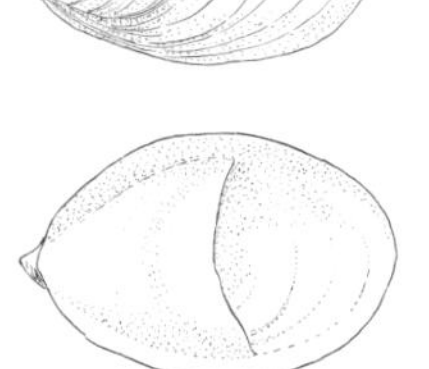
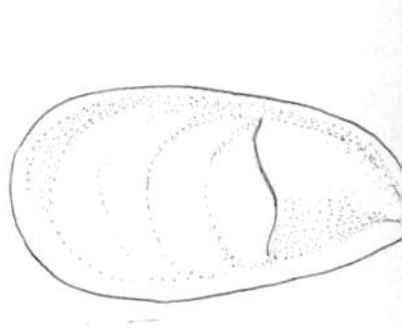

239. Slipper shells. On the extreme left is *Crepidula onyx* (§296), the southern California slipper shell, shown about twice natural size. The others, left to right, are *C. fornicata, C. glauca,* and *C. plana,* all east coast species that have been reported from northern bays on this coast.

240. *Ceratostoma nuttalli* (§296), a murex.

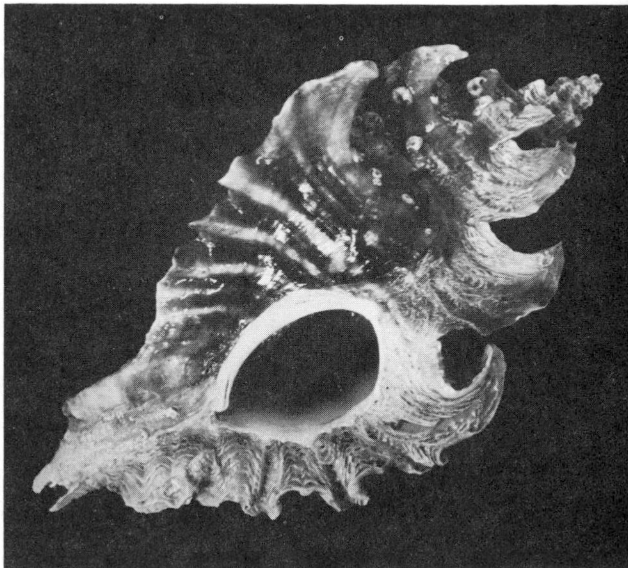

241. An especially gaudy murex, *Pterynotus trialatus* (§296).

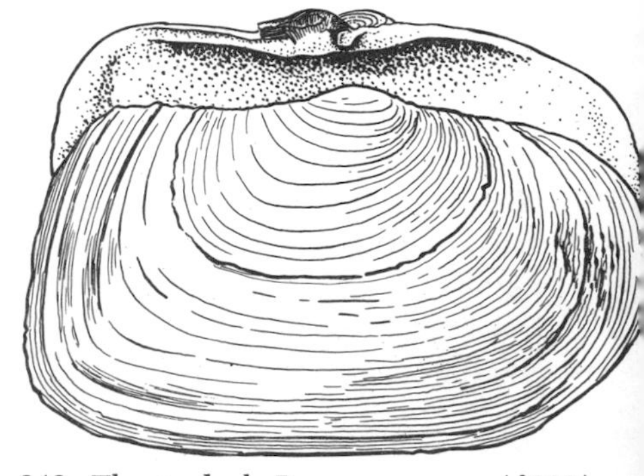

243. The geoduck, *Panope generosa* (§299).

242. Left: *Tetilla mutabilis* (§298).

nuttalli (Fig. 240), is especially characteristic of southern mud flats. The ambitious collector at Newport may also run across the flared *Pterynotus trialatus* (Fig. 241), prize of the conchologists in that region. The largest of the moon snails (*Polinices lewisi,* §252) is almost as characteristic an inhabitant of the mud flats as it is of the sand flats.

§297. The most noticeable animal of all, but one that is by no means obtrusively common, is a magnificently orange-plumed *Phoronis* that has been seen at Newport, its gelatinous body protected by a tube that is buried in the mud. This anomalous "worm," which, according to MacGinitie, proves to be *Phoronopsis californica,* extends down into the substratum, retracts immediately at the least sign of trouble, and is difficult to dig. Although occurring solitarily, it is a near relative of the green-plumed *Phoronopsis viridis* that grows in great beds at Elkhorn Slough, and of the similar *Phoronopsis harmeri* of the Vancouver region, which sometimes carpets the mud with green fuzz for many square yards. These are not at all difficult to dig, and colonies of the stringy worms in their stiff and upright, sand-covered, parchment-like tubes are very much a feature of the estuary fauna.

§298. *Tetilla mutabilis* (Fig. 242) is one of the most remarkable of sponges. It attaches loosely, now and then, to the roots of eelgrass, but for the most part it rolls aimlessly about in Newport Bay or lies around on the mud flats—an unheard-of practice for a sponge. The lightweight clusters, sometimes as large as a clenched fist, are sometimes dirty yellow to purple but are usually red with green glints. This sponge has been likened to the egg case of a spider; to another observer it suggests the gizzard of a chicken.

Burrowing Habitat

§299. The clams and worms are the obvious animals in the substratum of this low zone, and there is a varied assemblage of both. Largest of all our clams, if not one of the largest intertidal burrowing bivalves in the world, is the geoduck, *Panope generosa* (Fig. 243). (Weymouth, in his paper on the edible bivalves of California, says that the name was taken from the Indians of the north, which undoubtedly accounts for the pronunciation: the "eo" is pronounced like "oi" in "oil.") The geoduck lives in soft muck, or even in firm sandy mud; for, contrary to the popular belief, it is an extremely poor digger. Lest this statement should rouse the ire of many people who have exhausted themselves in a fruitless effort to reach one of the animals, we hasten to explain. The geoduck lives in a semipermanent burrow that is often 3 feet deep, sending his immense siphons upward to the surface. Any disturbance in his neighborhood causes him to partially retract his siphons, expelling contained water from them as he does so and thus giving the impression that he is digging down to greater depths. Continued disturbance causes continued retraction, although the siphons are much too large to be withdrawn completely into the shell.

The geoduck has been credited, too, with a vast amount of will power that he probably does not possess because inexperienced diggers often try to pull him out by his siphons. The siphons break, and the obvious conclusion is that the animal is hanging on with grim determination. As a matter of fact, he could not come out if he would, for his siphons are by no means strong enough to drag his bulging body through the substratum. Apropos of the difficulty of procuring the animal whole, Rogers remarks that a noted conchologist who, with the aid of two other men, had spent a long time in digging one out, referred to it feelingly as "a truly noble bivalve."

Individuals with shells more than 8 inches long are not uncommon, and the maximum weight of the entire animal is about 12 pounds. Some of the specimens we have seen would probably have tipped the scales at 10 pounds. The shells are relatively light and not even large enough to contain the animal's portly body, let alone the siphons; even in living specimens the valves always gape open an inch or so.

Some time ago a movie newsreel showed the digging of geoducks in Puget Sound, while the "spieler" stated that the animals were known only from Puget Sound and from Africa. The statement is incorrect, for the geoduck is taken at Elkhorn Slough and Morro Bay, although not very plentifully, and is known from many quiet-water areas as far south as southern California (it has been taken by divers at La Jolla), but not from Africa. Anyone who has ever tried to dig it will understand why it is known so slightly and is used so infrequently for food.

§300. Another large clam that resembles the geoduck in appearance, habits, and size and is often mistaken for it is variously known as the gaper, the summer clam, the rubberneck clam, the big-neck clam, the horse clam, the otter-shell clam, and the great Washington clam. After this appalling array of popular names it is almost restful to call the animal *Tresus* (once familiarly known as *Schizothaerus*!) *nuttalli* (Fig. 244). The individual gaper-etcetera is readily located by its siphon hole in the mud—"squirt hole" in the vernacular—from which, at fairly regular intervals, it shoots jets of water 2 or 3 feet in the air. These jets are particularly powerful when the clam is disturbed, and are often aimed

244. The gaper, *Tresus nuttalli* (§300).

with deadly accuracy; but at any time, a succession of the largest geysers produced by any clam betrays the presence of a bed. Digging them out, however, is no small job, for they lie from 1½ to 3 feet below the surface. Their average size is smaller than the geoduck's but the shell may be 6 to 8 inches long, and the clam may weigh 4 pounds. As with the geoduck, the easily broken shell is incapable of closing tightly over the large body, and the huge siphon must shift for itself. The siphon is protected by a tough brown skin and by two horny valves at the tip—a device not found in any other of the common clams. Nevertheless, the siphon, or part of it, must often be sacrificed, for the tips are commonly found in the stomachs of halibut and bottom-feeding rays.

Of the big clams, *Tresus* is the least used for food (although it is popular locally), partly because its gaping shell cannot retain moisture and therefore makes it unfit for shipment. Appearances are against it, too; the mud-flaked brown neck is unappetizing to look at. Weymouth says, however, that the Indians used to dry the siphons for winter use, and that at Morro Bay the siphons are skinned, quartered, and fried, although the bodies are discarded. As a matter of fact, the bodies are perfectly edible and, after being run through a meat grinder, make a not unsatisfactory chowder. The clam under discussion is protected by California law, 10 per person being the limit for one day.

Tresus nuttalli is interesting biologically (although it is perfectly safe) as the host of minute, flukelike tapeworm larvae in which the living flame cells (excretory organs) can be seen under the microscope. MacGinitie has determined that the adult tapeworms (tetraphyllids) are to be found in the intestines of sting rays. The large pea crab *Pinnixa faba* is quite likely to be present in the mantle cavity in California, in Puget Sound, and in British Columbia; and numerous specimens of *Opisthopus* (§76) have been found. Occasionally the smaller *P. littoralis* will be taken.

The clam ranges from Alaska to San Diego, and, although it decidedly prefers quiet bays, it may be found also on the outside coast; but outer-coast specimens are small and usually look rather battered. Specimens taken "outside," or even from those inlets directly adjacent to oceanic waters, ought never to be used for food during the summer because of the danger of mussel poisoning (§158).

§301. *Saxidomus nuttalli* (Fig. 245), the Washington clam, butter clam, or money shell (this last because the California Indians used the shells for money), ranges from Bolinas Bay to San Diego. Clam diggers however, do not distinguish between this clam and the very similar but smaller *S. giganteus,* which is the most abundant clam on suitable beaches in Alaska, British Columbia, and Puget Sound and occurs, but not commonly, as far south as Monterey. The same popular names are used for both. The shells of *S. nuttalli* sometimes rival those of the geoduck in size, but are characteristically only 3 to 5 inches long, while *S. giganteus*—again proving that there is not much in even scientific names—averages about 3 inches. Both appear in the markets, but not

245. *Saxidomus nuttalli* (§301), the Washington clam.

commonly except in the Puget Sound region, where the smaller form is used rather extensively. A pea crab, the small *P. littoralis,* may be found occasionally in the mantle cavity; but the Washington clam is, on the whole, remarkably free from parasites.

In a study of the distribution and breeding of *S. giganteus,* Fraser and Smith examined 2,600 British Columbia specimens. Tidal currents are an important factor in distribution because of the animals' pelagic larvae. With the exception of clams in their seventh year, the result of a particularly prolific season, the age distribution was fairly regular from the fourth year to the tenth. About half of the clams spawn at the end of the third year. At Departure Bay spawning took place in August, but at other beaches along the inner coast of Vancouver Island there was much variation. The larvae appeared as bivalved veligers in 2 weeks, and at the end of another 4 weeks, when still less than 3⁄16 inch long, they settled down on the gravel. It is reported that mussel poisoning (§158) may be present during the summer to a dangerous degree in Washington clams occurring (as they do rarely) on the open coast or in inlets directly adjacent to oceanic waters.

§302. The eastern soft-shell clam *Mya arenaria* (Fig. 246) was first noticed on the Pacific coast in 1874, when Henry Hemphill collected it in San Francisco Bay. There is no record of any deliberate introduction, but apparently the first oysters from the East were planted in San Francisco Bay in 1869, and the clam may have been introduced at that time. The second recorded introduction is 1874. If so, *Mya* may have become established in the short period of 5 years. However, one must remember that Henry Hemphill was an expert collector and seldom missed rarities, so a Hemphill record is hardly a criterion of abundance. There are *Mya* in the fossil record (Pliocene) of the Pacific coast, and in Alaska there is an apparently native species very similar to *Mya arenaria.* At this late date the record is so confused that we may never be sure whether *Mya arenaria* was introduced from the east coast or whether it has been here all the time as a widely distributed boreal species. In any event, *Mya* is one of the common bay clams of the Pacific coast from Vancouver Island to San Diego and is important commercially, especially in northern bays.

Mya is egg-shaped in outline, averages less than 5 inches in length, and is characterized by a large, spoon-shaped projection on the left valve at the hinge. Only one other species (*Platyodon,* §147) has this projection. The shell is light and brittle. The adults of the species are incapable of maintaining themselves in shifting substratum, having lost all

246. Piled shells of the eastern soft-shelled clam, *Mya arenaria* (§302), which is now important commercially on the Pacific coast.

power of digging, and hence require complete protection. They thrive in brackish water, however, if it does not become stagnant, and can stand temperatures below freezing. In the north *Mya* is likely to carry the same pea crab as the gaper.

In common with several other clams that can live in foul estuary mud, *Mya* exhibits, but to a striking degree, a facility for anaerobic respiration —the ability to live in a medium absolutely lacking in free oxygen. Experimentally, these clams have been known to live in an oxygen-free atmosphere for 8 days. They produced carbon dioxide continually but showed no subsequent effects beyond a decrease in stored glycogen (an animal "sugar") and a considerable increase in the metabolic rate after being replaced in a normal environment. Many bacteria are anaerobic, but such a revolutionary physiological process in an animal as specialized as a mollusk is somewhat startling. It has been suspected that one of its specialized organs called the crystalline style has a hand in the matter, for it is possible that the style may be capable of oxidizing certain of the products of animal metabolism, thus enabling the life processes to go on temporarily just as well as though free oxygen were available.

§303. The bent-nosed clam, *Macoma nasuta* (Fig. 247), is a small, hardy species (seldom reaching 2½ inches in length) that may be turned out of almost every possible mud flat between Kodiak and Baja California. At Elkhorn Slough it is the commonest clam. It can stand water so stale that all other species will be killed, hence it is often the only clam to be found in small lagoons that have only occasional communication with the sea. Also, it can live in softer mud than any other species. At rest the clam lies on its left side at a depth of 6 or 8 inches, with the bend in its shell turned upward, following the upward curve of the separate yellow siphons. When burrowing it goes in at an angle, sawing back and forth like a coin sinking in water.

The California Indians, and possibly those to the northward, made extensive use of the bent-nosed clam for food, and many of their shell piles, or kitchen middens, contain more shells of this species than any other. Later on, San Francisco Chinese dug these clams, kept them in boxes for a day or two and furnished them with several changes of water by way of an internal bath (to void the contained sand and mud), and then marketed them. Now, however, this species seems to be very little used.

In a nearly related small clam of the Atlantic, *Cumingia,* similar in form and habitat, a lunar periodicity of the breeding seasons has been

247. The bent-nosed clam, *Macoma nasuta* (§303); almost twice natural size. The slight bend is visible in the bottom shell.

detected, the sexual products being extruded most plentifully at full moon during the summer.

§304. South of Santa Barbara the common jackknife clam of mud flats is the active *Tagelus californianus* (Fig. 248), a nearly cylindrical form reaching a length of 4 inches. It has a long, flexible, and powerful foot that cannot be completely withdrawn, and it can dig in rapidly, even from the surface. It lives in a permanent burrow about 16 inches deep, within which it moves up and down at ease. If pursued, it will dig down still farther. Seen from the surface, the burrow appears to be double, for there are two separate openings 1½ to 2 inches apart for the separate siphons.

§305. In some estuaries there are regions of stiff clay or hard mud where relatives of the rock-boring mollusks may be found. One of these, *Zirfaea pilsbryi* (formerly *Z. gabbi*) (Fig. 249), the rough piddock, is the largest of the borers, reaching a shell length of 4½ inches. It occurs

248. Shell of the jackknife clam *Tagelus californianus* (§304).

249. The rough piddock, *Zirfaea pilsbryi* (§305), the largest of the intertidal boring mollusks; about ⅓ size.

from the Bering Sea to San Diego, and is found in outside rocky reefs as well as in protected mud and clay, but it is apparently more characteristic of the latter habitat. As with other borers, the valves gape at both ends and are roughened at the "head" end with grinding teeth. MacGinitie has watched the boring movements of this form in a glass tube. While the foot, which is modified to a sucking disk, holds on to the substratum, the shell is moved up and down and slightly rotated. Thirty-two turning movements make one complete revolution, requiring a minimum of 70 minutes. At approximately each revolution the direction of rotation is reversed. During the process the mantle is pulled back to allow the loosened material to be drawn up and shot out through the siphons. The shell is fragile, since the animal is amply protected in its burrow, and it is much too small to contain the long siphons even when they are fully contracted. Unlike most of the pholads, which quit work when their burrows are completed, *Zirfaea* remains active throughout its life of 7–8 years or more. Robert Menzies (or the late Colonel Miles?) found a specimen in Bodega Harbor that had bored a neat, large hole through a branch of redwood buried in the mud.

§306. A wormlike echiuroid is common in the mud flats. This is *Urechis caupo* (Fig. 250), aptly named the fat innkeeper, for it is the portly chief of as motley a crew of guests as one could hope to find. At the low-tide water's edge, or below it, *Urechis* constructs a burrow in the shape of a broad U, with entrances from 16 to 38 inches apart. The entrances are constricted like the muzzle of a shotgun, and for the same reason, so that they are only about ⅓ the diameter of the burrow, and around one of them is a pile of castings. Once completed the burrow is semipermanent, and needs to be enlarged only at infrequent intervals, for the animal is a slow-growing form. Our knowledge of the innkeeper's habits was gained by Fisher and MacGinitie, who found that it would live and carry on its usual activities in a glass burrow placed in an aquarium. There the strange animal's housekeeping, feeding, digging, and social life were closely watched for months.

The innkeeper itself is roughly cigar-shaped when resting, and grows up to 19½ inches long, although the average is perhaps nearer 8 inches. It is flesh-colored, with two golden bristles under the mouth and a circlet of bristles around the anus. When the animal is active, its shape is continually changing owing to the peristaltic movements that pump water through the body and through the burrow. These movements also provide locomotion, naturally, in either direction. To move forward, the animal swells the forward part of its body so that it presses firmly against the walls of the burrow. The remainder of the body is now contracted and drawn up, after which the swelling is transferred to the after part of the body. The contracted forward part is then pushed ahead and reanchored, and the cycle is repeated.

Digging is begun with the short proboscis, and the burrow is enlarged by scraping material from the walls with the bristles about the mouth end. The animal crawls over the loosened material and works it backward with the anal bristles. The material accumulates for a time, along with the castings, and is "finally blown out the 'back door' by a blast of accumulated respiration-water from the hind gut." This is where the constriction of the burrow entrance becomes useful, for it momentarily accelerates the velocity of the issuing water, thus enabling *Urechis* to expel rather large chunks of debris. Pieces too large to be so forced out are buried.

The animal's most remarkable trait is the spinning of a slime net, with

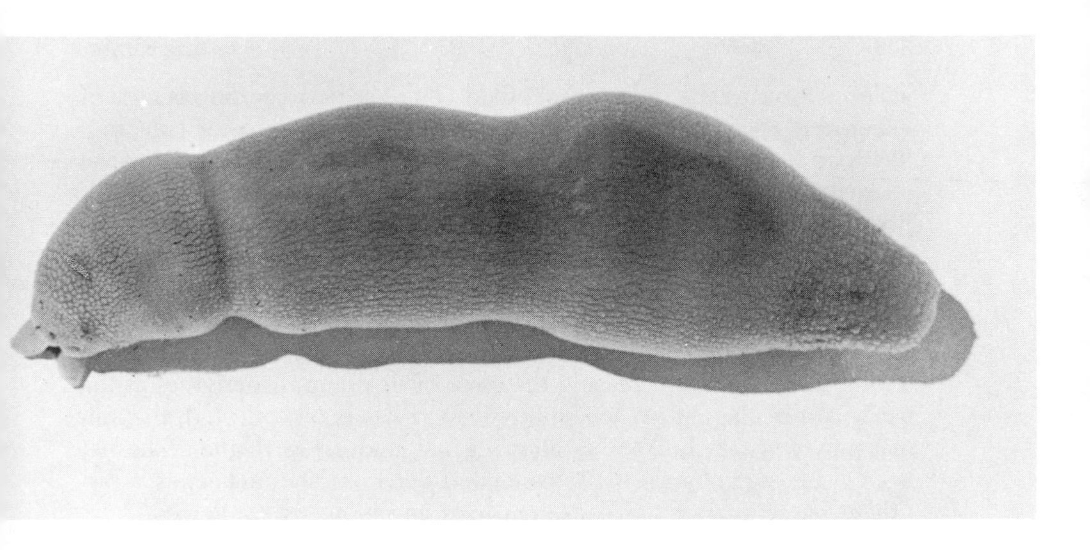

. The echiuroid worm *Urechis caupo* (§306), the fat innkeeper. Above: photo of a living animal, fairly
xed. Below: *Urechis,* with guests, in its burrow. (*a*) Two drawings of the innkeeper, showing the
iety of shapes it may assume. (*b*) A permanent guest, the scale worm *Hesperonoë adventor.* (*c*) The
crab *Scleroplax granulata.* (*d*) A transient, the goby *Clevelandia ios.* (*e*) Slime net used for feeding.

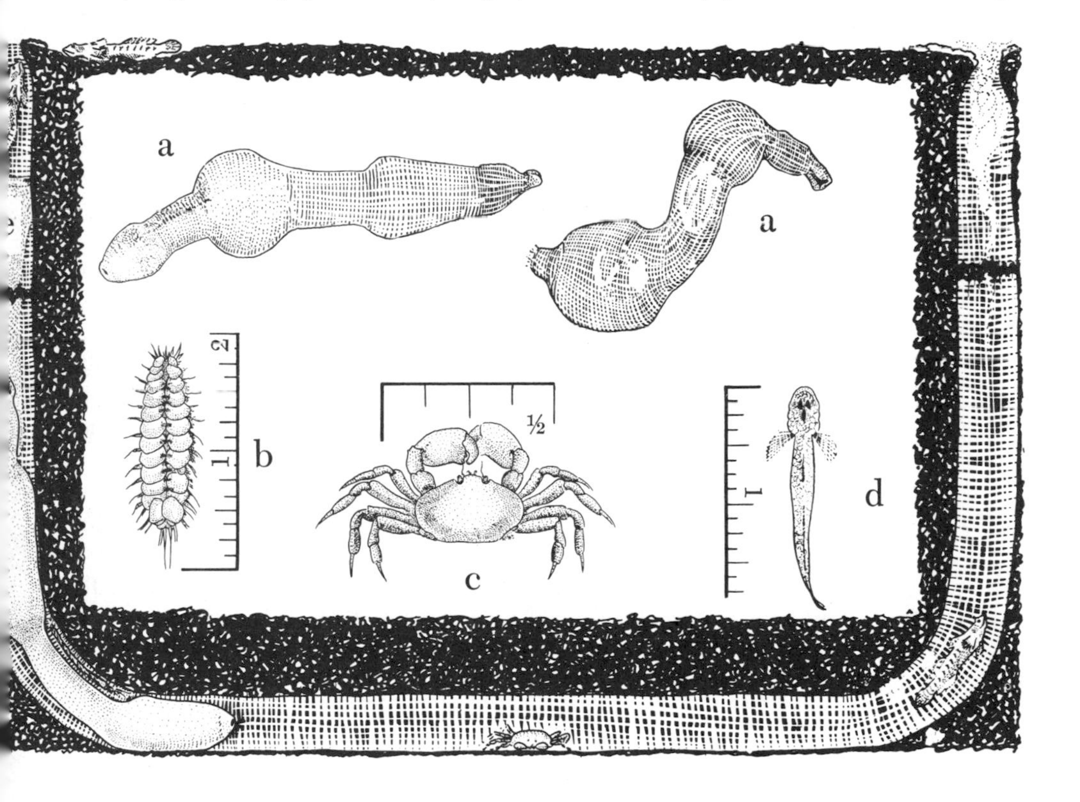

which it captures its microscopic food. The net permits the passage of water and yet is so fine that particles one micron in diameter (one-millionth of a meter) are caught. The openings are invisible even under a microscope. In preparing to feed, the innkeeper moves up the vertical part of the burrow at its head end and attaches the beginning of the net to the burrow walls close to the entrance. It then moves downward, spinning a net that may be from 2 to 8 inches long. At first the net is transparent, but as it collects detritus it turns gray and becomes visible. For about an hour *Urechis* lies at the bend in its burrow, pumping water through the net and increasing the force of its pumping activities as the net becomes clogged. When sufficient food has been collected, the animal moves up its burrow, swallowing net and all as it goes. The net is now digested along with its contained detritus. The innkeeper, a fastidious eater, discards all large particles as the net is swallowed.

It is scarcely to be expected that so ready a food supply would go to waste, and it does not. The innkeeper's three guests stand ready to grab all particles the instant they are discarded (Fig. 250). The most dependent of these is an annelid, the beautiful reddish scale worm *Hesperonoë* (formerly *Harmothoë*) *adventor,* from ½ to 2 inches long, which remains almost continually in contact with the body of its host. To quote Fisher and MacGinitie: "It moves from place to place with its host, making little runs between peristaltic waves, and turns end for end when *Urechis* does. After *Urechis* spins its mucus-tube *Hesperonoë* may crawl forward and lie with its palps almost touching the proboscis. As soon as *Urechis* starts to devour the tube *Hesperonoë* also sets to, making absurd little attacks on the yielding material with its eversible pharynx."

The second guest, of which there are sometimes a pair, is the pea crab *Scleroplax granulata,* usually not more than 5/16 inch across the carapace.* The third guest is a goby, *Clevelandia ios,* and there may be from one to five specimens in a burrow. This little fish is a transient guest, foraging outside much of the time and using the innkeeper's burrow chiefly for shelter. *Urechis* derives no apparent benefit from any of these commensals. On the contrary, "both crab and annelid interfere with the regular activities of *Urechis,* especially its feeding and cleaning reactions. A particle of clam dropped into the slime-net is immediately sensed by both commensals. Their attempts to reach it cause *Urechis* prematurely to swallow the tube when the clam morsel is stripped out the open end. It is usually snapped up by *Hesperonoë* and swallowed if small enough;

* Another pea crab, *Pinnixa franciscana* (§232), occurs rarely, and *P. schmitti,* which is the commonest *Echiurus* commensal, has also been taken here.

otherwise *Scleroplax* will snatch it away, when the annelid must be content with what remains after the crab's appetite is satisfied. . . . Enmity exists between crab and annelid in which the latter is the under-dog. This feud may account for the close association of *Hesperonoë* with *Urechis*." A more congenial relationship exists between the crab and the goby, for the goby has actually been observed to carry a piece of clam meat that was too large for it to swallow or tear apart to the crab and stand by to snatch bits as the crab tore the meat apart. Even the crab and the scale worm forget their differences, however, when danger threatens, for at such times both rush to their host and remain in contact with his body wall until the danger is past.

In *Urechis* the sexes are separate. MacGinitie (1935) has shown that in its breeding habits (the eggs and sperm are discharged into the water) *Urechis* exhibits a specialization as remarkable as that of its feeding. The eggs or sperm mature while floating freely in the coelemic fluid, where very young, immature, and fully ripe sexual products are mixed together indiscriminately. A remarkable set of spiral collecting organs—modified nephridia—pick up only the ripe cells and reject all other contents of the body cavity, and provide an instant means of disposal for sexual products as soon as ripe. After being removed from the body, these collecting organs function outside as autonomous organs. But even in cultures of blood or of seawater, the excised collectors refuse to pick up eggs other than those of *Urechis* (which have a characteristic indentation through which the organs operate), although the foreign eggs are identical in size. The same investigator has determined (1934) that the eggs can be fertilized artificially while still in the body cavity, and that the resulting development is more rapid than that of controls under simulated normal conditions, quite the opposite of what usually occurs with most sea animals, owing to the inhibitory effect of the animal's blood on the fertilization process.

There must be the usual tremendous mortality among *Urechis* larvae, but once the animals become established in burrows they are relatively safe, except from stingrays. At Newport Bay *Urechis* breeds during the winter and is usually spawned out before the end of the summer, but at Elkhorn Slough specimens are normally ripe throughout the year. The difference is caused by the difference in temperature at the two places; during the exceptionally warm summer of 1931, many of the animals at Elkhorn Slough were spawned out before August.

While embryologists go to great trouble to provide the proper conditions for fertilizing gametes and raising embryos, it would appear that nature is not so exacting. Some years ago, while preparing a motion picture, the late Alden Noble worked assiduously to fertilize *Urechis* eggs and have trochophore larvae ready for the photographer at the right moment, barely getting enough for the picture. After the photographer departed, Dr. Noble discovered that a 12-inch bowl left on the table in the sun was almost solid with trochophore larvae of *Urechis*. This had been the jar into which he had discarded all the slop and failures from his previous efforts!

Urechis is known chiefly from Elkhorn Slough, but has been found also in Newport, Morro, San Francisco, Tomales, Bodega, and Humboldt bays. A related, smaller form *Echiurus echiurus alaskensis,* with its pea crab *Pinnixa schmitti,* is exceedingly abundant in southeastern Alaska, occurring embedded in gravelly substratum where there is a clay admixture, as well as in the mud flats proper. The Alaskan form lacks the food net of *Urechis,* but the scoop-shovel proboscis achieves results just as effectively, if abundance of individuals is an index of successful food-getting. It is possible that *Urechis,* under duress, may be able to feed in the same manner if out of its burrow.

It was estimated by MacGinitie that *Urechis* may attain an age of 60 years. Even if this is on the generous side, it would certainly appear that a large *Urechis* cannot establish itself overnight, and one wonders how it will survive its principal intertidal enemy, the casual zoology student who must dig one up. Usually the digging process results in some false starts, in which a great hole is dug in the sediment; the *Urechis* is either lost in the caving walls of the hole, or chopped in half by the shovel, so that several may be sacrificed for the edification of scientific curiosity. Man is not the only enemy of *Urechis*: certain rays, e.g. the bat ray, eat echiuroids and other burrowing animals. Apparently they are able to pop potential prey out of its burrow by using their broad, flattened bodies as a sort of plumber's friend.

§307. Two species of echiuroid worms, besides *Urechis,* may occur in Newport Bay (at least they did years ago), the small green *Listriolobus pelodes* (found also in Tomales Bay) and the larger, also greenish, *Ochetostoma octomyotum. Ochetostoma* is similar in habit to *Urechis,* but it has a longer proboscis and feeds by sweeping detritus from the walls of its burrow onto the mucus-covered surface of this appendage. A concise summary of the domestic ways of the various echiuroids will be found in the MacGinities' book.

§308. The striking worm *Sipunculus nudus* (Fig. 251), cosmopolitan on warm shores, is common at Newport Bay and has been taken in Mission Bay and at Ensenada. The white skin is shining and iridescent,

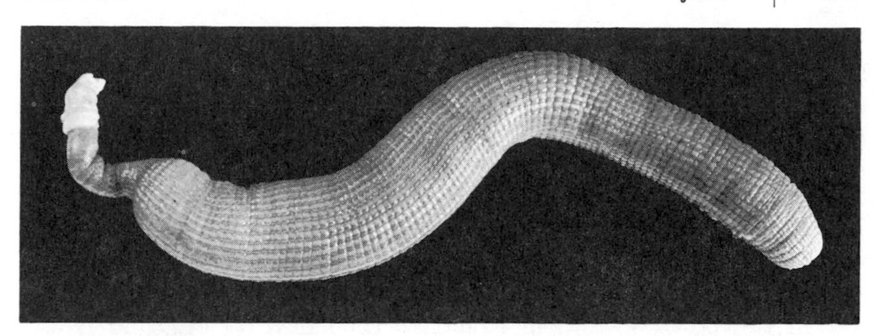

251. *Sipunculus nudus* (§308), a large sipunculid worm.

and shows the muscles in small rectangular patches. Like the "sweet potato" (*Molpadia,* §264), this animal passes great quantities of the substratum through its intestines and extracts the contained nourishment. More than half the weight of a test *Sipunculus* at Naples consisted of sand in the intestine. It has often been remarked that such animals play a large part in turning over and enriching the shallow bottoms of bays, just as earthworms function in the production of rich vegetable mold on land. *Sipunculus* finds its optimum conditions, on our coast, in a downward extension of the flats peopled by the burrowing anemone *Edwardsiella,* the brachiopod *Glottidia,* and the sea pens.

Sipunculus nudus is known as far north as Monterey Bay. To the north it is replaced by a similar sipunculid, *Siphonosoma ingens,* which differs principally (as far as external anatomy goes) in the structure of its extrovert, somewhat shorter and more flower-shaped than that of *Sipunculus nudus.*

§309. The true segmented worms have a large and obvious representative in the tubed *Chaetopterus variopedatus* (Fig. 252), which is known from Vancouver to San Diego, and is widely distributed elsewhere. It is found on the European coast, e.g., near Naples. In its U-shaped burrow it constructs a fairly thick and woody brown tube, which may be several feet long; the curiously shaped worm itself measures from 6 to 15 inches. Many generations of tube-dwelling have softened the worm's body to the point where it is helpless outside its tube, and it almost invariably registers its protest at being removed by breaking in two just behind the head.

The animal secretes a great amount of slime, which covers its body and lines the tube; this slime seems to have some connection with the worm's brilliant phosphorescence, for when the worm is touched, some of the slime will come off on one's fingers and continue to glow there.

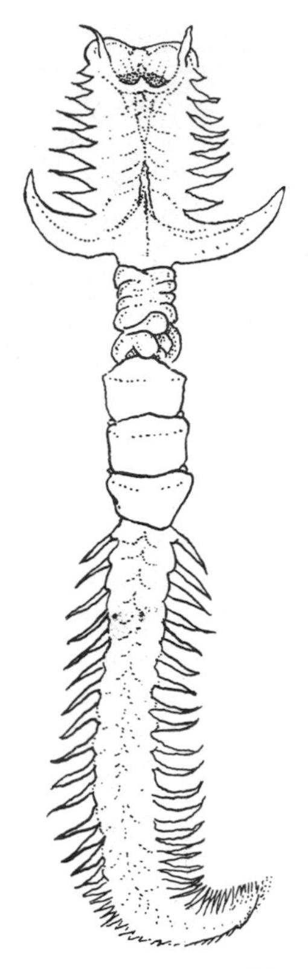

252. A phosphorescent annelid tube worm, *Chaetopterus variopedatus* (§309); natural size.

253. Two Pacific coast lugworms, genus *Abarenicola* (§310); the larger, darker animal is *A. vagabunda,* the other is *A. pacifica.*

§310. The low-tide deep sandy or muddy flats in the Puget Sound region are the favored environment of a black, rough-skinned lugworm, *Abarenicola vagabunda* (Fig. 253), which is popular with fishermen as bait and with physiologists for experimentation. A somewhat similar species, lighter in color, is found in muddier bottoms at higher levels; this is *Abarenicola pacifica* (Fig. 253). *A. vagabunda* ranges south to California. The species found from southern California to Humboldt Bay is *Arenicola cristata.* Along with this species, especially from San Francisco Bay south, another species has been recorded, *Arenicola brasiliensis* (it would take more ink than is desirable to explain why a northern hemisphere species has been saddled with the name *brasiliensis*). Thus the lugworm situation in California is similar to that in Puget

Sound, but less is known of the ecological requirements of the California species. Lugworms are very scarce in Tomales Bay, or at least rarely encountered; either the right place for them has not yet been found or they require a bit more fresh water drainage than is usually available in Tomales Bay.

Lugworms may attain a length of 1 foot, but 6 inches is about the average length for this coast. The animal's presence is indicated by little piles of mud castings, for, like so many other inhabitants of the mud flats, its adjustment to the food-getting problem involves eating the substratum. It usually lies in a U-shaped burrow of a size that will permit both its greatly swollen head end and its tail end to almost reach the surface. Sometimes, however, the head is buried deep in the substratum. In either case the worm maintains a current of water through its burrow by dilating successive segments. This current, which can be made to flow in either direction, aerates the animal and assists its delicate bushy gills in respiration. At low water the current is up, that is, from head to tail; but when the animal is covered, the current is reversed, thus being filtered through the sand in the tube—a process that neutralizes any toxic alkalinity caused by the nearness of certain seaweeds.

The whole mud-burrowing situation, as typified by the lugworm, offers an interesting example of adaptation to environment: by burrowing, the animals conceal themselves from enemies, escape wave action and drying, and place themselves in the midst of their food supply. The habitat requires, however, some provision against suffocation; and for the lugworm the protruded gills and the stream of water are the solution.

In the spring great transparent milky masses, containing possibly half a million eggs, can be seen extending up from the burrows below the tide line.

Arenicola is the only member of the bay and estuary fauna that can also live in the barren sand of the protected outer beaches, where very few animals seem able to exist. On those beaches the lugworm occurs, as previously noted, only where some silt is deposited around rocky outcrops.

§311. We have had biting worms and pinching crabs, stinging jellyfish and stinging sponges; in *Pareurythoë californica* we have a stinging worm. On mud flats like those in Elkhorn Slough the digger of *Urechis* or nemertean worms will sometimes see a long, pink, flabby-looking worm somewhat resembling *Glycera* but with white and glistening spicules about the "legs." If he picks it up, as he undoubtedly will the first time he sees it, he will regret the action instantly unless his hands are extraordinarily calloused. Once a specimen is preserved it may be handled with impunity; but for some unknown reason, the living spicules are capable of penetrating human skin with mildly painful consequences.

The annelid proboscis worm *Glycera* is fairly common in mud flats; the relatively gigantic *G. robusta* (Fig. 254) occurs in beds of black mud. Of other worms, there are the forms mentioned in §293 as occurring higher up, and, at Newport Bay, there is a plumed tube worm, *Terebella californica* (in appearance similar to *Thelepus*, §53). *Terebella* is interesting because it very commonly harbors a commensal crab, *Parapinnixa affinis*, formerly known from a single specimen only. Here is another instance in which a seemingly rare animal can be taken almost at will, once its ecological station is known.

Nereis virens occurs where there is an admixture of sand in the firm mud, building semipermanent tubes just as it does on the Atlantic coast. It appears that in *N. virens* we have a generalized and well-adjusted animal capable of meeting new situations as they arise, with no particular fixed pattern that must first be overcome. Yet its breeding habits are fixed in a pattern that may date back to the probably great tides of the

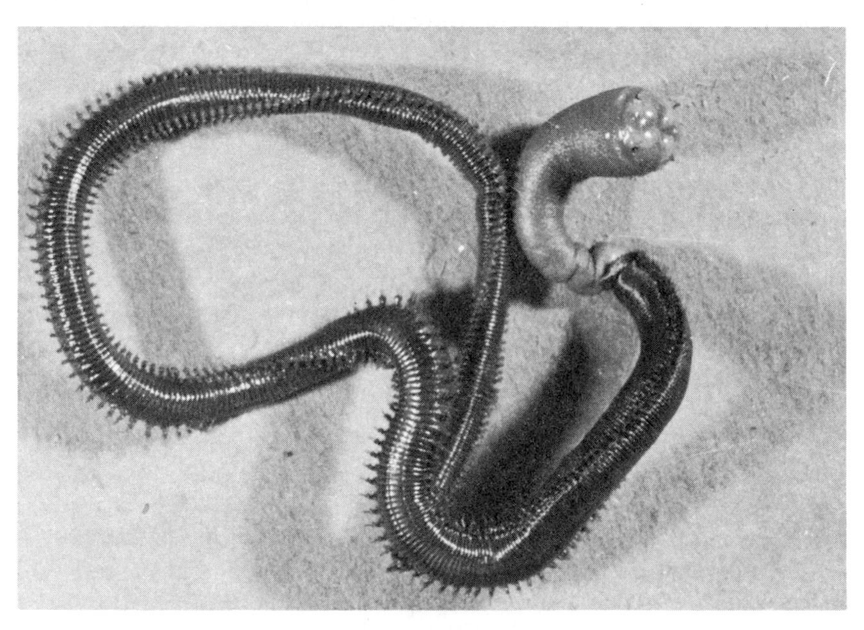

254. *Glycera robusta* (§311), with proboscis extended.

Cambrian period or before. The lunar rhythms in the breeding habits of polychaete worms, among them *Nereis*, are famous (see §151); and those of the Atlantic species have been investigated and plotted.

The giant *N. brandti* (§164) is frequently considered a subspecies of *N. virens*. Specimens up to 5½ feet long have been taken in the sandy mud of Elkhorn Slough (MacGinitie, 1935).

§312. As would be expected from the terrain, nemerteans (ribbon worms) are abundant, both in species and in number of individuals. Some of the larger species have extensible proboscises that can be shot out several feet; they are armed with venomed barbs or stylets, which serve to capture prey but are incapable of penetrating human skin. The animal's armchair method of hunting is to lie quietly in safety while extending its proboscis in search of food. Evidently the stylet is lost either regularly each time an attack is made or by accident at fairly frequent intervals, for in little pouches within the body the animal carries several spare stylets, much as a goodly yeoman carried spare arrows in his quiver. These stylets apparently make nemerteans much more formidable antagonists than their appearance would lead one to believe. We once put a nemertean and a mussel worm (*Nereis*) together in a dish of seawater, anticipating a speedy slaughter of the nemertean. Instead, the nemertean's questing proboscis touched the mussel worm amidships and took hold, evidently inflicting considerable pain. The mussel worm writhed for several minutes before shaking itself free, and then it showed no disposition to return the attack. Dr. Coe remarks that some nemerteans are known to be able to swallow living *Nereis* larger than themselves.

The most conspicuous trait of nemertean worms is their habit of breaking themselves into pieces when persistently disturbed. It has long been known that in some species the separate fragments would grow into complete animals, and Dr. Coe has recently demonstrated that a worm may be cut and the resulting portions recut until eventually "miniature worms less than one one-hundred-thousandth the volume of the original are obtained." The limiting factor is that the wound will not heal, nor will regeneration take place, unless the fragment is nearly half as long as the diameter of the body. In other words, a worm ⅛ inch in diameter could be cut into 1/16 inch slices, or a little less, and all the fragments, barring accident, would become new worms. Decidedly this is an item for "believe it or not" addicts. Coe also found that the animals can survive freezing, and that an adult worm, or a large regenerated fragment, can live for a year or more without food. The deficiency is compensated for by a continual decrease in the body size—a trait that we have noticed in captive and improperly fed nudibranchs as well, although the nudibranchs cannot continue the process for more than a few weeks. The species Coe used for his experiments was *Lineus vegetus* (§139).*

A wide white *Micrura* is very common in this habitat, as is a dirty-pink, bandlike *Cerebratulus*. The latter has a firm consistency and may be 12 feet or more long, but when captured it immediately breaks into many fragments. It is an efficient digger and swims well. An Atlantic *Cerebratulus* takes first prize in rapidity of digestion: Wilson (1900) records that one of the creatures ingested the tentacle of a squid. The squid itself was too large to be swallowed, and the tentacle could not be broken off, so the nemertean was compelled to regurgitate what it had eaten. This it did, but the portion of the tentacle that was spewed out, after only 5 or 6 minutes, was almost completely digested! MacGinitie has found our Pacific species associated with large annelid worms, which he believes to be its food supply.

§313. Among the crustaceans native to this lowest zone of the mud flats is the long-handed ghost shrimp *Callianassa gigas*, which differs from the related *C. californiensis* (§292) chiefly in being very much larger and in having a longer and more slender large claw. It is usually white or cream-colored, but is sometimes pink. Recorded from Vancouver Island to San Quintín Bay, it probably parallels the range of *C. californiensis* and reaches into Alaska.

Burrowing in the lowest areas of mud bared by the tide, between southeastern Alaska and San Quintín Bay at least, but with small individuals rather high up in the north, occurs the most striking shrimp in this zone—the blue mud shrimp *Upogebia pugettensis* (Fig. 255), whose actual color is usually a dirty bluish-white. These animals are firmer, larger, and more vigorous than the *Callianassa* found higher up. They are harder to dig out, also, because the ground in which they live is not only continually inundated by the tide but so honeycombed with burrows that the water pours in almost as fast as the mud can be spaded out. The blue mud shrimp's burrow is permanent and little branched, and has several enlarged places for turning around. It extends downward from the surface for about 18 inches, horizontally for 2–4 feet, and then to the surface again. It is inhabited, almost always, by one male and one female, and they are probably moderately long-lived.

* In a later paper Coe (1930) suggests that this extraordinary fragmentation constitutes the normal method of reproduction during warm weather, since the fragments (in an eastern *Lineus*) develop into normal adults, and since fragmentation can be induced experimentally at any season simply by raising the temperature.

255. *Upogebia pugettensis* (§313), the blue mud shrimp.

In burrowing, MacGinitie observed, the animal carries its load of excavated material in a "mud basket" formed by the first two pairs of legs. Four other pairs function as paddles to keep the water circulating through the burrow, while the mud basket, when it is not being used as such, serves as a strainer to catch the minute food particles in the water. For this latter purpose the mud-basket legs are provided with hairs. The fifth pair of legs terminates in brushes, and is used not only for walking but also for cleaning the body. An egg-bearing female, especially, spends a great deal of time in brushing her eggs to keep them clean of diatoms and fungus growths that would otherwise kill the larval shrimps. She also keeps the eggs aerated with her swimmerets. The shrimp's telson, or recurved tail, can be used to block a considerable head of water, or, on occasion, can be flipped powerfully so as to cleanse the burrow by blowing out a swift current of water. Like the related ghost shrimp, this mud shrimp will die if its body is not in contact on all sides with the walls of its burrow or their equivalent.

There are commensals, of course, one of them being distinctive in that it is a small clam, *Pseudopythina rugifera*, which attaches to the underside of the tail of its host. The little crab (*Scleroplax granulata*) that lives with the innkeeper *Urechis* and with the ghost shrimp *Callianassa* is another. There is also a scale worm similar to the ones that are commensal with the innkeeper and the ghost shrimp (§306), and the transient goby *Clevelandia ios* occasionally seeks protection in the burrows.

Upogebia as well as *Callianassa* is in ill repute with the oyster men of Puget Sound, for the mud that is dug up smothers many of the young oysters, and the burrows cause drainage from the areas that have been diked in to keep the oysters covered at low water.

Free-swimming Animals

§314. Several familiar shrimps will be found in pools on the mud flats, and, more conspicuously, such fish as flounders, skates, and rays. The very presence of these fish furnishes interesting sidelights on the food chains of estuary regions, since all were in search of food when the ebbing tide trapped them on the flats. The skates and rays were after clams, which they root out and eat after crushing the shells with their powerful "pavement" teeth. The stingray is to be avoided, and barefoot collectors should be particularly careful about stepping into pools. The wounds they inflict are never serious, but are said to be decidedly painful. Dr. Starks advises soaking the affected part in very hot water; and he adds, no doubt from experience, that the sufferer is perfectly willing to put his foot into water even hotter than he can painlessly tolerate.

Water brought into the laboratory from the Elkhorn Slough mud flats may contain free-swimming specimens of the active little amphipod *Anisogammarus confervicolus,* associated also with eelgrass, which is one of the most common amphipods of the north Pacific. It ranges from Sitka to Elkhorn Slough. It is more common in brackish water than in oceanic salinity, especially in the Puget Sound region. Water taken from the mud flats at Mission Bay near San Diego will sometimes be teeming with a very common opossum shrimp, *Mysidopsis californica,* similar to the form mentioned in §61 but apparently adapted to the stagnant water and fairly high temperatures of this nearly landlocked shallow bay.

12. *Wharf Piling*

PILING ANIMALS show variations apparently correlated with their degree of exposure to surf. Whether these differences are due actually to the presence or absence of wave shock, as is thought to be the case with rocky-shore animals, or to the different makeup of bay waters and open waters, cannot be stated. It is known that the waters of bays and harbors differ from oceanic waters materially in oxygen content, salinity, degree of acidity or alkalinity, temperature, and especially in nutrients. It may be that we must look to these factors for the explanation of the faunal differences; but until more definite information is available the degree of protection from surf is a convenient method of classification.

For obvious reasons, no wharves or piers are built in fully exposed positions. Nevertheless, the outer piles of long piers, such as those at Santa Cruz and Santa Barbara, get pretty well pounded during storms, and it is noticeable that the animals most frequent on these piles are barnacles and California mussels, both adapted to surf-swept rocky headlands and perfectly able to take care of themselves in rough weather. The piling fauna, therefore, will be considered under two headings, *exposed* (using the term relatively) and *protected.*

Piling offers a conspicuous means of observing depth zonation, although for some reason this zonation is not as obvious and clear-cut on the Pacific as it is on the Atlantic. Dr. Allee speaks of the piles at Vineyard Haven, Massachusetts, as having four distinct bands, recognizable by their color. Apparently very few west-coast workers have been interested in the the ecology of wharf pilings, and much of the data reported below is more or less tentative.

Exposed Piles

Zones 1 and 2. Spray Area and High Intertidal

§315. Very few *Littorina* are to be found on piling, and the highest animals encountered are very small barnacles, often under ¼ inch in diameter. They occur sparsely and grow almost flush with the wood. The statement in earlier editions that there were *Balanus balanoides* at Monterey appears to have been some sort of *lapsus calami*; this species has not been recognized since from Monterey, and the specimens in question may have been *Balanus glandula.*

Intermingled with these barnacles, and a little below them, are a few scattered clusters of California mussels. With the mussels, and a bit below them, some of the smallest examples of the rather prettily colored warty anemones (the aggregated phase of *Anthopleura*) are likely to occur. A few of the little rock crabs (*Pachygrapsus*) may be seen walking about here, retreating, when danger threatens, into the larger mussel clusters just below.

Zone 3. Middle Intertidal

§316. This zone (and the upper part of the lower zone) is characterized by lush growths of an almost white hydroid, *Obelia*; by large clusters of big California mussels; and by the great barnacle *Balanus nubilis*. The hydroid *Obelia longissima* (Fig. 256) and similar smaller species form long trailers or furry growths, usually on barnacle or mussel shells but also on the bare wood where the area is not already occupied —a rare circumstance. *O. longissima* is the commonest hydroid of the piling environment in British Columbia, and is almost as common in Puget Sound (§332), ranging northward to Alaska and southward to San Pedro. The stalks of *Obelia* branch and rebranch in every plane. Each little "branchlet" is terminated by a cuplike hydrotheca that protects the retracted tentacles. When other, larger sacs, the gonangia, are swollen, they should be watched for the birth of minute jellyfish (Fig. 256). The young medusae, which usually have 26 tentacles, are given off most abundantly during the summer, but reproduction continues throughout the year. The newly born jellyfish swim away with all the assurance in the world, propelling themselves with little jerks. It is a pity that they are so small, for they are delicate, diagrammatic, and beautiful; they can barely be seen with the unaided eye if the containing jar is held against the light.

The growth of hydroids is very rapid—so rapid that many of them

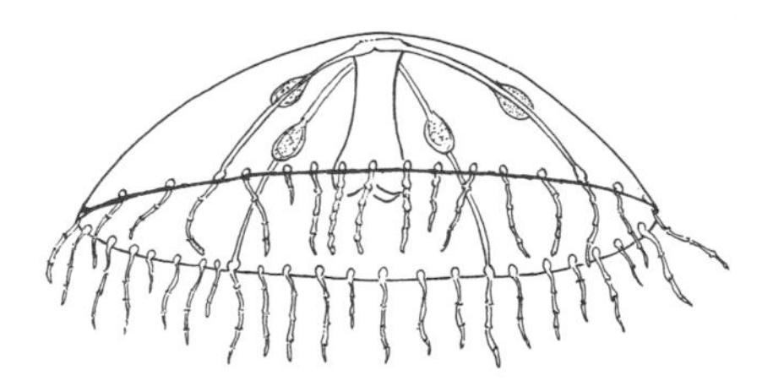

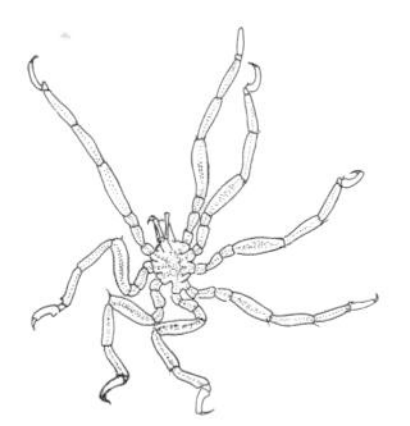

56. Top: the hydroid *Obelia longissima* (§316); the inset shows a greatly magnified branch of *O. dubia* §332), a similar species. Bottom, left: a jellyfish (medusa) from the hydroid *Obelia*; greatly enlarged. ottom, right: a sea spider common on *Obelia* colonies, *Halosoma viridintestinale* (§316); enlarged.

pass through several generations in a year. In one instance, 6 weeks sufficed for *Obelia* to cover a raft moored at sea, and a single month is sometimes long enough for a newly hatched larva to become a hydroid and release its own medusae.

Associated with these *Obelia* colonies there is almost invariably a minute sea spider, *Halosoma viridintestinale* (Fig. 256), not much larger than the hydranths on which it probably feeds. As its specific name implies, this pycnogonid has green guts, which show plainly through the legs; the significance of this color has not been investigated. A careful look at *Obelia*, especially in late spring, may reveal the presence of immature pycnogonids developing in the hydranths. The species occurs from Monterey to Tomales Bay, where it is common.

In this zone and lower, especially on pilings and floats in San Francisco Bay, the entoproct *Barentsia gracilis* occurs in brown, furry mats, often covering piling and barnacles alike. Anemones may attach to this matting; when they do so, they may easily be removed. Another ectoproct, or bryozoan, *Alcyonidium mytili,* likewise brown and furry but close-cropped, very often encrusts the insides of empty barnacles and mussel shells.

§317. *Balanus nubilis* is the largest barnacle on our coast, and probably one of the largest in the world. It ranges on this coast from southern Alaska to San Quintín, Baja California. It is commonly 2½ to 3 inches high, with a basal diameter exceeding the height, and in some places specimens grow on top of each other until they form great clusters. We have such clusters from the Santa Cruz wharf that are more than a foot high. The animal's great size and accretionary habits are not entirely helpful to it, since clusters get so large and heavy that they often carry away the bark or part of the disintegrating wood and sink to the bottom, where it is likely that the barnacles cannot live. The shell of *B. nubilis* is frequently covered with *Barentsia* (§316), and may also provide an attachment base for anemones and tube worms. The mantle through which the cirri, or feeding legs, are protruded is gorgeously colored with rich reds and purples. *B. nubilis* is another of the long series of barnacles originally described by Darwin. His description still stands correct—a tribute to the carefulness of the work devoted to this difficult group by a thoroughly competent mind.

A smaller but even more striking barnacle that is characteristic of boat bottoms and is occasional on wharf piling (and even on rocks) is *Balanus tintinnabulum* (Fig. 257). The nearly cylindrical ridged shell is a pinkish red with white lines, and the "lips" of the mantle, as with *B. nubilis,*

258. *Pachycheles rudis* (§318); natural size.

257. A red and white barnacle, *Balanus tintinnabulum* (§317).

are vividly colored. It is characteristically a southern California species.

§318. The big-clawed porcelain crab *Pachycheles rudis* (Fig. 258), ranging from British Columbia to Baja California, seeks crevices, nooks, and interstices, and is often at home in the discarded and *Alcyonidium*-lined shells of barnacles. It is about the same size as the porcelain crab of rocky shores (*Petrolisthes*) but may be distinguished from the latter by the granulations on the upper surface of the big claws. We have found ovigerous specimens in February, March, and July, and probably eggs are borne by at least a few individuals in every month of the year.

§319. Crawling about on these same empty barnacle shells are specimens of *Pycnogonum stearnsi* (Fig. 259), the largest and most ungainly of all local sea spiders. A border design of these grotesque yet picturesque animals might surround the pen-and-ink representation of a nightmare. Most sea spiders spend part of their tender youth in close juxtaposition to a coelenterate—the larvae, in fact, usually feed on the juices of their hydroid or anemone host. This sea spider, white to pale salmon pink in color, is found in a variety of habitats—*Aglaophenia* fronds, *Clavelina* clusters and the like, and around the bases of the large, green *Anthopleura*. It is especially common among the anemones of the caves and crevices of Tomales Bluff, Marin County; sometimes half a dozen occur on a single anemone.

Like the rest of the genus, *Pycnogonum stearnsi* feeds primarily on

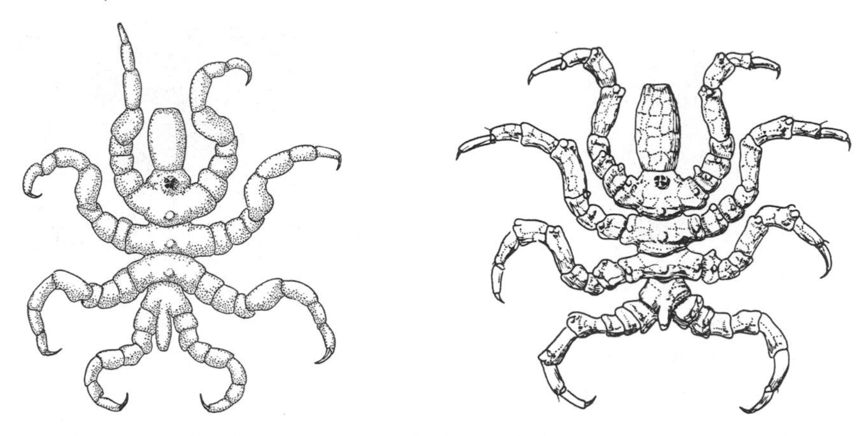

259. Two central California sea spiders (§319): left *Pycnogonum stearnsi*; right, *P. rickettsi*. Both enlarged about three times.

anemones and will often be found with its proboscis sunk into the column of its host. Although *P. stearnsi* is supposed to occur as far south as San Diego, it has not been found there. It is common among anemones in the Monterey area, where it may also be found hidden among the accumulation around the bases of anemones set in pockets or crevices. It does not appear to occur north of Sonoma County in any abundance, and only one specimen has been collected in Oregon—at Cape Arago some years ago. In the Friday Harbor area the common *Pycnogonum* is *P. rickettsi* (Fig. 259), a "reticulated" species that also occurs on Duxbury Reef in the same association as *P. stearnsi* (but not, as far as is known, on the same individual anemones); it was originally collected from anemones at Monterey by Ed Ricketts.

§320. There are a few clusters of goose barnacles on the outer piles, and some large solitary specimens of the great green anemone *Anthopleura*. Under the mussel clusters there are many acquaintances from other environments, among them the common scale worm (§49) of many loci and *Nereis mediator*, the ubiquitous pile worm of the bait gatherer; *N. mediator* resembles *Nereis vexillosa* (§163) but is smaller, rarely reaching more than 2 to 4 inches.

Zone 4. Low Intertidal

§321. The anemone *Metridium senile* (Fig. 260), usually white but occurring in several definite color patterns, prefers bare piling. Small specimens, however, may be found on dead shells (or even on the shells of living barnacles) and on the club-shaped tunicate *Styela.* Several times we have taken, from piling, kelp crabs that carried *Metridium* on their backs and claws. The largest specimens occur subtidally, extending clear to the bottom in water 20 to 25 feet deep, but small specimens extend into the intertidal, where, at low tide, they hang fully relaxed and pendulous. Expanded, they are delicate and lovely.

Entire clusters of this anemone may be colored a rich brown, and reddish-yellow specimens are not uncommon. The fact that specimens of one color seem to segregate themselves is accounted for by the common method of reproduction via basal fragmentation. This asexual method, actually a form of division, explains the presence of the many little specimens that seem to have been splattered about near larger individuals. *Metridium* also reproduces itself sexually, unisexual individuals discharging eggs and sperm from the central cavity through the mouth. The chance unions of these products result in free-swimming larvae that carry the fair breed of *Metridium* far and wide—pretty well throughout the world in the northern hemisphere. On this coast we know them to extend from Sitka to Santa Barbara on wharf piling. Gigantic exhibition specimens are dredged from deep water, one specimen, from 60 fathoms, filled a 10-gallon crock when expanded. Piling specimens will average around 2 inches in diameter, but 6-inch specimens have been taken.

Metridium is a fascinating animal to physiologists, who have used it in many experiments. Pantin (1950) leveled a movie camera at this seemingly stolid and motionless animal. When speeded up 60 times, his

260. Top and side views of the white anemone *Metridium senile* (§321); about natural size for specimens found on wharf pilings.

motion pictures reveal that *Metridium* is seldom motionless, and that its movements follow a well-defined pattern. When it is stimulated by a whiff of food, the movement is accelerated and the pattern changed. "We may conclude," writes Dr. Pantin, "that the very slow responses of *Metridium* involve a complex neuromuscular pattern of inherent activity involving reciprocal inhibition and the successive activities of two antagonistic muscular systems. The system is normally active in varying degrees in different animals. . . . The whole pattern is already there in the normal animal, though usually only released by the presence of food." So much for *Metridium*'s "instinctive" reaction to the promise of a square meal!

§322. Where it occurs at all, the stalked simple tunicate *Styela montereyensis* (Fig. 261) may be present in great numbers. Although it was described originally from rocky-shore specimens a couple of inches long, the most luxuriant growths are found on piers like those at Santa Cruz and Santa Barbara, where the water is fairly uncontaminated. Here occasional specimens may be more than a foot in length, and some are gorgeously festooned with growths of the ostrich-plume hydroid, with other tunicates, and even with anemones. They are dark red in color, except on the lower part of the body, which is lighter.

Styela montereyensis occurs from British Columbia to San Diego; it is not rare on rocky shores throughout its range. In bays and sheltered places south of Santa Barbara another *Styela, S. barnharti* (Fig. 261), is more common. It is somewhat thicker and has shorter siphons, the tips of which are deep maroon or magenta instead of reddish orange. *Styela barnharti* occasionally occurs as far north as Monterey, whence larvae have probably been carried by the Davidson Current; but it does not establish breeding populations there, according to Donald P. Abbott. The common *Styela* of the Puget Sound area is *S. gibbsi,* which has a much shorter stalk than the other species.

Slabs of *Amaroucium* (§99), the "sea pork" of children, occur in varying sizes and shapes, and their reds, yellows, or browns add touches of color to the piling assemblages. Locally there may be numbers of a simple tunicate, probably *Corella,* related to the *Molgula* so common on east-coast piling. The single animals are almost circular but are usually so crowded together that they form polymorphic clusters; and the test, where not covered by sand or bits of seaweed, is water-clear.

All these tunicates are the sedentary adults of free-swimming, tadpole-like larvae that indicate the group's relationship with the vertebrates. The free-swimming stage is very brief, and those tadpoles that do not find a suitable attachment surface quickly must perish. It has been determined that Atlantic *Amaroucium* larvae at first swim toward the light and away from the pull of gravity; later they swim away from the light and toward gravitational influence. Just how this affects their ultimate position in the low intertidal on piling is uncertain. Those that settle in the high-tide area must die of exposure, and those that drift down to the bottom of the piles, where conditions would seem to be satisfactory, die for some other reason, adults rarely or never being found there. Apparently, then, only those that happen, in the course of their vertical migrations, to touch a suitable spot on the right level have any chance of survival.

Adult tunicates feed on microscopic particles in the water, keeping currents of water circulating through the body by the waving action of great numbers of cilia. The food particles become entangled in mucus and are then transferred by other cilia to the digestive part of the gut.

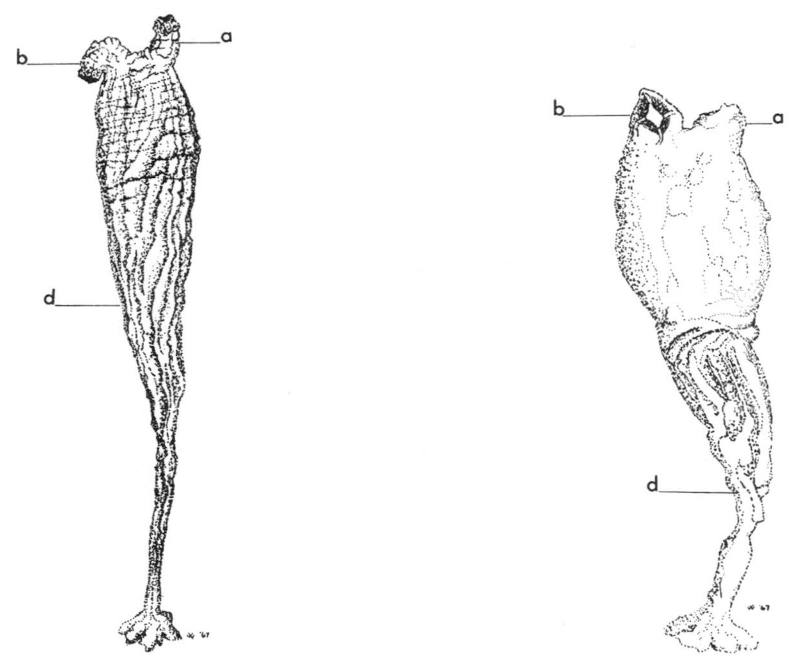

261. Common stalked tunicates of wharf pilings (§322): left, *Styela montereyensis*; right, *S. barnharti*. Both animals about natural size. The two siphons are at *a* and *b*; *d* is the division between body and stalk (determined by internal organs).

§323. The kelp that grows on piling in this zone has the usual concomitants of life in a crowded environment. Most of it is plastered and encrusted with unbidden guests: *Aglaophenia, Obelia,* and sertularians among the hydroids; the omnipresent *Membranipora* and tiny colonies of the tubed *Tubulipora* among bryozoans. Even the single-celled *Ephelota gigantea,* a stalked suctorian protozoan a fraction of an inch tall, furs the blades of kelp with minute vases of white protoplasm, finding these swaying plants an ideal attachment site. These protozoans are related to the animals occurring in the respiratory trees of the cucumber *Stichopus,* more distantly to the barely visible infusorians that haunt stagnant swamps and hay infusions, and still more distantly to the host of protozoans that are infamous as agents of malaria, tropical dysentery, sleeping sickness, etc.

§324. The infrequent bare spots on piling are furred and plumed with the usual hydroids—the same that occur on kelp and the hard tests of tunicates. *Bugula neritina,* a red or yellow-purple bryozoan in sparse palmate clusters, seems to be characteristic of bare spots on piling or on boat bottoms. We have often found it on submerged wood, but never on rock. Another bryozoan, the encrusting *Cryptosula pallasiana,* also seems to be specific to wood. It occurs in thin, white, watery sheets, sometimes slightly separated from their support, and will be recognized as a near relative of the bryozoans (such as *Eurystomella bilabiata,* §140) that encrust the underside of rocks. There are a good many others, some characteristic of piling on this coast but many of them cosmopolitan. This universal distribution is easily explained by the fact that ships always, unless they are fresh from drydock, carry growths of attached animals on their hulls. Deep-water vessels take care of the intercontinental aspects and smaller coastwise vessels carry on the distribution, for there is always a chance that some of the attached animals will be discharging sexual products or larvae while the vessel lies at the wharf.

§325. The seastars found on piling all attain their maximum abundance elsewhere. Considerable numbers of the common seastar (*Pisaster ochraceus*) are attracted by the abundance of mussels, and in the lower levels a few of the quiet-water *Pisaster brevispinus* will be found. The brilliantly colored sea bat *Patiria* occurs also, and the little, six-rayed *Leptasterias* may be abundant.

Another echinoderm occurs, the white cucumber *Eupentacta quinquesemita* (§235), but it is an almost exclusively subtidal form. Also chiefly subtidal is the rock oyster *Pododesmus* (§217), which belies its popular name by preferring piling to rocks in the regions south of Puget Sound.

§326. The chunky little shrimp *Betaeus setosus* (Fig. 262), some ¾ inch long, may often be found on piling. Like its relative of the southern tide pools (§46), this species has a carapace that extends over its eyes. No doubt this device affords protection to the eyes, but it must also be a considerable handicap to the animal's vision. *B. setosus* has been reported from Point Arena to Laguna Beach.

262. *Betaeus setosus* (§326), a shrimp often found on wharves.

§327. The kelp crab *Pugettia producta* (§107) is quite as characteristic of piling as it is of rocky tide pools. At low tide it will usually be found at least a few inches below the surface, and even when the tide rises it seems to prefer this low zone. This fairly large, active spider crab is the most likely of all the crabs on this coast to be infected with the strange parasite *Heterosaccus californicus*. This creature is a crustacean, actually a specialized barnacle whose free-swimming larval stage looks very much like any other crustacean nauplius. It is incapable of developing independently, however, and must find a crab host in order to complete its life cycle. It attaches itself to a hair on the crab's body or legs, penetrates the base of the hair with its feelers, and then enters the crab by the nightmarish method of slipping through its own hollow feelers. This is possible only because the degenerative process is already well advanced: the legs, the bivalve shell, and some of the inner organs have been shed, and the body is consequently reduced to a fraction of its larval size. Once inside, *Heterosaccus* migrates through the blood stream to the gut, attaches itself near the crab's stomach, and begins to grow. Avoiding the vital organs, the roots of its tumorlike sac extend throughout the crab's body, even into the claws; and the crab, its energies drained to feed the intruder, becomes sluggish. The crab molts just once after the parasite has gained entrance, and during that molt the parasite pushes out through the temporarily soft shell and assumes its final shape and position as a brownish mass under the crab's abdomen. Thereafter, the host is unable to molt until the parasite has completed its life cycle and died—a matter of years. Usually the crab is unable to survive; but if it does, it becomes normal again after the parasite disappears.

Sacculinids attack both males and females, and one of their strangest effects is to modify the sex of the afflicted crab. Both sexes are rendered sterile, but the only other effect on the female is to speed up her assumption of adult sex characters. The male, however, develops various female characteristics, such as a broad curved abdomen and smaller claws; and if he survives the life cycle of the parasite, he may then produce eggs as well as sperm, having become a male hermaphrodite. If he has been only slightly feminized by the parasite, he may regenerate normal male organs. The mechanism of this transformation is still incompletely understood. Reinhard (1950) suggested that the normal male crab is the result of a balance between the male and female hormones, and that the parasite upsets this balance. The effect of this is the removal of the inhibition on "femaleness" in the male and the stimulation of "femaleness" in the female.

On the California coast usually less than 10 per cent but more than 1 per cent of the kelp crabs are infected. Infestation becomes more abundant with the increase in latitude on this coast. On the Atlantic coast, sacculinids are rare except on the Gulf coast, where infestation is common on mud crabs in Louisiana and on the blue crab in Texas waters.

§328. To many well-informed persons the biological connotations of wharf piling all center around the word "teredo." These wood borers have been the bane of shipping for at least 2,000 years, for they were known, unfavorably, by the ancient Greeks and Romans. It seems quite possible that the early Mediterranean custom of hauling boats ashore when they were not in use was motivated, in part at least, by the knowledge that frequent drying would protect them from attack by the shipworms. The species of *Teredo* and their near relatives are still, after many years of scientific research on preventives, the cause of tremendous destruction to marine timbers—destruction that often runs into millions of dollars a year in a single seaport.

The term "shipworm," which is popularly applied to the whole group, is a misnomer, for the animal is actually a clam, although its small, calcareous shell—its boring tool—covers only the "head" end of its long, wormlike body. *Teredo*s are highly efficient woodworkers, but countless generations of protected life in timber have made them helpless outside of that element, and once removed from their burrows they cannot begin a new one. In these days of steel hulls they have had to confine their activities to wharf piling and other marine timbers, but that has affected neither their aggressiveness nor their numbers.

In open water the true *Teredo* rarely occurs, but the one shipworm that is native to this coast, the giant *Bankia setacea* (Fig. 263), known as the northwest shipworm, operates in relatively exposed piles from at least as far north as Kodiak Island to as far south as San Diego, although below Monterey Bay it does little damage. In San Francisco Bay proper it has been known since the earliest days of shipping; but since it is less resistant to lowered salinity than other shipworms, it does not extend its activities into the brackish waters of the adjoining San Pablo and Suisun bays, or into similar enclosed bays in other regions. Neither has it ever been as destructive, according to a report by the San Francisco Bay Piling Committee, as the true *Teredo* that suddenly appeared there about 1910–13. However, if some of the plans discussed by engineers are carried out—all the fresh water shunted south to the San Joaquin Valley and Los Angeles, with the tainted fluids carried off elsewhere in a giant sewer—the salinity will increase up the river, and if there is any wood left, the shipworms will chew it up.

The work of the shipworm is begun by a larva, which drills a hole that is barely visible to the naked eye. For the next few months the animal grows rapidly, enlarging the burrow as it grows but leaving the entrance very small. A few inches inside the wood the burrow turns, usually downward, so as to follow to some extent the grain of the wood. The animal turns aside, however, to avoid obstructions like nails, bolts, and knots, and to avoid penetrating the burrows of neighbors. Thus the actual course of the burrows is nearly always sinuous. *Bankia* concentrates its attack at the mud line, digging out burrows that are sometimes 3 feet long and almost an inch in diameter. The shipworm's method of boring is very similar to that employed by the rock-boring clams (§242); see §337 for an account of *Teredo navalis*. A heavy attack will reduce a new, untreated pile to the collapsing point in 6 months, and survival for more than a year is unlikely. The life of a pile may be prolonged to 3 or 4 years by chemical treatment, such as creosoting, but nothing keeps the animals out for long. Copper sheathing, which did good service in protecting ships for so many years, is impractical on piles because of its expense, the likelihood of its being stolen, and the ease with which it is damaged by contact with boats and driftwood. Also, the sheathing cannot extend below the vulnerable mud line, and the mud line is likely to be lowered by eroding currents. Numerous other jacketings have been tried, but not even concrete jackets are entirely satisfactory, for when the wood borers are thus thwarted, the concrete borers (§242) come to their assistance. Steel and iron piling has numerous disadvantages, aside from its great expense; so the battle between man and the shipworm goes on, with the shipworm still getting somewhat the best of it.*

Bankia setacea likes low water temperatures, and extrudes its eggs, in San Francisco Bay at least, at that time of the year when the water temperature is lowest—from February or March to June. In southeastern Alaska it breeds about a month earlier, as indicated by the attack on a test block at Petersburg. Fertilization is external to the burrow and a matter of chance. It has been suggested by Hill and Kofoid, the editors in chief of the *Report of the San Francisco Bay Marine Piling Committee,* that this inefficient reproductive method may account for the inferiority as a pest of *Bankia* to *Teredo.* Further considerations relating to the similar *Teredo* will be found in §§337 and 338.

§329. Dampier, the English buccaneer, who was something of a naturalist, is said to have written that his wormy ship was also being attacked by small white animals resembling sheep lice. Undoubtedly he was referring to a wood-boring isopod, the gribble, of which there are three species on this coast: *Limnoria lignorum, L. quadripunctata,* and *L. tripunctata.* Shipworms perform their work on the inside of wood; *Limnoria* works, with almost equal efficiency, from the outside, attacking piles at all levels from the middle intertidal down to depths of 40 feet or more (Fig. 264).

These animals are so tiny—scarcely ⅛ inch long—that hundreds of them may occur in a square inch of heavily infested wood. Their legs

* Although he was not much of a poet, H. D. Thoreau of Concord concisely stated the matter of the shipworm's ultimate supremacy some years ago:

Though all the fates should prove unkind,
Leave not your native land behind.
The ship, becalmed, at length stands still;
The steed must rest beneath the hill;
But swiftly still our fortunes pace,
To find us out in every place.

The vessel, though her masts be firm,
Beneath her copper bears a worm;
Around the cape, across the line,
Till fields of ice her course confine;
It matters not how smooth the breeze,
How shallow or how deep the seas,
Whether she bears Manilla twine,
Or in her hold Madeira wine,
Or China teas, or Spanish hides;
In port or quarantine she rides;
Far from New England's blustering shore,
New England's worm her hulk shall bore,
And sink her in the Indian seas,
Twine, wine, and hides, and China teas.

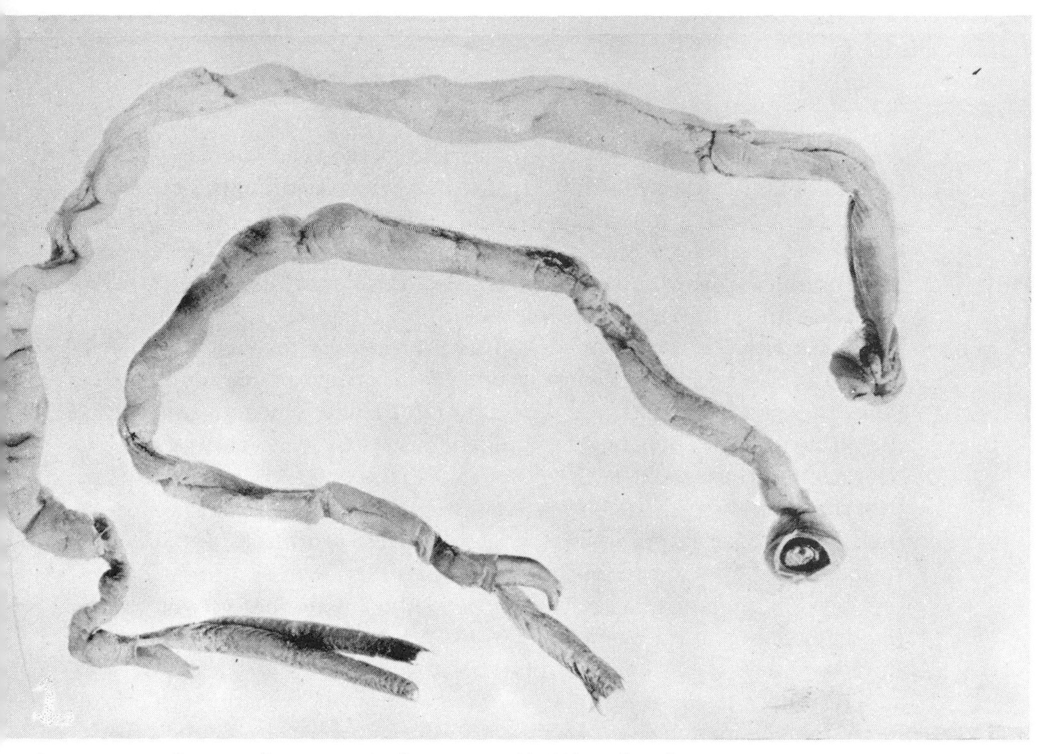

. The native Pacific coast shipworm, *Bankia setacea* (§328); reduced.

The gribble *Limnoria quadripunctata* (§329), a wood-boring isopod; the photograph, enlarged about
mes, shows the living animals in position in their burrows.

are adapted to holding to wood in the face of severe currents. The digging is done by the mouth parts, and the wood is eaten as fast as it is gouged out, passing so rapidly through the intestine that a given particle remains in the digestive tract only an estimated 80 minutes. This speed of passage is what would be expected in an animal that derives all its nourishment from wood, as *Limnoria* apparently does, for the food content is very low. There is no mechanism for filtering the contained plankton out of the water, and nothing but wood has ever been found in the digestive tract. This is a unique situation, for cellulose is highly indigestible and remains unchanged by animal digestive juices unless, as with the termites, there are intestinal protozoa to aid in the conversion. In one experiment, however, specimens of *Limnoria* ingested pure cellulose in the form of filter paper and lived nearly twice as long as controls that were unfed.

When digging, gribbles jerk their heads backward and forward, turning them slowly at the same time. They can completely bury themselves in from 4 to 6 days, starting their burrows at a rather flat angle. Presumably because of the difficulty of obtaining oxygen, they seldom go more than ½ inch below the surface. Breathing in a solitary burrow would probably be difficult even at that depth, but the burrows of different individuals are usually connected by small openings that permit freer circulation of water than could be obtained in an isolated burrow. There can be little doubt about the function of these openings, for they are not large enough to permit the passage of the animals themselves.

The females of *Limnoria* carry, on the average, only 9 or 10 eggs, but breeding is continuous throughout the year. The newly hatched animal, unlike the larvae of shipworms, does not swim at all, but it gradually acquires some swimming ability as it grows. Experiments recorded in the *San Francisco Bay Marine Piling Report* show that at no stage of its life is the animal positively attracted to wood. Presumably, gribbles do not fasten on wood unless they accidentally touch it; but since they are never very good swimmers, those that fail to make this contact probably die. At any stage, however, unlike shipworms, they may be transferred to new locations without harm. Thus driftwood floating among piling is assumed to be an important factor in their distribution, the work being begun by adult animals instead of by free-swimming larvae. The taking in and discharging of ballast water by ships may aid *Limnoria* in spreading from port to port.

Limnoria lignorum is primarily a cold-water animal, occurring on the Scandinavian coast, within the Arctic Circle, and on this coast from

Kodiak Island to as far south as Point Arena. *L. quadripunctata* occurs on this coast from Tomales Bay to San Diego, and *L. tripunctata* from San Francisco to Baja California. The various species are responsible for damage to piling in most of the great harbors of the world, and the warmer water species, *L. tripunctata,* is undeterred by the creosote with which most piling is treated to protect it from these depredations.

Over the last 20 years or so, the Navy has financed much research on the lives and times of borers, both worms and gribbles. It has been established that gribbles do have cellulase, the enzyme necessary to digest wood; and that they attack almost any kind of wood. Some exotic and expensive hardwoods are less susceptible, but creosote does not deter one species at least from attacking treated timbers. One investigator for the Navy has suggested that perhaps the Navy should consider breeding a resistant strain of trees to provide piling unappetizing or inedible to borers, since all other attempts to keep borers at bay (and since concrete pilings are cheerfully attacked by rock-boring piddocks) have been unsuccessful.

§330. Whoever frequents wharves, piers, and floats, especially at not-quite-respectable hours, will be certain to see some of the more obvious free-swimming invertebrate animals that haunt the harbors. While pelagic animals are not within the scope of this work, it would be a pity not to mention the bell-shaped jellyfish *Polyorchis penicillatus* (Fig. 265), whose transparent white against the dark of under-wharf waters makes it very conspicuous. It swims beautifully, and rather well, by kicking its manubrium—a kneelike veil within the bell. The average

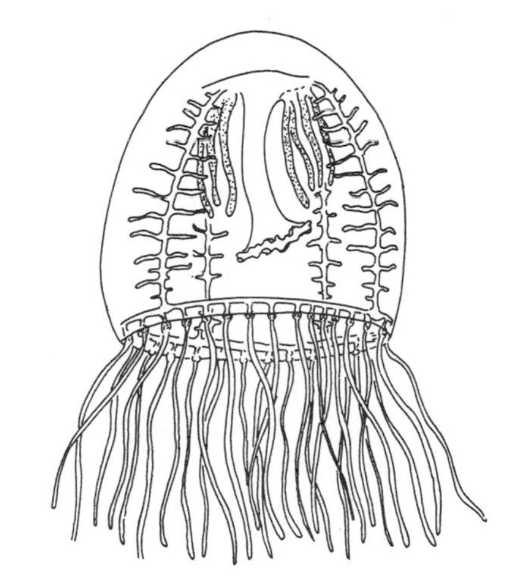

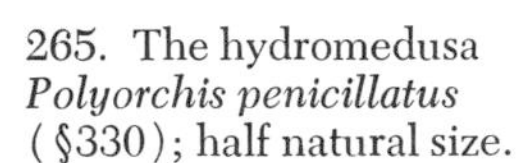

265. The hydromedusa *Polyorchis penicillatus* (§330); half natural size.

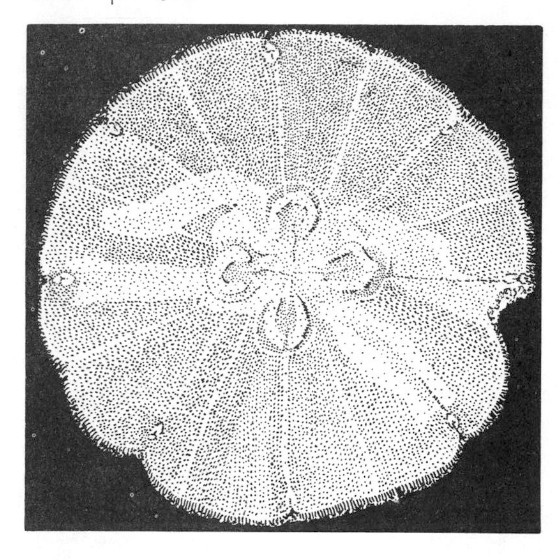

266. The scyphozoan *Aurellia aurita* (§330); about 1/10 natural size.

diameter is an inch or more. The gonads and some other internal organs are variably colored from yellow-brown to purple.

In quiet bays in Puget Sound and British Columbia the jellyfish *Aurellia aurita* (Fig. 266), nearly colorless except for its four horseshoe-shaped brown gonads, sometimes occurs in such immense numbers that it is impossible to dip an oar without striking several of the beautiful, pulsating animals. A boat seems to glide through a sea of jellyfish rather than through water. These tremendous aggregations are but a brief phenomenon, usually lasting during one flood tide only. A few hours later it may be difficult to find a dozen specimens. Similar swarms occur in Tomales Bay during the summer months.

The ubiquitous *Nereis vexillosa* (§163) occurs in this environment also whenever clusters of mussels extend so far down; but a smaller worm, probably *Platynereis agassizi,* seems to be both abundant and characteristic, especially in its free-swimming (heteronereis) stage. We discovered the interesting periodicity of this form quite by accident, in connection with a proclivity for roaming about at odd hours that is no doubt very annoying to conventional night watchmen. Swarming takes place in Monterey harbor during the spring and summer months when, on moonless nights, the extreme high tides occur about midnight. At first only one or two of the vividly white and wriggly worms will be in sight; by the end of half an hour there may be two or three score, apparently attracted to the area of observation by the rays of the flashlight. The male comes first—a tiny animal ½ inch long or less—followed by the female, sometimes more than an inch long and having red after-segments. When the two sexes are together, the heteronereis go completely crazy, swimming about in circles at furious speed and shedding eggs and sperm. Only one who has tried to capture them with a dip net knows how rapidly and evasively they swim. Specimens brought into the laboratory have continued this activity all night, although in their natural environment the swarming seems to last only a few hours at the height of the tide. In the morning, captive specimens look withered and worn out, no doubt an accurate reflection of their condition. In this connection, see also §163.

Finally there will be dozens, possibly hundreds, of smaller, free-swimming organisms of various types—micro-crustaceans, minute jellyfish, larval stages of worms, mollusks, and echinoderms—all outside the province of this book. It might be mentioned that one of the usually pelagic euphausiids, *Thysanoëssa gregaria,* or perhaps *T. longipes,* distantly related to the opossum shrimps (§61), has been taken abundantly in Monterey harbor within a stone's throw of the shore.

Protected Piles

As previously stated, we can draw only a vague line between relatively exposed and fully protected piling. Between the obvious extremes there is probably great overlapping, and we have to remind the reader that this classification is offered tentatively. Further work may justify it as a working hypothesis or may show that other factors than exposure to waves are the primary ones.

Depth zonation is as apparent here as on more exposed piling, but the distinctive animals are so few in number that they will be treated without formal zonal classification.

§331. The common small barnacles of protected piling in Berkeley and Newport Bay are the omnipresent *Balanus glandula* (§4). For some reason, the small barnacles of this coast have been almost entirely overlooked by collectors, so that until someone works them over carefully it is impossible to make any general remarks that further investigation may not refute.

§332. The California mussels of the more exposed piling are replaced in these quiet waters by the smaller and cosmopolitan bay mussels *Mytilus edulis* (§197). In favorable environments they often form great bunches that double the diameter of the piles on which they grow, and they may be found in probably every suitable port between the Bering Sea and northern Mexico. They attain their maximum development in the middle zone, so they are very obvious, even at half tide.

The small mussel worm *Nereis mediator* (§163) will be found here no less frequently than in open-shore mussel clusters, only the mussels are different. The quiet-water shield limpet, *A. pelta olympica* (§195), occurs on these pilings just as on rocks and gravel, minute specimens occupying the highest populated zone along with the small barnacles.

Interspersed among the mussel clusters are several bay and harbor hydroids. The cosmopolitan *Obelia commissuralis* was reported by Torrey (1902) from San Francisco Bay, where it was adapted to the estuary conditions of dirty water and lowered salinity. Clusters of *Obelia* may still be found in these situations about the wharves in West Oakland, in water so filthy that all animal life would seem to be precluded, and *O. commissuralis* is presumed to be the species involved. There is an excellent illustration of a living Atlantic coast colony in Nutting (1915, Pl. 21).

The sea spider *Achelia nudiuscula* may occur in hydroid clusters of this sort, and was once abundant on *Obelia* on pilings and junk along the east shore of San Francisco Bay—in very filthy water. It is known only from this, the type locality.

In cleaner stretches of quiet water, particularly up north, the tenuous *Obelia longissima* (§316) is known to be the predominant shore hydroid, avoiding pollution and direct sunlight but tolerating some fresh water. Fraser (1914, p. 153) says that *O. longissima* is "the commonest shallow water campanularian in the region. It grows throughout the year on the station float, Departure Bay, and medusae are freed at many times during the year." On continually submerged floats, the situation seems to be that the very common and bristling *Clytia edwardsi* (illustrated in Nutting, 1915, Pl. 4) occurs in considerable sunlight, but that the lush clusters of *O. longissima* are either underneath the float or on portions protected from direct sunlight by some of the superstructure. In late July 1935, the *Clytia* in southeastern Hood Canal were producing round, four-tentacled medusae. *Obelia* and *Clytia* comprise up to 90 per cent of the littoral hydroid population of this region (the rest is made up mostly of *Gonothyraea*, and of the smaller and more hidden *O. dichotoma*, §271).

On floats in the Puget Sound region (as at Port Madison, releasing medusae in July 1932), and on bell buoys, boat bottoms, etc., along the open coast at Monterey, the small and sparkling *Obelia dubia* (Fig. 256) occurs, so similar to *O. dichotoma* (§271) that some authorities consider them identical. We have never knowingly taken *O. dubia* on fixed piling, where *O. dichotoma* frequently occurs. *O. dubia* seems to be intolerant equally to stagnation, lowered salinity, pollution, and tidal exposure—a "touchy" form, difficult to preserve expanded unless narcotized within a few minutes after capture. This is the species that we have been calling *O. gracilis* in Monterey Bay, according to determinations by Dr. Fraser, who says (1914, p. 151) that the juveniles are sometimes impossible to distinguish from juvenile *O. longissima.*

Four of the five commonest hydroids here release free-swimming medusae, thus increasing many times the opportunities for wide distribution of the race. Each hydroid colony produces thousands of minute, active jellyfish, which can be carried far from the parent hydroid by currents before reaching sexual maturity. The union of the jellyfish's eggs and sperm produces planulae, free-swimming larvae that can drift still farther before they settle and develop into the polyp stage.

Gonothyraea clarki resembles *Obelia,* but it develops medusae that are perfectly functional but, for some strange reason, are never released. Four or five of these little captives pop out of each vaselike gonangium and pulsate ineffectually at the ends of their stalks, releasing germ cells and eggs to be united by chance currents. In British Columbia and Puget Sound, *Gonothyraea* is plentiful in spring and early summer, growing on rocks as well as on floats and piling, but it dies out later and is not to be found. It extends as far south as San Francisco.

§333. In the extreme low-tide zone enormous bushy clusters of the naked hydroid *Tubularia crocea* (Fig. 267) will be found banding the piles or floating docks with delicate pinks or reds. Its nearest relative is the small, solitary *T. marina* of rocky shores, but this clustered form, which may extend 6 inches out from the piles, is far more spectacular. It is the "heads" that provide the color; the supporting stems look like small flexible straws. The heads, or hydranths, autotomize readily; they seem to break off as a regular thing if the water gets warm or the conditions are otherwise unfavorable, after which, if conditions improve, the stem regenerates a new head within a few days. The sexual medusae are never liberated. We have found *T. crocea* in Newport Bay, Elkhorn Slough, and San Francisco Bay; it is also known from many regions to the northward and from both sides of the Atlantic.

Hilton found the *Tubularia* clusters at Balboa heavily populated with a small sea spider, *Anoplodactylus erectus* (which we have found rarely on compound tunicates at Pacific Grove). He worked out the life history. The eggs, which are produced in the summer, develop into larvae that

267. **A cluster of the naked hydroid *Tubularia crocea* (§333); natural size.**

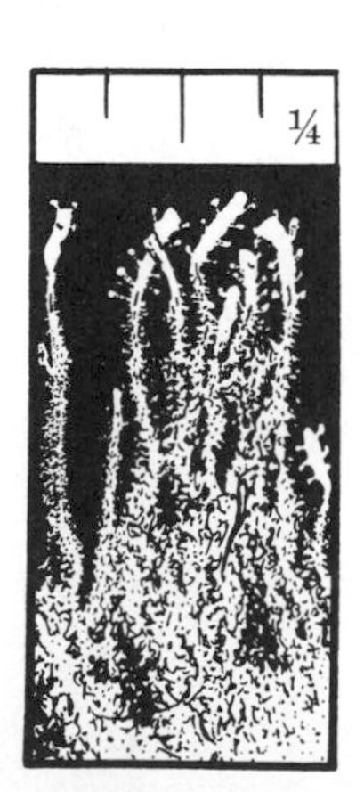

268. The hydroid *Syncoryne mirabilis* (§333): left, a typical cluster; right, a greatly enlarged view of the clublike polyps.

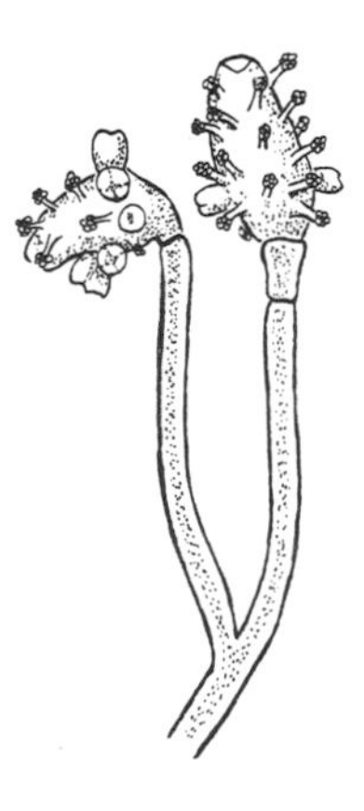

pierce the body wall of *Tubularia* and enter the digestive tract, there to live parasitically until further grown.

Flatter, tangled clusters of another naked hydroid are often interspersed with *Tubularia,* but the two will be differentiated easily. The tiny, bulbous heads of *Syncoryne mirabilis* (Fig. 268) bear clubbed tentacles and bud off the free-swimming medusae from their lower sections. Agassiz found this form in San Francisco Bay in 1865. There is at least a distinct possibility that it is a relic of the days of wooden ships, for the same species occurs on the east coast, and it seems unlikely that its natural distribution would account for its occurrence on this coast also. If it was so imported it probably became established at several different places, since it now ranges along the entire coast as far south as Chile.

§334. Any of several crabs may be seen crawling about on the piling throughout the intertidal zone but keeping fairly well under water, whatever the height of the tide. Practically all of the rocky-shore forms have been reported, and two spider crabs seem to be characteristic. In Monterey Bay and southward to Panama a thoroughly attenuated, gangly form, *Pyromaia tuberculata* (formerly *Inachoides tuberculatus*), ½ inch or less in carapace width, is fairly common. Sponges and seaweeds mask this sluggish little crab, so that it will seldom be noticed until what appears to be a bit of the fixed piling growth moves slowly away from the observer.

Oregonia gracilis is very similar, but twice the size of *Pyromaia.* It may be seen occasionally on wharf piling in Puget Sound and in eelgrass beds, whence it ranges northward to the Bering Sea and southward, but in deeper water, to Monterey. The masking habit of this group of crabs is discussed in §131.

Several cucumbers occur, in the Puget Sound area especially. Huge *Stichopus californicus* (§77) crawl about on the pilings frequently enough to be considered characteristic here but keep pretty well below the low-tide line.

§335. On these fully protected piles we find a good many of the animals already mentioned as occurring on more exposed piling: the seastar *Pisaster,* both the common *P. ochraceus* (§157) and *P. brevispinus* (§209), and the usually white anemone *Metridium* (§321). There are the usual tunicates; *Styela barnharti* (§322) is common in southern California harbors especially. The elongate simple tunicate *Ciona intestinalis* reaches a length of from 4 to 5 inches. Its translucent green siphons are long and glassy, the basal portion being often covered with debris. It is well distributed throughout the northern hemisphere but requires pure water. Some of the small rock-loving tunicates, notably the red *Cnemidocarpa* (§216), are found on Puget Sound piling, and the gelatinous clusters of the compound *Botrylloides diegensis,* in reds, yellows, and purples, are features of the piling in San Diego Bay.

Referring to another tunicate, *Pyura haustor johnsoni,* which is almost identical in appearance with the Puget Sound *P. haustor* (§216), Ritter remarks: "A striking thing about this species is its great abundance in San Diego Bay, and the large size reached there by individuals, as compared with what one finds on the open shores. Its favorite habitat appears to be the piles of wharves where, at times, it makes almost a solid coating. Although it must be counted as a native of the whole littoral zone, we have found only occasional specimens at outside points."

§336. Boring forms are as common and destructive here as on less sheltered piling. The boring isopods *Limnoria* (§329) operate quite indiscriminately in these quiet waters, assisting the shipworms to make life miserable for harbor engineers. Another isopod of ill fame is the larger *Sphaeroma pentodon* (up to nearly ⅜ inch long), which cannot, however, compete with *Limnoria* in destructiveness. Wood is for it apparently only a secondary habitat, for the animal (which resembles *Cirolana,* Fig. 24) is found in great numbers in clay and friable rock. It rarely seems to attack piling until the wood has been riddled by shipworms, hence it is not of much economic importance. *Sphaeroma* bores for protection only, making oval openings up to nearly ½ inch in diameter that give the pile a pitted appearance. Boring is accomplished with the mouth parts, as with *Limnoria,* but the loosened material (at least in the case of an observed specimen boring in chalk) is passed backward by the feet and washed out by the swimmerets. The fact that very little wood is found in the intestines of wood-boring specimens indicates that wood particles are handled in the same manner. Algae and other growths are the food supply.

Eggs are carried under the abdomen of the female for a time, and young have been found with the adults in the spring, summer, and fall in San Francisco Bay; breeding probably lasts all year.

§337. Sixty years ago the European shipworm, *Teredo navalis,* was unrecognized on this coast. Just when or how it gained a foothold is not known, but in 1914 it was discovered that piling at Mare Island, in the northern part of San Francisco Bay, had been extensively damaged. The attack was repeated in 1917, and within 3 years the damage reached unprecedented and catastrophic proportions. Ferry slips collapsed, and warehouses and loaded freight cars were pitched into the bay. All piling was attacked, and most of it destroyed. Since then *Teredo navalis* has spread to all parts of the bay. The animals can stand much lower salinity than other shipworms, which accounts for their devastating attacks in the northern parts of the bay, where piling was largely untreated because it had been supposed that the influx of fresh water from the Sacramento and San Joaquin rivers afforded immunity from borers. The fresh water does kill them off during seasons of heavy rainfall, but when conditions improve, they come back in force.

These animals are usually hoist by their own petard within 6 months, for their honeycombing of timber is so complete that by the end of that time it begins to disintegrate, the resulting exposure causing their deaths. They can reach a length of more than 4 inches in 4 months, however, and one 14-month-old specimen was nearly a foot long. The rate of boring is more rapid in comparatively young animals; a 3-month-old specimen bores about ¾ inch a day. Thus the animals can ruin a pile during their short lives, making ample provision in the meantime for another generation to take care of the next set of piles.

Unlike *Bankia*, this *Teredo* breeds during late summer and early fall, the last lot of larvae being attached by December. The sexes are separate, but the same individual is first male, then finally female.* Eggs are retained in the gill cavity until fertilized by spermatozoa contained in the incoming water. They are then stored in a brood pouch that forms in the gills until they are ready to be discharged through the parent's siphons as free-swimming bivalves known as "veliger" larvae. Apparently the breeding period is continuous for some weeks, for the brood pouch contains larvae in various stages of development, while at the same time the ovaries are enlarged with unfertilized eggs. Since the ovaries may contain an estimated million or more eggs at one time, the tremendously rapid spread of the animals is easy to understand.†

* Coe (1933) discovered at Woods Hole that although the primary gonad is bisexual, *T. navalis* is "essentially protandric, nearly all females passing through a functional male phase before reaching the definitive phase of sexuality." This means that when the current year's brood is young, early in the season, there is a preponderance of males (any specimens left over from the previous season, however, being females). Later, as this brood ages, practically all the individuals in the colony will be females. Still later, when the new generation liberates its larvae and the second filial generation attaches and matures, there will be ample extruding males to fertilize all the available eggs. Some oysters are known to reverse sex still a second time, but so far as known the first female phase is definitive with *Teredo*.

† Working with this species at Woods Hole, Massachusetts, Grave (1928) found that spawning, during the summer and early fall, was a function simply of water tempera-

In the first century Pliny the Elder conjectured that teredos bored with their shells. His theory went unchallenged until 1733, when the Dutch naturalist Sellius announced that since the shells appeared to be inadequate for the work, the boring must be accomplished by the foot. Recent investigations have vindicated Pliny. Dr. R. C. Miller, working for the San Francisco Bay Marine Piling Committee, succeeded in cutting into a burrow and covering the opening with a piece of glass. Through this window observers watched *Teredo navalis* at work. The animal rasps the wood with its shell by rocking on its foot as do the rock-boring pholads, repeating the motion rhythmically 8 to 12 times per minute. Also like the pholads, it rotates, so that a perfectly cylindrical burrow is cut. Interestingly, there are two distinct grades of rasping surface on the shell, and the animal's intestinal contents show two correspondingly different sizes of wood particles.

All the wood removed from the head of the burrow passes rapidly through the animal's stomach into its intestine and is finally removed via the siphons. *Teredo* probably extracts some nourishment from the wood en route, but certainly not enough to keep it alive. Its chief food supply comes, as with all bivalves, from the plankton contained in the respiratory water.

All other means of controlling *Teredo* having failed, there may be some hope in selective tree breeding, as mentioned in §329.

§338. In Los Angeles and San Diego harbors, and in Hawaii, the havoc-working shipworm is the small *Teredo diegensis* (Fig. 269). Reaching a length of 5 inches but averaging nearer 2 inches, it is the smallest shipworm occurring on the Pacific coast. Breeding apparently takes place throughout the year, but the heaviest infections by larvae take place when the water is warmest, from April to October. This is just the opposite of what takes place with *Bankia,* and differs somewhat from the breeding of *T. navalis.* As with *T. navalis,* the young are retained in a brood pouch until well along in their development, but they are larger and fewer in number. In the brood pouch of an average female 490 were counted. The brown-shelled, nearly globular larvae are plainly visible through the distended body wall of the parent.

This *Teredo* has been found (1920) at one spot in San Francisco Bay—

---

ture, 11° to 12° C. or more. There was no evidence of tidal influence or of stimulation by the presence of the other sex. In its life history, *Teredo* spends 5 weeks developing, the first half of this period in the gill chamber of the mother and the next as a free-swimming veliger. The largest specimens found were 40 by 1 cm. Sexual maturity is attained at a length of 4 to 5 cm., about 2 months after settling. The normal adult size is reached the first year, and death comes in the second.

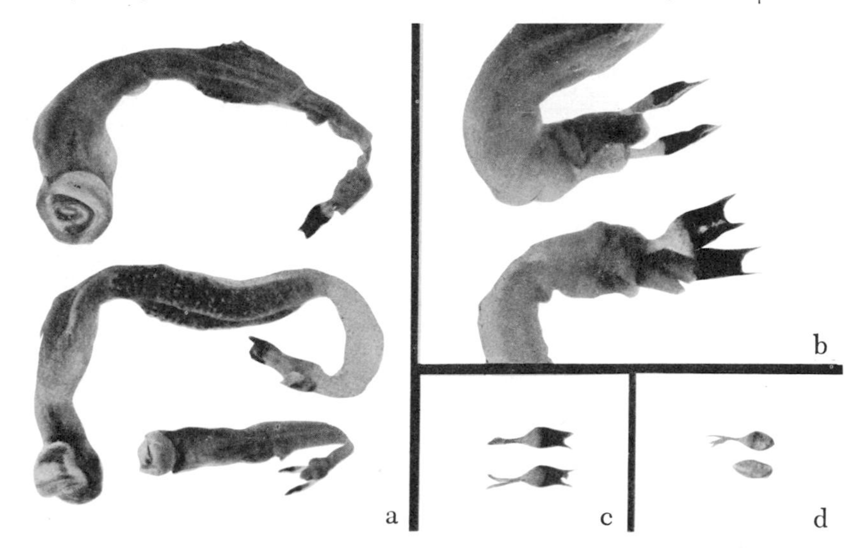

269. *Teredo diegensis* (§338): (*a*) entire animals, showing enlarged brood sacs with larvae; (*b*) posterior ends, showing the siphons and the flaplike pallets used to close the entrance to the burrow; (*c*) normal pallets; (*d*) pallets with horny tips removed.

a small inlet near South San Francisco, where, apparently, it found survival conditions after being accidentally introduced from the south or from Hawaii. Within four years after its discovery there its numbers were considerably depleted, and it is possible that it has already disappeared. Even if it survives, it is unlikely to become an important destructive factor in San Francisco Bay's relatively cold waters.

13. Recent Developments

ON ROCKY SHORES it is not always easy to make out the groupings of animals into communities (with the obvious exception of the mussel beds), for the animals are often jumbled together. A tidal flat on sandy or muddy bottom, however, is a striking contrast—the animals are so characteristically grouped that it is possible to say, here is an area of worms or phoronids, and there a colony of mud shrimps. These groupings are called communities, and their arrangement is a result of the complicated interplay of the physical factors of the environment: the substrate, currents and tidal action, the influx of fresh water, salinity and temperature, and the biological factors imposed by the animals on their environment (such as the binding of the sandy bottom by tube builders and the retention of water in burrows) and upon each other. In this complex system of interrelationships, a change in any one factor may set in motion a whole train of changes, which will be reflected by a perceptible change in the limits of the various communities. Thus the communities of an area like upper Tomales Bay will change from year to year in details, although the general aspect, as well as a characteristic arrangement of zones, may remain the same (Fig. 270).

In recent years, our attention in Tomales Bay—as elsewhere—has turned to analysis of the communities of the shallow seas. Many of the same species that occur on tidal flats live in—and on—these shallow bottoms, but other members of these communities are rarely if ever seen between the tides. These animals of the shallow bottoms are a major source of fish food, and many of them feed in ways that make them of

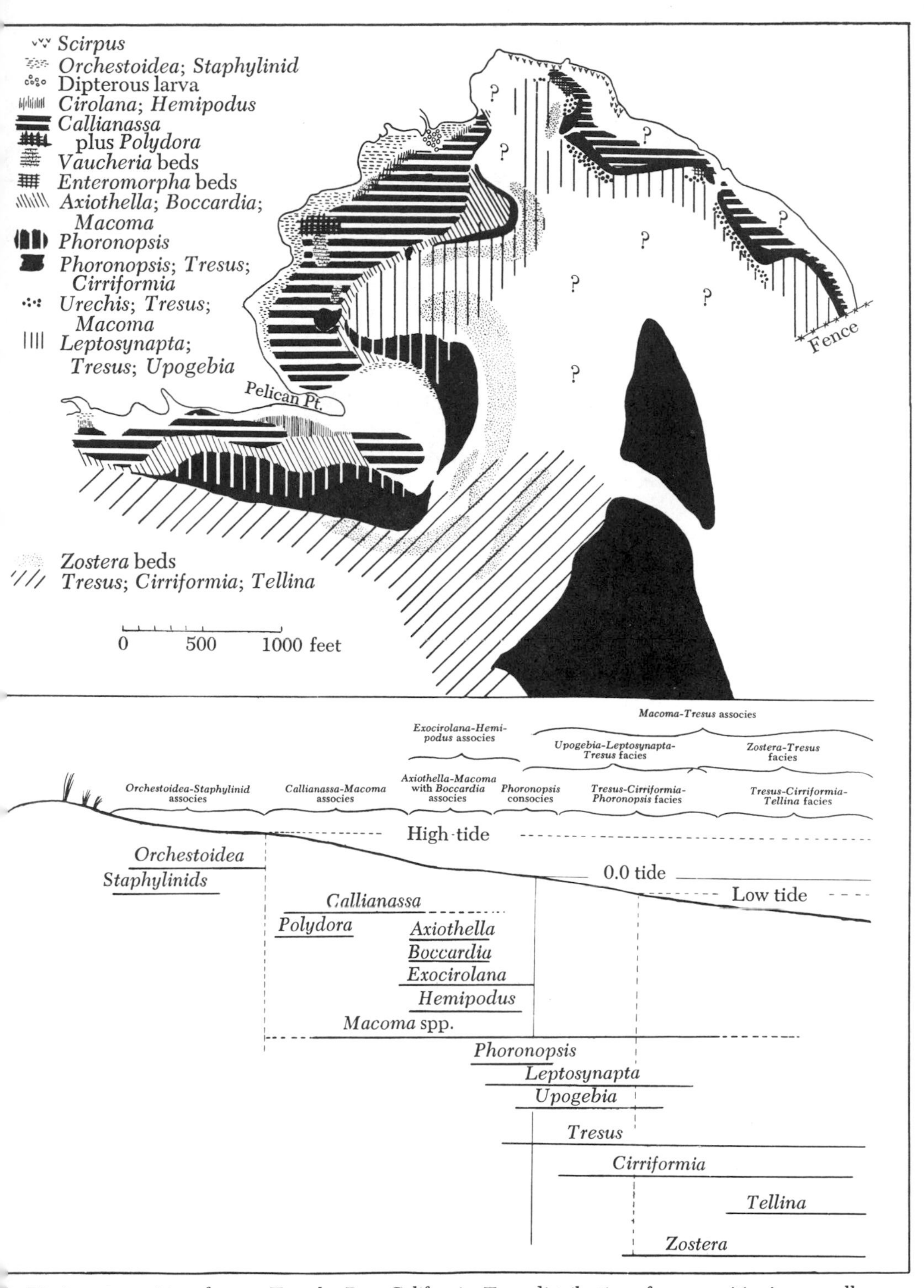

). Biotic communities of upper Tomales Bay, California. Top: distribution of communities in a small 'e just east of Sand Point, near Dillon Beach. Bottom: the range of representative genera through the 'erent intertidal zones, showing the ecological associations and faciations in which they may be grouped. 'h charts are based on the pilot survey by F. A. Pitelka and R. E. Paulson in spring 1941.

critical interest for study of the fate of radioactive materials that may be dumped into the ocean. But there are also problems of pure science involved: are these communities entities in themselves, or is their composition governed only by the physical conditions of types of bottom sediment and patterns of water currents, and how may such groupings be used to understand the incomplete record of fossil assemblages?

To examine these questions, a program of long-term studies of communities and environmental changes has been under way in Tomales Bay for some years. Phases of this study go back as far as the Pleistocene, for at such places as Tom's Point the fossils in the unconsolidated banks along the shore contain some of the same species present today, as well as others that suggest a somewhat warmer temperature in the past. Intensive sampling of the existing fauna and flora and studies of such environmental variables as the temperature and salinity in the bottom sediments have been conducted. For the last several years much of this work has been carried out by teams of high school biology teachers, with Ralph G. Johnson, from the University of Chicago, as "Big Daddy." One of the unexpected by-products of this effort has been the production of an inexpensive guidebook to seashore life of the area by a group calling themselves the "Tide-pool Associates."* Many papers have been published concerning this work, including one on the geology of Tomales Bay, which is cited in textbooks as a "typical" fault-controlled bay (although it may be the only bay quite like it in the world). Yet there is still much to learn before it can be said that we understand even this small part of the marine environment.

Newcomers to the Pacific Coast: The Estuarine Itinerants

One problem of particular interest to students of the geographical distribution of plants and animals is the colonization of new areas by organisms that have somehow strayed from their native haunts. A man named Guppy, for whom a certain little fish was named, devoted several large volumes to the subject of plant dispersal, one of them concerned entirely with the floating leguminous seeds of the West Indies, the famous "sea beans." On the Pacific coast the most noteworthy immigrants are from estuaries of other parts of the world; hence they have settled in our bays and estuaries. Most of these have hitchhiked their way with oysters, either from the Atlantic coast or from Japanese waters. Usually

* See *Exploring Tidal Life along the Pacific Coast, with Emphasis on Point Reyes National Seashore,* by Tierney, Ulmer, Waxdeck, Foster, and Eckenroad. This little booklet is now in its second edition.

these immigrants have become well established before their presence is noticed, which, as in the case of undesirable aliens, is too late.

On the other hand, efforts to establish desirable invertebrates from other waters have been on the whole unsuccessful. The most spectacular failure was that of the Atlantic lobster, which was brought across the continent in the 1870's and dumped into the Pacific off the Golden Gate. It has since been determined that the larvae of this lobster require a minimum temperature of about 15° C., although the adults can do very well in low temperatures and sometimes stray into the waters around Iceland. Since temperatures off the Golden Gate rarely average above 13° C. during the summer months, it is not surprising that the lobsters planted off the Golden Gate have never been seen again. On this coast the temperature conditions most closely resembling those of the lobster's native haunts appear to be in the sheltered waters of southern Alaska. Currently (1968) scientists of the Fisheries Research Board of Canada are attempting to naturalize the Atlantic lobster in the waters of British Columbia.

§339. Efforts to establish the eastern or Virginia oyster, *Crassostrea virginica,* have for the most part failed (there are some small colonies in Willapa Bay); but the Japanese oyster, *Crassostrea gigas* (now ambiguously called "Pacific oyster" by the trade), is now well established, under supervision, in Hood Canal and has recently settled in Drake's Estero. About once every three years there is a successful spatfall in Willapa Bay. Since 1947 *Crassostrea gigas* has become naturalized and abundant in British Columbia waters and forms massive trottoire-like growths in Departure Bay and the Georgia Straits area. The bulk of the Japanese oyster production in American waters, however, is still from spat imported from Japan and raised in our coastal waters. Importation of spat has become increasingly more expensive, and experiments are now under way to introduce the European flat oyster, *Ostrea edulis,* in Pacific coast bays.

§340. With the Japanese oysters has come the Japanese oyster drill, *Ocenebra japonica,* now established in Puget Sound as a serious pest. It has also become naturalized in Tomales Bay. This unwelcome guest was preceded by the related Atlantic drill *Urosalpinx cinereus,* reported to be common in San Francisco and Tomales bays, and by the dog whelk, *Ilyanassa obsoleta,* in San Francisco Bay. A possibly more sinister immigrant from Japan is the parasitic copepod *Mytilicola orientalis,* reported to be a minor oyster pest in Puget Sound. Recently copepod-infested oysters were found in Tomales Bay. This copepod, or a similar one, has wiped out the mussel industry of France and the Low Countries in the last few years. Since 1947 spat inspectors have been stationed in Japan to prevent the continued introduction of Japanese drills and other pests.

§341. A number of "neutral" immigrants from Japan have probably arrived with oysters in past years and have only recently been noticed. Two of these have become abundant in certain areas. A mussel, *Volsella* (or *Modiolus*) *senhousei,* has been identified from Samish Bay, Puget Sound, Tomales Bay, San Francisco Bay, Moss Landing, and Mission Bay; in lower Tomales Bay near Millerton it occurs in clusters, with densities of several hundred per square meter. It is thin-shelled, and greenish with zigzag patterns. A horn snail, *Batillaria zonalis,* occurs in dense aggregations at the mouth of Walker Creek and near Millerton in Tomales Bay. It is superficially similar to *Cerithidea,* but is easily distinguished because it lacks the weltlike ribs of *Cerithidea* and because its aperture tends to form a short canal at the base. *Batillaria* has been recognized in various bays from Puget Sound to Moss Landing.

§342. Although San Francisco Bay no longer produces market oysters, there are several souvenirs of the halcyon days of Jack London, the oyster pirates, and the French restaurants of San Francisco living in the Bay. The Virginia oyster, *Crassostrea virginica,* was once well naturalized in San Francisco Bay, and there are probably still a few stragglers, as there are in Tomales Bay; but progress and pollution have destroyed the commercial fishery in San Francisco Bay. Since the early transplantations were made in days of ecological innocence, all sorts of animals came with the eastern oysters. Today there are dense stands of the horse mussel *Volsella demissa* (*Modiolus demissus*) in the southern part of San Francisco Bay, and consideration has been given to harvesting these mussels for pet food, chicken meal, or even some extract for space travelers. *Volsella* is brownish to black, with prominent radial ribs on the shell. Shore birds are apparently fond of dining on this mussel at low tide, and the mussel sometimes protects itself the only way it can—by gripping the bird by the foot until the tide comes in and the bird is drowned. Clapper rails seem to be especially easy victims of this tactic. The large predatory gastropod *Busycon canaliculatus* also appears to be well established as a subtidal species in San Francisco Bay. It was not observed during the intensive survey of San Francisco Bay carried out by the United States Bureau of Fisheries and the Department of Zoology at Berkeley in 1912–14, so it has probably been carried in with oysters since that time (active transplantation of eastern oysters in San Francisco Bay ceased some time after 1924). This reminds us that there has been no comparable study of San Francisco Bay in all the 60 years since the old *Albatross* survey, although recent developments indicate the critical need for information about the present physical and biological status of California's largest bay. There have been a few nibbles here and there, especially in the Delta area, but no concerted effort.

§343. Somehow, perhaps with the oysters and their various associated mollusks, a small crab has made its way from Atlantic to Pacific waters. This is *Rhithropanopeus harrisi,* a crab smaller than the native estuarine *Hemigrapsus oregonensis* and recognizable by its three spurs on the sides of the carapace and its whitish claws. It was first noticed in San Francisco Bay about 1940, and it is now abundant in the sloughs of northern San Francisco Bay and up the river as far as Stockton. In 1950 Dr. James A. Macnab found this crab in the sloughs of southern Coos Bay in Oregon. A small, dingy Atlantic tunicate, perhaps unnoticed in most places, appears to have made the crossing of the continent as well. Dr. James E. Lynch found *Molgula manhattensis* on the piling at Marshall in Tomales Bay in 1949.

§344. Recently (1966) Dr. G. Dallas Hanna has revised his paper on introduced mollusks (including both terrestrial and freshwater species). Some 30 species of marine mollusks have been recognized. Some of these evidently have not survived, or are dubiously known from dead shells of uncertain provenance. Some species not already mentioned are the slipper limpets *Crepidula fornicata* (abundant in Puget Sound and sparingly in Tomales Bay) and *Crepidula plana* (in Puget Sound). The case for *Mya arenaria* as an introduced species may be moot (see §302), as well as the case for the small clam *Gemma gemma,* which is sometimes confused with a similar small clam of the sand flats, *Transenella tantilla.* Recently (1963) efforts have been made to establish the quahog, or hard-shelled clam, *Mercenaria mercenaria* (alas for the lovely Linnaean name *Venus mercenaria*!), in various bays from Washington to California. The returns are not yet in on this experiment. The most spectacular introduction of an edible clam (*Protothaca semidecussata*) was, as usual, unintentional.

§345. The most recent unwelcome invader to come to notice is the Atlantic green crab, *Carcinides maenas,* which turned up in Willapa Bay in 1961. This crab attacks young oysters and could become a serious pest. Its progress is being watched with apprehension (Fig. 271).

It is pleasant to conclude our roster of undesirable and indifferent invaders with a mention of two of some value, at least by our homocentric standards. First is the Japanese "littleneck" clam, *Protothaca semidecussata,* now widespread and common enough in Puget Sound to appear on the market. It was first noticed in British Columbian waters about 1939,

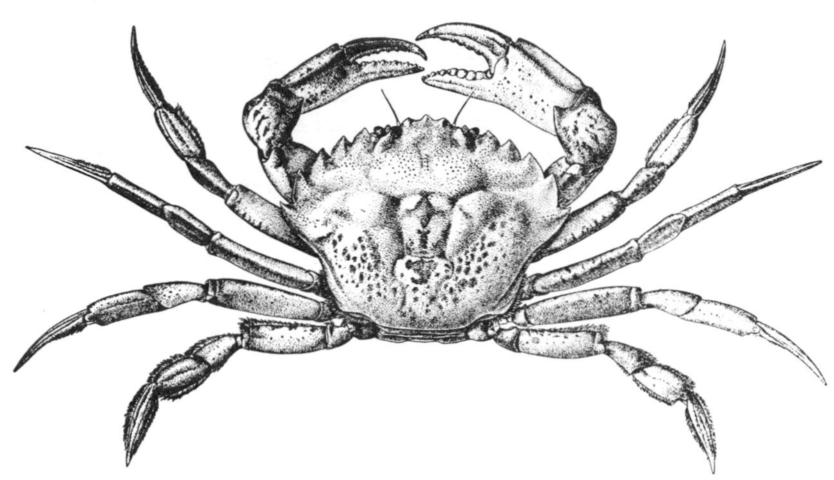

271. *Carcinides maenas* (§345), a sinister new arrival from the Atlantic.

when it was described as a new species, *Paphia bifurcata*. It has become abundant in Puget Sound within the last 10 or 12 years. Its presence in California waters was noticed some years earlier; there is a note by Bonnot (*Nautilus*, Vol. 49, pp. 1–2) indicating its arrival in central California at least as early as 1930. It is now abundant in San Francisco and Tomales bays. Somewhat smaller than its relative *Protothaca staminea*, it is considered better flavored than the native species. In appearance it is more elongate, with brownish to bluish banding on one end, indicating the source of its specific name, *semidecussata*.

An unexpected result of the Korean War (apparently) is the naturalization in San Francisco Bay of *Palaemon macrodactylus*, a brackish-water shrimp about 2 inches long. It was first noticed about 1954 in San Francisco Bay, and is now common in small streams in Marin County flowing into San Francisco Bay and in the main river as far as Collinsville and Antioch. It now seems to be more abundant than the native *Crago*, from which it may easily be distinguished by its longer and apparently more numerous antennae and its long, toothed rostrum. Introductions can rarely be dated as closely as this one has been, and its progress into the Delta is being followed with interest. Its obvious vigor and success may make it an important food resource for the fish populations of the region. It appears to have been introduced in ballast water from ships returning from the Korean peninsula, its native haunts. It thrives well in captivity, and will probably become an important experimental animal in our halls of learning.

PART IV. BETWEEN AND BEYOND THE TIDES

Daring also should he be, and dauntless and temperate; and he must not love satiety of sleep but must be keen of sight, wakeful of heart, and open-eyed. He must bear well the wintry weather and the thirsty season of Sirius; he must be fond of labor and must love the sea.

Oppian, *Halieutica,* III, 42–48

14. *Intertidal Zonation*

STUDYING THE ARRANGEMENT of plants and animals of the shore and shallow sea in horizontal belts is the oldest and most "durable" concern of marine biology, as Rupert Riedl reminds us in his centennial tribute to Josef Lorenz. Perhaps the phenomenon was discussed in some lost book of Aristotle's; in our time, at least, the scientific literature on the subject has been accumulating since the 1840's. Perhaps it was indeed Charles d'Orbigny, in 1820 or earlier, who made the first statements in "the literature" about zonation and suggested a classification according to tidal range. As Max Doty (1957) put it in his review of the subject for the *Treatise on Marine Ecology*: "Even earlier someone else may have said essentially the same things; many have since. Indeed, reporting the intertidal distribution of organisms has been an occupation attractive to many authors."

In any event, the first formal description of zonal patterns seems to be that of J. V. Audouin and Henri Milne-Edwards in 1832. They recognized five zones: (1) *Balanus,* (2) seaweed, (3) corallines, (4) *Laminaria,* and (5) oysters. Interestingly enough the first critical study of the factors that influence zonation was made by Josef Lorenz (1863) in the Adriatic, where the tides are not the most significant influence on zonation. The specific relation of zones to tidal levels is often attributed to one L. Vaillant (1873), especially by the Russians, who speak of "Vaillant's principle."

By 1929, when Torsten Gislèn summarized the work on seashore ecology in a masterful review with an exhaustive bibliography (Gislèn,

1930), there was already a vast literature. Since then another 5-foot shelf of books and papers has appeared, and it has become a labor of Sisyphus to keep up with it all (and Sisyphus was not required to read Russian or Chinese!).

I

In a sense, *Between Pacific Tides* is an expanded paper on intertidal zonation. It is the first book for general readers to discuss the distribution of animals on the seashore according to their levels of occurrence and to attempt to use this zonal distribution as an aid to identification. Although the original edition of this book was essentially completed in 1936 (but not published for 3 years after that), Ricketts, as early as 1930, was working on a paper on intertidal zonation. This paper, titled "The tide as an environmental factor, chiefly with reference to intertidal zonation on the Pacific coast," was never published. It was freely discussed with anyone available, and when Torsten Gislèn, from Lund, Sweden, spent the winter of 1930–31 at Hopkins Marine Station, the manuscript was, as he acknowledged, at his "free disposal." It was also made available to Willis Hewatt, who studied the zonation along a transect at Hopkins Marine Station from 1931 to 1934 for his doctoral dissertation, and who was permitted to use one of the illustrations from the paper (Figure 3 of this edition). It seems apparent that Ed Ricketts was developing his own ideas of zonation independently; and since he was blissfully oblivious of the academic need to publish or perish, he gave freely of his ideas.

In spite of all this ferment at Hopkins Marine Station, and the publication of *Between Pacific Tides* in 1939, there has been remarkably little study of intertidal zonation as such on the Pacific coast of North America. Victor Shelford and his students studied bottom communities in the vicinity of Friday Harbor and published the results in 1935; but these, with their verbal thickets of associations and faciations, had little pertinence to zonation problems. During the summers of 1936, 1937, and 1938 Rigg and Miller studied the zonation at Neah Bay, Washington; but they did not publish their findings until 1949. Beginning in 1937, an impressive series of papers on the intertidal zonation of the shores of South Africa began to appear; and in 1947 the Stephensons arrived in western North America to continue their worldwide studies of zonation. In a sense, this invasion deterred other students, and everyone waited for the papers. But only one of them, on Vancouver Island, was published (by T. A. and Anne Stephenson, in 1961). The work at Pacific Grove and La Jolla will never be published as part of the series, since T. A. Stephenson died in April of 1961.* Instead, we have been led to expect a posthumous work on the shores of the world, in which the unpublished parts concerning our own coast will be included as chapters.

Although many references to intertidal studies in various parts of the world will be found listed in the Bibliography, we will not attempt to call the roll for all of them here. The important review papers since Gislèn (1930) and Doty (1957) are: Riedl (1962, 1964); Southward (1958); Knox (1963) for the southern hemisphere; and Gurjanova (1958) for work on Asiatic shores. We understand that Oleg Mokyevsky is working on a book on the subject that already occupies 800 typescript pages, and doubtless somewhere in the world someone is working on another book or paper. The most recent large contribution is that of J. R. Lewis, *The Ecology of Rocky Shores* (London, 1964), a zone-by-zone and limpet-by-littorine account of the rocky shores of the British Isles. Anyone who reads this book faithfully and carefully should be at home almost anywhere on the rocky shore of Britain; if he is gifted with a photographic memory, he need only remember that the seascape before him was discussed on page so-and-so.

Well, almost. If things change, it may be a little difficult to recognize the zones, even if the shape of the rocks in the background remains the same. My complaint about the lack of dates on photographs, first made in the 1952 edition, must again be reiterated. A photograph of a recognizable location is an ecological document; it should be treated as such and dated. Curiously, although Lewis has no dated photographs in his otherwise fine book, he has a number of dated diagrams. Somehow, we doubt that the situations encountered on the seashore are as eternal and immutable as many of the dateless discussions in the literature imply.

II

That zonation occurs is obvious to all, and enough had been published on it by 1938 to cause MacGinitie to express an opinion that the subject had "been decidedly overdone" (1939, p. 45). In view of what has been

* Perhaps, some bright day, the book will finally appear. It will not, of course, be quite the same. Many have been impatient with the Old Master and have felt that having once determined a universal scheme of zonation, he managed to fit everything into it "like a glove." Perhaps Alan Stephenson was indeed more artist than scientist; nevertheless (or because of it), he was the author of one of the great monographs on sea anemones, a very difficult group. He was also a perceptive gentleman, who knew, when I presented myself at his door on the hillside above Aberystwyth with a harp, that I had not come to Wales to see him alone, and obliged me, in the tradition of a people he was not always in sympathy with, according to their hospitable motto: *Deuwch pan fynnoch, croeso pan ddeloch.*

published since then, it would seem that discussion of the subject has hardly started.*

MacGinitie, at least in 1938, recognized three zones, somewhat equivalent to the three basic zones of Stephenson; but in *Natural History of Marine Animals* discussion of the subject is conspicuously lacking. This represents one extreme; the other is that of the Shelford school, with its array of biomes, associations, and faciations, "a positively terrifying outburst of terminology referring to communities" (in the Stephensons' words) in which the zones have been submerged in a high tide of related but not exactly synonymous subdivisions.

The number of zones that may be recognized on a seashore depends not only on the complex variables of tide, climate, and the life subject to these variables, but also on the degree of refinement the student seeks to attain. In Figure 272 some of the various ways in which zonation has been recognized on the Pacific coast are represented. At first glance, this appears hopelessly complex. Even Hewatt and Ricketts, working in the same locality, do not quite agree. The samples from Doty's study, however, show the differences that may occur in algal zonation in the same locality under different conditions of exposure. Rigg and Miller set up two schemes of zonation, one for algae and the other for invertebrates. These zones show an essential similarity to central California, despite the much greater tidal range at Neah Bay: "The major difference is one of nomenclature, the present writers having regarded mid-tide not as a zone but as the boundary between the upper and lower intertidal regions. A very little vertical displacement would bring the two sets of concepts into harmony." To these various zonal arrangements we have added, principally for historical interest, the community terms of Shelford *et al.*

A few words of caution about these diagrams (Figs. 272, 273) are in order. First, they are, especially as presented in this interpretation, highly schematic, and it is, in a way, doing the authors an injustice to fix these zones on a scale of tidal heights. We are actually dealing with proportions related to tidal heights at specific localities, which are quite different from the fixed levels based on a standard established for a gauging station. Hence the measurements of Hewatt, Ricketts, and Doty, as referred to tidal levels at San Francisco, are abstractions or idealizations of conditions that may not even occur on the rocks near the gauging station. The water levels of the schemes of Yonge and Stephenson in Figure 273

* One eminent ecologist of the editor's acquaintance refuses to recognize the existence of zonation; this of course may simply be a way of emphasizing his interest in other problems of ecology, such as the heretical practice (for an ecologist) of upsetting the ecosystem of his personal garden with some potent bug killer or another.

Height (ft)	La Jolla, Calif. — RASMUSSEN (in Shelford, *et al.*, 1935)	Pacific Grove, Calif. — HEWATT, 1937 (roman numerals); R. & C., 1939 (arabic numerals)	Pacific Grove, Calif. — DOTY, 1946: *MUSSEL POINT*	Pacific Grove, Calif. — DOTY, 1946: *HEWATT STRIP*	SHELFORD, 1935 — Puget Sound	"Postelsia Point," Neah Bay, Wash. — RIGG & MILLER, 1949: *ALGAE*	"Postelsia Point," Neah Bay, Wash. — RIGG & MILLER, 1949: *ANIMALS*
14′–13							
13–12							A SPLASH ZONE
12–10	"*BALANUS-LITTORINA* BIOME"					*Ralfsia* *Prasiola*	*Littorina sitkana*
10–8			(arrow upward to 10)		*Littorina-glandula* faciation	*Endocladia* *Gigartina*	*Acmaea digitalis*
8–7							B UPPER INTER-TIDAL
7–5	*Balanus-californianus* Association: *Littorina planaxis* Faciation	I *Littorina planaxis* 1; *Acmaea digitalis*	*Prasiola*			*Postelsia*	*Balanus glandula* *Mytilus*
5–4		II *Balanus glandula* 2; *Littorina scutulata*	crustose reds	*Gigartina cristata*	*Pollicipes-Mytilus* fac.	*Halosaccion*	*Pollicipes*
4–3			*Endocladia* *Porphyra*	*Pelvetia* *Fucus*			C LOWER INTER-TIDAL
3–1	*Mytilus californianus* Faciation	*Pelvetia*; III *Pollicipes* 3; *Mytilus*; *Tegula funebralis*; *T. brunnea*	*Mytilus*; *Gigartina agardhii*; *Iridaea flaccida*	*Cladophora trichotoma*		*Alaria*	
1–0			*Egregia* *Ulva lobata*	*Gigartina canaliculata* *Iridaea splendens*	*Anthopleure* fac.	*Lessionopsis*	D DEMERSAL
0 to –2		IV hydroids, sponges, etc. 4	*Zanardinula*; *Gigartina corymbifera*; *Calliarthron*	*Zanardinula*; *Gigartina corymbifera*; *Laminaria andersonii*			*Strongylocentrotus* sponges, etc.
–2 to –3 and below			*Cystoseira* *Laminaria andersonii*		*Strongylocentrotus-Pugettia* fac.	*Laminaria* *Nereocystis*	

2. Zonation on the Pacific coast, as observed by various authors. The numbers at either side represent the height (in feet) above or below mean lower low water.

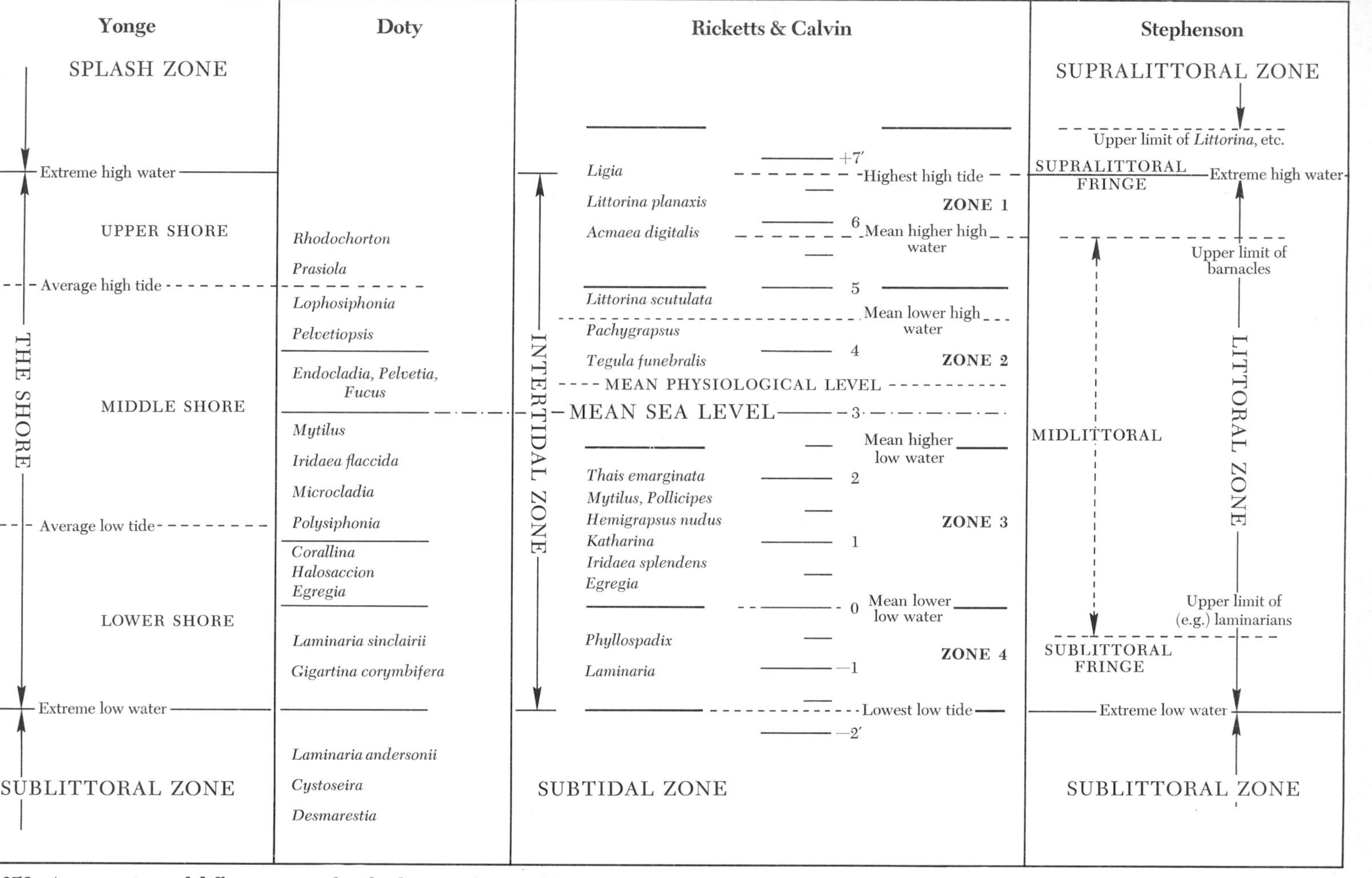

273. A comparison of different generalized schemes of intertidal zonation. The actual heights given apply only to the Ricketts and Calvin system; other systems are measured from the points shown (high-tide average, upper limit of barnacles, etc.).

are not strictly comparable with those at San Francisco in the center of the figure, an indication of the difficulties inherent in reconciling the semidaily and mixed tidal patterns. Zonation, like everything else in nature, is not a simple pattern, although we would sometimes like to have it so. Zones may overlap: they are quite different on a vertical face and a gently sloping reef, other things being equal; and, in the case of some algae, they may change with the seasons. The Stephensons provided a detailed account of some of these variations in their Nanaimo papers.

Nevertheless, the zones in the rocky intertidal region look pretty much the same in England, South Africa, Australia, and California, as anyone who examines the excellent (but undated) photographs in various books and papers will discover. Coniferous forests in Russia and in the United States also look the same in photographs. These climax developments, the most stable sort of growth that can be produced under given climatic or physiographic conditions, are known in ecological terminology as biomes. Beveridge and Chapman (1950) seem to have been the first to apply this term clearly to the rocky intertidal: "The biome of the rocky seashore represents a physiographic climax since it is dependent upon the tide rather than upon the climate" (p. 188). We might argue a bit about "physiographic" here—after all, the tide is a component of what might be termed the "hydrographic climate." Application of the term biome (i.e., as equivalent to the terrestrial "climatic climax") to the sea is as yet undecided; it may perhaps apply to coral reefs, but the term means more than simply a larger, more inclusive rank in the hierarchy of communities. Perhaps "ecosystem" would be a better word for the three-dimensional conditions in the sea.

Although we cannot afford a detailed discussion of the nature of tides and tidal differences, it is necessary to call the reader's attention to the different types of tides occurring in various parts of the world, since some of the controversy about terminology is based on experience with different tidal regimes. Our English terminology for tides has been developed for the tides of northern European waters, the more or less even procession of high and low tides twice a day known as semidiurnal or semidaily (Fig. 274).* A tidal pattern of this type is associated with the phases of the moon, and during the full and new moon the ranges are greater than during the quarters of the moon. The tides of greater range

* This difference between tidal characteristics has had unanticipated legal complications. Lawyers, always reluctant to concede the validity of facts of nature, have applied the concept of mean high tide level developed in English common law and U.S. Constitutional law to property ownership on the Pacific coast, where the mean high tide is actually below the highest high tide (because of the differences in daily highs in our mixed tidal cycle) at least 20 per cent of the time. The effect of this is

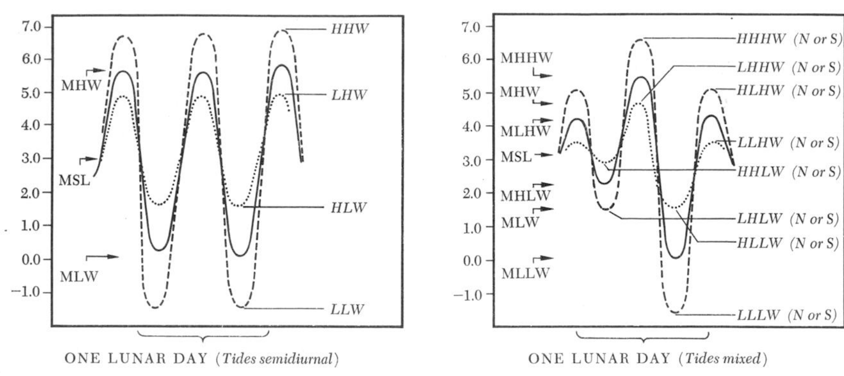

274. Types of tidal curves, after Doty. Solid line = mean range of tides; broken line = maximum range; dotted line = minimum range.

are known as spring tides (which has to do not with the spring of the year but with the Anglo-Saxon word "to jump," which we also use for certain mechanical devices). The tides of minimum range are known as neaps. The neap tide is therefore not the low tide, but the tide cycle of lesser range; a low during the neap cycle may actually be higher than the low of the spring cycle. It is simple ignorance of English to call the low tide the neap.

The tides on the eastern side of the Pacific (which means California to Alaska) are also semidaily tides, but the highs and lows of the two cycles are quite uneven (Fig. 274); these are known as mixed tides. The variations of mixed tides tend to coincide more with the declination of the moon (i.e., its angle above or below the celestial equator), the tides of greater range being associated with the tropic declination of the moon and those of minimum range with the equatorial declination; these tides are often called declinational tides, and the proper terms are "tropic" for the springs and "equatorial" for the neaps. Figure 275 shows the predicted tide for a lunar month at San Francisco, together with the positions of the respective heavenly bodies. The progression of highs and lows is uneven, and periodically a tide is skipped, as on July 4 and 18, so that the curve gets back into phase, so to speak.

---

to establish property rights to tidelands property at lower levels than those on the Atlantic coast, and to prevent access to the shore even more. At the same time the states proclaim the right of all citizens to access to the shore between high and low tides, often blissfully unaware that there are indeed two high tides (and two low ones). One may, of course, take consolation in the thought that in the fullness of time (or sooner, if the "greenhouse effect" melts the glaciers) the sea will rise again and wash away all the private estates, fancy motels, and condominiums—as well as all the marine laboratories.

Although T. A. Stephenson insisted that zonation is not caused by the tides, the patterns of zonation are nevertheless associated with tidal patterns. A number of workers have noticed that in regions of mixed tides there are more discrete zonal divisions than in regions of equal semidaily tides. The subzones recognized by the Stephensons at Nanaimo correspond to what the Russian workers recognize as various "stories," or subzones, associated with complicated tidal patterns (Gurjanova, 1958). O. B. Mokyevsky, however, considers that the actual type of tide is not important in the "littoral bionomy," and that the differences merely cause "a certain shifting of some critical levels in the vertical distribution of the organisms." Nevertheless, it is this "certain shifting" that complicates zonation patterns. Alan Southward's review of intertidal studies (1958) also concludes that different types of tide influence zonation patterns.

Virtually all students of zonation agree that one break is well defined. This is the approximate 0.0 of tide datum, or mean lower low water, below which is the zone of laminarians, and of *Phyllospadix* on our coast (see Figs. 276 and 277). There is also agreement on the zone above the average high tide (or mean higher high water of the Pacific coast), frequented by certain littorines and limpets. These two regions are (roughly) the sublittoral and supralittoral fringes of Stephenson's terminology, the *D* and *A* of Rigg and Miller, the sublittoral and upper-littoral of Beveridge and Chapman, the littoral-sublittoral fringe and supralittoral zone of Dakin *et al.*, and, of course, the "4" and "1" of this book (Fig. 3).*

There remains the intervening zone or zones, the multilittoral zone of Stephenson's proposed universal system. Because of the dual application of "zone" to both the inclusive littoral (= intertidal) and the subsidiary midlittoral in this classification, we have reduced "midlittoral zone" to "midlittoral." Since it separates the fringes, and since one can hardly call the zone between two fringes a fringe, it is simpler to leave the term an unadorned adjective. The gap between Stephenson's supralittoral and littoral zones is, according to him, an inconsistency of nature rather than of his classification.

* J. R. Lewis (1966) expresses a strong dislike for the term supralittoral, and denounces it as "being entirely superfluous and illogical for a belt of marine organisms . . . it has, historically at least, a supratidal meaning." Perhaps so, but language is not always logical, and we may soon have to adjust ourselves to discussions of the "geology" of the moon. In any event, supralittoral is a universally used term, and if we wish to be logical, we should also dispense with subtidal. As for historical meanings, one should by this token restrict a certain word referring to arches to architecture and ignore what has been going on under them since Roman times.

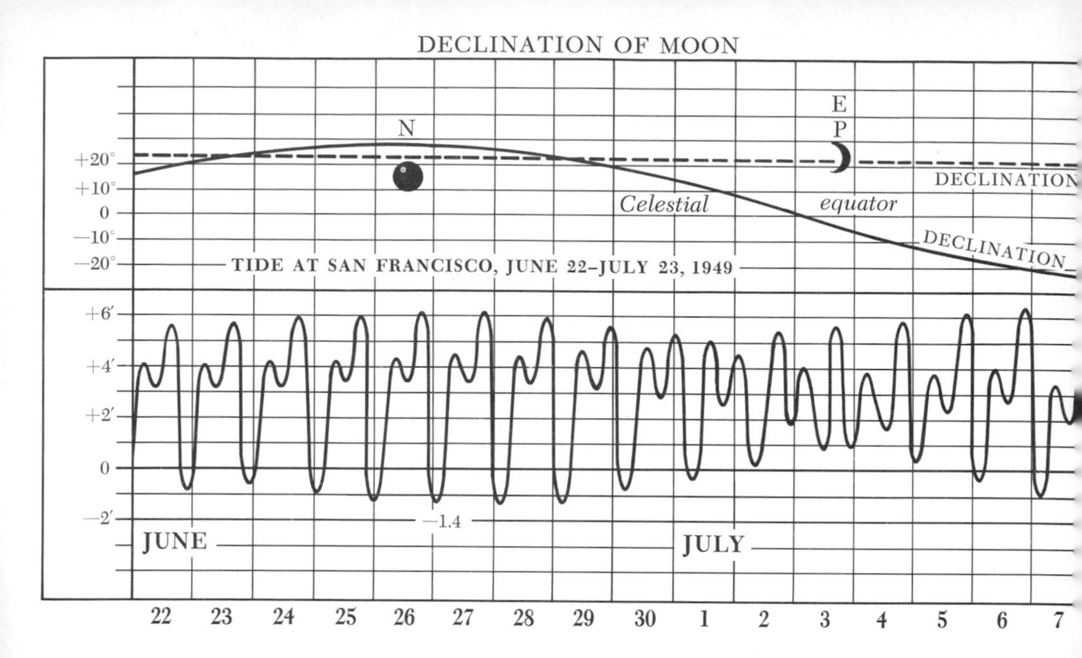

275. The relation between the transits and phases of the moo

276. The *Laminaria* Zone: laminarians ("kelp"), which grow in the low intertidal, expose at low water.

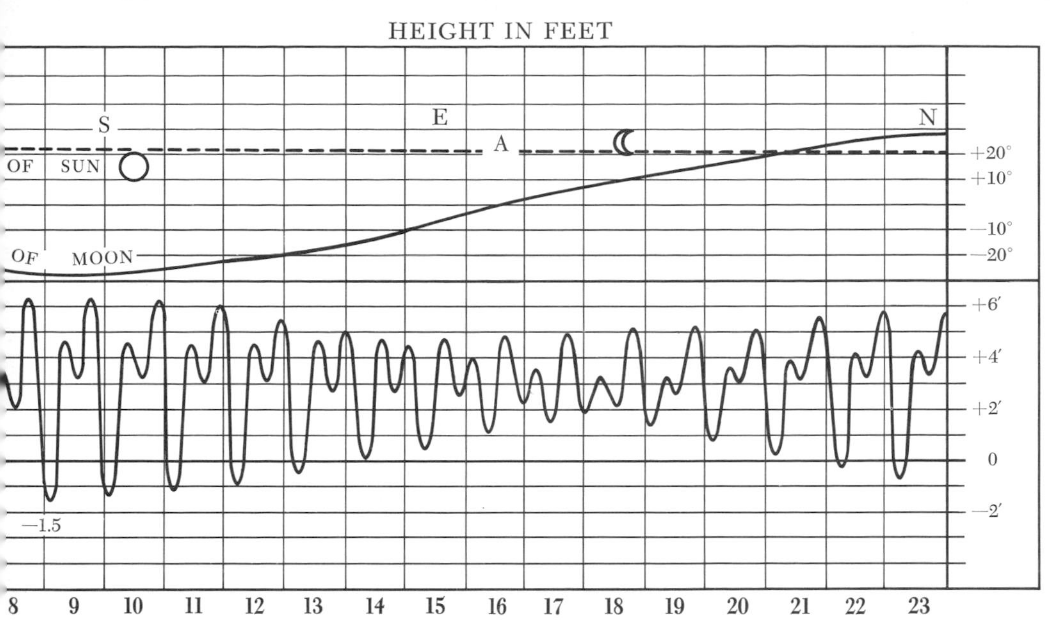

nd the ranges of the tides in a mixed tidal pattern.

7. The eelgrass *Phyllospadix* growing on rocks in the low intertidal.

As can be seen from Figure 273, the broad midlittoral of Stephenson has at least two natural divisions on this coast, the break occurring at mean sea level (3.0 feet) according to Hewatt and at 2.5 feet according to Ricketts. Stephenson, during his trip to the Pacific coast in 1947, recognized this division and tentatively called the two parts the upper and lower "Balanoid zones" in his privately printed *Report on Work Done in North America during 1947–48* (1948). The terminology is now abandoned; and he would now call them upper and lower midlittoral. Doty's scheme can be reconciled with the four-zone system of this book by considering his Endocladia-Pelvetia-Fucus zone a lower subdivision of Zone 2, 2*b* perhaps; and similarly, his zone of Corallinas, etc., becomes 3*b*.

Difficulty is sometimes encountered in distinguishing Zones 2 and 3 on the basis of animal distribution. Rigg and Miller, for example, found no animals in their Zones *C* and *D* abundant enough to serve as indicators (we have added *Strongylocentrotus* to the diagram); algal growth is so heavy that it obscures animal life at this locality. We must resort to the algae, and here Doty's work is especially useful. Although Figure 273 does not list all the algae that are restricted in their vertical distribution, it does name the most conspicuous species or genera. These should be recognizable with the aid of some plates from Gilbert M. Smith's *Marine Algae of the Monterey Peninsula,* which have, for this purpose, been rearranged approximately according to their vertical distribution at the end of this chapter (pp. 407–15). Hopefully, no one will suppose that he can now get along without a handbook to the algae by using this handful of plates; they are intended simply as a rough guide to the more pronounced zonation to be found on this coast, and nothing more.

In his exhaustive book on British shores, Lewis (1966) uses the term Littoral Zone, which he divides into the Littoral Fringe, between the upper limit of *Littorina* and *Verrucaria* and the upper limit of barnacles (all of Zone 1 and perhaps a bit of Zone 2 in our terminology), and the Eulittoral Zone, between the upper limit of barnacles and the upper limit of laminarians (all of our Zones 2 and 3). The region below the *Laminaria* is termed the Sublittoral Zone.

Perhaps it should be pointed out that our consideration of zonation is basically biological in approach. As Eupraxie Gurjanova emphasizes, this approach does not always take into consideration the relations between the occurrence of species and measured tidal levels; and she states that it cannot be applied without consideration of the physical levels, which she terms "Vaillant's principle" (as contrasted with "Stephenson's principle"). The idea of classifying zones according to physical factors might just as well be called "Lorenz's principle," since he first treated this problem in an exhaustive, objective manner at least a decade before Vaillant. However, the "Vaillant" system does not lend itself easily to problems of wave-induced differences on exposed coasts; indeed, all systems seem to refer best to some abstract situations that neatly correspond to the diagrams.

The Stephensons were probably correct in postulating that there is a universal scheme of zonation applicable to all parts of the world, in the sense that there are generally three divisions that may be recognized on a seashore at low tide, the upper, middle and lower parts. But when it comes to saying that a zone of algae in one part of the world is equivalent to a zone of ascidians in another because they are at about the same tidal level, we are not discussing biological equivalents. We cannot ignore the question: what is essentially different about a shore that supports a dense growth of filter-feeding animals, as opposed to one that is characterized by seaweeds at an equivalent tidal level? There is the danger, as E. P. Hodgkin (1960) observed, that preoccupation with universal schemes can lead one to fit observations into the scheme to the extent that what is really going on may be overlooked.

III

Zones are not only bands of organisms along the shore; they are also groups of biotic aggregations, or communities. When we speak of zonation, we are thinking of the vertical distribution of various plants and animals as related to tidal levels. Within these vertical limits we find horizontal differences as well. The plants or animals that are considered characteristic of certain zones do not occur universally in the zones they indicate. Sometimes their patchiness indicates unfavorable physical conditions (aside from the obvious fact that when there is no rock, there can be no rocky shore organisms except as sporadic strays); at other times particular combinations are repeated in many places and suggest some interdependency. If we accept E. W. Fager's "operational definition" of a community as a group of organisms found living together, there are, of course, many such groups on the seashore. Some of the terminology developed in the past by eminent ecologists like Shelford and proposed for the Puget Sound communities is, as Rigg and Miller remarked, "more complex than is required for the present purpose."

Nevertheless, this terminology did emphasize the community aspects of zonation, and we find that Ricketts used some of it now and then—notably in speaking of the "*Pisaster-Mytilus-Mitella* association." This is equivalent to the "*Mitella-Mytilus* faciation" of Shelford—in simpler terms, the mussel beds of the outer coast. Nowadays, if we wish to use the old terminology, we shall have to speak of "*Pisaster-Mytilus-Pollicipes*" or "*Pollicipes-Mytilus*" because of the procedural change in names.*

A mussel bed is far from a simple grouping of mussels, goose barnacles, and hungry seastars, as the name "*Mytilus-Pollicipes-Pisaster* association" might imply. Figure 278 represents the occurrence of some of the more conspicuous members of a mussel-bed community at Pacific Grove, as compiled by Frank A. Pitelka for a marine zoology lecture. Several of the animals found in this mature mussel bed could not live at the various levels embraced by the mussel zone on bare rock—for example, the little flat crab *Petrolisthes,* the isopod *Cirolana,* and the worm *Nereis.* Without the mussels they must live elsewhere, and some of them occur "normally" at different levels. The dense growth of mussels provides shelter for them, and the spaces between the mussels and their byssal threads serve as traps for the detritus on which some of them feed. Others find their food in the film of algae that grows on the mussel shells and the sheltered areas of rock beneath them. As the mussel bed grows, more and more crevices and sheltered spots are developed beneath and between the shells until some animals, like sipunculid worms and sea cucumbers, which usually live under rocks in sheltered places nearer shore, find living quarters in the mussel bed. Because of this shelter we will find, as under trees and bushes, that the temperature is lower than on bare, exposed surfaces, at least when the tide is out.

The mussel not only alters the physical environment, the substrate, on which it settles, it also affects the water around it by extracting plankton and detritus from it in tremendous quantities. According to Fox and Coe (1943), a mussel 70 mm. long filters (at about 60 per cent efficiency) the plankton and suspended detritus out of 60 liters of water a day, or 22,000 liters a year. This is the activity of a single mussel; the turnover

* Names—often very widely used ones—are changed in zoology not because zoologists made misidentifications but because of reclassification by a specialist involving reallocation of old concepts, or, more often, because it is found that an older name has priority and the later, possibly more familiar name must be abandoned. Sometimes the rule of priority may be suspended in favor of an established name, but the process is too involved to be convenient. Three familiar names have fallen by the wayside in the last few years: *Mitella* has been abandoned in favor of *Pollicipes,* *Tethys* for *Aplysia,* and *Schizothaerus* for *Tresus.*

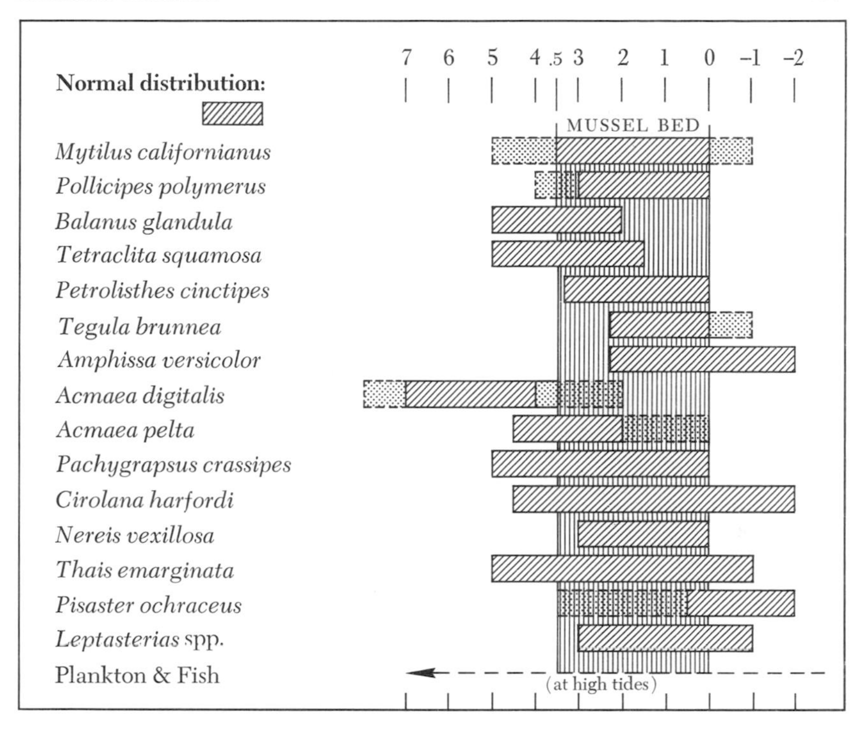

278. Vertical distribution of animals in a mussel bed.

of water and accumulation of body and shell weight by a million mussels, as calculated by Dr. Fox at the Scripps Institution, is enormous. For those with a taste for figures, we present part of his table (somewhat modified) here:

	Weight (in metric tons)		Increment (in metric tons)
	At 1 year	At 2 years	
Total	51.2	117.6	66.4
Dry weight of tissues	2.3	6.4	4.1
Weight of shell	26.0	77.7	51.7

To produce this growth, the 1 million mussels had to filter at least 22 million metric tons of seawater, from which they extracted at least 121 tons of organic matter and about 400 tons of inert material. The mussels were able to retain 4.1 tons out of the 121 tons of organic matter as tissues in their bodies, so their efficiency in making the conversion was about 3.4 per cent.

Ecologists have a useful shorthand system, called the "food chain," for illustrating the interrelationships of a community. The food chain of a mussel bed would look something like Figure 279.

Contrary to what might be expected at first guess, the mussel bed does not start with young mussels settling on a bare rock. It is not exactly a matter of "getting thar fustest with the mostest." Hewatt, as part of his work at Pacific Grove, scraped a square yard of mussel bed bare in November, counting the animals he removed, and made subsequent counts every few months for a year. His results are best summarized in the accompanying table.

REESTABLISHMENT OF A MUSSEL BED

	1931	1932						
	Nov. 23	Feb. 19	Mar. 4	Apr. 25	June 3	July 18	Oct. 13	Dec. 26
Algae		xxx						
Acmaea spp.	319	61	232	314	353	331	997	792
Amphissa versicolor	39							
Balanus glandula	872		95	2126	1967	1904	2175	2374
Cirolana harfordi	926							
Leptasterias aequalis	6							
Lottia gigantea	10	1	7	8	11	11	15	15
Mytilus californianus	1612			55	51	7	53	89
Nuttallina californica					2	34	14	9
Pachygrapsus crassipes	14							
Petrolisthes cinctipes	416							
Phascolosoma agassizi	327							
Pollicipes polymerus	356		76	109	109	109	109	108
Tegula brunnea	17							
Tetraclita squamosa	78		423	416	378	371	359	329
Thais emarginata	218			2	7	367*	47	59

* Average size, 5 mm.

As can be seen, the first event was the growth of a film of algae, accompanied by a vanguard of limpets grazing on it. Then came the barnacles. At the end of the observation period, the barnacles were still in the majority, and none of the shelter required by the crevice-loving animals was yet available. Unfortunately, Hewatt was unable to follow this through until the test square had returned to "normal." Other studies on denuded surfaces suggest that it may take 5 years for a mussel bed to become established. Recent Japanese work by Hoshiai with *Mytilus edulis* zones indicates that this mussel can reestablish itself completely in about 3 years, but that it may also be susceptible to greater fluctuations in abundance from year to year than *M. californianus*.

The mussel bed is not, of course, the only significant community on

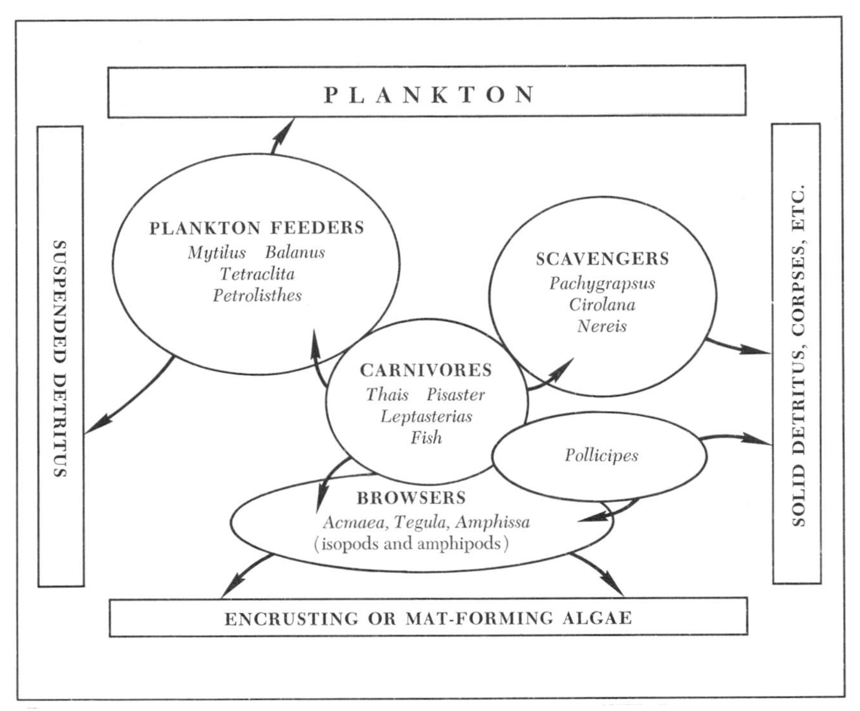

279. The food chain in a Pacific coast mussel bed.

the rocky shore. There are many others, and even an apparently insignificant association of organisms may involve appreciable amounts of energy and a standing crop (or "biomass") of organic matter. For example, there are the tufts of the red alga *Endocladia* that are found interspersed with *Balanus glandula* on the rocks above the *Mytilus* zone. This *Endocladia muricata–Balanus glandula* association (Fig. 280) was studied for some years at Pacific Grove by Peter Glynn (1965). He found 93 species of plants and animals in this association, but only 28 of the animal species occurred more than 98 per cent of the time. Of these, a small viviparous clam, *Lasaea cistula,* was among the most abundant, sometimes occurring by the thousands in comparatively small areas. In the most concentrated part of the zone Glynn found more than 210,000 individuals per square meter, representing a total dry weight of 2,640 grams per square meter (roughly equivalent to 5 pounds per square yard). Of course, this may not represent much protein; only about 4 or 5 per cent protein remains after all the material is reduced to ash.

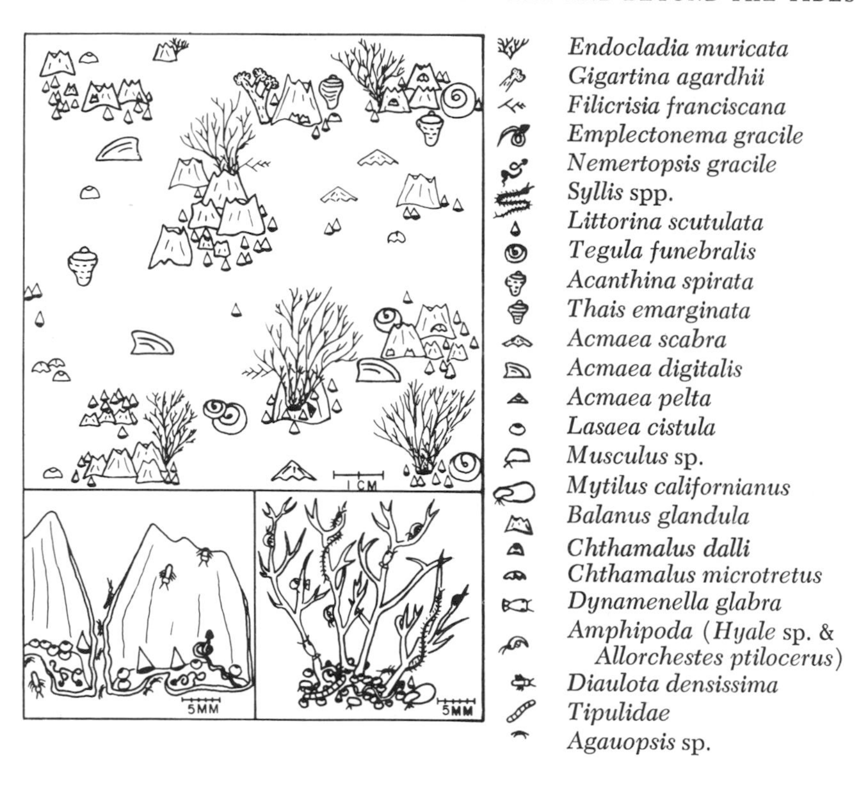

280. Spatial arrangement of some of the more common species in the *Endocladia-Balanus* association. The relative quantities of organisms shown at the top are proportional to those found in a test patch. Lower left: interior of two empty *Balanus* shells. Lower right: an enlarged thallus of *Endocladia muricata.*

Nevertheless, this elaborate microcosm thrives in a zone that is under the sea only 28 per cent of the time and accordingly receives suspended food material from the sea at the rate of 570 milligrams (dry weight) per square meter per year. Figure 281 shows Glynn's interpretation of the ecological energetics of this system.

Our study of these "aggregations of animals living together" is still in its early stages, for there has been very little of this repeated observation from year to year. The apparently orderly progression in the mussel bed cleared by Hewatt (and in similar patches elsewhere) has suggested to some that there is indeed a true ecological succession here, a situation in which the first group of animals to settle prepares the way for the next by modifying the substrate or providing suitable shelter. This succession leads, according to ecological theory, to the "climax" stage,

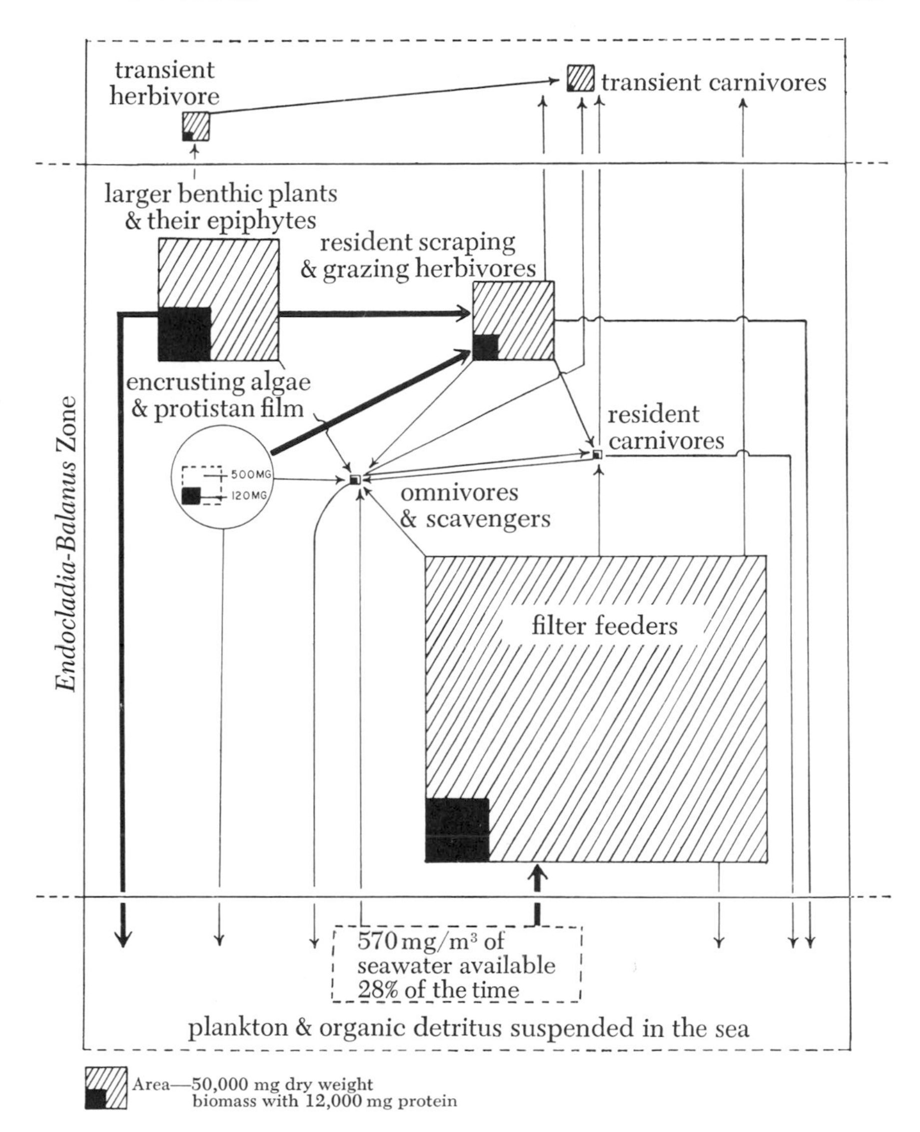

281. Diagram showing average dry weight biomass and protein content of the organisms of the *Endocladia-Balanus* association, grouped in categories according to food relations. The areas of the shaded and solid block for each group are proportional to its dry-weight biomass and protein content in a square meter of test area (key at lower left). Arrows show major food pathways. Dotted lines indicate estimated quantities. For the encrusting algae and protistan film, a greatly magnified view is shown.

or most stable condition; thus conditions leading to change are generated within the community by members of the community. But it also appears just as logical that in some places at least, things settle at approximately the same place on top of whatever has settled before, and the most vigorous animals—and plants—will crowd out the earlier arrivals.

This view of the state of affairs has been most concisely stated by Hoshiai (1964): "Therefore, it may be said conclusively that the zonation is formed through the process that the zone-forming species settle according to their characteristic distribution range, that the early settlers were subsequently replaced by the later arrivals in the overlapping part of the distribution range of each of the said zone-forming species and that the later occupants prevent the early ones from reappearing in all or in a part of their zones."

We may, of course, regard this changing of the guard as phases of the same thing, with all the phases extant at some point in the community. As E. W. Fager (1965) puts it, *Mytilus* beds on exposed rocks show "a buildup from bare rock to a crop of *Mytilus* with its associated organisms which is too heavy to withstand winter storms; practically bare rock exposed by the breaking off of sections of the bed and a sequence of stages leading again to the *Mytilus.* Although the different phases might be separated for the convenience of study, it seems more reasonable to consider them all as parts of the *Mytilus* community."

Many—if not most—of the inhabitants of an area or zone like the mussel bed have free-swimming larval stages and could therefore settle almost anywhere in the intertidal region, depending on the age of the larvae at a given stage of the tide. Vast numbers of them are probably eaten by the hordes of filter-feeding creatures already established on the rocks, and presumably those able to travel after they settle on the rocks have a difficult time reaching the level where they may thrive. For those who cannot move after once settling, such as barnacles, it may be just too bad if they settle where others do better. Joe Connell (1961) has shown that the barnacle *Chthamalus stellatus* in Europe may indeed settle almost anywhere between tides, but that it thrives only at the highest levels. When it settles in the zone occupied by *Balanus balanoides,* the faster-growing, more vigorous *Balanus* simply crowd *Chthamalus* out. *Balanus* itself does not have an easy time, for lower down there are voracious snails, *Thais,* who eat the stragglers as far up as the snails can comfortably graze. Thus it appears that the lower limits of some, if not many, animals on the seashore may be controlled either by competition with other species or by predation (Fig. 282). Algae that have settled

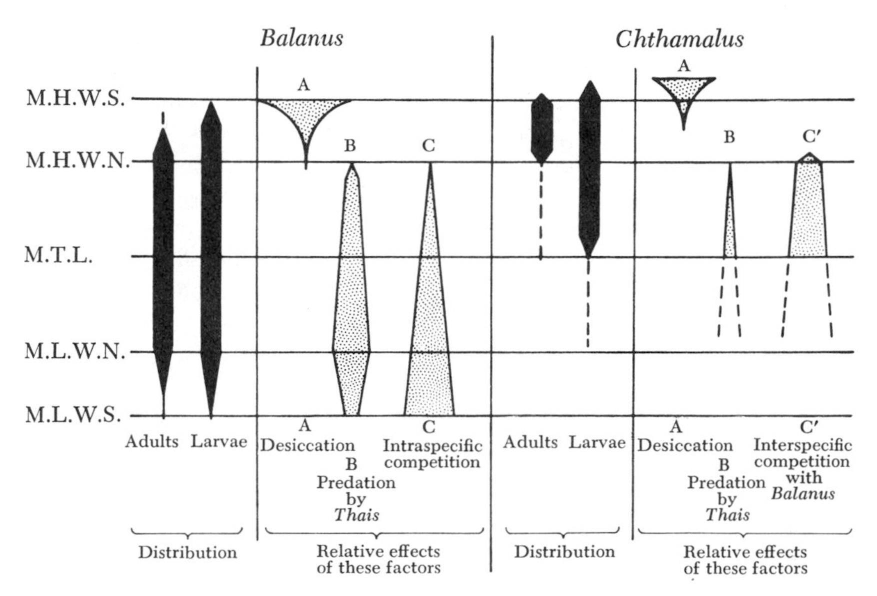

282. The intertidal distribution of adults and newly settled larvae of *Balanus balanoides* and *Chthamalus stellatus*, with a diagrammatic representation of the relative effects of the principal limiting factors.

too low will naturally be grazed on by herbivores, much as shrubbery may be undercut by browsing animals. Indeed, the competition for a place to settle down is the most desperate of all the many contests on the seashore.

IV

Undoubtedly, as Connell suggests, one of the controlling factors for the lower limits of such characteristic zonal indicators as barnacles is the biological circumstance of competition or predation. In these cases the upper limit is set by the ability of the organisms to resist desiccation: *Chthamalus,* for example, is a small barnacle that resists desiccation better than *Balanus*; therefore, it can live high up, sometimes well above average high tide level on an exposed shore. *Thais* is either less resistant or cannot feed if exposed too long, so its upper limit governs the lower limit of *Balanus*. And so on. J. S. Colman (1933) suggested that competition between species was the selective agent 30 years before, but Connell has proved the case.

Nevertheless, exposure is also one of the prime causes of zonation, for many organisms are evidently arranged in order of their ability to resist the desiccation or wave action attendant upon exposure. This is obvious from the "spreading" of zones toward increased wave action, as indicated in Figure 4. As Figure 283 shows, where wave action is severely limited, and exposure is governed by the tidal cycle primarily, the uppermost limits of intertidal organisms may actually be below the highest tides of the year; and where there are no tides (or essentially none), the limit may be something like mean sea level instead of the maximum sea level.

There are still unexplained problems, such as the almost geometrical cutoff of zones in some places, with bare, unoccupied space above or below. Another unexplained phenomenon is the alternation between zones of plants and animals that may often be observed in the upper parts of the intertidal region.

There is a pronounced difference in the velocity of the tidal current—the ebb and flood—through the cycle, resulting in a velocity through the middle part of the tidal cycle twice, or more than twice, the velocity of the tidal current near the high or low of the cycle. Observing the

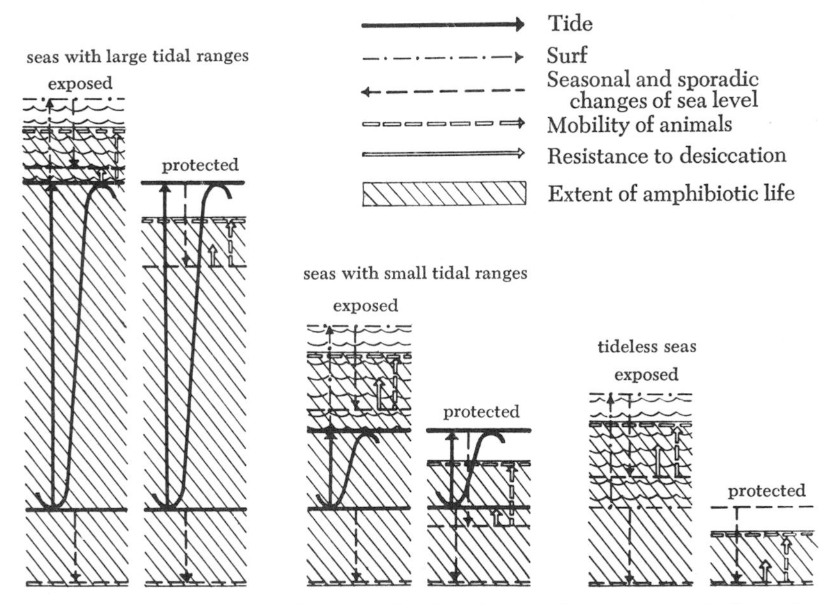

283. Interaction of factors influencing the distribution of organisms in the littoral (after Mokyevsky).

establishment of organisms on the concrete slopes of a new seawall, Moore and Sproston (1940) found that the middle zone, most subject to the tidal currents, was the last to reach the climax condition, whereas the extremes developed first. Again, as with wave action, we are at a loss to explain how this difference in tidal flow could produce the sharp breaks observed.

The degree to which organisms, especially plants, may require light is undoubtedly an influential factor. Brown algae are more efficient in water of certain depths than green algae, since their pigment stops more light than the unaided chlorophyll of such greens as *Ulva,* and transfers the energy to the chorophyll of the brown seaweeds. This gives them a competitive advantage over the greens in the lower levels, at least in the temperate zone. It may be some such relationship as this that influences the sharp but narrow zonation observable in comparatively tideless regions. Colman (1943) has commented on this zonation at Tortugas, Florida. It is also conspicuous on jetties in Texas, where the upper zone consists of seasonal greens (which die off in winter), followed by a zone of reds about 6 inches to 1 foot wide, and finally by a slightly broader zone of browns. Then the algae end abruptly, and the next zone is onc of encrusting sponges. The tidal range in this region is about 2 feet, but the pattern is complex.

The idea of critical levels was first introduced by Lorenz in 1863, but intensive study of this aspect of the intertidal environment was not undertaken until comparatively recently. Colman (1933) and Evans (1947) have determined these levels for various algae in England, and Beveridge and Chapman have done so in New Zealand. Doty (1946) attacked the problem of critical levels in a slightly different manner. He considers that the vertical distribution of algae, in particular, is correlated with the sudden increases in maximum time of single exposure that occur at certain parts of the tidal cycle or, conversely, to sudden increases in the duration of submergence of the algae. Ricketts recognized the striking difference in exposure on either side of the 3.5 foot level in central California (cf. Gislèn, 1943, p. 23), where maximum single exposure below this level is only half as long as exposure immediately above it. This coincides roughly with the "mean physiological level" of his diagram in this book (Fig. 3). Ricketts was considering "mean" or "average" conditions; Doty is concerned with the differences in exposure during a given time, and his tide factors are combinations of both tidal levels and time. They become "critical levels" when they coincide with other, secondary, factors to act upon the organisms. Doty has found, on the basis of his study of the vertical distribution of algae and the tidal curves, that these sudden changes in the tidal cycle occur at LHHW (4.7 feet), LLHW (3.5 feet), HHLW (3.0 feet), LHLW (1.0 feet), MLLW (0.0 feet), and LLLW (−1.5 feet). The figures in parentheses refer to predicted San Francisco tide levels for 1945 and are simply conventions, since these levels vary from place to place and with time.

An experimental test of this change in gradient, or "tide factor hypothesis," was made by Doty and Archer (1950), who submerged algae for prolonged periods. These experiments consisted of exposing algae to temperatures 5 degrees above normal, and indicated, according to the authors, that when they doubled or tripled the time "at which only just significant injury took place, death occurred sufficiently frequently to indicate acceptability of the tide factor hypothesis." The critical level concept is further discussed by Doty in his 1957 review of intertidal problems (written to provoke as well as to inform).

A still unexplored aspect of zonation is the effect of temperature upon the organisms *in situ,* although Lorenz (again!) emphasized that one of the important factors to be considered was the rapidity of the temperature change, both at the surface and in the upper water layers, from day to day and from year to year. In Lorenz's day there was no adequate instrumentation to measure this, and the problem is still a formidable one owing to the difficulty of maintaining accurate instruments for any period of time between the tides, except perhaps on docks in harbors. Yet we suspect that the actual temperature variations experienced by many intertidal animals may be much greater, and perhaps more severe, than the temperatures of the sea a few miles offshore or at some station where the records were compiled on the basis of dipping a thermometer in a bucket once a day. Much of our information on temperature and reproductive periods may be less than approximate, at least as far as intertidal organisms are concerned. Obviously, those animals living on comparatively bare rock experience a very different temperature regime from those under seaweeds or mussel beds. Some intertidal animals may be where they are because they actually need exposure to comparatively high air temperatures for short periods of time.

Behavior may also have some influence on zonation. It certainly affects the distribution of some animals in mud flats—*Phoronopsis,* for example, is evenly spaced, so that the lophophores, when expanded for feeding, do not interfere with each other. Connell (1963) has observed a tendency toward even spacing in a tubicolous amphipod whose holes are spaced so that the inhabitant of each hole has a feeding range around him. Perhaps this sort of thing is not possible on a turbulent rocky shore, but we cannot be certain without looking into it more thoroughly.

In any event, the subject of intertidal zonation is far from exhausted, and it is worthy of the concentrated attention of the subtlest ecologist, as well as that of the current breed of electronic physiologists with bright new oscilloscopes and multiple-channel data recorders. We still agree with C. M. Yonge's comment of 1949: "We still have much to learn about the interplay of forces on the shore."

Algal Zonation on the Pacific Coast

On this page and the following eight pages we have arranged a series of plates (from G. M. Smith's *Marine Algae of the Monterey Peninsula*) to show the characteristic plants of each intertidal zone. Below are two upper-zone algae: (*a*) *Prasiola meridionalis,* a small green alga growing high on inaccessible rocks (×18); (*b*) *Pelvetiopsis limitata,* one of the brown rockweeds from a lower level (×1).

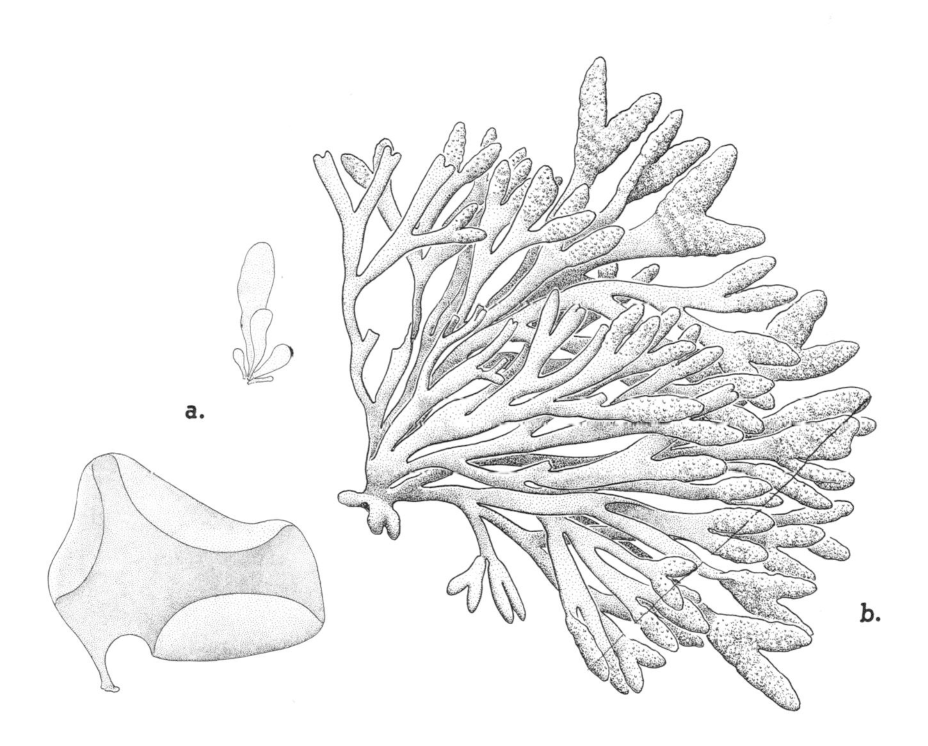

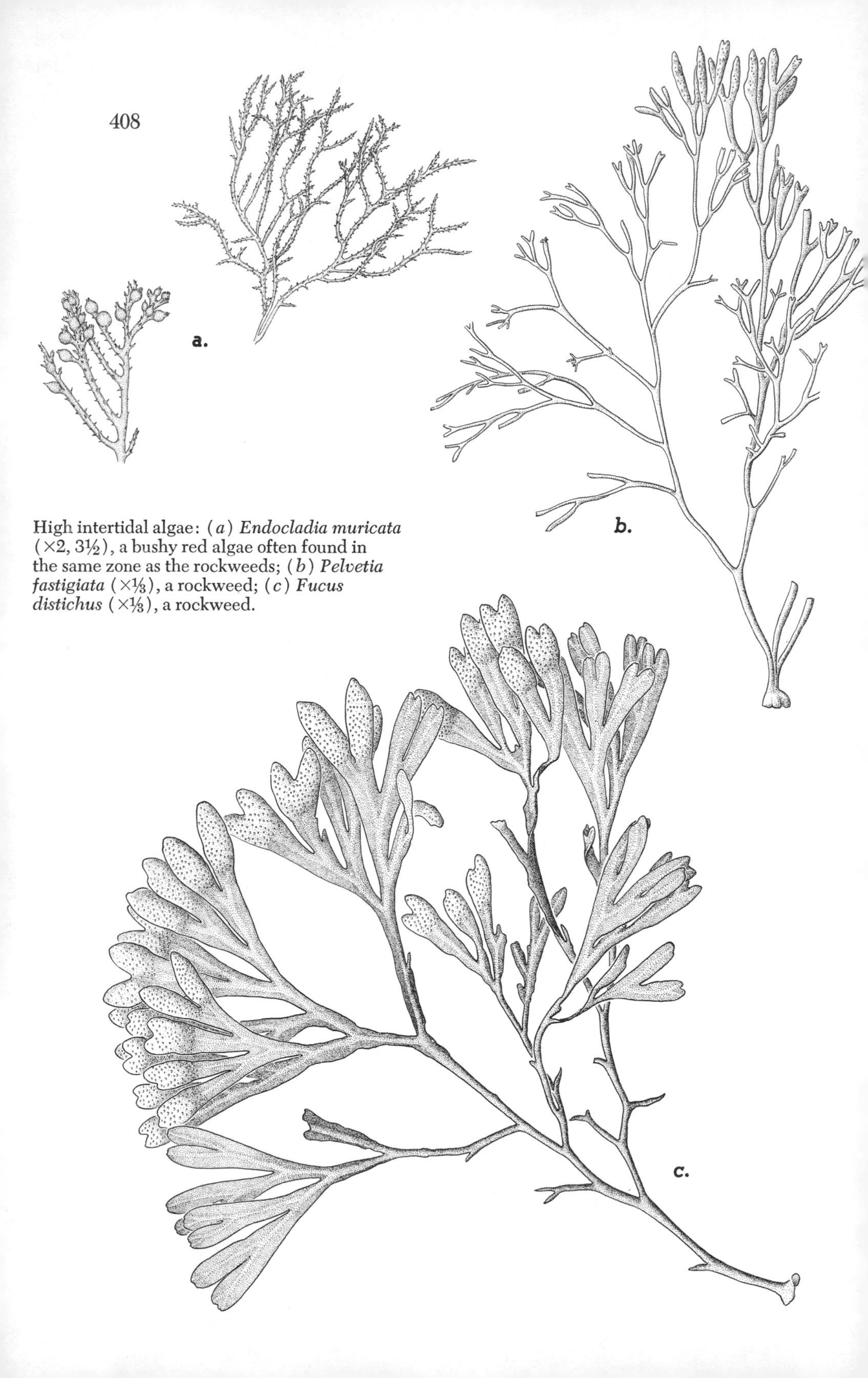

High intertidal algae: (*a*) *Endocladia muricata* (×2, 3½), a bushy red algae often found in the same zone as the rockweeds; (*b*) *Pelvetia fastigiata* (×⅓), a rockweed; (*c*) *Fucus distichus* (×⅓), a rockweed.

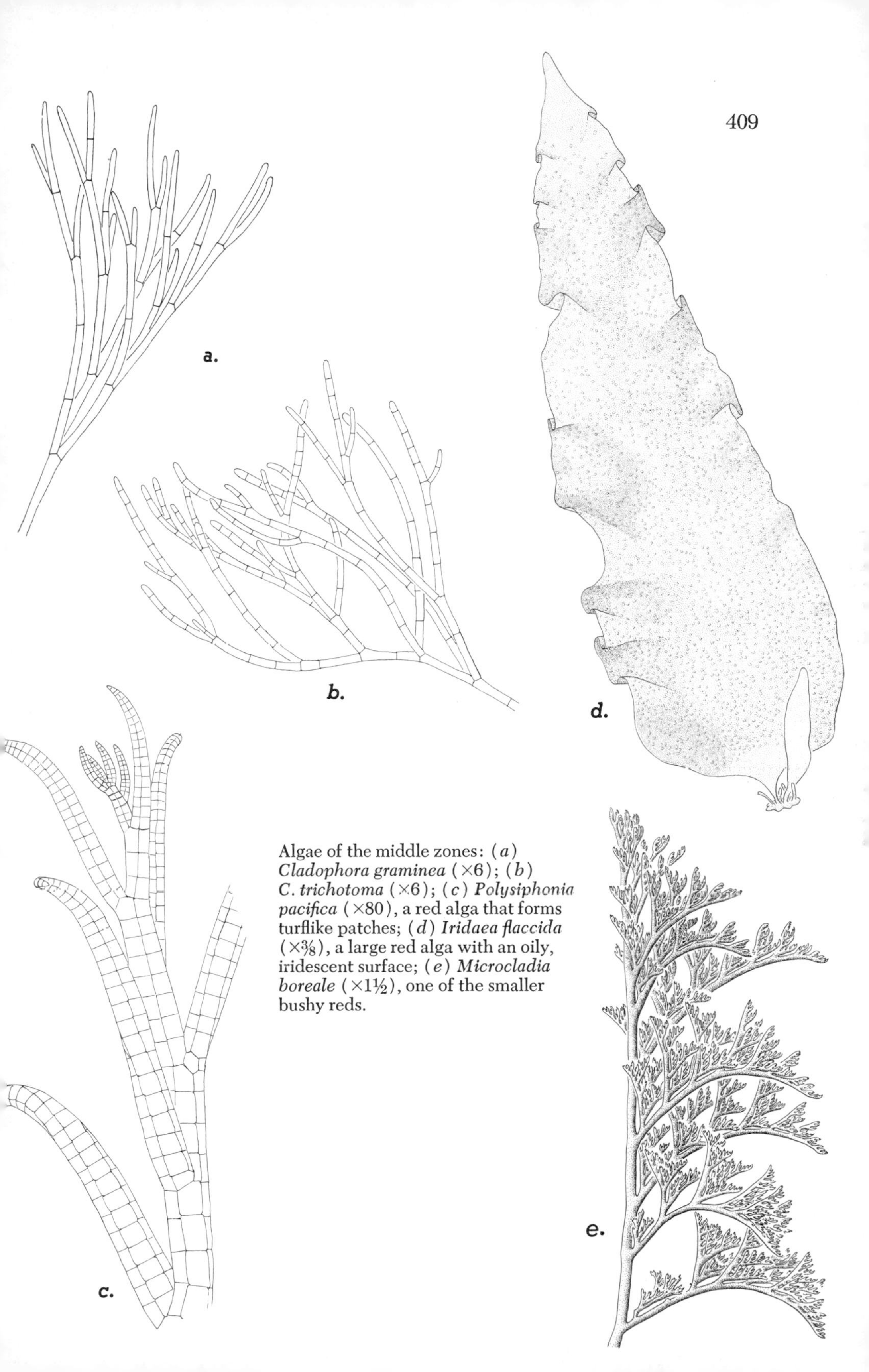

Algae of the middle zones: (*a*) *Cladophora graminea* (×6); (*b*) *C. trichotoma* (×6); (*c*) *Polysiphonia pacifica* (×80), a red alga that forms turflike patches; (*d*) *Iridaea flaccida* (×3/8), a large red alga with an oily, iridescent surface; (*e*) *Microcladia boreale* (×1½), one of the smaller bushy reds.

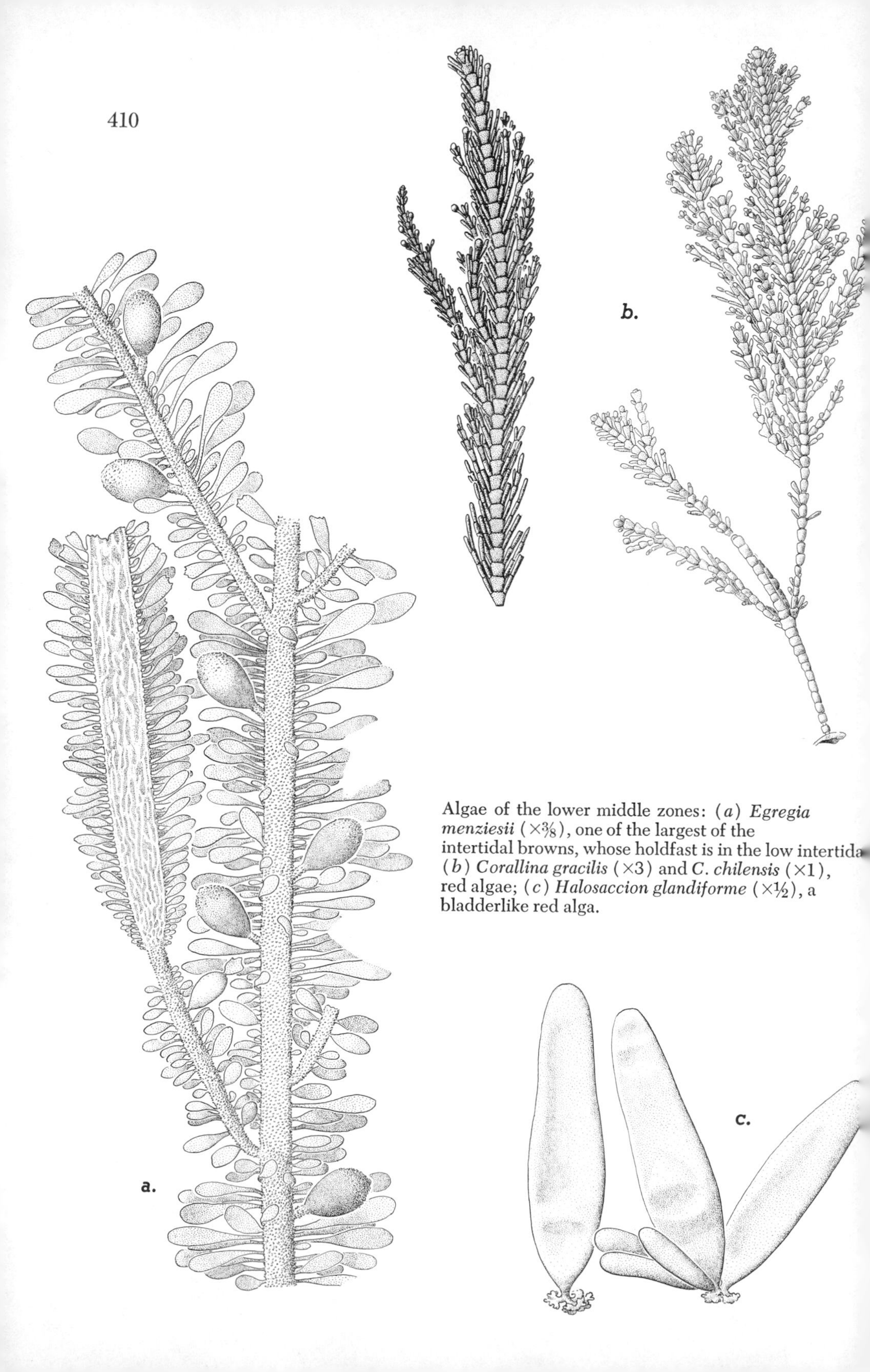

Algae of the lower middle zones: (*a*) *Egregia menziesii* (×3⁄8), one of the largest of the intertidal browns, whose holdfast is in the low intertida[l;] (*b*) *Corallina gracilis* (×3) and *C. chilensis* (×1), red algae; (*c*) *Halosaccion glandiforme* (×½), a bladderlike red alga.

Algae of the *Laminaria* Zone: (*a*) *Prionotis lanceolata* (×⅓); (*b*) *P. australis* (×½); (*c*) *P. andersonii* (×⅕), more characteristic of sheltered coves than open rocks; (*d*) *Iridaea splendens* (×¼).

Two of the laminarians: (*a*) *Laminaria andersonii* (×¼); (*b*) *L. farlowii* (×¼). These are brown algae, or "kelp."

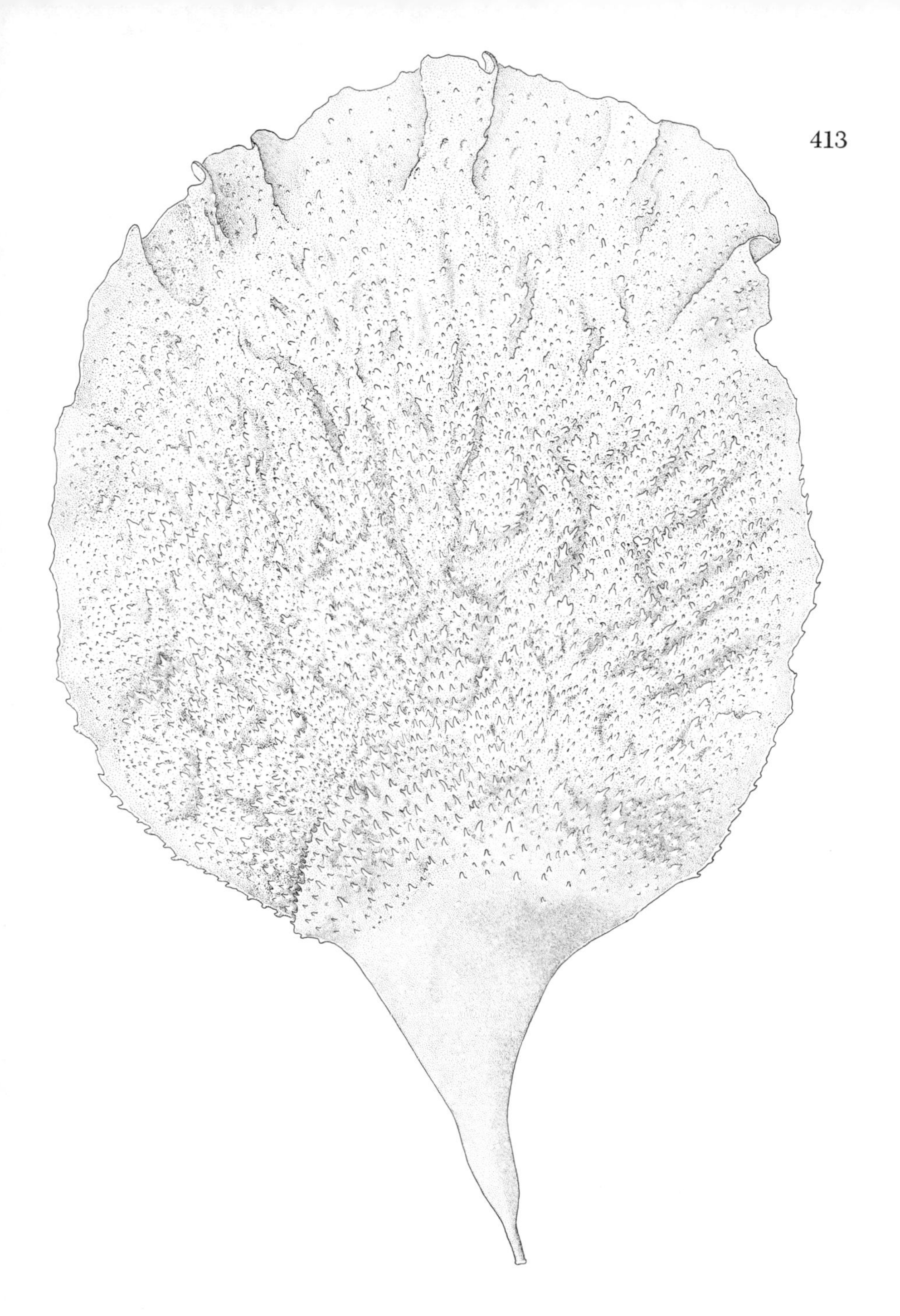

Gigartina corymbifera (×½), a large red alga of the lower zone.

Cystoseira osmundacea (×3/8), a brown seaweed growing just below the lowest tide level.

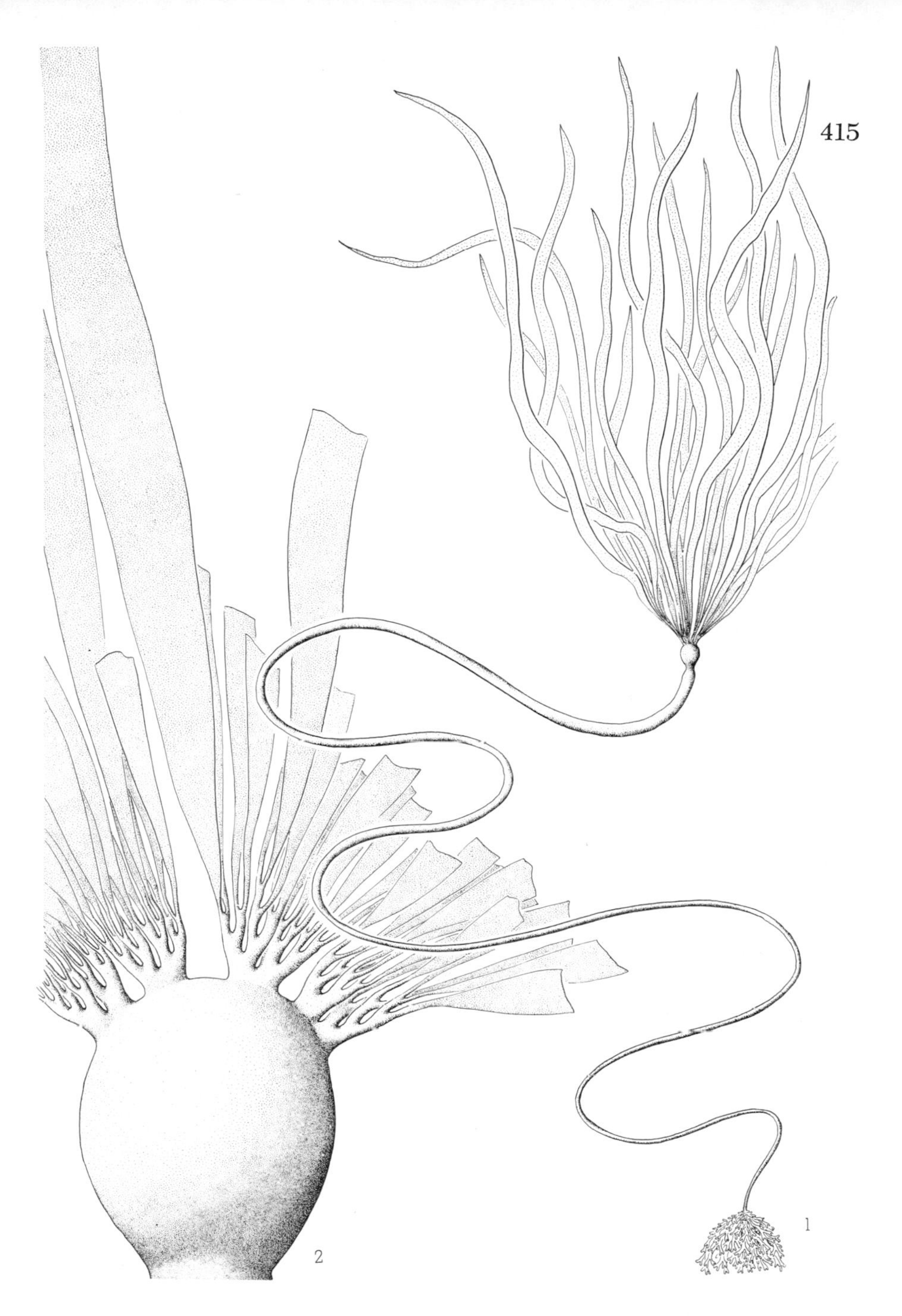

The whiplike *Nereocystis luetkeana,* a subtidal species that is often cast up on sandy or rocky beaches (usually minus its holdfast): (1) the entire plant, greatly reduced; (2) the hollow float, or pneumatocyst, and the bases of the blades (×½).

15. *Beyond the Tides: The Uncertain Sea*

THIRTY YEARS AGO the scientific waterfront of the Pacific coast was calm if not serene. To the north, at Friday Harbor, nothing much was happening except during the summer; and to the south, at La Jolla, the Scripps Institute for Biological Research was not yet accustomed to its new name of Scripps Institution of Oceanography. In the middle, at Pacific Grove, Hopkins Marine Station had not changed much since it began in the 1890's. At the same time, out in the sea, one of the greatest fisheries in history was being pursued. In the peak year of 1936 almost 800,000 tons of sardines were landed at Pacific coast ports, most of them at Monterey under the noses of the populace. Cannery Row was in its heyday. Nobody knew very much about the sardines, however, although some of the professional fisheries people were beginning to worry a bit about the future of this vast fishery. Today there are almost no sardines, but there are quite a few people who know something about the sardines. After the last flurry of moderately good fishing (about 1950) the sardine fishery came to a virtual end, and Cannery Row began to rust away into a ghost town.

Today—three decades from the 21st century—there are three major oceanographic establishments, more than a dozen seaside laboratories associated with colleges and universities, and several government-sponsored mission-oriented laboratories from Alaska to San Diego. Hundreds of people are studying all aspects of the ocean, and in a few years there may be hundreds more. Congress has passed a Sea Grant College Act designed to assist certain universities and laboratories somewhat after the manner of the old Land Grant College Act. It is expected that this Sea Grant program will develop and increase our utilization of the resources of the sea. At the outset, it has stimulated a scramble for funds. Increasing quantities of information about the sea have come—and will continue to come—from all this activity. But if anyone should think we ought to have enough information to explain everything by now, we must disabuse him of the notion. Although we now have a pretty good idea of what may have happened to the sardines, we cannot claim factual certainty. And we have no idea of the effect of the intensive Russian and Japanese fishery off our shores—especially the so-called "pulse" concept of fishing, which is, in effect, to saturate the fishing grounds with boats, securing the entire year's catch in a few days, and then to move on.

In addition to the scientific information that is filling up journals, bulletins, and mimeographed progress reports, there is another form of literature that is accumulating exponentially: the committee reports on what we should do; the industry brochures, expensively printed, on what industry should or will do; the house magazines from well-heeled concerns, the newsletters that seem to thrive on taking in each other's news and regurgitating it; and the stodgily printed proceedings of Congressional hearings. Anyone on a few mailing lists will accumulate mountains of this material in a short time. There are abstracts of abstracts, analyses of journal contents, and similar devices to help the harried scientist keep up with all this—if he or his department can afford the several hundred dollars per year needed to subscribe to them. It is indeed a far cry from those gentle, bygone days when the most one read about oceanography was William Beebe's adventures in Bermuda, or an article on the Galápagos Islands, illustrated with "autochromes," in the *National Geographic*.

How—or why—did everyone get so interested in the oceans? The original National Research Council committee of the 1930's recommended that we establish some oceanographic institutions, and the Rockefellers thereupon started Woods Hole on its way and funded a building for the University of Washington; but the first flurry had long since subsided. Oceanography was still a gentleman's science until the grim business of landing on coral reefs in World War II stimulated research in the processes of the shore, and, more recently, in the properties of the unspecified ocean depths at which the atomic submarines move. In any event, because of a combination of military necessity, decline of fisheries, and the appeal of free-diving apparatus, oceanography has become a matter of national policy. Politicians give inspirational addresses on the future of oceanography, and the TV exudes all sorts of scenes, from programs *au* Cousteau (who conveys the impression he invented oceanography) to the adventures of Flipper and such episodes as a man-eating seaweed proliferating out of a 250 cc. beaker to overwhelm the crew of the computerized submarine in defiance of all ecological verities. There is money in oceanography, although still not enough for anyone's liking, especially as compared with what is expended on rockets to the moon; but oceanographers do not give up hope of getting what they consider to be their share.

All of this activity produces information—if not knowledge—about what is going on in the sea in such quantities that it is no longer possible to prepare a brief summary (there are dozens of brief books now) for the general reader; the best that can be done is to attempt an introduction to the work carried on in one locality or branch of the diverse agglomeration of sciences known as oceanography.

It is hoped that this concluding chapter will give the reader some idea of what is being done and thought about the life of the ocean a few miles beyond our shores; it is hoped that the reader will also understand that this is, at best, a fragmentary introduction.

I

What did happen to the sardines? This was the question along the waterfront in 1949. Very little is said in the popular press about the sardines anymore; most of the people who dine at the fancy restaurants on Cannery Row probably do not give the great, silent canneries a thought, except to wonder, perhaps, why they are not torn down. Then, for a while, in the late 1950's, we had sharks. They came close to shore, divers were attacked, and some were killed. For a while the ocean was a few degrees warmer than it used to be, currents were perhaps just a bit different, and there seemed to be more sharks. Perhaps the sharks had been out there all the time, and more people were exposing themselves to attack. Sharks seem to be curious creatures, and certainly the modern diving gear—the dark suit, often trimmed with orange, and the bright air tanks with their streams of bubbles—makes an object worthy of curiosity. Now, of course, we wonder whether *our* fish can survive the spectacular mass fishing tactics of the Russians and Japanese. Unfortunately, we are not sure, for we do not have the kind of information we would like to have to answer that question. It is possible, for example, that this kind of fishing will constitute an experiment in selective breeding, by catching the densely aggregated stocks, for stocks or races of fishes that tend to be more widely dispersed; this would in turn require more fishing effort in space as well as time. But who can be sure?

We who live on the Pacific coast of North America live along the eastern edge of the largest ocean on the globe. The general pattern of movement of the surface layers in the northern part of this ocean is that of a great clockwise rotation: northward along the coast of Japan, westward across the open space below the Aleutians, and southward along the coast of California. The water is warm as it moves northward past Japan, cools off as it crosses the northern Pacific (but, since it is relatively warmer than the waters of the Gulf of Alaska, the counterclockwise eddy that turns back toward Alaska is considered a "warm" current), and gradually warms up until it finally swings to the west far to seaward off Baja California. Beneath this surface current, some 600 feet (or 200 meters) deep, the water flows southward offshore, but near shore there is a deep northward countercurrent along the coast at all times of the year (Fig. 284). During the winter months the surface current near the coast also flows northward. The pattern of a generally southward surface current, as presented in an atlas, is characteristic of only the summer months.

The surface water of the ocean does not move directly before the

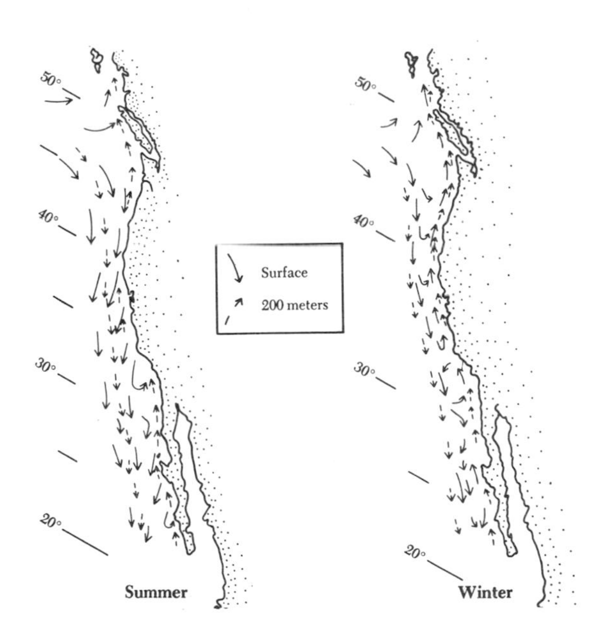

284. Seasonal variations in current patterns along the Pacific coast of North America.

wind, but slips off at an angle of as much as 45 degrees to the right. Thus the prevailing northwest winds that blow parallel to our coast push the surface water away from the coast (Fig. 285). To replace the water that is shifting seaward, cold water moves to the surface from depths of a few hundred feet, especially to the south of headlands and along sheer coasts. Although this upwelling is not restricted to the summer months, it is most intense during that period. As a result, the coast along California and Oregon usually has very cold sea temperatures during the summer (Fig. 286). In winter and early spring the cold waters near the coast are shifted north as a warmer northward current develops near shore (Fig. 287). The northward surface current near shore during winter (known as the Davidson Current) seems to be the result of the weakening of the southward surface current. It does not appear to be directly related to the wind system, for the change to the north may precede the actual shift of the wind. Whatever the cause, the water near shore off northern California moves with considerable velocity to the north. Drift bottles released just north of San Francisco have been recovered on Washington beaches after about a month, indicating a northward drift of about ½ mile per hour at the least. Bottles are not the only things that move northward; good years for the Davidson Current may be marked by the settling of *Emerita* larvae on beaches as far north as Vancouver Island and the establishment of *Styela barnharti* in Monterey Bay.

This, in brief, is the California current system, although the name "California Current" applies only to the principal southward surface current. There are all sorts of variations in the system from year to year and from season to season. In the winter, the wind is associated with the low-pressure area over the Pacific, and as the storms move out of the northwest from the Aleutians and beyond, their counterclockwise rotation results in the familiar southerly gales. As a result, the prevailing strength of the northwest wind is reduced, the California Current is weaker, and the countercurrent becomes more pronounced. As the summer high forms over the ocean, the winds from the northwest blow more from the northward and the general southward movement of surface water is strengthened. Often, as this wind pattern develops, drifts of the little blue *Velella* (§193) are stranded on the beaches. In 1961—a warm summer with little fog and noticeable mortality of limpets along the coast at least as far north as Oregon—few *Velella* were seen in the spring and only a handful came ashore in late June.

It is difficult for oceanographers to measure the actual movement of waters in the ocean because there are no fixed reference points. What

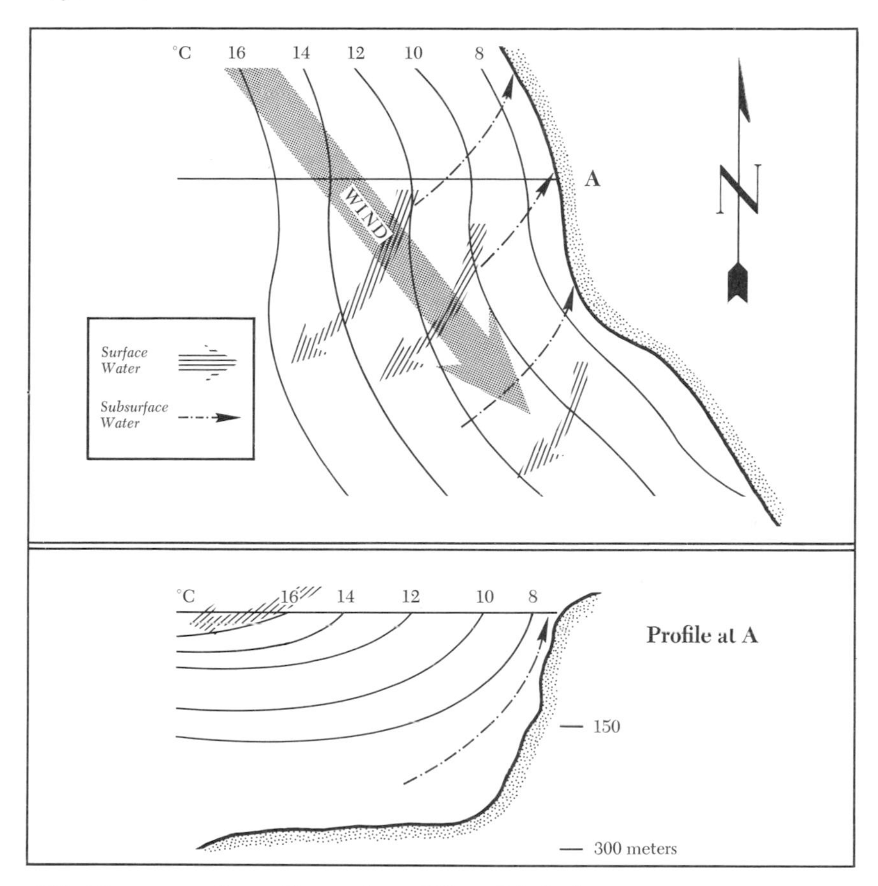

285. The process of upwelling at about 38° north latitude.

is known about the oceanic circulation is inferred for the most part from measurements of the water's temperature and salinity, and for exact knowledge about just where the water is going we have had to rely on one of mankind's oldest oceanographic instruments, the drift bottle. Thousands of these bottles, each containing a return postcard, are cast overboard every year. Only about 2.5 per cent of those set adrift off northern California are ever seen again. In winter they move northward, some as far as the Queen Charlottes, but by April they start to drift southward. Bottles cast off in southern California waters are often caught in the eddy in the Los Angeles basin. If the bottles are released 50 miles offshore, however, they are never seen again. Somewhere out in the ocean, there must be a lot of bottles drifting around.

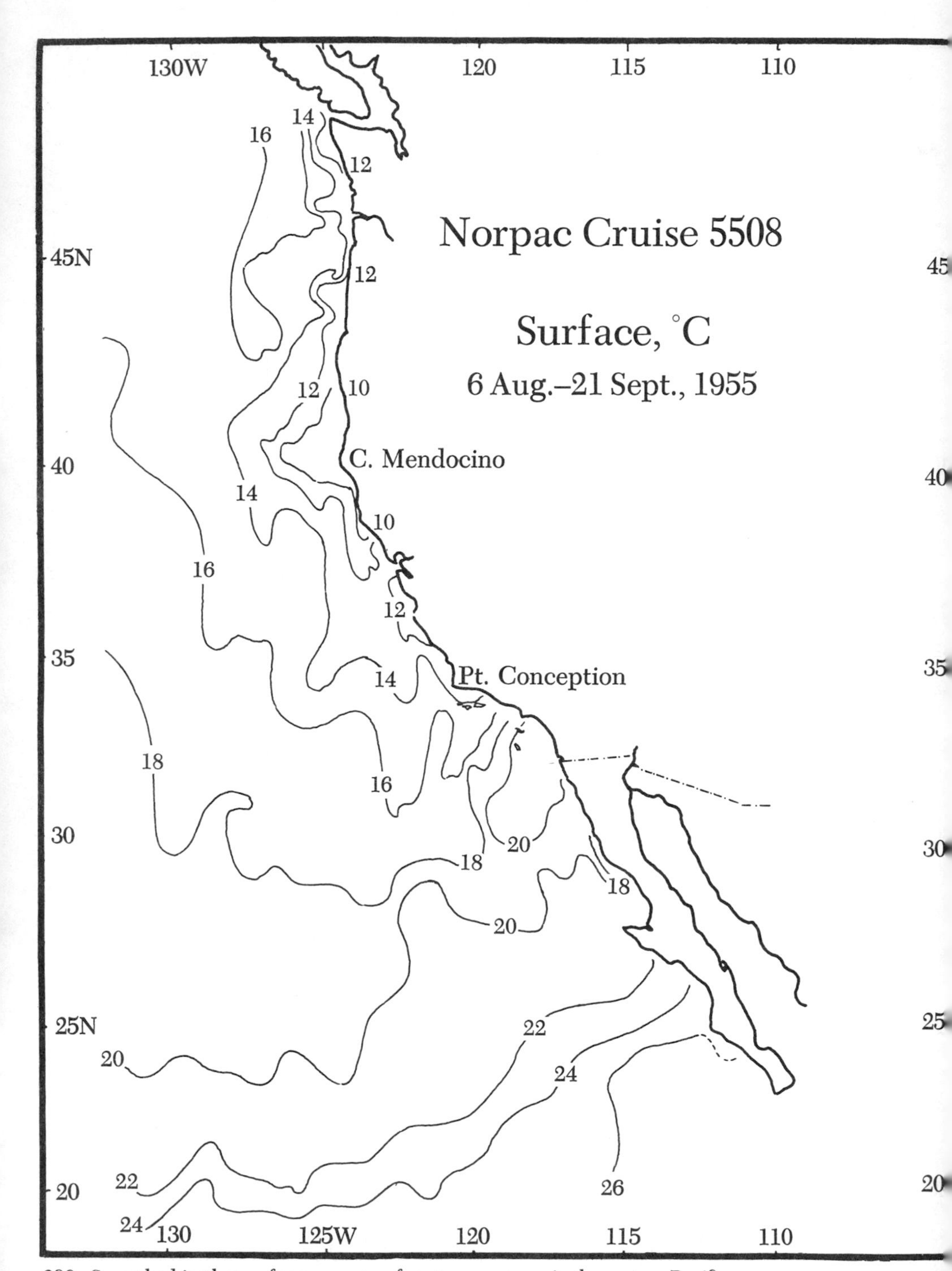

286. Smoothed isotherms for summer surface temperatures in the eastern Pacific.

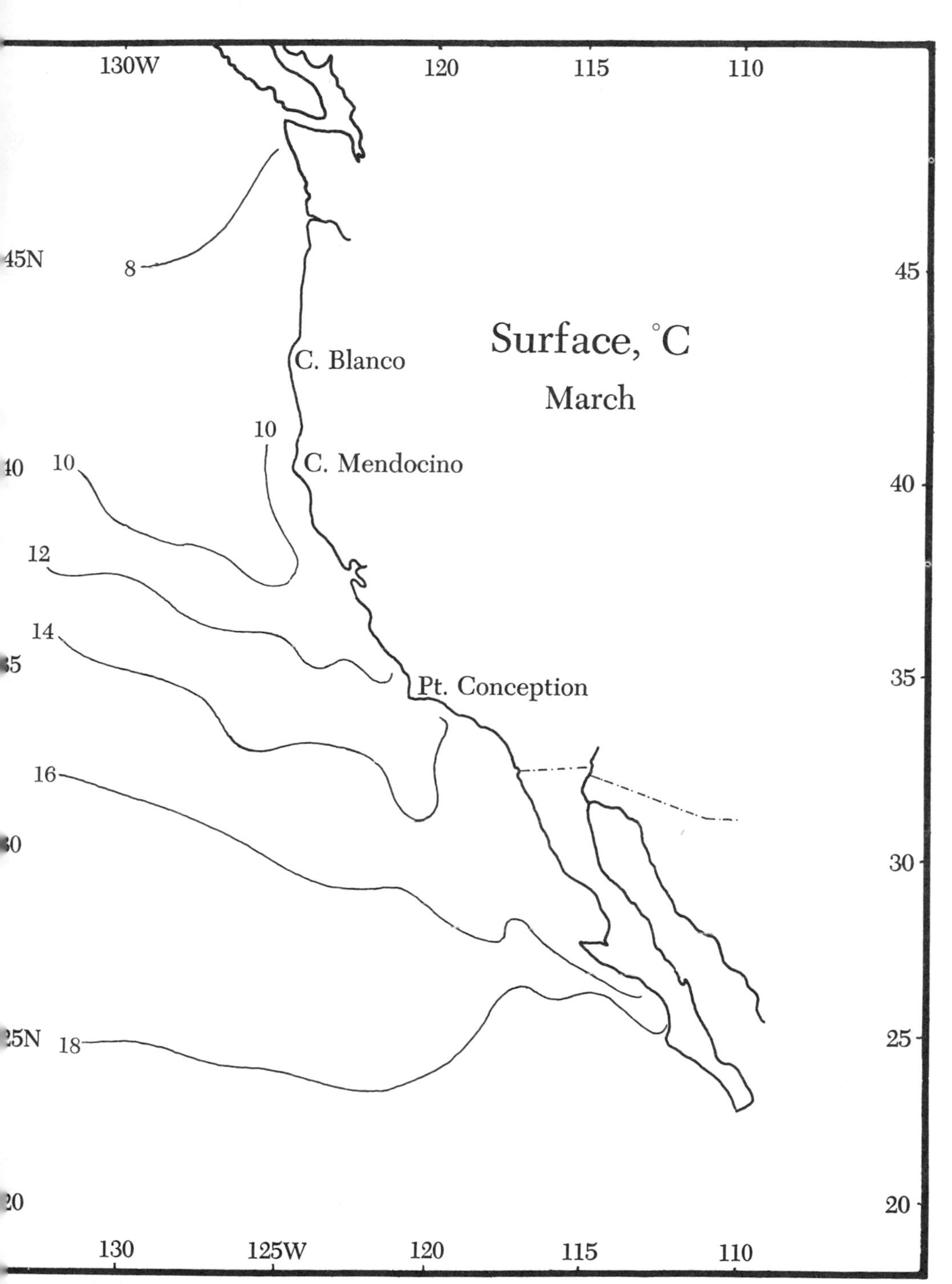

7. Smoothed isotherms for March surface temperatures in the eastern Pacific.

Recently the bottles have been supplemented with drogues, or floating, counterweighted, buoylike contraptions, some of them with battery-powered signals. The drogues make it possible to follow the smaller movements of currents in more detail. Some of these are simple devices that look as if they were lost from a beach party: bright yellow disks about 6 or 8 inches in diameter, with holes in them. Some have counterweighted tails so that they will maintain a certain level or drift along the bottom. The mute plea to notify the sender is printed on them, and their bright color makes them one more mysterious object to be seen from a distance and quicken the beachcomber's hopeful pace.

During the First World War a new fishery was developed along the coast of California, and in a few years California sardines were found in every grocery store in the land, usually in large oval tins. By the middle 1930's the annual catch of sardines was reckoned in millions of tons and the economy of Cannery Row was in its palmiest days. Fisheries experts told us we ought to be calling this fish, *Sardinops caerulea,* a pilchard instead of a sardine, but this affectation did not take. Following the record catches of sardines in 1936, there was a sharp drop; the catches began to oscillate, and fisheries experts began to be worried. Were we overfishing these apparently inexhaustible stocks of fish? Then, in 1947, came the catastrophic drop. The sardines were disappearing and something must be done. Thus the California Cooperative Sardine Research Program was organized, and in 1949 the data began to accumulate. This program, which we now call the California Cooperative Oceanic Fisheries Investigations (CalCOFI), was originally organized and conducted at Scripps Institution of Oceanography, but it involves all interested agencies and institutions, including the California Fish and Game Department, the Federal Fish and Wildlife Service, Stanford University, and the California Academy of Sciences. In some years other organizations, including Pacific Marine Station and Oregon State University, have been participants.

One memorable project involved the Japanese and Canadians as well. This was Operation NORPAC, in which 19 research vessels (11 of them Japanese) sailed 55,000 miles during July to September of 1955 to make simultaneous oceanographic observations over the entire North Pacific. This is an area about 1½ times as large as North America, and from it come about 18 billion pounds of fish a year. This was the largest single oceanographic operation in history, and the data that resulted are still only partially digested. Figure 286 presents a sample of the results.

As a result of the first 9 years of work, certain average conditions were established. The sardines had shifted their spawning well to the south, off the coast of Baja California, and they were none too abundant anyhow, dropping to almost zero in 1952 and 1953. The average temperatures along the coast were lower than those of the period of the sardine's ascendancy (Fig. 288), and the winds were blowing just a bit stronger during the years 1949–56, as compared with 1920–38. It looked as if we were in for a long cold siege, and perhaps it would take years for conditions to return to "normal," and for the sardines to return to something like their former abundance. (The term "normal" in relation to weather is, of course, illusory; our personal ideas of what normal weather is supposed to be are governed by the first few years of our awareness of weather, so that the climate of the place we grew up in is the "normal" climate. In the broader sense, we who live in this interglacial period can hardly be expected to accept a glacial stage as "normal.")

Then, in 1957, something happened to the ocean. The first typhoon in memory visited Hawaii; the ice broke up in the Arctic earlier than ever before recorded; oceanic temperatures were higher in many parts of the North Pacific. Along the coast, southern species of fish turned up in the anglers' catches, and collectors between the tides saw many ani-

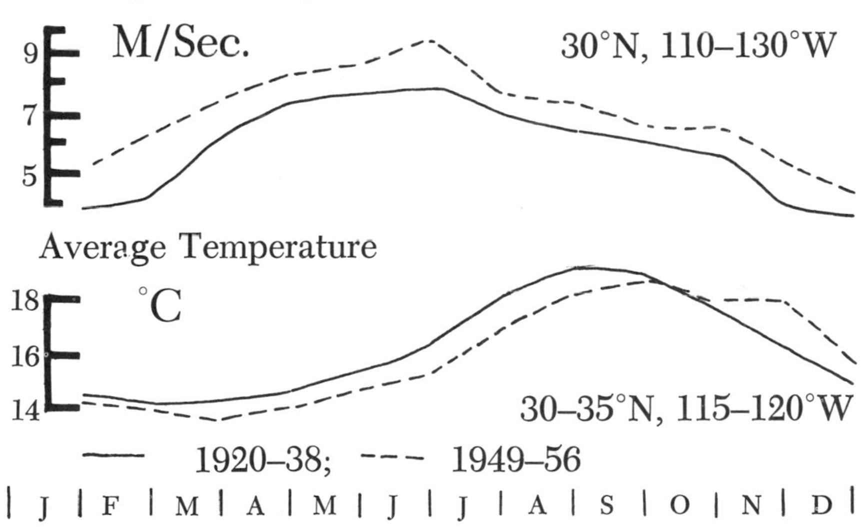

288. Average wind and temperature conditions for 1920–38 and 1949–56.

mals farther north than they were expected. Episodes involving sharks became more frequent, although fatalities from shark attacks were still an infinitesimal fraction of the fatalities on the highways, even on a "normal" weekend. Along the Pacific coast the winds weakened, sea level rose by nearly half a foot (low-tide collecting was poor after 1957 in many places along the coast, and some places are still "not the way they used to be"), and the water got warmer by several degrees. Last, but not least, there was a slight upsurge in the sardine catch. On Figure 289 are plotted the average departures from temperature and sea-level means for the entire Pacific coast; Figure 290 shows the temperature changes in 1957 against the long-term averages for several places along the coast. It will be noted that these temperature changes are not really very large. However, in terms of the possible heat that would be required to cause this rise in temperature, the changes would have required four times the solar heat that actually reached these waters. Obviously, then, these changes are not local, but the comparative suddenness of the events of 1957 indicated that the mechanisms by which atmosphere and ocean affect each other are much more complex than we had suspected. Perhaps it all began with events in the southern hemisphere; at least there seems to have been a connection between changes in the South Pacific and in the North. Possibly the events were set in motion by something in the Antarctic regions. It is said that somewhere in North America, at one time, the position of a pebble determined the course of the Missis-

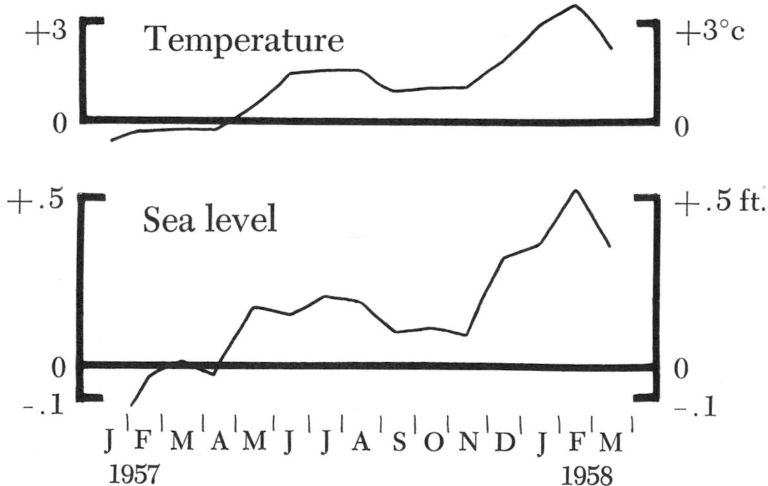

289. Mean differences (1957) from the long-term averages for temperature and sea level on the Pacific coast (La Jolla to Sitka).

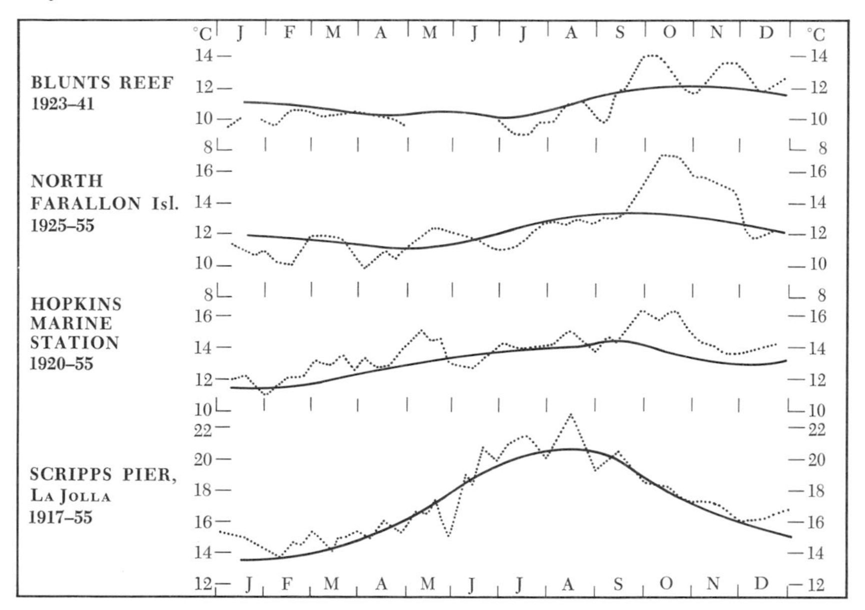

290. Temperatures for 1957 (dotted lines) compared with long-term averages for several localities on the California coast.

sippi. We would not want to say that the stumbling of a penguin in the Antarctic could have such far-reaching effects, but it is possible that some comparatively insignificant perturbation, in the Antarctic regions, of the complex system of air and ocean could induce the changes of the last few years, or that the lack of this perturbation was responsible for the monotonous decade that preceded them.

From the viewpoint of a terrestrial organism like man, the changes in temperature do not seem very large. However, the total variation in sea temperatures in the eastern Pacific is comparatively narrow, and a delicate balance of adjustments has become established. The sardine eggs hatch most rapidly at 17°C.; decreasing the temperature 1° adds 6 hours to the hatching period, and at 15° the hatching period is prolonged from 54 to 77 hours. This, of course, increases the period of time during which the eggs may be eaten by something; and if, as seems logical, the development of the young fish is similarly retarded, its period of vulnerability is increased. Garth Murphy (1961) calculates that a 3° decrease in temperature during the period of hatching and development could decrease survival by as much as 10 times.

The events of 1957 and following years may not be as unique as they seem; they could be, for the most part, reflections of better record-keeping. Something like this seems to have happened 100 years before, when many warmer-water fish now uncommon in southern California were recorded by the early railroad survey expeditions (Hubbs, 1948). In 1859 there was a noteworthy abundance of the pelagic red "crabs" (actually a galatheid) *Pleuroncodes planipes* at Monterey Bay, and in 1960 great numbers of this same animal reappeared (Glynn, 1961). The year 1926 was also notable for the occurrence of many warm-water fish north of their usual range, and temperatures south of San Francisco were noticeably higher. But, from all the available data, there is no clear indication of regularity; no "cycles" can be demonstrated.

If we examine a number of records all at once, as presented in Figure 291, some interesting things become apparent. First, we must bear in mind that the record of the sardine catch does not, especially in the 1920's, indicate the size of the sardine stock accurately. After 1940 there is perhaps a closer relation between the actual population and the size of the catch. Nor is there any clear relation between the diatom counts—so faithfully made by W. E. Allen over so many years—and the sardine catch, or any connection with environmental changes. What is apparent from this composite graph is the unusual nature of the period from about 1945 to 1956. It seems to have been a period of monotony, with temperature fluctuations consistently—but of less magnitude—below established averages and wind consistently a little stronger. The sea level at San Francisco was somewhat higher for the period, while the salinity at Scripps Pier was remarkably uniform. In all, nothing like this had happened since records had been accumulated, and the real question may be not what happened in 1957 but what was going on during the decade before. Two things are plain: the averages set by the Marine Life Research program are not the same as the averages of the entire 40-year period, and the decline of the sardines was not all the fault of man. The implications of this, both for research in oceanography and for our hopes for increase and management of oceanic fisheries, are also obvious.

One could, of course, blame this all on sunspots, which increased during this period, but that is begging the question. What we really want to understand is the mechanism—the means—by which these changes have been effected over vast areas of the ocean—perhaps, in some degree or another, everywhere. Nothing is more revealing of the limitations of our knowledge about processes in the ocean than the record of oceanographers thinking out loud about the changes of 1957, and the patient

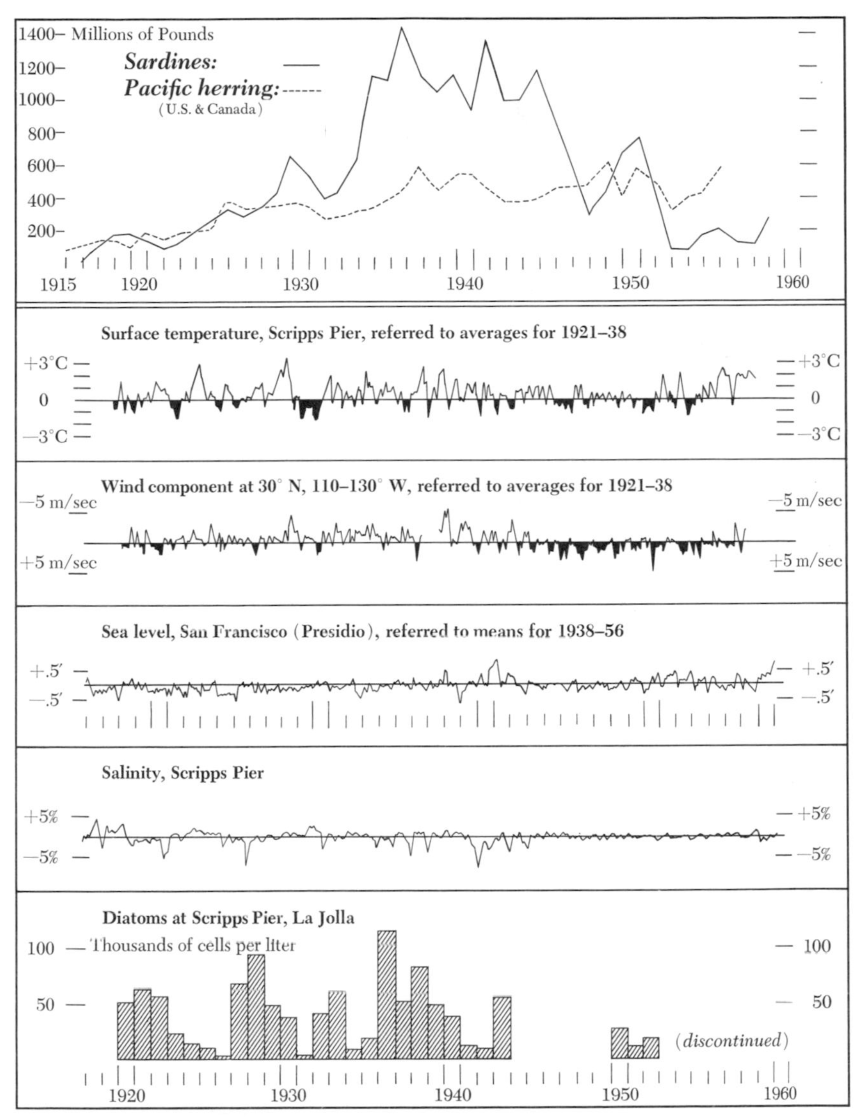

291. Composite graph of variations in the sardine and herring catch and variations in oceanographic conditions on the Pacific coast, compiled from Isaacs, Reid, Sette, and Stewart (in CalCOFI VII); data for 1959 from Roden, 1961 (CalCOFI VIII); the salinity records of 1944–55 for Scripps Pier are suspect, according to Roden. The diatom count at Scripps Pier has been resumed, but with different methods; it will be some years before we have a good record period to compare with the other data.

reader will find these deliberations, as set down in the seventh CalCOFI report, a fascinating document. The oceanographers could see what had happened, but how it happened was beyond the adequacy of their theories to explain; and, without a theory, predictions cannot be made with confidence.

II

The complex of organisms and environment is called an ecosystem, and we have so far been discussing the physical part of a natural system that includes the sardine. It includes many other animals, plants, and processes as well, and it is this total ecosystem that we set out to understand when we began the sardine research program in 1949. Some critics maintain that the idea of an ecosystem is too broad and complex for practical treatment by the present techniques of ecology, that what we are trying to say is that we are studying nature as we find it. That is indeed what we have to do in the sea. The concept of an ecosystem means something more to an ecologist than just a fancy name for nature, however. It asks for numbers, for quantities, rates, and time; in fact, for all that is on Figure 291 and a great deal more. These matters are now fairly well understood for ponds and small lakes, and the extension of the ecological techniques and concepts of the ecosystem to the ocean is simply a matter of scale. True, the scale is vastly greater, and "unexpurgated nature in all its complexity"* may be too much for mere mortals to take in all at once.

However, a start must be made somewhere. In Figure 292 we present a diagram of some of the things in the ecosystem of a part of the shallow sea. As with every other living system, we begin with the sun, the nucleus of our atom. The fate of this insignificant nucleus is our fate, but that is beyond our control to worry about, and no one yet has any accurate idea about the half-life of our system; if someone did, we might not like the news. Anyhow, solar energy is the moving power, and the initial step in the system is the activity of plant life in fixing carbon by sunpower in a form that can then be utilized by animals. The rate at which this process goes on is called productivity, or, more specifically, primary productivity. Some people feel that if we know this, we can understand the system. The beginning of this process may be seen in very shallow water on any sunny day, by the bubbles of oxygen produced by the active brown film of diatoms on the bottom (Fig. 293).

* A. J. Lotka, *Elements of Physical Biology* (1925), p. 301.

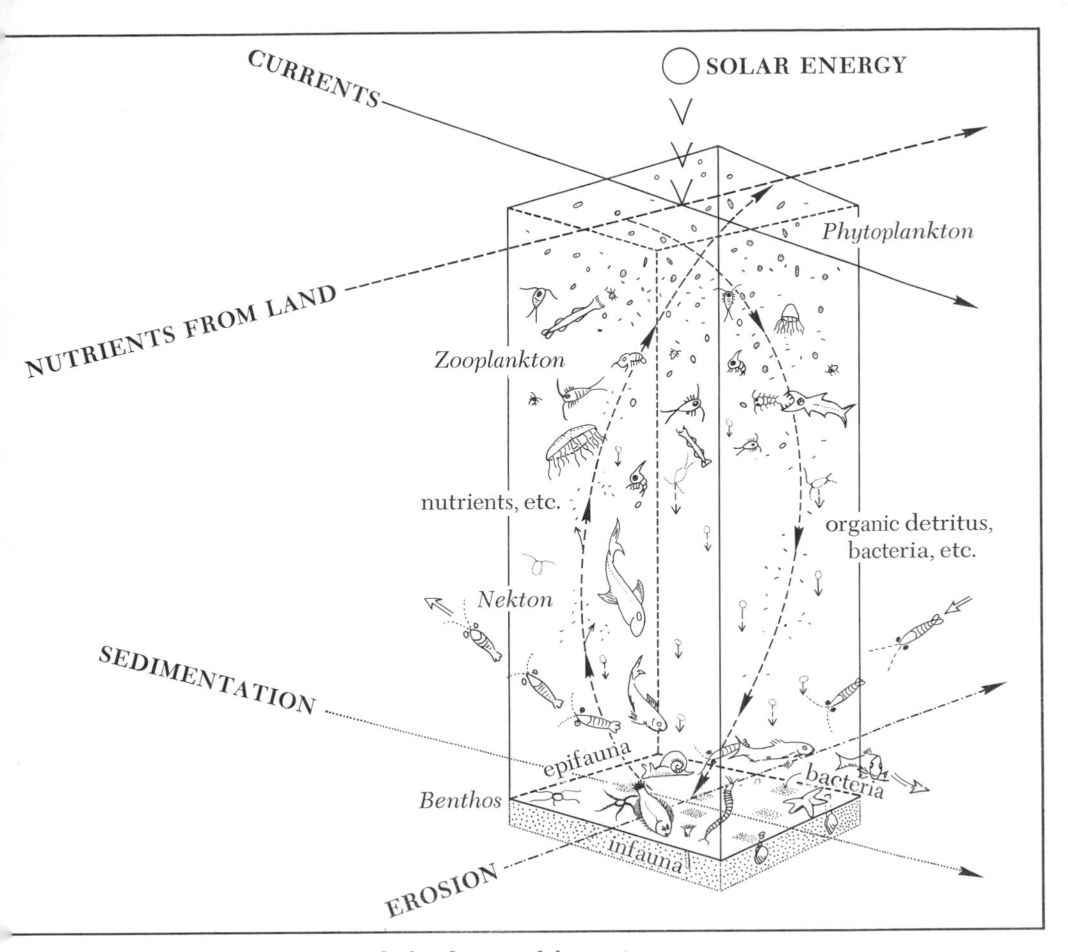

292. An idealized prism of the marine ecosystem.

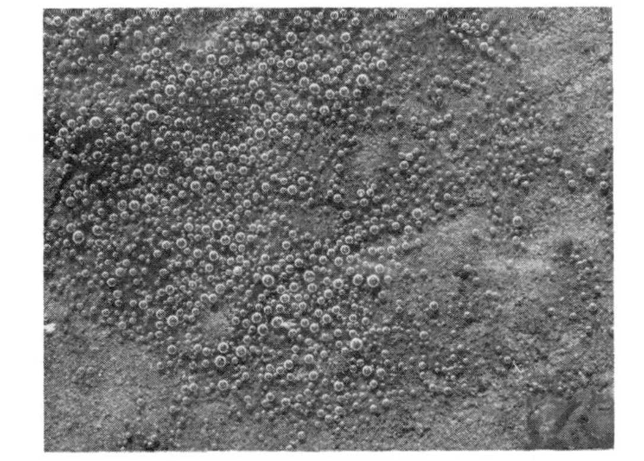

293. Bubbles of oxygen escaping from the diatom film on the bottom of a shallow pool.

However, this is only the beginning, and it is equally essential to understand how much of the activity of the phytoplankton of the sea (the activity of seaweeds along the shore is comparatively insignificant) is carried over to the next step, and the next. It is now thought that the total amount of carbon fixed by plants in the sea is roughly equal to that fixed on land, but there are indications that the process in the sea goes on at something much nearer the maximum possible rate than that on the land. Although W. E. Allen's painstaking work may not have provided the basis for a good index of the activity of a large area of the ocean beyond the Scripps Pier, he did demonstrate how some of the common species of diatoms and dinoflagellates fluctuated from almost nothing to spectacular blooms, while at the same time some of the less common species remained at about the same abundance. These blooms, or outbreaks, are in themselves indications of a system that is capable of violent fluctuations. Small changes in sunlight and in available supplies of phosphates, nitrates, and trace elements (including vitamin B_{12}) can have spectacular effects. If the system is going at a high rate, it is easy to see how this can happen—like an automobile that can be wrecked by hitting an obstacle at 50 miles an hour that it could safely run over at 5 miles an hour. The small organisms that comprise the phytoplankton reproduce by dividing: thus their populations double by reproduction, and some variation that can accelerate or retard the number of divisions by only one per unit of time will thus double or halve the population, if all of them divide at the same rate.

The most spectacular symptoms of oscillation in the marine ecosystem are "red tides." These are blooms of various dinoflagellates, usually yellowish or red in color, occurring so abundantly as to discolor the water. Whereas diatoms grow best in shaded water and do not reproduce well during periods of strong sunlight, some of the dinoflagellates thrive during periods of bright, sunny days. When sunlight is combined with other circumstances, such as perhaps a good supply of B_{12} or some chemical from the land, dinoflagellates reproduce at an exponential rate. Some of them, like *Gonyaulax* of mussel-poisoning fame, are extremely poisonous to many organisms; others seem to be innocuous, but the effects of great numbers of them are not fully understood.

It is a rare sample of seawater that does not have at least a few diatoms in it, although dense populations of them may be scarce or may occur in patches in the ocean. On these microscopic plants and dinoflagellates depend a succession of creatures—the herbivores, or the first level of consumers (the plants are the producers), some of them for only their early stages, others throughout life. An intricate pattern of baby fish eating diatoms, of copepods eating diatoms, and of many other herbivores—some of them never changing their diets, others turning to animal food after they have overtaken it in size and sometimes eating each other—is set up. If we look at what is found in the water and what is found in the stomachs of sardines (Fig. 294), we see that virtually everything included in the plankton is also eaten by sardines, although it is obvious that quite a few things that are occasionally eaten by sardines may be taken in accidentally, or perhaps experimentally. In estimating how much of this material constitutes the regular food of the sardine, we must also consider the sizes involved. Diatoms are small, but they are taken in bulk, especially by the youngest fish, and are consumed very rapidly. Also, as the fish grow larger, the total food consumed increases, even if there are losses in the population. But the copepods live on diatoms as well; so, in one way or another, most of the primary production of the sea is put to use. If not directly eaten by the zooplankton of the upper layers, it sinks deeper into the sea and becomes the food of animals in midwater, and eventually feeds the filtering clams and mud grubbers on the bottom, or adds to the nutrient supply of the sediments. Some of this, sooner or later, is returned to the surface to complete the cycle.

Obviously, there is sometimes competition for all this food, since there are so many things in the sea. However, it does not appear that this is as serious as might be thought at first, at least in a system not tampered with by man. For example, the diagram reproduced as Figure 295 shows the competition for food of the same size range in three common juvenile fishes (the adults have noncompetitive food habits). The original caption for this graph suggests that the jack mackerel prefers somewhat larger sizes, so that the overlap is not so real as it looks. However, if we examine other data, we find still other exceptions (Fig. 296). For the year 1952, for example, the jack mackerel were most abundant in areas where there were neither sardines nor anchovies. Although the sardine and anchovy overlapped in areal distribution, about 30 per cent of the area occupied by anchovies was also occupied by sardines. They were somewhat differently distributed according to depth, with more than 65 per cent of the sardines occurring in the uppermost layer and with less than 50 per cent of the anchovies in the same layer. Furthermore, only a few days' difference in rates of development (which has not been considered in these graphs) could separate these species even further in time and space. Thus, Figure 295 is an excellent example of what might be called ecological overstatement; it is true only if all three species are

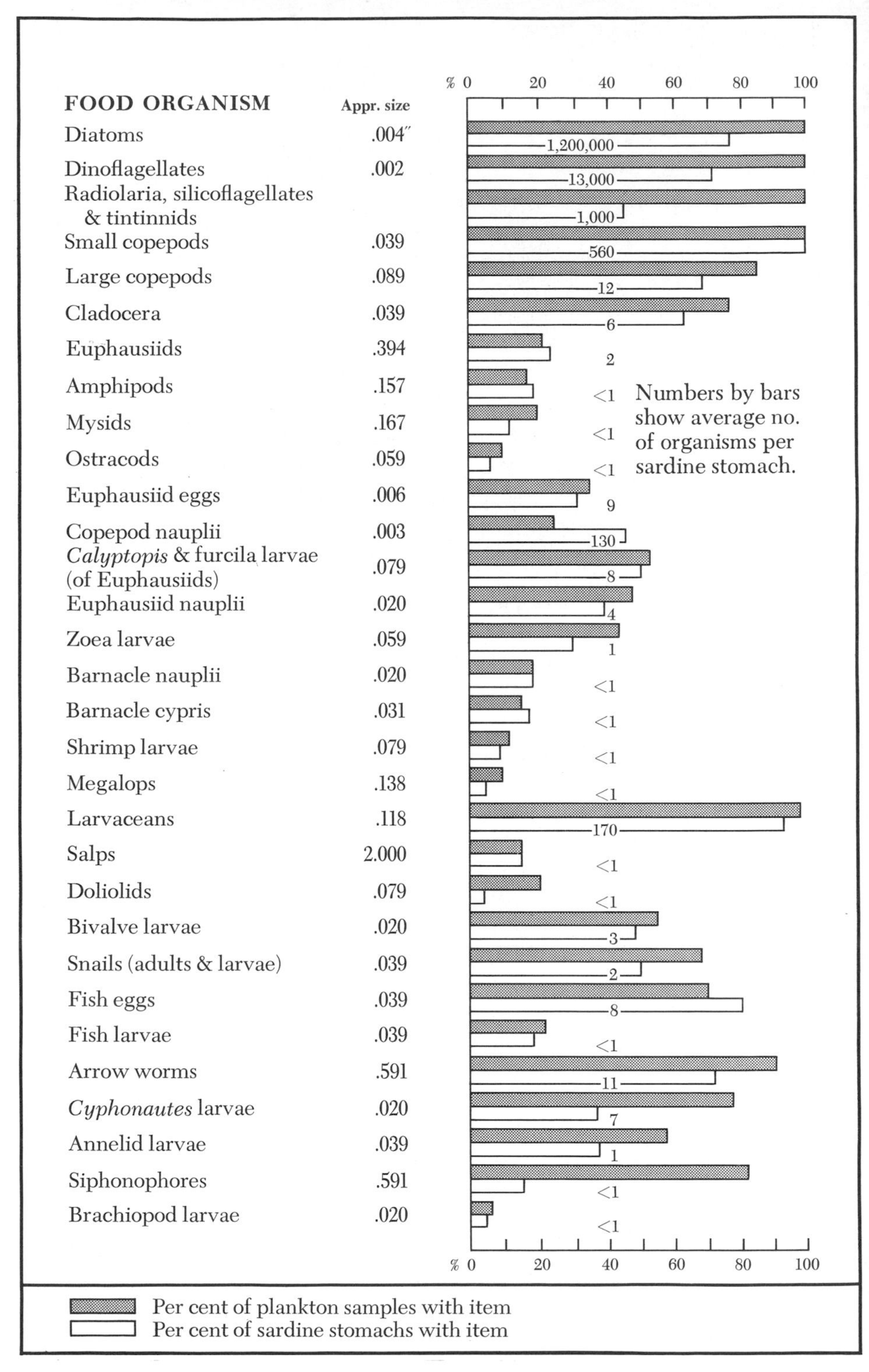

294. The abundance of various common plankton organisms in sardine stomachs, compared with their abundance in the total plankton.

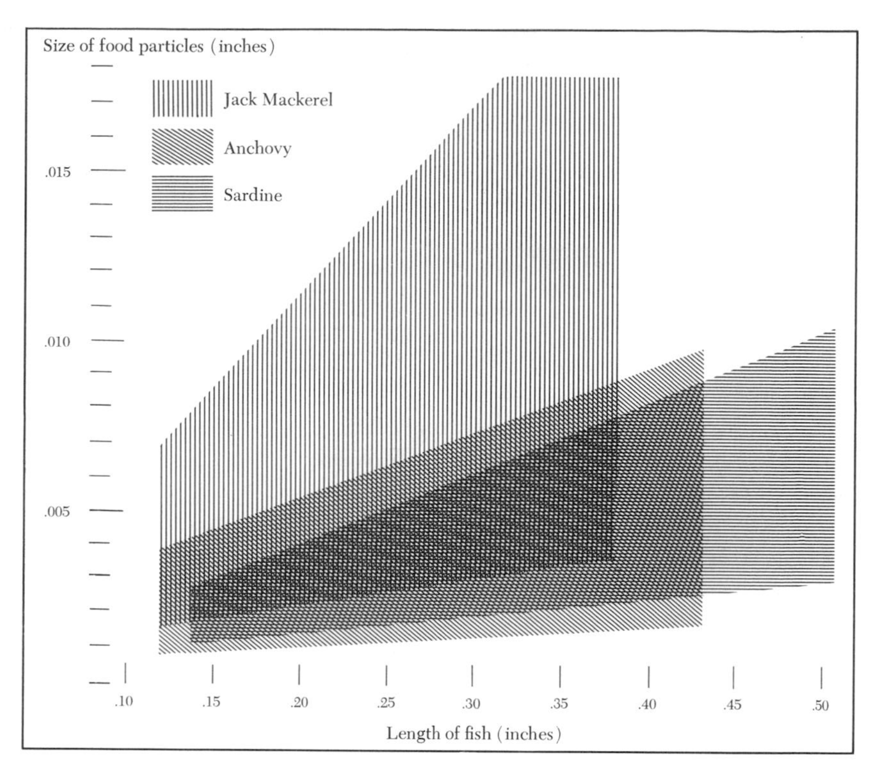

295. Food particle size preferences in the young stages of three fish.

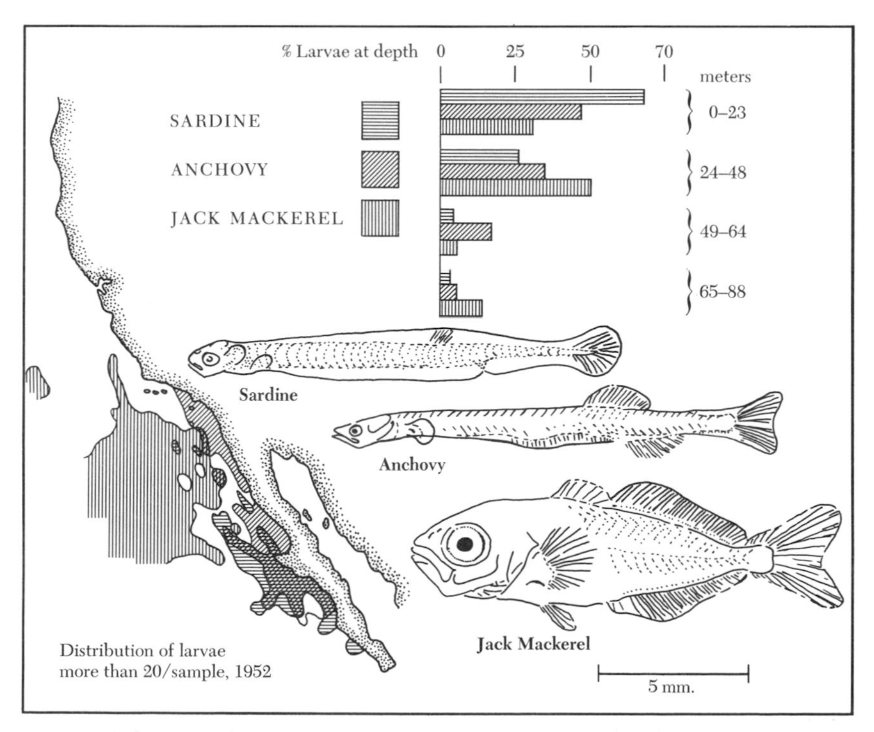

296. Distribution of larval sardines, anchovies, and jack mackerel.

of the same age, rate of development, and geographical distribution in depth and area. What evidence we have suggests that shifting and replacement of populations with changes in the climate of the sea are more important in determining the abundances of fish in the sea than competition between species.

One indication of this is the greater abundance of zooplankton during cold-water years in California waters. Many of these organisms are predators on fish eggs and larval fish, and their abundance was sometimes 20 times higher during the colder periods (Reid *et al.*, 1958). Accordingly, not only would the lowered temperature affect survival of eggs and larvae directly; it would also add to the hazards of being eaten by providing conditions for the rapid increase of zooplankton. On top of this, there is also the physical effect of current action and upwelling in dispersing eggs and larvae to possibly less favorable regions (Murphy, 1961).

Competition between species is difficult to prove to the satisfaction of an ecologist. Certainly the sardine and anchovy are "ecologically equivalent" species, and when times are bad for one species and not the other, the opportunity will be seized. According to Garth Murphy's latest analysis (1966), the sardine had at best an uncertain future because it is subject to population fluctuations associated with a widely varying reproductive success from year to year. The great decline of 1949 was a crash brought on by an intensive fishery that removed the older fish from the population, combined with a poor reproductive year. Sardines were overfished for many years before 1949; and even in recent years, at greatly reduced catches, they are being overfished, for the more sardines we fish, the better it is for the anchovies. And because we are not fishing anchovies significantly, it would not help the sardines even if we stopped catching them, since the anchovies now have the upper hand. Murphy concludes: "It appears that judicious utilization of all ecologically similar species within a trophic level offers the only hope for sustained yields."

Besides everything that goes on in the overlying water, the bottom of the sea must be considered. The sea bottom, one of the classical grounds of marine biology and one from whose study have come some of our ideas of ecological communities, has been neglected by the CalCOFI program. This is in part due to its emphasis on the study of a pelagic fishery. Some of the gaps have been filled by the work of the Allan Hancock Foundation, although this work has been for the most part restricted to the basins of southern California between the mainland

and the Channel Islands. Along the shores of this basin lies one of the great megalopolitan concentrations of our civilization. One hesitates to call it a center of population, for there is no center, only a vast urban sprawl. This is an ecological and social cancer, drawing resources from other parts of the country and fouling both sea and air with its offal. Every day more than 500 million gallons of water containing sewage and wastes are released into the sea between Los Angeles and San Diego. The only immediate beneficiaries of this, besides those organisms that find nourishment in the sewage, have been the staff of the Hancock Foundation, whose studies of the sea bottom have in large part been financed by the State Pollution Board.

Like the CalCOFI program, the Hancock Foundation's study of the benthos is the largest survey of its type to be conducted. More than 500 bottom samples have been taken since the project began in 1956 (see the various papers by Hartman and Barnard, in Section II of the Bibliography). Hundreds of species of worms, crustaceans, mollusks, and other animals have been identified, counted, and weighed. In spite of all this, the percentage of bottom fauna sampled may be less than that of plankton sampled by CalCOFI because it takes more time to take the sample from the bottom and to sort and identify the material. Quantitative bottom surveys of this type have always been more static, with less attention to seasonal changes, than plankton surveys. In part, this is justified, since many of the bottom animals live several years, and their "standing crop" is much larger than their annual turnover. There appear to be amazing standing crops of some of the bottom animals. The green echiuroid *Listriolobus pelodes* occurs over an area of about 46 square miles of the bottom between Santa Barbara and Ventura, with a gross weight of 825,000 tons. This figure, for 1957, was somewhat larger than the total weight of the 1956 California fish landings, including tuna. Of course, there are some extrapolations, as Figure 297 shows, because this estimate is based on less than one sample per square mile (however, marine biologists estimate the biomass of the deep sea on the basis of a sample or two in hundreds of square miles). Further, there is quite a difference between the nutrient value of a squishy echiuroid and that of a succulent tuna. What we need to know about these bottom aggregations is the turnover—how old these animals really are, and how fast they grow. However, we must do what we can with what data we can get; and even our Russian colleagues are not without a grain of skeptical humor concerning their efforts to obtain quantitative information about the sea bottom (Fig. 298).

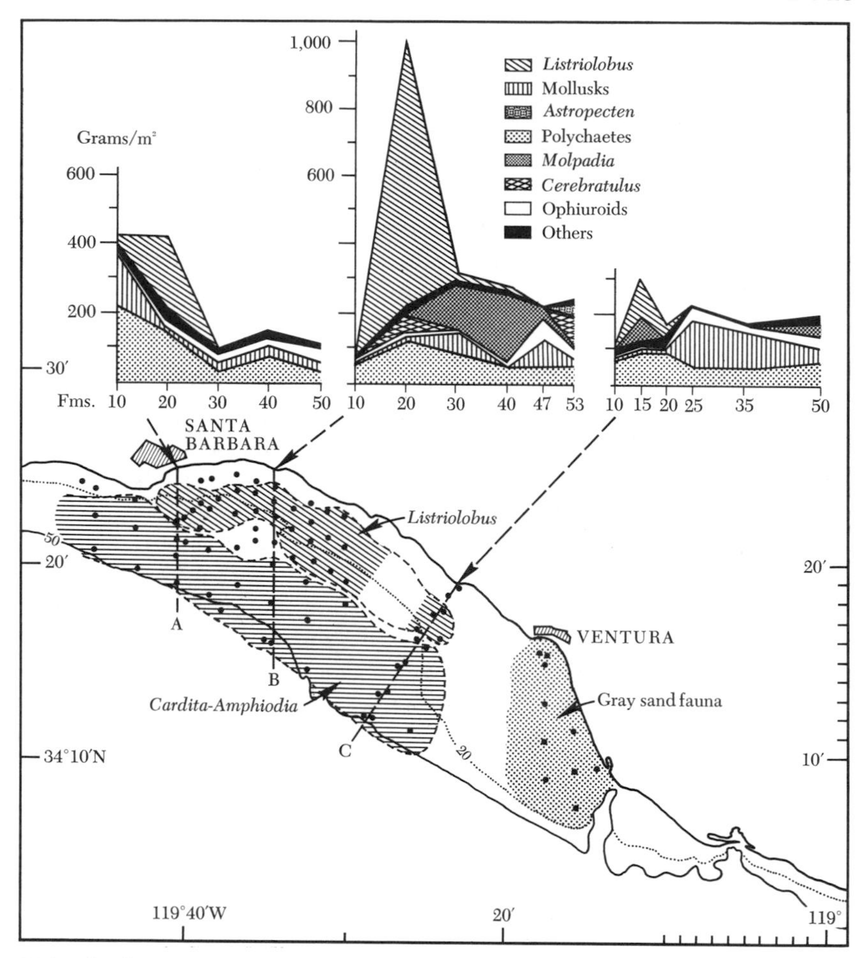

297. The density of bottom organisms at different depths along three transects off Santa Barbara. The map at bottom shows the location of the sampling points and the areal distribution of some dominant types of animals.

What is the balance sheet for all of this complex of plankton, fishes, and bottom organisms? No one really knows, and the data are too incomplete for more than guesses.

Unfortunately, there are neither funds nor people to do the job adequately, and the only attempt so far at estimating the annual organic budget for our area is that by K. O. Emery of the Hancock Foundation (1960, pp. 175–79). This estimate (based on dry weights—that is, with-

298. The Soviet deep-sea bottom dredge in action, by Zina Filatova.

out water content) is for the southern California region only. In the first place, only about 2 per cent of the solar energy is converted to plant material, which is about twice the average efficiency for the ocean as a whole. The annual production of plant matter is estimated at around 44 million tons, and that of zooplankton at 3.4 million tons, or about 7.5 per cent of the plant production. (This annual production amounts to about 4,500 pounds per acre, which is roughly the same as that on good agricultural land in the United States.) The fish, in turn, are estimated to amount to 0.1 million tons—3 per cent of the zooplankton, or about 0.2 per cent of the plants. The bottom organisms fare somewhat better, with about 1.5 million tons, or 3.4 per cent of the plant production. Ultimately, about 7 per cent of the organic matter produced reaches the bottom, but most of this is recycled, with perhaps 0.6 per cent permanently lost in the bottom. These are guesses based on all sorts of assumptions, and freely admitted to be such.

Obviously, nature is not as efficient as we would like it to be, and the suggestion is often made that we should go directly to the plankton, short-circuiting this elaborate system. At the present time, however, our methods for collecting plankton are even less efficient than those of the sardine, and we would be put to the additional expense of converting the plankton to something usable. Usually these suggestions are made without realization of the uncertainties of the plankton cycles or of the difficulties of obtaining plankton in significant quantities. It would not appear that direct utilization of oceanic phytoplankton will occur in our time, although the day when bays and lagoons may be converted to intensive plankton culture may not be far off. At the present time, something between 1 and 2 per cent of the food consumed by man comes from the sea. Surely we can draw more heavily than we have upon the sea's food resources without severely upsetting the ecosystem, although the moral of the sardine episode should now be plain: in times of natural disturbances of the environment, an intensely fished species may be more vulnerable to these changes.

Harvesting the zooplankton, the smaller animals in the feeding level (trophic level) just below the fishes, may be another matter. Some of these animals, like the pteropods and euphausiids that have been the principal food of certain whales, are probably the most abundant creatures on earth. Their numbers may be in the billions. Now that we have about exterminated the blue whales of the Antarctic seas, the vast populations of krill (which may be nutritionally not much different from shrimp) are unharvested (at least by men). Russian fisheries specialists are experimenting with ways to harvest and process these euphausiids; the processing may present more problems than the harvesting, however, for they are too small to be economically handled except by machinery. In the meantime there are indications that some of the penguins, who also depend on euphausiids, are on the increase. Perhaps we had better be studying ways to can penguins, since the euphausiids are the mainstay of much of the larger sea life of Antarctic seas, and drastic reduction of their numbers would disrupt the entire system.

It is possible that we are altering the productivity of both sea and land by our inadvertent tampering with the basic carbon cycle of our planet. The content of carbon dioxide in the atmosphere has increased perceptibly in the twentieth century, either from the use of coal and oil or because our agricultural practices reduce the net efficiency of plants on the surface of the earth (Hutchinson, 1948). As yet, these effects are slight, and are overcome by the stabilizing effect of the oceans, which absorb most of the atmospheric CO_2. In any event, the carbon cycle (Fig. 299) is the basic pathway of our terrestrial ecosystem, and evidently it has always been so. In Palaeozoic seas somewhat more carbon was accumulated in sediments than in Mesozoic or Tertiary seas, which suggests greater productivity in those distant times, according to Evelyn Hutchinson. Whatever the long-term variations in this basic cycle may be, the most noticeable oscillation of the cycle is not a matter of centuries, but of days and years, since the cycle depends on rates of growth of organisms; these, in turn, depend on rates of photosynthesis, which in turn depend on the rate of return of CO_2 to the atmosphere.

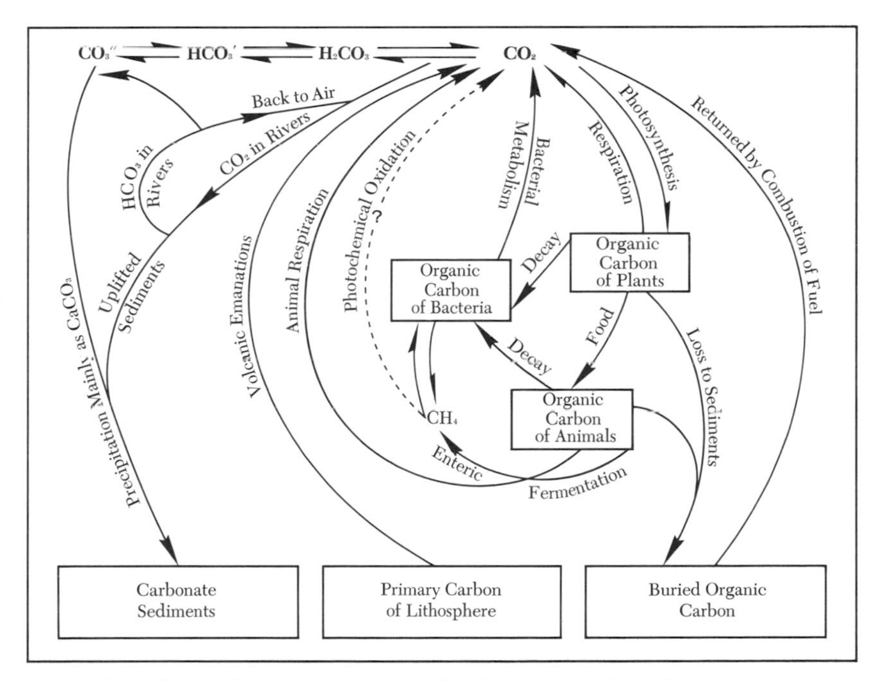

299. The carbon cycle. Exact quantities of carbon passing through various stages are hard to determine, since consumption of fossil fuel, smog production, etc., are changing each year. (Modified from Hutchinson, 1948.)

III

Many of these intricate interrelationships might have remained matters of special interest to theoretical ecologists, had the world remained in a state of comparative innocence, but the Pandora's box of the atom has been opened. Now we are beginning to realize other implications of this movement of elements through natural community systems, and the ability of humble invertebrates to concentrate materials by factors of several thousand times is no longer a biochemical curiosity; it is a warning of what may happen if highly toxic chemicals and radioactive materials are thrown indiscriminately into the sea. At first the physicists and engineers were unaware of these implications, and looked to the sea, as mankind has always done, as the great universal dispose-all system, into which all manner of things could be dumped and forgotten. Perhaps this would be possible in a completely lifeless sea, although the oceanographers are not so sure of this either. The circulation of the oceans is still incompletely understood, with too many unresolved themes and slippery differentials that cannot be integrated.

One thing about radioactive isotopes is that they can be measured easily in extraordinarily minute amounts. Because of this, isotopes have become one of the major tools of science: we can trace the fate of certain chemicals by "tagging" them with radioactive isotopes—in plants or animals, or almost anything else. We usually reckon an isotope's activity by its "half-life," the time it takes for half the radioactivity to dissipate. Some isotopes, like that of xenon, which leaks from atomic power plants, may have a half-life of seconds; others have a half-life of several thousand years. Carbon is an essential part of all living systems, and its radioactive isotope C^{14} has a half-life of 5,580 years but is at the same time relatively innocuous in very small concentrations; hence C^{14} has become one of the most important of the isotopes for biological research. One of the ingeniously simple uses of C^{14} is a technique for determining the activity of phytoplankton in the sea. "Labeled" sodium carbonate ($Na_2C^{14}O_3$) is added to samples of seawater. The plants in the water will take up carbon, and a proportion of the C^{14} will be fixed in them. The tagged phytoplankton is then filtered out and the increase in radioactivity over a known period is measured. From this measurement it is possible to calculate the amount of carbon fixed per unit of time and per unit of volume. (See Doty and Oguri, 1959, for an excellent description of this technique.)

Radioactivity, both from the elements of the earth and as ionizing radiation from space, has been since the beginning a part of the environment of life. Quite probably, life as we know it would not have evolved or be possible without radiation, and possibly we need a certain amount of radiation. Some idea of the present range of natural radioactivity can be obtained from a study of Figure 300. The units of this diagram are millirads—that is, thousandths of a rad. The rad stands for 100 ergs per gram of matter and is a slightly larger unit than the roentgen unit,

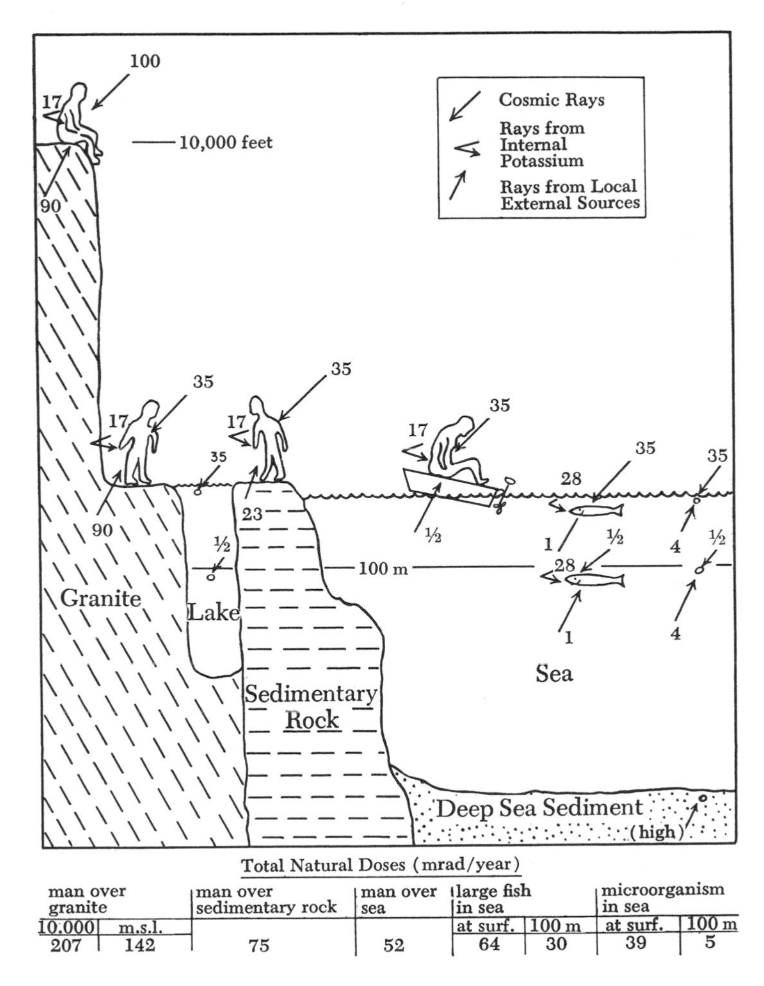

300. Comparative natural radiation dosages (Folsom and Harley, 1957).

but it applies to radiation received internally, whereas the roentgen unit refers to absorption of external radiation. In any event, the annual natural dosage for people living on high granite mountains is near the maximum level of 10 roentgens during 40 years, which is considered the limit of tolerance before man may sustain genetic damage. Few of us live on the top of a granite peak for 40 years, and man does seem to be comparatively radio-resistant—radio-resistant, that is, to natural conditions; the potential effect of introducing artificially produced isotopes in the natural system, whose course through the human body is quite different from what man has normally been exposed to, is the critical problem of our civilization. The analogy of gross quantities of natural radioactivity in this context is meaningless, and the argument of some physicists that fallout is comparatively harmless should not be accepted at face value when isotopes of elements important in biological systems may be involved.

Thanks to the insulating effect of water, the penetration of cosmic rays in the sea is reduced by 99 per cent in the first 100 meters (about 300 feet). Exposure to all sorts of radiation decreases with depth, and is at its lowest in deep lakes. But when the deep-sea sediments are reached, the exposure to radiation increases to roughly the same as that experienced by an organism living in a hole in granite, owing to the natural radioactivity in the sediments. For a very small animal, life becomes more complicated; most organisms with shells or having some bulk are partly protected from the short-ranged alpha and beta radiations; but a very small organism, of perhaps a millimeter or less, is vulnerable to these densely ionizing radiations as well as to the gamma radiations that concern us larger animals. In this context, one must remember that the ratio of surface to volume is largest in the smallest organisms, so that adsorption on the surface, or concentration in the shell, would increase exposure to radioactivity. It is not surprising, therefore, that such things as viruses, bacteria, and protozoans are less radiosensitive than larger, higher organisms.

The problem of radioactivity in the sea is complicated by many more uncertainties, not only because of our ignorance of much of the circulation system in the ocean, but because we know so little about the genetic tolerances of marine organisms. Our concern, sharpened by these uncertainties, has led us to discontinue the disposal of radioactive wastes in the Pacific Ocean, at least for the time being, although the British are deliberately disposing of radioactive waste in the Irish Sea through a long pipe from the atomic works at Windscale. We have a similar isotope sewer, whose maintenance is less defensible. This is the Columbia River on the Pacific coast, into which 2,000 curies are discharged at Hanford every day. Of this amount, 1,000 curies reach the sea, and some of this accounts for the radioactive zinc found in oysters at Willapa Bay.

In some circles, this steady buildup of radioactivity in the sea is complacently accepted because the concentrations are considered well below the safety limits we have set for our own species. In setting our standards for the disposal of atomic wastes, we have as yet given little consideration to the possible effects on other organisms, except that we do know we must take into consideration the capacity of other organisms to concentrate these materials. As the British say of their Windscale release into the Irish Sea: "We place most of our emphasis on the ultimate level in the food chain—that is man." We can study uptake, transfer through the food chains, and potential effect on man; yet all this will tell us nothing about the effects of added material in an environment on the life cycles and well-being of the organisms living within that environment. It is very possible, for example, that the levels we might eventually build up in the sea by continued fallout and waste disposal would have more effect, ecologically, than the same levels on land. This implies, of course, that there would have to be much more radioactivity added to the ocean, because of dilution in its vast volume, to add to the total radioactivity of the sea in proportion to the increase on land. Granted this assumption, it would follow that the percentage increase in the sea would be much greater in terms of present natural activities on land and sea. Maurice Ewing suggested that some archaic types persist in the deep sea because there is less radioactivity (this could not apply to animals living in deep-sea sediments, where there is enough radium to give rather high total dosages).

In some of the published literature, there have already been suggestions that we cannot determine, on the basis of present knowledge, *any* level of activity that will not affect life in some way. We have, for example, indications that the manner in which radioactive materials become associated with organisms may not always be a simple linear function of what they may be exposed to. Radiosensitivity has been demonstrated to decrease with age in mammals, trout, and freshwater snails, and the highest sensitivity to external radiation is at periods of rapid division and growth (Donaldson and Foster, 1957). It has been suggested that if we know more about the chemical composition of an organism, we will be able to predict what the organism will do with radioactive isotopes of elements making up the internal system, but this does not seem likely if sensitivity varies with age, since high dosages might change the pattern of uptake in later stages. Donaldson and Foster suggest that the damage effects of radiation have a similar pattern throughout the animal kingdom, depending on the dose, but we really do not know this; all we know, so far, is based on a handful of animals. In the sea we have many organisms that grow actively for most of their lives and have comparatively short adult lives, or whose life cycles are of short duration. These organisms may be vulnerable to radiation for the majority of their lives, since most of their lives are spent in those stages of rapid growth and reproduction that may be most affected by radiation.

Then, there are indications that uptake may be influenced by environmental or ecological factors. Oysters fed on radioactive *Chlorella* assimilated very little radioactive phosphorus (P^{32}), but when fed on other phytoplankton of less or similar P^{32} content, the oysters assimilated much more radioactivity (Krumholz, Goldberg, and Boroughs, 1957).

All of this work so far has been done with relatively high amounts of radioactivity. Concerning the possible effects of long-term, low-level exposure in the sea, we have no information. Ketchum (1960) has suggested that a comparatively low radiation dose might have a disproportionately high genetic effect on a marine animal, because the great natural wastage of larvae leaves the possibility of a generation being maintained by comparatively few survivors, which will be exposed to radiation during the stages of rapid growth.

In addition to the actual uptake, or incorporation of materials by organisms, the magnitudes of surfaces available for physical recirculation by adsorption in a water column are immense. The great Russian marine biologist Zenkevich has provided an interesting diagram of this process (Fig. 301), which suggests surfaces of several thousand square meters a year available for this process of "biocirculation." Accumulation of isotopes on these surfaces would also increase the potential radiation dose to the organism involved.

Obviously we need to know much more about the physiology, ecological transfer, and genetics of marine organisms than we do, and we also need to know what the natural conditions are in the sea before going further with our use of radioactive materials. In the first monographic treatment of the effect of radionuclides on marine organisms, the Russian radioecologist G. G. Polikarpov emphasizes over and over again how incomplete our knowledge is; nevertheless, he concludes, we do know enough to state that "*further radioactive contamination of the sea is in-*

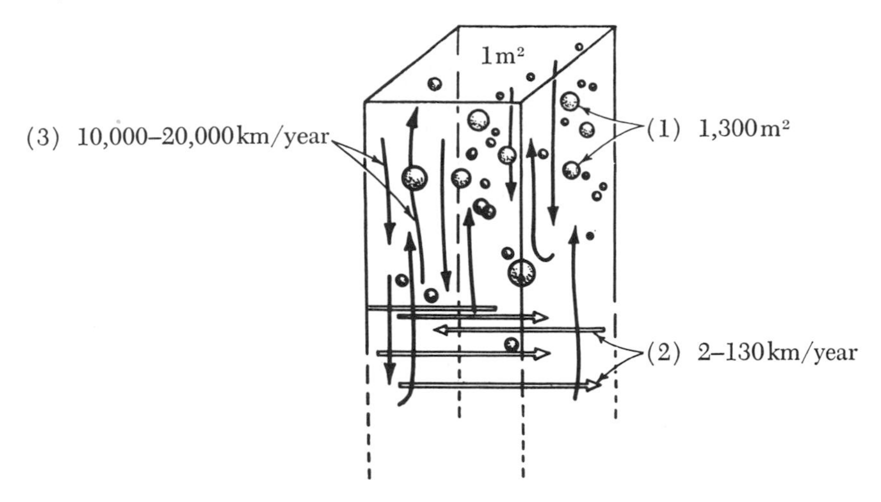

301. Adsorptive surfaces and biocirculation in a column of water beneath one square meter in the temperate zone: (1) potential adsorptive surfaces; (2) biocirculation by fish; (3) biocirculation by zooplankton.

admissible" (his italics). This is considered an extreme position by some specialists on this side of the Iron Curtain, and it has even been insinuated that Polikarpov's position is politically motivated. But all pollution is inadmissible; at some point we must improve our environmental hygiene.

Instead, we have been assaulted by public-relations hirelings, who exude sweetness and light at service club luncheons and impugn the integrity of concerned scientists by suggesting they are taking advantage of their position and their knowledge to advance their own misguided causes.*

Yet, rather than freely conceding their obligation to make adequate study of the potential effects of nuclear reactors, pulp mills, or other environment-disrupting activities, the organizations concerned too often resist all suggestions for careful study as interfering with the economics of the project. They will plead that they serve the greatest good of the greatest number (a shockingly cynical slogan), even if the process kills a few fish, or even people. For 6 years an effort was made to install a

* We hear that one of these ardent apologists, no longer in the service of a certain power company, has expressed a wish to join forces with his former opposition and fight their cause against his former masters. Apparently miracles are still possible along El Camino de los Kilowattos, if not on the road to Damascus.

nuclear power station on Bodega Head, in simple defiance of the obvious circumstance of geology—until an earthquake in Alaska finally discouraged the project. The monument to this folly, the famous "Hole in the Head," is now full of water, and makes an excellent place for divers to test apparatus (Fig. 302).

Atomic energy is only one improvement of modern civilization that threatens to destroy our environment for ourselves and for the creatures we share it with. At about the same time that the atomic bomb was demonstrated to be practical, we began to develop powerful pesticides. The first of these was the famous DDT, now somewhat old-fashioned and mild compared with some of the newer substances, which may kill animals—or plants—in concentrations of a few parts per billion. We also discovered penicillin, introducing antibiotics into our culture. It is an ideal combination: penicillin and all the other antibiotics that have followed it will cure or prevent our ills, so that most of us will survive longer than ever; atomic energy will provide the power to run our civilization; and the pesticides will kill all undesirable competitive plants and insects, so that we may raise more food for all. This will all be to the good: no more diseases, abundant power, and no more bugs. And, we will clean it all up, using our fine new detergents by the ton.

Well, not quite. We may be inadvertently selecting more virulent strains of disease with our drugs and antibiotics; we may not really have enough raw material for all the atomic power plants we would like to have; and our pesticides may be ruining the land and sea for all the other creatures we share this earth with. The detergents used so liberally to clean up the oil from the *Torrey Canyon* wreck have caused more devastation than the oil alone, and some of them may last for years.* Los An-

* At least this lesson was not lost on the authorities in January 1969, when a vast "black hemorrhage" broke loose off the coast near Santa Barbara (*San Francisco Chronicle,* Feb. 7, 1969), although some detergents were spread from planes in an effort to disperse the surface oil slick. The Santa Barbara episode, caused by offshore drilling in a dangerously faulted region of the ocean bottom, released at least 230,000 gallons of crude oil, forming a slick over 800 square miles. The damage may take years to assess. The State of California is bringing suit against the Department of the Interior and the oil companies, demanding a compensation of more than two billion dollars for damage to property and marine life and "loss of tourist potential." Even so, the natural conditions upset by the oil may not be reestablished for many years; and there is always the possibility, now that the oil pool has apparently been breached, that even a minor earthquake may permanently rupture the ocean bottom and create a state of chronic disaster. This affair is one more demonstration that we are reaching the limits of tampering with our environment. It is difficult to predict where the next disaster will be; but undoubtedly there will be many more before it is generally agreed that it is we who must come to terms with our environment, not the environment with us.

302. Bodega Head, showing remains of the abandoned reactor project.

geles needs more water, but of course we cannot deliver the water from the fields of northern California there because of the potentially deadly pesticides. The ocean, some think, is the best place for this stuff—to say nothing of all the sludge and gunk from our factories and cities. One scheme to solve this problem is the "Big Drain." This would be an enormous pipe into the ocean (perhaps near beleaguered Bodega Head, perhaps near the quiet solitude of Pigeon Point), which would convey all the contaminated water from the Central Valley into the ocean. A project of this type could mean no more seashores worth visiting for miles along the central California coast. Already we suspect that some pesticides draining into the ocean via San Francisco Bay may have something to do with the poor crab fishing around San Francisco in recent years. Larger quantities, poured directly into the ocean, could be carried far to the north by the Davidson Current in winter, possibly in concentrations dangerous to fisheries for hundreds of miles. In the summer the stuff could drift along the pleasant shores of Carmel and Point Lobos.

It seems incredible that a project of this type should be considered even for a minute; yet the indignant denials by some state officials, the virulent memoranda that somehow leak out of official files by way of the easily accessible copy machine, and the furious flurry of telephone calls to hopefully cooperative people suggest that there is entirely too much smoke for no fire at all. Somehow it sounds to an old campaigner like the Battle of Bodega all over again.

There will be many more battles before those motivated by economic concerns fully realize that there is more to this dissent than the interests of bird watchers and lily pickers, and that opposition is based on the growing concern for the integrity of the environment of which man as a species is a part and on which he depends for continued existence. Seashore property is so valuable to investors because it seems to them (if they think about it at all) that the seashore will always remain, in spite of everything that may be done, a pristine, renewable environment: there will always be fresh air, clean, sandy beaches, and clear water. But there is increasing evidence that the ocean itself may be vulnerable to change, that we cannot treat it as our global septic tank without affecting ourselves in the long run, and that we cannot interfere with processes of the shore without setting in motion a train of undesirable events. Wesley Marx, in his book *The Frail Ocean,* brings together many of these disquieting indications, especially for southern California.

Among the matters as yet unmentioned here is the problem of the vast amounts of heat we must waste in generating power: water, by the hundreds of thousands of gallons per minute, is required to cool down steam and atomic plants alike. So much of this heat will be released that engineers are talking of "thermal pollution." More optimistic people prefer to call it "thermal enrichment." It may be impossible to heat up whole areas of the ocean (or even warm up beaches for swimmers, as some entrepreneurs have suggested); but if this did come to pass, it could change our flora and fauna in the spots so affected, not perhaps so much by warming up the water, as by depriving organisms of necessary cold water periods in their life or damping out oscillations they may need. We do not know enough about the temperature requirements for intertidal organisms to be sure of what might happen when there are rows of atomic plants along the coast, spewing out millions of gallons of hot water every hour. The engineer's approach is to try it out and see; the biologist's reaction is that after it is done it will be too late to undo the damage.

If the naturalist has anything to contribute to human culture and well-being beyond the aesthetic and emotional delight that comes from pretty books and bird-watching, it is the knowledge that nature is an interrelated system that should not be lightly tampered with. This ought to mean more to man's survival on this planet than towering pyramids of technology and a knowledge of what moon dust is made of. To discard lightly the admonition of the naturalist is to admit that we are not very far from the caves, and perhaps to hasten the day when we will join the dinosaurs in their fossil beds. Perhaps, if we overload our environment with radioactivity for a few thousand years, evolution will take a different course (especially from the one envisioned for us by Sir Julian Huxley), and our descendants will become something quite unlike us. Such a people might well be happy with their fate, but of course we do not see this as a happy prospect for our own descendants. Perhaps that is a selfish view in terms of the whole long history of the earth, but the danger of a less favorable outcome is too great for us to wish this deluge lightly upon the future. As a species, we must not forget that our continued well-being is in turn dependent on the well-being of our ecosystem. It would be illogical, if not stupid, to talk about the greater resources of the sea and the brave new Sea Grant colleges to be financed at government expense while at the same time we use the sea as a septic tank for our most dangerous wastes. Perhaps it would be wiser to finance colleges for the development of ways to process and utilize the wastes of our civilization before doing anything else. Certainly it is shortsighted to talk of increasing our consumption of fish while at the same time carelessly dumping pesticides into the environment to be concentrated in the fish that we plan to eat. As the editor's old English professor, George R. Stewart, puts it in his book on the garbage problem (*Not So Rich as You Think,* Houghton-Mifflin, 1968), we have run out of space for this sort of thing.

In some ways we are not worried enough about what we may do; in other ways we may err on the conservative side. Nevertheless, our concern for the potential effects of the atomic age is perhaps the most encouraging aspect of our brave new world. At least this seems to be the first time in history that man has expressed any concern at all about pollution of any kind at its outset. If the big red button in some deep underground command post is not pressed, there may be some hope for us—and for the oceans.

Annotated Systematic Index

THIS INDEX has several purposes. First, for those who wish to know, it will give some idea of where a certain animal stands in the scheme of classification, at least to the level of orders and suborders (families have been avoided for the most part). The classification used here may not be universally accepted (none is anyhow), but it is not notably divergent from what might be called standard zoological opinion. Classification is, after all, simply an effort to present our ideas of how things should be arranged and how they may possibly be related; in this sense, it reflects our ideas of evolution. In any event, the reader should be aware that there is no standard number of phyla—despite the insinuations of textbook advertisers who say that their text considers "all 37 phyla," or whatever other number that particular author decided to use. Having once heard an instructor inform his class that the specimen before them was an example of "Phylum Ten" in their text, I feel it necessary to belabor the point, since this book is widely consulted, if not read, in biology classes on this coast.

This Index should also enable an interested reader to track down more information on any animal that especially interests him, or to locate closely related species that are discussed in other sections of *Between Pacific Tides.* (For books and articles of broader scope, the reader should consult the General Bibliography, pp. 535–65.) All levels of classification, as well as such substantive material as occurs in the annotation, will be found listed in alphabetical order in the General Index.

All names of plants and animals are listed according to their formal designation: the genus (plural, "genera") and specific name in italics, followed by the author's name in roman. This is the designation by which a species (a pure Latin noun of the fifth declension, and hence the same form in both singular and plural) is known. These names are supposed to look like Latin and be treated as such grammatically, so the specific name, if an adjective in the nominative singular, must agree in gender with the generic name. However, the gender of the genus may be derived from the original language (when that language has grammatical gender) and may not be governed by the artificial Latin ending, which means that not all words ending in *-us* are masculine nouns of the second declension. Nor are all Latin nouns of the first declension ending in *-a* to be treated as feminine, for when the subject is obviously masculine they are treated as masculine, so we could have *Nauta superbus*. Names based on nonsense, outlandish languages, or unrecognizable anagrams are to be treated as undeclinable. All of this creates uncertainty, since few people now know enough Latin grammar to know which ending should be used, and some rebellious dissidents have suggested that all names be arbitrarily considered masculine. It is obvious that problems of nomenclature do not have much to do directly with biology, but they do constitute a branch of international law. In zoology the guide for this sort of thing is the International Code of Zoological Nomenclature (approved by the XV International Congress of Zoology; published by the International Trust for Zoological Nomenclature, London, 1961). The text of this book is in French and English; of its 176 pages more than 60 are devoted primarily to these problems of formation of names.

There are some differences in botanical names. Obvious patronymics are often capitalized in botany, but never in zoology. The transfer of a name to another genus in zoology is indicated by enclosing the author's name in parentheses without indicating the reviser responsible for the change, whereas botanists include the reviser's name.

PLANTS

This is, of course, a book about animals; and a surprisingly small number of them are directly related to the intertidal algae as grazers or browsers, although the seaweeds are one of the major sources of organic detritus in the intertidal and provide refuge in one way or another for many species of animals. The literature on Pacific coast marine algae is large and is steadily growing; certain papers dealing principally with algal ecology are cited in Section III of the Bibliography. The following references are useful as guides or background reading, and Smith's book in particular contains a comprehensive bibliography of systematic literature on Pacific coast algae.

CHAPMAN, V. J. 1964. Coastal vegetation. New York: Pergamon Press and Macmillan Co., Commonwealth and International Library, Botany Division, Vol. 3. 245 pp., illus.

A paperback. This is essentially an ecology of intertidal and maritime plants, including those of dunes and marshes. It is based primarily on British examples, but ecological concepts are not tied to species or localities.

CUPP, EASTER E. 1943. Marine plankton diatoms of the west coast of North America. Bull. Scripps Inst. Oceanogr., **5** (1): 1–238, 5 pls., 168 figs.

Seventy years or so ago, it was the fashion to collect diatoms and mount them in designs on slides, which were then admired at the meetings of the local microscopical society—an amiable pastime now gone the way, it seems, of the free lunch and the horse-drawn trolley. This book went out of print, and was for a while scarce; it has now been reprinted, along with other numbers in this series, by one of those firms that makes reprinting of serials its business—at a price. In any event, it does not answer our need for a comprehensive handbook to the benthic and sessile diatoms of the intertidal regions.

DAWSON, E. YALE. 1946. A guide to the literature and distributions of the marine algae of North America. Mem. S. Calif. Acad. Sci., **3** (1). 134 pp.

A useful checklist.

——— 1966. Marine botany: An introduction. New York: Holt, Rinehart, and Winston. 371 pp., illus.

I first met Yale Dawson in W. A. Setchell's laboratory; the Old Master had found a student worthy of the torch and was in a mellow mood. I suppose he hoped that his disciple would eventually inherit his chair; but that did not come to pass, and Yale finally wound up in the Smithsonian, which, alas, was to be his last job. He died in action, appropriately enough, but that does not mean we shall miss him less. We can ill spare our good naturalists.

This is a good general text for nonbotanists, with a strong flavor of the author's sense of the history of his subject.

——— 1966. Seashore plants of southern California. Berkeley: Univ. Calif. Press. Paperback, 101 pp., 64 figs., 8 col. pls.

——— 1966. Seashore plants of northern California. Berkeley: Univ. Calif. Press. Paperback, 103 pp., 76 figs., 8 col. pls.

Salt-marsh and dune plants are included in addition to the seaweeds. The southern California guide goes as far north as Gaviota. The northern California guide is concerned mainly with the Monterey and mid-California area; it does include plants found in Oregon, but only perfunctorily.

DOTY, MAXWELL S. 1947. The marine algae of Oregon, I: Chlorophyta and Phaeophyta. Farlowia, **3** (1): 1–65, 10 pls. II: Rhodophyta, *ibid.*, **3** (2): 159–215, 4 pls.

GUBERLET, MURIEL LEWIN. 1956. Seaweeds at ebb tide. Seattle: Univ. Wash. Press. 182 pp., illus.

About the seaweeds of the Puget Sound area, with the usual hoked-up common names, such as "loose color changer" and "red eyelet silk." Although there may be problems with some species because of lack of detail, the work as a whole is useful, and is available as a comparatively inexpensive paperback.

HOLLENBERG, GEORGE J., and ISABELLA A. ABBOTT. 1966. Supplement to Smith's *Marine algae of the Monterey Peninsula.* Stanford: Stanford Univ. Press. 130 pp., 53 figs.

Smith brought up to date: new records, new species, and name changes; records 55 species as new records for the area. Cross-referenced to be used with the earlier work.

KIKUCHI, TAIJI. 1966. An ecological study on animal communities of the *Zostera marina* belt in Tomioka Bay, Amakusa, Kyushu. Pub. Amakusa Mar. Biol. Lab., Kyushu Univ., **1** (1): 1–106, illus.

An excellent start for a new journal! Our *Zostera* beds have not yet been studied with such loving care and critical attention to statistical procedures.

MUNZ, PHILIP A. 1964. Shore wildflowers of California, Oregon, and Washington. Berkeley: Univ. Calif. Press. 122 pp., 178 figs., 96 col. pls.

Includes almost everything that grows within sight of the sea or on the dunes and marshes, from cactus to Sitka spruce.

ROUND, F. E. 1965. The biology of algae. London: Edward Arnold (New York: St. Martin's Press). 269 pp., 68 figs., 8 pls.

An excellent, solid introduction to the algae, both fresh-water and marine; as the dust jacket says, it is "scholarly and general."

SANBORN, ETHEL I., and MAXWELL S. DOTY. 1947. The marine algae of the Coos Bay–Cape Arago region of Oregon. Ore. St. Monogr. Bot., **8**. 66 pp., 4 pls.

SCOFIELD, W. L. 1959. History of kelp harvesting in California. Calif. Fish and Game, **45** (3): 135–57, 7 figs.

SMITH, GILBERT M. 1944. Marine algae of the Monterey Peninsula. Stanford: Stanford Univ. Press. 622 pp., 98 pls.

Because the majority of the algae known from the Pacific coast of the United States and Canada occur at Monterey, this becomes a handbook for the whole coast.

———, ed. 1951. Manual of phycology. Waltham, Mass.: Chronica Botanica. 375 pp.

A symposium treatment of the algae in the broadest sense, including the flagellates, with a long chapter on intertidal algal ecology by Feldman. Excellent bibliographies.

In addition to the algae, there are three species of higher plants commonly found intertidally on this seacoast, the eelgrasses. Rather than refer the reader to a flora, which would be devoted primarily to terrestrial plants, we offer the following key, or synopsis, of the Zosteraceae:

1. Flowers unisexual but borne on the same spike; fruit ovoid, hornless. Plants growing on sand or mud in bays and similar protected situations; leaves 5–7-nerved (*Zostera*) 2

   Flowers unisexual but borne on different spikes; fruit heart-shaped with prominent horns at the base. Plants growing on rocky, exposed, or semiprotected situations; leaves 3-nerved (*Phyllospadix*) 3

2. Leaves usually 2–8 mm. broad; fruit subsessile; seed distinctly ribbed; Atlantic and Pacific Oceans, known from San Pedro to Alaska on the Pacific coast of North America .... *Zostera marina* L.

   Leaves usually 6–12 mm. broad; fruit distinctly stipitate; seed ribless; Santa Barbara to Puget Sound.............*Zostera marina* var. *latifolia* Marong

3. Leaves usually 1–2 mm. broad, thin, flat, indistinctly 3-nerved; several seed-bearing spikes, scattered along the stem; Lower California to Mendocino County ....*Phyllospadix torreyi* Wats.

   Leaves usually 2–4 mm. broad, elliptical in cross section, distinctly 3-nerved; seed-bearing spike, usually one (rarely two spikes), attached near the base of the stem; Santa Barbara to Vancouver Island, also Japan ....
   .... *Phyllospadix scouleri* Hook.

Since fruiting stages of these plants are seldom seen, at least by unobservant zoologists, we must rely on habitat and general appearance for field determination. The varieties of *Zostera* are of "doubtful distinctness," according to one botanist; only the variety *latifolia* is recognized, from Marin County, in Howell's flora, and is probably the common phase of the central California coast. Of the *Phyllospadix*, *P. scouleri* seems to prefer more exposed, wave-beaten conditions than *P. torreyi.*

Phylum "PROTOZOA"

Acellular, or single-celled, animals (and some groups that botanists claim as plants) are ordinarily too small for unaided eye observation, and have been for the most part omitted from the text. However, there are some large enough to be seen with a hand lens, and microscopes of one kind or another are becoming less expensive. A small sample of papers on various protozoans is offered below; there is no comprehensive reference, and probably never will be, with the possible exception of the Foraminifera, for which there is a large literature because oil companies have been convinced of the significance of the tests of these creatures as possible stratigraphic and environmental indicators.

Kozloff, Eugene N. 1961. A new genus and two new species of ancistrocomid ciliates (Holotricha: Thigmotricha) from sabellid polychaetes and from a chiton. J. Protozool., **8** (1): 60–63, 11 figs.

——— 1965. *Colligocincta furax* gen. nov., sp. nov., an ancistrocomid ciliate (Holotricha: Thigmotricha) from the sabellid polychaete *Laonome kröyeri* Malmgren. J. Protozool., **12** (3): 333–34, 5 figs.

——— 1966. *Phalacrocleptes verruciformis* gen. nov., sp. nov., an unciliated ciliate from the sabellid polychaete *Schizobranchia insignis* Bush. Biol. Bull., **130** (2): 202–10, 15 figs.

LYNCH, JAMES E. 1929. Studies on the ciliates from the intestine of *Strongylocentrotus,* I. Univ. Calif. Pub. Zool., **33**: 27–56 (*et seq.*).

MATSUDO, HITOSHI. 1966. A cytological study of a chonotrichous ciliate protozoan, *Lobochona prorates,* from the gribble. J. Morphol., **120** (4): 359–90, 9 pls.

MOHR, JOHN LUTHER. 1948. *Trichochona lecythoides,* a new genus and species of marine chonotrichous ciliate from California, with a consideration of the composition of the order Chonotricha Wallengren, 1895. Occ. Paps. Allan Hancock Found., **5**. 20 pp., 1 pl.

Ciliates attached to pleopods of amphipods.

——— 1966. On the age of the ciliate group Chonotricha, pp. 535–43 in *Some contemporary studies in marine science.* London: Allen and Unwin, 1 fig.

———, JEAN ANTHONY LEVEQUE, and HITOSHI MATSUDO. 1963. On a new collar ciliate of a gribble: *Lobochona proprates* n. sp. on *Limnoria tripunctata.* J. Protozool., **10** (2): 226–33, 11 figs.

MYERS, EARL H. 1940. Observations on the origin and fate of flagellated gametes in multiple tests of *Discorbis* (Foraminifera). J. Mar. Biol. Assoc., **24**: 201–26, 3 pls.

Beautifully illustrated paper on a group of benthic tide-pool Foraminifera well represented on the California coast. Species occurring at La Jolla, Monterey, and Moss Beach are discussed.

STEVENS, N. M. 1901. Studies on ciliate infusoria. Proc. Calif. Acad. Sci., Ser. 3, Zool., **3**: 1–42.

PICKARD, E. A. 1927. The neuromotor apparatus of *Boveria teredinidi* Nelson, a ciliate from the gills of *Teredo navalis.* Univ. Calif. Pub. Zool., **29**: 405–28.

NOBLE, A. E. 1929. Two new species of . . . *Ephelota* from Monterey Bay. Univ. Calif. Pub. Zool., **33**: 13–26.

ARNOLD, ZACH M. 1951. Occurrence of *Gromia oviformis* Dujardin in California (Testacea). Wasmann J. Biol., **9** (3): 351–53.

JEPPS, MARGARET W. 1936. Contributions to the study of *Gromia oviformis* Dujardin. Quart. J. Micr. Sci., n.s., **70**: 701–19, 3 pls.

This large (2–3 mm.) foram, mistaken by some for "fecal pellets," is common, especially among *Phyllospadix* holdfasts, from Pacific Grove to Dillon Beach and probably elsewhere. It is ivory-white and (for a protozoan) conspicuous.

Phylum (?) **MESOZOA**

Curious animals of uncertain position, possibly highly modified representatives of some larger group (flatworms, perhaps). The dicyemid mesozoans are parasites—or guests—in the kidneys of octopods; the orthonectids occur associated with various animals, and only recently have they been found on the Pacific coast.

KOZLOFF, EUGENE N. 1965. *Ciliocincta sabellariae* gen. et sp. n., an orthonectid mesozoan from the polychaete *Sabellaria cementarium* Moore. J. Parasit., **51** (1): 37–44, 18 figs.

MCCONNAUGHEY, BAYARD H. 1949. Mesozoa of the family Dicyemidae from California. Univ. Calif. Pub. Zool., **55** (1): 1–34, 8 pls.

——— 1957. Two new Mesozoa from the Pacific Northwest. J. Parasit., **43** (3): 358–64, 3 figs.

Species described from "*Octopus apollyon* or *O. hongkongensis*" from Friday Harbor and Cape Arago.

——— 1960. The rhombogen phase of *Dicyema sullivani* McConnaughey (Mesozoa: Dicyemidae). J. Parasit., **46** (5): 608–10, 1 pl.

———, and Evelyn McConnaughey. 1954. Strange life of the dicyemid mesozoans. Sci. Month., **79** (5): 227–84, 11 figs.

Popular account, with short bibliography.

Stunkard, H. W. 1954. The life-history and systematic relations of the Mesozoa. Quart. Rev. Biol., **29** (3): 230–44.

Phylum **PORIFERA.** Sponges

The systematics of sponges is a difficult matter, and things have not been simplified by the circumstance that perhaps more have been called to attempt the problem than should have been chosen; not the least of these at times was the late Dr. de Laubenfels. In any event, the paper by Bakus is the first significant contribution to Pacific coast sponge taxonomy in more than 30 years, and more work is obviously needed.

Bakus, G. J. 1966. Marine poeciloscleridan sponges of the San Juan Archipelago, Washington. J. Zool. Lond., **149**: 415–531, 52 figs.

Concerns 23 species from the Friday Harbor area; proposes four new genera and seven new species. Discusses zoogeography of Pacific coast species and reviews the development of poriferology on this coast.

Laubenfels, M. W. de. 1928. Interspecific grafting, using sponge cells. J. Elisha Mitchell Scient. Soc., **44**: 82–85.

——— 1932. The marine and fresh-water sponges of California. Proc. U.S. Nat. Mus., **81**: 1–140.

——— 1935. Some sponges of Lower California (Mexico). Amer. Mus. Novit., **799**. 14 pp.

——— 1936. A discussion of the sponge fauna of the Dry Tortugas . . . with material for a revision of the families and orders of the Porifera. Carnegie Institution of Washington, Pub. No. 467. 225 pp.

——— 1948. The order Keratosa of the phylum Porifera—a monographic study. Allan Hancock Pub., Occ. Paps., **3**. 217 pp., 30 pls., 31 text figs.

Concerns the bath sponges and their relatives; mostly tropical and warm-water forms.

Parker, G. H. 1914. On the strength and the volume of the water currents produced by sponges. J. Exp. Zool., **16**: 443–46.

Phylum **COELENTERATA.** Jellyfishes, Corals, etc.

Some zoologists prefer to classify the coelenterates together with the ctenophores as the phylum Cnidaria, and the recent confirmation that nematocysts occur in at least one ctenophore lends support to this arrangement. However, this is not a textbook of zoology, and we need not abandon categories of convenience. In any event, ctenophores (comb jellies) are not encountered alive on our shores, although moribund specimens of the sea walnut or sea gooseberry *Pleurobrachia* are sometimes encountered on sheltered shores of outer bays and sand spits, cast there by the waves.

MOORE, RAYMOND C., ed. 1953. Treatise on invertebrate paleontology, Part F: Coelenterata. Geol. Soc. Amer. and Univ. Kansas Press. 498 pp., 358 figs.

Excellent treatment of major features and classification of living as well as dead groups.

REES, W. J., ed. 1966. The Cnidaria and their evolution. Published for the Zoological Society of London by Academic Press. 449 pp., illus.

The proceedings of a symposium held in London with the idea of not only bringing workers up to date with each other, but also providing, in the published proceedings, a useful reference work for students. In essence, this is therefore a supplement to Hyman's treatment of coelenterates in her first volume.

UCHIDA, TOHRU. 1963. The phylogenetic lines of coelenterates from the viewpoint of symmetry. J. Fac. Sci. Hokk. Imp. Univ., Zool., **15** (2): 276–82, 4 figs.

Class HYDROZOA. Hydroids and Hydromedusae

For years the works of C. McLean Fraser have gathered dust on library shelves, unused and unusable, perhaps conveying by their bulk alone the impression that there is not much left to be done with hydroids. Yet the study of hydroids is now entering the critical stage of detailed work on complete life cycles, with identification of both polyp and medusa stages, and we are only at the beginning of what needs to be done. Fraser's work was dry and sterile; too often he used as illustrations of Pacific coast species the drawings made for his Atlantic coast monograph; and seldom, if ever, did he indicate that he even knew that many hydroids had medusa stages. Those with the need to know can find this work; citing it here would imply endorsement.

HAND, CADET. 1957. The systematic affinities and hosts of the one-tentacled, commensal hydroid *Monobrachium,* with new distributional records. J. Wash. Acad. Sci., **47** (3): 84–88, 2 figs.

———, and MEREDITH L. JONES. 1957. An example of reversal of polarity during asexual reproduction of a hydroid. Biol. Bull., **112** (3): 349–57, 5 figs.

HYMAN, L. H. 1947. Two new hydromedusae from the Californian coast. Trans. Amer. Micr. Soc., **66**: 262–68.

JHA, RAJ K. 1965. The nerve elements in silver-stained preparations of *Cordylophora.* Amer. Zool., **5**: 431–38, 5 figs.

REES, W. J. 1950. On *Cladonema myersi,* a new species of hydroid from the Californian coast. Proc. Zool. Soc. Lond., **119** (4): 861–65, 4 figs.

——— 1957. Evolutionary trends in the classification of capitate hydroids and medusae. Bull. Brit. Mus. (Nat. Hist.), Zool., **4** (9): 455–534, 56 figs., 2 pls.

RUSSELL, F. S. 1953. The Medusae of the British Isles: Anthomedusae, Leptomedusae, Limnomedusae, Trachymedusae, and Narcomedusae. Cambridge, Eng.: Cambridge Univ. Press. 530 pp., 319 text figs., 34 pls. (incl. col.).

A useful reference even on the Pacific coast.

STECHOW, E. 1923. Zur Kenntnis der Hydroiden Fauna des Mittelmeers, Amerikas, und andere Gebiete, II. Zool. Jahrb., Abt. f. Syst., **47**: 29–270.

Mentions only four Monterey and Vancouver forms, one of them new. A previous paper (Pt. I, 1919, *42*: 1–172) describes a new species of *Orthopyxis* (*Eucopella*) from Vancouver (p. 69).

UCHIDA, TOHRU. 1963. The systematic position of the Hydrozoa. Jap. J. Zool., **14** (1): 1–14, 5 figs.

Also in Dougherty, *et al., The lower metazoa, (q.v.).*

Order GYMNOBLASTEA

SKOGSBERG, TAGE. 1948. A systematic study of the family Polyorchidae (Hydromedusae). Proc. Calif. Acad. Sci., Ser. 4, **26** (5): 101–24, 2 figs.

HAND, CADET, and JOHN R. HENDRICKSON. 1950. A two-tentacled, commensal hydroid from California (Limnomedusae: *Proboscidactyla*). Biol. Bull., **99** (1): 74–87, 5 figs., 2 pls.

This quaint little creature lives around the rim of the tube of the sabellid *Pseudopotamilla ocellata,* and like a bureaucrat, dips its hands into the food grooves of its host. It occurs in the Monterey area. The young medusae are also described in this excellent paper. Two other species, including the northern one of the following reference, occur on this coast.

UCHIDA, TOHRU, and SHIRO OKUDA. 1941. The hydroid *Lar* and the medusa *Proboscidactyla.* J. Fac. Sci. Hokk. Imp. Univ., Zool., **7** (4): 431–40, 11 figs.

Describes hydroid and medusa stages of *Proboscidactyla flavicirrata,* the species in northern Japan; it also occurs in the Puget Sound region.

Order CALYPTOBLASTEA

C. urceolata Clark occurs abundantly on the open coast from Monterey Bay to Sitka, and *H. annulatum* Torrey is common at Monterey; both are epizoic on larger hydroids. Many other species are listed for the Pacific coast, but all are small; their differentiation requires careful microscopic examination, and no attempt will be made to consider them here, despite their abundance.

This is the commonest Monterey Bay *Obelia.* Formerly considered to be *O. gracilis,* it is found on bell buoys, on boat bottoms, and on stipes of *Macrocystis.*

STECHOW, E. 1926. Einige neue Hydroiden. Zool. Anz., **48**: 96–108.

Order HYDROCORALLINA. Hydrocorals

FISHER, W. K. 1938. Hydrocorals of the North Pacific Ocean. Proc. U.S. Nat. Mus., **84**: 493–554.

FISHER, W. K. 1931. California hydrocorals. Ann. Mag. Nat. Hist., Ser. 10, **8**: 391–99.

The Alaska hydrocoral that occurs littorally on the exposed shore near Sitka turns out to be *A. petrograpta* Fisher.

Order TRACHYMEDUSAE

RUGH, R. 1929. Egg-laying habits of *Gonionemus murbachi* in relation to light. Biol. Bull., **57**: 261–66.

Order CHONDROPHORA

EDWARDS, C. 1966. *Velella velella* (L.): The distribution of its dimorphic forms in the Atlantic Ocean and the Mediterranean, with comments on its nature and affinities. In *Some contemporary studies in marine science,* pp. 283–96. London, Allen and Unwin.

With complete bibliography.

Mackie, G. O. 1960. The structure of the nervous system in *Velella.* Quart. J. Micr. Sci., **101** (2): 119–31, 5 figs.
——— 1962. Factors affecting the distribution of *Velella* (Chondrophora). Internat. Rev. Ges. Hydrobiol., **47**: 26–32.

Class SCYPHOZOA. The Scyphomedusae, usually large jellyfish

Order Stauromedusae

Haliclystus stejnegeri Kishinouye .... §273

Gwilliam, G. F. 1960. Neuromuscular physiology of a sessile scyphozoan. Biol. Bull., **119** (3): 454–73, 10 figs.

Order Semaeostomeae

Aurellia aurita (Lamarck) .... §§193, 330
Chrysaora melanaster Brandt .... §193
Cyanaea capillata (Eschscholtz) .... §193

Hedgpeth, Joel W. 1948. Re-examination of "The adventure of the Lion's Mane." Baker Street Journal, **3** (3): 285–94. (Previously published in *Scientific Monthly,* March 1945.)
Welsh, John H. 1955. On the nature and action of coelenterate toxins. Deep-sea Res. Suppl., **3** (Bigelow volume): 287–97, 2 figs.

Scyphistoma of scyphozoan .... §228

Galigher, A. E. 1925. On the occurrence of the larval stages of Scyphozoa. Amer. Nat., **59**: 94–96.

Class ANTHOZOA

Subclass ALCYONARIA

Hickson, S. J. 1915. Some Alcyonaria and a Stylaster from the west coast of North America. Proc. Zool. Soc. Lond. **1915** (4): 541–57, 1 pl., 5 text figs.
Mostly dredged material from Oregon, British Columbia, and Alaska.
Kükenthal, W. 1913. Über die Alcyonarian Fauna Californiens. Zool. Jahrb., Abt. f. Syst., **35**: 219–70.
Corrects and revises the paper of Nutting cited below, which is reported to have incorrectly determined more than half of its species.
Nutting, C. C. 1909. Alcyonaria of the California coast. Proc. U.S. Nat. Mus., **25**: 681–727.

Order Stolonifera

Clavularia sp. .... §144
Clavularia sp. (formerly listed as *Telesto ambigua* Nutting) .... §144

Order Alcyonacea

Gersemia rubiformis (Pallas), Sea Strawberry .... §173

Order Pennatulacea

Acanthoptilum gracile (Gabb) .... §289

Known from shallow water in Tomales Bay (12–20 feet) and Monterey Bay (to 50 fathoms). In both localities it occurs with *Stylatula elongata.* The southern species, found near Corona del Mar, is *A. scalpelifolium* Moroff.

Leioptilus guerneyi (Gray) (as *L.* or *Ptilosarcus quadrangulare* in 1948 ed.) ....not treated

Reported to occur in the low intertidal from Puget Sound northward to Prince Rupert, B.C. The expanded animal is a translucent, pale orange color, and may attain a length of perhaps 18 inches. Undoubtedly an ideal experimental animal, to judge from the following on a similar species:

MORI, SYUITI. 1960. Influence of environmental and physiological factors on the daily rhythmic activity of a sea pen. Cold Spr. Harb. Symp. Quant. Biol., **25**: 333–44, 16 figs.

Renilla köllikeri Pfeffer, Sea Pansy, formerly considered to be the more southern *R. amethystina* Verrill ....§248

PFEFFER, GEORG. 1886. Neue Pennatuliden des Hamburger Naturhistorischen Museum. Jahrb. Wist. Aust. Ham., III: Beilage Jahresb. Naturh. Mus. Ham., **1885**: 53–61.

This species ranges from Wilmington to Cedros Island.

Stylatula elongata (Gabb), Slender Sea Pen..........................§289

The known range of this species is from Tomales Bay to San Diego (possibly farther south) from low tide to 35 fathoms. It is abundant in Tomales Bay.

Virgularia bromleyi Kolliker and *V. galapagensis* Hickson are reported by Deichmann (*in litt.*) as occurring in the vicinity of Corona del Mar in shallow water (less than 12 fathoms); two species of *Balticina, B. californica* (Moroff) and *B. septentrionalis* (Gray), are shallow-water forms (20 fathoms to deep water); *septentrionalis* possibly occurs at lesser depths in Burrows Inlet near Vancouver, B.C.

Order GORGONACEA

Muricea californica Aurivillius....................................§218

AURIVILLIUS, MAGNUS. 1931. The Gorgonarians from Dr. Sixten Bock's expedition to Japan and the Bonin Islands, 1914. K. Sven. Vet., Ser. 3, **9** (4): 1–337.

Subclass ZOANTHARIA. Anemones and Corals

STEPHENSON, T. A. The British sea anemones. London: Ray Society. Vol. I (1928), 148 pp., 14 pls. Vol. II (1935), 426 pp., 19 pls.

The *vade mecum* for anemone students. Several of the forms considered also occur on the Pacific coast. Excellent color plates.

TORREY, H. B. 1902. Anemones: Papers from the Harriman Alaska expedition, XXX. Proc. Wash. Acad. Sci., **4**: 373–410.

There is no up-to-date general account of Pacific shore Zoantharia, a most obvious need. There are recent considerations of Alaska and California forms in the following.

CARLGREN, OSKAR. 1931. Zur Kenntnis der *Actiniaria Abasilaria.* Arkiv. Zool., K. Sven. Vet., **23a** (3): 1–48 (esp. pp. 22, 33, and 39).

——— 1934. Zur Revision der Actiniarien. Arkiv. Zool., K. Sven. Vet., **26a** (18): 1–36 (p. 16, *Evactis artemisia* from Wrangell, Alaska).

——— 1934. Some Actiniaria from Bering Sea and Arctic waters. J. Wash. Acad. Sci., **24** (8): 348–53.

——— 1936. Some West American sea anemones. J. Wash. Acad. Sci., **26** (1): 16–23.

Corynactis californica and two other new species, dredged from Monterey Bay.

——— 1949. A survey of the Ptychodactiaria, Corallimorpharia, and Actiniaria. K. Sven. Vet., (4) **1** (1). 121 pp., 4 pls.

Diagnoses and keys to the genera and families of anemones; an indispensable reference, but requiring a specialist's knowledge to use. Lists most of the described species.

HAND, CADET. 1955. The sea anemones of central California, Part I: The corallimorpharian and athenarian anemones. Wasmann J. Biol., **12** (3): 347–75, 7 figs. (Journal dated "Fall 1954" but pub. Jan. 7, 1955.)

——— 1955. The sea anemones of central California, Part II: The endomyarian and mesomyarian anemones. Wasmann J. Biol., **13** (1): 37–99, 17 figs. (Journal dated "Spring 1955" but pub. Oct. 3, 1955.)

——— 1956. The sea anemones of central California, Part III: The acontiarian anemones. Wasmann J. Biol., **13** (2): 189–251, 13 figs. (Journal dated "Fall 1955" but pub. Feb. 14, 1956.)

——— 1957. Another sea anemone from California, and the types of certain California anemones. J. Wash. Acad. Sci., **47** (12): 411–14.

Nematostella vectensis, a small burrowing anemone in shallow water of *Salicornia* flats in the San Francisco and Tomales Bay areas.

———, and RALPH BUSHNELL. 1967. A new species of burrowing acontiate anemones from California (Isophelliidae: Flosmaris). Proc. U.S. Nat. Mus., **120** (3554): 1–8, 2 figs.

MARTIN, EDGAR J. 1963. Toxicity of dialyzed extracts of some California anemones (Coelenterata). Pac. Sci., **17** (3): 302–4.

PANTIN, C. F. A. 1964. A problem in muscular excitation in the Anthozoa. In *Essays on physiological evolution*, pp. 224–32, 6 figs. Pergamon Press.

——— 1965. Capabilities of the coelenterate behavior machine. Amer. Zool., **5**: 581–89, 4 figs.

Order CERIANTHARIA

TORREY, H. B., and F. L. KLEEBERGER. 1909. Three species of *Cerianthus* from southern California. Univ. Calif. Pub. Zool., **6** (5): 115–25, 4 text figs.

Order ACTINIARIA

FORD, CHARLES E. 1964. Reproduction in the aggregating sea anemone, *Anthopleura elegantissima*. Pac. Sci., **18** (2): 138–45, 3 figs.

POWELL, DAVID C. 1964. Fluorescence in the sea anemone *Anthopleura artemisia*. Bull. Amer. Littoral Soc., **2** (3): 17.

Quite a few invertebrates, especially coelenterates, fluoresce under ultraviolet. They are very pretty.

MCMURRICH, J. P. 1913. A new species of *Edwardsiella* from southern California. Proc. U.S. Nat. Mus., **44**: 551–53.

This widespread anemone has been variously called *Sagartia, Diadumene*, or *Aiptasiomorpha luciae* (Carlgren prefers the last name). Although its occurrence at such places as Puget Sound, Hokkaido, and Woods Hole suggests a cold-temperate distribution pattern, records from such localities as the Suez Canal and Port Aransas, Texas, suggest that it is a eurythermal species as well as a euryhaline one.

HAUSMAN, LEON AUGUSTUS. 1919. The orange-striped anemone (*Sagartia luciae* Verrill): An ecological study. Biol. Bull., **37**: 363–71.

TORREY, H. B. 1904. On the habits and reactions of *Sagartia davisi*. Biol. Bull., **6**: 203–16.

From San Pedro and San Diego bays; its occurrence here in the 1900's suggests other means of dispersal than traveling with oysters.

UCHIDA, TOHRU. 1932. Occurrence in Japan of *Diadumene luciae,* a remarkable actinian of rapid dispersal. J. Fac. Sci. Hokk. Imp. Univ., (6) **2** (2): 69–82, 4 figs., 1 pl.

Suggests that the original home of this widespread anemone is the coast of Japan and Asia, where sexually reproducing phases are found. Elsewhere, reproduction is usually by budding.

Harenactis attenuata Torrey . . . . §257

Metridium senile (L.), (*M. dianthus, marginatum*) . . . . §321

BATHAM, E. J., and C. F. A. PANTIN. 1950. Muscular and hydrostatic action in the sea anemone *Metridium senile* (L.). J. Exp. Biol., **27** (3&4): 264–89, 9 figs., 1 pl. Inherent activity in . . . . *Ibid.,* 290–301, 6 figs. Phases of activity in . . . and their relation to external stimuli. *Ibid.,* 337–99, 11 figs.

——— 1951. The organization of the muscular system of *Metridium senile.* Quart. J. Micr. Sci., **92** (1): 27–54, 12 figs., 2 pls.

PANTIN, C. F. A. 1950. Behavior patterns in lower invertebrates. Symp. Soc. Exp. Biol., **4**: 175–95, 3 figs. (New York: Academic Press).

WESTFALL, JANE A. 1965. Nematocysts of the sea anemone *Metridium.* Amer. Zool., **5** (3): 377–93, 8 pls.

——— 1966. The differentiation of nematocysts and associated structures in the Cnidaria. Zeitschr. Zellf., **75**: 381–403, 20 figs.

Electron microscope studies; the more we look at these things, the more complicated they appear to be.

Tealia crassicornis (Mueller) . . . . § 65

T. lofotensis (Danielssen) . . . . § 65

Order MADREPORARIA. Stony Corals

DURHAM, J. WYATT. 1947. Corals from the Gulf of California and the North Pacific coast of America. Geol. Soc. Amer. Mem., **20**. 68 pp., 14 pls., 2 figs.

Well-illustrated monograph; based on hard parts.

———, and J. LAURENS BARNARD. 1952. Stony corals of the eastern Pacific collected by the *Velero III* and *Velero IV*. Allan Hancock Pac. Exp., **16** (1): 1–110, 16 pls.

VAUGHAN, T. W., and J. W. WELLS. 1943. Revision of the suborders, families, and genera of the Scleractinia. Geol. Soc. Amer. Spec. Pap., **44**. 363 pp., 51 pls., 39 figs.

Astrangia lajollaensis Durham . . . . § 39

BOSCHMA, H. 1925. On the feeding reactions . . . of *Astrangia.* Biol. Bull., **49**: 407–39.

Balanophyllia elegans Verrill . . . . § 39

YONGE, C. M. 1932. A note on *Balanophyllia regia.* J. Mar. Biol. Assoc., **18**: 219–24.

Caryophyllia alaskensis Vaughan . . . . § 39

Phylum **CTENOPHORA.** Comb Jellies

Class TENTACULATA

Pleurobrachia bachei (A. Agassiz) . . . . §193

Phylum **PLATYHELMINTHES**

Class TURBELLARIA. Flatworms, Planarians

The recent revival of the pleasant parlor game of theoretical phylogeny has stimulated research in turbellarians, since they suggest a possible or plausible line of evolution from ciliates. Many of them are very small, and must be observed alive as well as studied in preparations. We have hardly begun an inventory of our fauna, as evidenced by Karling's bag of about 80 species, mostly undescribed, during a short stay in California. The classification of the free living forms is in a state of flux.

AX, PETER. 1963. Relationships and phylogeny of the Turbellaria. In Dougherty, *et al., The lower metazoa,* pp. 191–224, 16 figs.

Cited here as an example of recent work. Ax is a prolific contributor to that specialized branch of study of small creatures of uncertain affinities familiarly known in Germany as "Ramenetierchen."

KARLING, TOR G. Marine Turbellaria from the Pacific coast of North America. I: Plagiostomidae, Arkiv. f. Zool., (2) **15** (6): 113–41, 47 figs. (1962). II: Pseudostomidae and Sylindrostomidae, *ibid.,* (10): 181–209, 44 figs. (1962). III: Otopanidae, *ibid.,* **16** (26): 527–41, 38 figs. (1964).

Order POLYCLADIDA. Polyclad Worms

BOCK, SIXTEN. 1925. Planarians, Pt. IV. No. 28 in Papers from Dr. Th. Mortensen's Pacific expedition, 1914–16. Vidensk. Medd. fra Dansk Naturh. Foren., **79**: 97–184.

Describes *Kaburakia excelsa* sp. nov., from Nanaimo littoral, on p. 132.

BOONE, E. S. 1929. Five new polyclads from the California coast. Ann. Mag. Nat. Hist., (10) **3**: 33–46.

FREEMAN, DANIEL. 1930. Three polyclads from the region of Pt. Fermin, San Pedro, Calif. Trans. Amer. Micr. Soc., **49**: 334–41.

Described as *P. molle,* the southern polyclad mentioned by Johnson and Snook as a new species, probably of *Prosthiostomum.*

——— 1933. The polyclads of the San Juan region of Puget Sound. Trans Amer. Micr. Soc., **52**: 107–46.

Describes 12 new species, most of them littoral and several of them common.

HYMAN, LIBBIE H. 1953. The polyclad flatworms of the Pacific coast of North America. Bull. Amer. Mus. Nat. Hist., **100** (2): 265–392, 161 figs.

——— 1955. The polyclad flatworms of the Pacific coast of North America: additions and corrections. Amer. Mus. Novit., no. 1704, 11 pp., 8 figs.

——— 1959. Some Turbellaria from the coast of California. *Ibid.,* no. 1943, 17 pp., 19 figs.

Much critical work remains to be done with Pacific coast polyclads; there are many undescribed species, and life histories are unknown:

Order ACOELA

KOZLOFF, EUGENE N. 1965. New species of acoel turbellarians from the Pacific coast. Biol. Bull., **129** (1): 151–66, 30 figs.

Parotocelis luteola gen. nov., sp. nov.; *Raphidophallus actuosus* gen. nov., sp. nov.; and *Diatomovora amoena* gen. nov., sp. nov. All from San Juan Island, intertidally; the last species also collected at Charleston, Oregon.

ARMITAGE, K. B. 1961. Studies of the biology of *Polychoerus carmelensis* (Turbellaria: Acoela). Pac. Sci., **15**: 203–10.

COSTELLO, H. M., and D. P. COSTELLO. 1938. A new species of *Polychoerus* from the Pacific coast. Ann. Mag. Nat. Hist., (11) **1**: 148–55.

GARDINER, E. G. 1895. Early development of *Polychoerus caudatus* Mark. J. Morphol., **11**: 155–71.

SCHWAB, ROBERT G. 1967. Overt responses of *Polychoerus carmelensis* (Turbellaria: Acoela) to abrupt changes in ambient water temperature. Pac. Sci., **21** (1): 85–90.

The beast moves more rapidly up to 17°C. as temperature is increased from 5°, but slows down when it gets warmer. The author concludes that temperatures below 5° and above 29° "are not suitable." Apparently it seeks refuge below rocks and gravel, "should the tide pool environment warrant such behavior," where the temperature conditions are yet to be adequately investigated.

Order TRICLADIDA

HOLLEMAN, JOHN T., and CADET HAND. 1962. A new species, genus, and family of marine flatworms (Turbellaria: Tricladida, Maricola) commensal with mollusks. Veliger, **5** (1): 20–22, 3 figs.

Describes *Nexilis epichitonis*, about 3 mm. long, occurring on mantles of *Mopalia hindsi* and *M. muscosa*; also found on *Thais emarginata*. A very strange beast in its internal anatomy. This is the second triclad described from California (the other is *Procerodes pacifica* Hyman from San Diego); it has also been found in Oregon.

Order RHABDOCOELIDA

LEHMAN, H. E. 1946. A histological study of *Syndisyrinx franciscanus* gen. et sp. nov., an endoparasitic rhabdocoel of the sea urchin *Strongylocentrotus franciscanus*. Biol. Bull., **91** (3): 295–311, 8 figs.

Phylum **NEMERTEA.** Ribbon Worms

COE, WESLEY R. 1904. Nemerteans. Harriman Alaska Series, **11**. 220 pp.

——— 1905. Nemerteans of the west and northwest coasts of America. Bull. Mus. Comp. Zool., **68**. 318 pp.

——— 1930. Asexual reproduction in nemerteans. Physiol. Zool., **3**: 297–308.

——— 1938. A new genus and species of Hoplonemertea having differential bipolar sexuality. Zool. Anz., **124**: 220–24.

——— 1940. Revision of the nemertean fauna of the Pacific coasts of North, Central, and northern South America. Allan Hancock Pac. Exp., **2** (13): 247–323, 8 pls. (2 col.).

——— 1944. Geographical distribution of the nemerteans of the Pacific coast of North America, with descriptions of two new species. J. Wash. Acad. Sci., **34** (1): 27–32.

CORRÊA, DIVA DINIZ. 1964. Nemerteans from California and Oregon. Proc. Calif. Acad. Sci., (4) **31** (19): 515–58.

No pictures; no new species; range extensions, clarifications of descriptions, and keys to *Paranemertes* (Pacific coast species) and *Malacobdella* (all species). Does this mean that our nemertean fauna is well known and no new surprises are to be ex-expected? The editor's skepticism is aroused.

Amphiporus bimaculatus Coe. . . . . § 48

Carinella rubra Griffin. . . . . §239

Cephalothrix major Coe. . . . . §139

COE, WESLEY R. 1930. Two new species of nemerteans belonging to the family Cephalotrichidae. Zool. Anz., **89**: 97–103.

Cerebratulus montgomeryi Coe. . . . . §239

Cerebratulus sp. (of Elkhorn Slough). . . . . §312

WILSON, C. B. 1900. The habits and early development of *Cerebratulus lacteus* (Verrill). Quart. J. Micr. Soc., n.s., **43**: 97–198.

Emplectonema gracile (Johnston). . . . . §165

Lineus vegetus Coe. . . . . §139

COE, WESLEY R. 1929. Regeneration in nemerteans. J. Exp. Zool., **54**: 411–60.

——— 1931. A new species of nemertean (*Lineus vegetus*). Zool. Anz., **94**: 54–60.

Malacobdella grossa (Müller). . . . . §189

GUBERLET, J. E. 1925. *Malacobdella grossa* from the Pacific coast of North America. Pub. Puget Sd. Biol. Sta., **5**: 1–14.

Micrura sp. (of Elkhorn Slough). . . . . §312

M. verrilli Coe. . . . . §139

Paranemertes peregrina Coe. . . . . §139

Tubulanus polymorphus Reneir. . . . . §139

Phylum **BRYOZOA (POLYZOA)**

BASSLER, RAY S. 1953. Treatise on invertebrate paleontology, Part G: Bryozoa. Geol. Soc. Amer. and Univ. Kansas Press. 253 pp., 175 pls.

The most recent general summary outside of comprehensive zoological treatises; strongly necrological.

LAAGAAIJ, R. 1963. *Cupuladria canariensis* (Busk)—portrait of a bryozoan. Palentology (London), **6** (1): 172–217, 21 figs., 2 pls.

Cited here as a model paper; a synthesis of the present knowledge of the living creature with its fossil record.

OSBURN, RAYMOND C. Bryozoa of the Pacific coast of America. Part I: Cheilostomata–Anasca, Allan Hancock Pac. Exp., **14** (1): 1–269, 29 pls. (1950). Part II: Cheilostomata–Ascophora. *Ibid.*, **14** (2): 271–611, 34 pls. (1952). Part III: Cyclostomata, Ctenostomata, Entoprocta, and addenda. *Ibid.*, **14** (3): 613–841, 18 pls. (1953).

POWELL, N. A. 1967. Sexual dwarfism in *Cribrilina annulata* (Cribrilinidae: Bryozoa). J. Fish. Res. Bd. Canad., **24** (9): 1905–10, 2 figs.

ROBERTSON, ALICE. 1902. Some observations on *Ascorhiza occidentalis* Fewkes, and related Alcyonidia. Proc. Calif. Acad. Sci., (3) **3** (3): 99–108, 1 pl.

——— 1903. Embryology and embryonic fission in the genus *Crisia*. Univ. Calif. Pub. Zool., **1** (3): 115–56, 4 pls.

GRAVE, B. H. 1930. The natural history of *Bugula flabellata.* J. Morphol. Physiol., **49**: 355–84.

O'DONOGHUE, C. H. 1926. Observations on the early development of *Membranipora villosa* Hincks. Contr. Canad. Biol. and Fish., n.s., **3**: 249–63.

Phylum **ENTOPROCTA**

Although systematic reports on bryozoans usually include these animals, they are very different in structure and organization. The species are included in the third part of Osburn's monograph (*see under* Bryozoa).

ROBERTSON, ALICE. 1900. Studies in Pacific coast Entoprocta. Proc. Calif. Acad. Sci., Ser. 3, **2** (4): 323–48, 1 pl.

MARSICAL, RICHARD N. 1965. The adult and larval morphology and life history of the entoproct *Barentsia gracilis* (M. Sars, 1835). J. Morphol., **116** (3): 311–38, 12 pls.
Variations with age and station in this population in San Francisco Bay suggest that several species attributed to this genus may actually be one variable species; good description of morphology.

Phylum **BRACHIOPODA**

DALL, W. H. 1920. Annotated list of the recent Brachiopoda in the collection of the United States National Museum. Proc. U.S. Nat. Mus., **57**: 261–377.
See also the Smith and Gordon (1948) reference under Mollusca.

MOORE, RAYMOND C., ed. 1965. Treatise on invertebrate paleontology, Part H: Brachiopoda. Geol. Soc. Amer. and Univ. Kansas Press. 2 vols., 927 pp., illus.
The fact that this work is in two volumes is adequate testimony to the greater significance of these animals in the past; this book is essentially a necrology, although most of the first 138 pages (by Alwyn Williams and A. J. Rowell) concerns the anatomy of extant forms.

Glottidia albida (Hinds), Tongue "Clam" . . . . §291

CRAIG, GORDON Y. 1952. A comparative study of the ecology and palaeoecology of *Lingula*. Trans. Edin. Geol. Soc., **15**: 110–20, 2 figs., 1 pl.

Also an account of *Glottidia*, in danger of being buried in this journal.

JONES, GILBERT F., and J. LAURENS BARNARD. 1963. The distribution and abundance of the inarticulate brachiopod *Glottidia albida* (Hinds) on the mainland shelf of southern California. Pac. Nat., **4** (2): 27–52, 14 figs.

Terebratalia transversa (Sowerby), Lamp Shell . . . . §227

Phylum PHORONIDA

MARSDEN, JOAN RATTENBURY. 1957. Regeneration in *Phoronis vancouverensis*. J. Morphol., **101** (2): 307–24, 8 figs.

——— 1959. Phoronidea from the Pacific coast of North America. Canad. J. Zool., **37**: 87–111, 31 figs.

What we have been calling *Phoronopsis viridis* should be *Phoronopsis harmeri*. In all, six species of phoronids are recognized from the west coast of North America.

Phoronis vancouverensis Pixel . . . . §229

Phoronis sp. (the orange-plumed form of Newport Bay) . . . . §297

Phoronopsis harmeri Pixell, or *P. viridis* Hilton . . . . §297

JOHNSON, RALPH G. 1959. Spatial distribution of *Phoronopsis viridis* Hilton. Science, **129** (3357): 1221.

ZIMMER, RUSSEL L. 1967. The morphology and function of accessory reproductive glands in the lophophores of *Phoronis vancouverensis* and *Phoronopsis harmeri*. J. Morphol., **121** (2): 159–78, 14 figs.

Phylum NEMATODA

Continued recognition of this group as a phylum may not be compatible with some of the better opinion on invertebrates (see Hyman, Vol. III). However, they are significant but too often ignored members of the intertidal complex; some of the largest free-living nematodes are found on the shore. Their systematics is difficult, and only recently has any work been done with them.

CHITWOOD, B. G. 1960. A preliminary contribution on the marine nemas (Adenophorea) of northern California. Trans. Amer. Micr. Soc., **79** (4): 347–84, 4 pls.

INGLIS, WILLIAM G. 1964. The marine Enoplida (Nematoda): A comparative study of the head. Bull. Brit. Mus. (Nat. Hist.), Zoology, **11** (4): 263–376, 194 figs.

Phylum ANNELIDA

Subclass ARCHIANNELIDA

HILTON, W. A. 1922. The occurrence of *Polygordius* adult at Laguna Beach. Pomona J. Entom. and Zool., **14**, No. 4.

Very active, like an attenuated roundworm, but with characteristic antennae at head region. In sand with amphioxus.

WIESER, WOLFGANG. 1957. Archiannelids from the intertidal of Puget Sound. Trans. Amer. Micr. Soc., **76** (3): 275–85, 4 figs.

Describes two new species and mentions others; provides keys for the species of *Protodrilus* and *Nerilla* discovered up to that date.

Class OLIGOCHAETA

ALTMAN, LUTHER C. 1931. *Enchytraeus pugetensis* (n. sp.), a new marine enchytraeid from Puget Sound. Trans. Amer. Micr. Soc., **50**: 154–63.
Includes natural history notes.

Class POLYCHAETA

For better or worse—depending on one's inclination, interest, and ability—the polychaetes, or bristle worms, are among the most numerous and significant animals of the sea. Some are pelagic, living on the high seas, and a few are in the great deeps. There are many in the shallow sandy and muddy bottoms of the seas everywhere, and on intertidal mud flats there may be dozens of species. Since it takes a bit of hard work, aided by literature at times obscurely couched in a rather special terminology, to identify polychaetes, they are considered difficult subjects for study. But as Olga Hartman says, "We must look at the setae." If a diligent observer of the life of mud flats keeps at it for a while, he may find perhaps a hundred species on an acre or so of flats. Many of them are beautifully iridescent.

Thousands of papers have been published about polychaete worms, but surprisingly few books. Indeed, there is only one book in English devoted for the most part to these worms, the little book by Dales in the Hutchinson University Library (R. Phillips Dales, *Annelids*. London, 1963. 200 pp., 19 figs.). We wish it were longer and had more pictures.

Olga Hartman, doyenne of Pacific coast polychaetology, has done yeoman service in trying to keep us up to date with the literature. Her *Catalogue of the polychaetous annelids of the world* is a basic guide to the systematic literature. Parts I and II of this appeared in 1959 (Occasional Papers of the Allan Hancock Foundation, No. 23, Parts I and II) and the supplement and index appeared as Part III (197 pp.) at the end of 1965. Her summary of the literature, in which many topics are indexed so as to provide a key to the unwieldy mass of papers on polychaetes, appeared in 1951 and included more than 1,300 authors and 4,000 titles (*The literature of the polychaetous annelids, Part I: Bibliography and subject analysis.* Los Angeles: privately printed, 290 pp.). This work, of course, is now somewhat out of date, since the flood of papers has not subsided; indeed, it seems to increase as physiologists and biochemists find more things to look into; and of course there seems to be no end to the making of new species. The authors most concerned with systematics of northeastern Pacific polychaetes include: E. and C. Berkeley, K. J. Bush, R. V. Chamberlin, C. Essenberg, J. E. Guberlet, O. Hartman, H. P. Johnson, M. W. Johnson, G. E. and N. MacGinitie, J. P. Moore, T. Skogsberg, and A. L. Treadwell. Of first importance to Pacific coast workers are the many revisionary papers by Dr. Hartman, published principally in the "Allan Hancock Pacific Expeditions" series from 1939 on. Now, as we go to press, the first volume of her comprehensive key to California species has appeared: *Atlas of the errantiate polychaetous annelids from California* (Los Angeles: Allan Hancock Foundation, 1968. 828 pp., illus.).

In addition, monographs or manuals of major importance include: P. Fauvel, "Polychètes errantes" (1923) and "Polychètes sedentaires" (1927) in the *Faune de France,* Volumes 5 and 16; and W. C. McIntosh, "Monograph of the British Marine Annelids," Ray Society Monographs (1900–1924). There is also the recent manual by Berkeley and Berkeley: *Canadian Pacific fauna,* **9**, "Annelida"; **9b** (2), "Polychaeta sedentaria." Univ. Toronto Press, 139 pp., 292 figs. (1952).

There is a large literature in Russian, much of it essential. One major work has been translated, and copies should be at all Pacific Coast marine laboratories: *Polychaeta of the Far Eastern seas of the U.S.S.R.*, by Pavel V. Ushakov; Israel Program for Scientific Publications, published for the Smithsonian Institution and the National Science Foundation (Office of Technical Services, United States Department of Commerce—if available at all). Translation published in 1965; Russian original, 1955.

Most of the above sources are systematic; they will tell the reader, if he can master the jargon of polychaetology, what species he may or may not have. Worms are also animals, and rather fascinating ones, and it is possible to base a scholarly career in academic zoology upon them, as the papers below should attest:

ÅKESSON, B. 1967. The embryology of the polychaete *Eunice kobiensis*. Acta Zool., **48**: 141–92, 22 figs.

There is a large literature on polychaete anatomy and development, much of it of broad interest to zoologists and developmental biologists, and perhaps even to molecular biologists; this paper, based on studies of a species at Friday Harbor, is a good example.

CLARK, R. B. 1965. The integrative action of a worm's brain. Symp. Soc. Exp. Biol. **20** (Nervous and hormonal mechanisms of integration): 345–79, 17 figs.

——— 1967. Zoology: The study of animals. Inaugural lecture, University of Newcastle-upon-Tyne, Monday, November 21, 1966. 18 pp., frontispiece.

DAVENPORT, DEMOREST. 1950. Studies in the physiology of commensalism, I: The polynoid genus *Arctonoë*. Biol. Bull., **98** (2): 81–93, 2 figs.

———, and JOHN F. HICKOK. 1951. Studies in the physiology of commensalism, II: The polynoid genera *Arctonoë* and *Halosydna*. Biol. Bull., **100** (2): 71–83.

By the use of an ingenious arrangement of tubing, leading to tanks with and without the echinoderm hosts, the authors found that the worms will more often than not travel toward their hosts, provided the hosts are uninjured. They conclude that "a rather tenuous bond exists between host and commensal," and have not determined the source of the chemical agent involved; preparations of the echinoderm test or viscera fail to attract the worms. Worms are attracted for only a very short period to a tank from which the host has been removed. These experiments indicate that the attractive agent—or agents—is very unstable, and that it may be masked by other substances released when the host is injured or disturbed.

HEALY, EUGENE A., and G. P. WELLS. 1959. Three new lugworms (Arenicolidae: Polychaeta) from the North Pacific area. Proc. Zool. Soc. Lond., **133** (2): 315–35, 3 figs., 4 pls.

Three species are recognized for our area: *Abarenicola pacifica* from California to Japan; *A. vagabunda*, ssp. *vagabunda*, restricted to San Juan Islands region; *A. v.*, ssp. *oceanica*, from California through Alaska to Japan but not to the San Juan Islands.

WELLS, G. P. 1945. The mode of life of *Arenicola marina*. J. Mar. Biol. Assoc., **26**: 170–207, 10 figs.

——— 1962. The warm-water lugworms of the world (Arenicolidae: Polychaeta). Proc. Zool. Soc. Lond., **138** (3): 331–53, 4 pls., 3 figs.

The California species are *Arenicola cristata* and *A. brasiliensis*. "It would be most interesting to have full ecological information for the various forms to be found on the Californian coast."

Hesperonoë adventor (Skogsberg), as *Harmothoë adventor*. . . . .§306

FISHER, W. K. 1946. Echiuroid worms of the Pacific Ocean. Proc. U.S. Nat. Mus., **96**: 277–78 and Pl. 37.

Gives an account of the relations of this worm with its host and other commensals.

H. complanata (Johnson), as *Harmothoë complanata*. . . . .§292

Hesperonoë sp., commensal with *Upogebia*. . . . .§313

Hololepidella tuta (Grube), as *Polynoë tuta*. . . . .§234

Hydroides norvegica (Gunnerus) . . . . .not treated

REISH, DONALD J. 1961. The relationship of temperature and dissolved oxygen to the seasonal settlement of the polychaetous annelid *Hydroides norvegica* (Gunnerus). Bull. S. Calif. Acad. Sci., **60** (1): 1–11, 4 pls.

This warm-water species is found in southern California harbors, where it encrusts piling and other objects and may foul yacht rudders and propellers with masses of calcareous tubes.

Lubrineris erecta (Moore), as *Lumbriconereis erecta*. . . . .§153

L. zonata (Johnson), as *Lumbriconereis zonata*. . . . .§293

Marphysa stylobranchiata Moore. . . . .§129

Mesochaetopterus taylori Potts. . . . .§265

MacGinitie and MacGinitie (1949), in *Natural history of marine animals*, p. 205, have described the methods of feeding in a related species, *M. rickettsi*.

Nainereis dendritica (Kinberg), formerly *N. laevigata*. . . . .§191

Nephtys caecoides Hartman and *N. californiensis* Hartman. . . . .§265

BANKS, RICHARD C. 1962. Observations on the polychaete genus *Nephtys* near Bolinas, California (Annelida: Nephtyidae). Wasmann J. Biol., **20** (1): 107–14.

CLARK, R. B. 1962. Observations on the food of *Nephtys*. Limnol. and Oceanogr., **7** (3): 380–85.

Most species are carnivores; one in Long Island Sound, however, must be a facultative detritus feeder.

———, and E. C. HADERLIE. 1962. The distribution of *Nephtys californiensis* and *N. caecoides* on the Californian coast. J. Anim. Ecol., **31**: 339–57, 7 figs.

The second of these coincidental papers is a more comprehensive and careful treatment, with broader geographic data and sediment analyses, although the conclusions are the same. It appears that while the professionals were taking care to get all the information together, a student rushed to press with his summer project. Such things happen.

CLARK, MARY E. (Mrs. R. B.). 1964. Biochemical studies on the coelomic fluid of *Nephtys hombergi* (Polychaeta: Nephtyidae), with observations on changes during different physiological states. Biol. Bull., **127** (1): 63–84, 4 figs.

Nereis (Neanthes) brandti (Malmgren) . . . .§164

N. (Neanthes) succinea Frey and Leuckart. . . . . .not treated

N. (Neanthes) virens Sars. . . . .§311

N. limnicola Johnson (*Neanthes lighti* Hartman) . . . . .not treated

JONES, MEREDITH L. 1967. On the morphology of the nephridia of *Nereis limnicola* Johnson. Biol. Bull. **132** (3): 362–80, 17 figs.

SMITH, RALPH I. 1950. Embryonic development in the viviparous nereid polychaete *Neanthes lighti* Hartman. J. Morphol., **87** (3): 417–65, 38 figs.

——— 1953. The distribution of the polychaete *Neanthes lighti* in the Salinas River estuary, California, in relation to salinity, 1948–52. Biol. Bull., **105** (2): 335–47, 2 figs.

——— 1959. The synonymy of the viviparous polychaete *Neanthes lighti* Hartman (1938) with *Nereis limnicola* Johnson (1903). Pac. Sci., **13** (October): 349–50.

Phylum **SIPUNCULIDA.** Sipunculid Worms

BANG, FREDERIK B., and BETSY G. BANG. 1962. Studies on sipunculid blood: Immunologic properties of coelomic fluid and morphology of "urn cells." Cahiers Biol. Mar., **3**: 363–74, 7 figs., 1 pl.

FISHER, W. K. 1952. The sipunculid worms of California and Baja California. Proc. U.S. Nat. Mus., **102**: 371–450, 1 fig., 22 pls.

Essential for identification of the species in this region.

RUBIN, HARVEY, and FREDERIK B. BANG. 1964. *In vitro* studies of antibacterial activity of *Golfingia gouldi* (Pourtales) coelomic fluid. J. Insect Pathol. **6**: 457–65.

Though the paper may not be of world-shaking importance, we feel it necessary to call attention to the circumstance that papers on sipunculids may appear in an entomological journal.

FISHER, W. K. 1928. New Sipunculoidea from California. Ann. Mag. Nat. Hist., Ser. 10, **1**: 194–99.

PEEBLES, FLORENCE, and DENIS L. FOX. 1933. The structure, functions, and general reactions of the marine sipunculid worm *Dendrostoma zostericola*. Bull. Scripps Inst. Oceanogr., Tech. Ser., **3**: 201–24.

FISHER, W. K. 1950. The sipunculid genus *Phascolosoma*. Ann. Mag. Nat. Hist., (12) **3** (30): 547–52.

Phascolosoma, as recently used, is erroneous; this name properly belongs to the animals referred to under *Physcosoma*. For the animals known as *Phascolosoma*, the available name is Lankester's *Golfingia*, in honor of an afternoon on the St. Andrews links with Prof. McIntosh! This does not mean, as a bystander might suppose, that zoologists cannct make up their minds about the name of an animal, but that some people have not played according to the rules.

GEROULD, J. H. 1906. The development of *Phascolosoma*. Zool. Jahrb., Abt. f. Anat., **23**: 77–162.

Phylum **ECHIURIDA.** Echiurid Worms

FISHER, W. K. 1946. Echiuroid worms of the North Pacific Ocean. Proc. U.S. Nat. Mus., **96**: 215–92, 18 pls., 19 text figs.

GISLÈN, TORSTEN. 1940. Investigations on the ecology of *Echiurus*. Lunds Universitets Årsskrift, N F Avd. 2, Bd. 36, Nr. 10, pp. 3–39. 6 pls.

Also as: K. Fysiografiska Sällskapets Handl., N F, Bd. 51, Nr. 10.

NEWBY, W. W. 1940. The embryology of the echiuroid worm *Urechis caupo*. Mem. Amer. Phil. Soc., **16**. 219 pp., 85 text figs.

Urechis caupo Fisher and MacGinitie.................................§306

FISHER, W. K. 1947. New genera and species of echiuroid and sipunculid worms. Proc. U.S. Nat. Mus., **97**: 351–72, 8 pls.

———, and G. E. MACGINITIE. 1928. A new echiuroid worm from California. Ann. Mag. Nat. Hist., Ser. 10, **1**: 199–203; also, The natural history of an echiuroid worm. *Ibid.*, 204–13.

LAWRY, JAMES V. 1966. Neuromuscular mechanisms of burrow irrigation in the echiuroid worm *Urechis caupo* Fisher and MacGinitie. I, Anatomy of the neuromuscular system and activity of intact animals, J. Exp. Biol., **45** (2): 343–56, 9 figs., 1 pl. II, Neuromuscular activity of dissected specimens. *Ibid.*, 357–68, 17 figs.

Urechis shows no periodicity or regular rhythm, but pumps more or less all the time.

MACGINITIE, G. E. 1935. Normal functioning and experimental behavior of the egg and sperm collectors of the echiuroid, *Urechis caupo*. J. Exp. Zool., **70**: 341–54.

——— 1935. The fertilization of eggs and the rearing of the larvae of *Urechis caupo* within the blood cavity of the adult animals. J. Exp. Zool., **71**: 483–87.

REDFIELD, A. C., and M. FLORKIN. 1931. The respiratory function of the blood of *Urechis caupo*. Biol. Bull., **61**: 185–210.

TYLER, ALBERT. 1965. The biology and chemistry of fertilization. Amer. Nat., **99** (907): 309–34, 15 figs.

Principally about fertilization in *Urechis caupo*.

Phylum **ARTHROPODA**

CARTHY, J. D. 1965. The behaviour of arthropods. W. H. Freeman, San Francisco. 148 pp., 41 figs.

Mostly about insects.

MOORE, RAYMOND C., ed. 1959. Treatise on invertebrate paleontology, Part O: Arthropoda 1. Geol. Soc. Amer. and Univ. Kansas Press. 560 pp., 415 figs.

About arthropods in general, but mostly concerning Trilobites; although they are all dead, they are still worth knowing something about.

SHAROV, A. G. 1966. Basic arthropodan stock: With special reference to insects. New York: Pergamon Press. Monographs in Pure and Applied Biology, Vol. 30. 271 pp., illus.

The innocent is hereby warned that what may sound like fact in this book is quite often the author's assertion, and that the phylogeny is often on the wilder fringe. Why does it follow that because pycnogonids feed on coelenterates, they are the oldest of arthropods? Perhaps because they descended from coelenterates, and therefore had nothing to eat but their ancestors?

SNODGRASS, R. E. 1952. A textbook of arthropod anatomy. Ithaca, N.Y.: Comstock. 363 pp., 88 figs.

Appendages and muscles, for the most part. Very little about the squishier anatomy.

TIEGS, O. W., and S. M. MANTON. 1958. The evolution of the Arthropoda. Biol. Revs., 33(3): 255–337, 18 figs.

Subphylum **TARDIGRADA**

The tardigrades, small, eight-legged organisms found in damp moss and similar places, belong to the supercategory that includes the annelid worms, mollusks, and arthropods—sometimes called the superphylum Articulata. They have been associated with the Onychophora (*Peripatus*) as late worms or early arthropods. Most tardigrades are found in damp terrestrial situations, but a few species are marine, and one of these is common on our shores.

Crisp, D. J., and J. Hobart. 1954. A note on the habitat of the marine tardigrade *Echiniscoides sigismundi* (Schultze). Ann. Mag. Nat. Hist., Ser. 12, **7**: 554–60, 1 fig.

On this coast, this bear animalcule occurs in the green algal film on *Balanus glandula*, at Dillon Beach and also in the Newport Bay region. It was simultaneously discovered in these widely separated localities, and simultaneously reported by Pequegnat and Hedgpeth at the Western Society of Naturalists meeting in Santa Barbara in 1956. It is very hard to find your first tardigrade, but it is sometimes easier the second time.

Schuster, Robert O., and Albert A. Grigarick. 1965. Tardigrada from western North America: With emphasis on the fauna of California. Univ. Calif. Pub. Zool., **76**: 1–67, 43 figs.

Echiniscoides sigismundi, p. 45, also from filamentous algae on rocks; illustrated, fig. 36. Also reports occurrence of a small marine species, *Styraconyx* or *Bathyechniscus sargassi* "from California."

Subphylum **PYCNOGONIDA.** Sea Spiders

Benson, Peter H., and Dustin C. Chivers. 1960. A pycnogonid infestation of *Mytilus californianus*. Veliger, **3** (1): 16–18, 1 pl.

Infestation by *Achelia chelata* is so far known only from a restricted part of Duxbury Reef, Marin County, and along the San Francisco shore, although the species has been taken in other localities.

Cole, L. J. 1904. Pycnogonida of the west coast of North America. Harriman Alaska Exp., **10**: 249–98, 16 pls.

Still an essential reference.

Hedgpeth, Joel W. 1941. A key to the Pycnogonida of the Pacific coast of North America. Trans. San Diego Soc. Nat. Hist., **9** (26): 253–64, 3 pls.

Key paper for this group within the area involved—Alaska to southern California. Species described in later preliminary notes by Hilton require further description and figures before they can be recognized. It should be pointed out, however, that the genus *Endeis* is still unknown from this coast; his *E. compacta* from Dillon Beach is actually an abnormal specimen of *Halosoma viridintestinale*.

——— 1947. On the evolutionary significance of the Pycnogonida. Smiths. Misc. Colls., **106** (18): 1–53, 1 pl., text figs.

Bibliography lists most papers to 1947, supplementing the comprehensive bibliography of the Bronns Tierreich monograph (Helfer and Schlottke, 1935). A few papers were missed, notably the Okuda reference below.

——— 1948. Report on the Pycnogonida collected by the *Albatross* in Japanese waters in 1900 and 1906. Proc. U.S. Nat. Mus., **98**: 233–321, 33 figs.

Includes general discussion of the zoogeographical relations of this group in the North Pacific.

——— 1951. Pycnogonids from Dillon Beach and vicinity, California, with descriptions of two new species. Wasmann J. Biol., **9** (1): 105–17, 3 pls.

——— 1961. Taxonomy: Man's oldest profession. Stockton, University of the Pacific, 11th Faculty Research Lecture, 18 pp., 2 figs.

About the discovery of *Achelia chelata* and other matters that led to fame and fortune as a student of pycnogonids. It was also through these animals that the acquaintance of Ed Ricketts was made.

Okuda, Shiro. 1940. Metamorphosis of a pycnogonid parasitic in a hydromedusa. J. Fac. Sci. Hokk. Imp. Univ., (6) **7** (2): 73–86, 10 figs.

Describes development of *Achelia alaskensis* in the Japanese *Polyorchis karafutoensis*. Although local medusae have not been observed to harbor pycnogonids, they have nct been adequately examined.

Schmitt, Waldo L., 1934. Notes on certain pycnogonids. J. Wash. Acad. Sci., **24** (1): 61–70.

Pycnogonum rickettsi, from dredged *Metridium*, and a list of Monterey Bay species, littoral and dredged.

Ziegler, Alan C. 1960. Annotated list of Pycnogonida collected near Bolinas, California. Veliger, **3** (1): 19–22.

HILTON, W. A. 1916. Life history of *Anoplodactylus erectus* Cole. J. Entom. and Zool., **8**: 25–34.

SANCHEZ, SIMONE. 1959. Le développement des pycnogonides et leurs affinités avec les arachnides. Arch. Zool. Exp. Gen., **98** (1): 1–101, 29 figs.

There are still some uncertain points and some excellent, unworked material on our shores, especially *Halosoma viridintestinale,* common in Tomales Bay on (and in) hydroids on the eelgrass.

FRY, WILLIAM G. 1965. The feeding mechanisms and preferred foods of three species of Pynogonida. Bull. Brit. Mus. (Nat. Hist.), Zool. **12** (6): 195–224, 8 text figs., 5 pls.

Pycnogonum stearnsi is one of the species studied; basal circular muscles were formerly thought to be nerve rings.

A curious burrowing form found in coarse sand at Tomales Bluff (Hedgpeth, 1951). Probably to be found elsewhere.

Subphylum **CHELICERATA**

Class ARACHNIDA

SAVORY, THEODORE. 1964. Arachnida. London and New York. Academic Press. 291 pp., 109 figs.

An introductory summary.

Order ACARI

The minute but supramicroscopic animals of the seashore are almost unexamined on our coast; the two papers cited below are the second and third (the first, in 1912, cited therein) concerning them. The 1949 reference proposes two new genera and three species from the coasts of Oregon and California; the second redescribes one of the 1912 species from the California coast and adds five new species from Alaskan shores. The field is open for the hand-lens naturalist.

NEWELL, IRWIN M. 1949. New genera and species of Halacaridae (Acari). Amer. Mus. Novit., **1411**. 22 pp., 63 figs.

——— 1951. *Copidognathus curtus* Hall, 1912, and other species of *Copidognathus* from western North America (Acari: Halacaridae). *Ibid.,* **1499**. 27 pp., 91 figs.

Order CHELONETHIDA. Pseudoscorpions

Several species of these quaint creatures, especially of the genus *Garypus,* frequent the high beaches, living among stones. The common beach species on this coast appears to be *Garypus californicus* (Banks), found from Ensenada to Monterey, on San Nicolas Island, and also in the Tomales Bay area, especially on Hog Island. The body is about 3–4 mm. long, making it fairly conspicuous for a pseudoscorpion. A monster, *Garypus giganteus* Chamberlin, occurs on Baja California beaches, and another species occurs on beaches in the Gulf of California. These animals build little inverted, dome-shaped huts of sand grains on the undersurfaces of rocks. Another pseudoscorpion, *Halobisium occidentale* Beier, occurs on *Salicornia* flats and debris-littered flats of bays and estuaries from San Francisco Bay to Alaska, possibly intergrading with *H. orientale* of the Siberian coast. The curious "mating waltz" observed in some pseudoscorpions has not been reported for these halophilous forms, but would be well worth confirming. It is a ceremony that would have delighted the late Ed Ricketts.

CHAMBERLIN, J. C. 1921. Notes on the genus *Garypus* in North America. Canad. Ent., **53**: 186–91, 1 pl.

——— 1924. The giant *Garypus* of the Gulf of California. Nature, Sept. 1924: 171–72, 175, 6 figs.

——— 1931. The Arachnid order Chelonethida. Stanford Univ. Pub. Biol. Sci., **7**: 1–284, 71 figs.

Morphology and phylogeny, with key to genera and bibliography. *Garypus californicus* figured (Fig. 3); toto-figure of *Halobisium,* fig. 56.

WEYGOLDT, PETER. 1966. Moos- und Bücherskorpione. Wittenberg: A. Ziamsen. 84 pp., 76 figs.

A concise little book, packed with information about pseudoscorpions, including maritime species.

Class CRUSTACEA

CALMAN, W. T. 1909. Crustacea. In Lankester's *Treatise on zoology,* **7**: 1–346.

GREEN, JAMES. 1963. A biology of Crustacea. Chicago: Quadrangle Books. 180 pp., 58 figs., 4 pls.

Primarily biological in treatment, by contrast with the more systematic emphasis of Schmitt (below).

SCHMITT, WALDO L. 1965. Crustaceans. Ann Arbor: University of Michigan Press. 204 pp., illus.

Based on the original version in the Smithsonian Scientific Series, but brought up to date. A successor to the popular treatments by Stebbing and Calman, now long out of print. This and the Green book (above) are somewhat complementary in coverage.

SMITH, GEOFFREY, and W. F. R. WELDON. 1909. Crustacea. Cambridge Natural History, **4**: 1–217, 135 figs.

There has been no single volume treatment on Crustacea for zoologists since 1909, when both the *Cambridge Natural History* and Lankester's *Treatise* volumes were published. Though we might prefer Calman's, the two treatments are very similar, and it is the Cambridge volume that has been reprinted (along with the rest of the series).

WATERMAN, TALBOT H., ed. 1960. The physiology of Crustacea, I: Metabolism and growth. New York: Academic Press. 670 pp., illus.

——— 1961. The physiology of Crustacea, II: Sense organs, integration, and behavior. New York: Academic Press. 681 pp., illus.

One of those indispensable references, which, once purchased by a student, constitutes a substantial investment in his graduate career.

WHITTINGTON, H. B., and W. D. I. ROLFE. 1963. Phylogeny and evolution of Crustacea. Spec. Pub. Mus. Comp. Zool., Cambridge, Mass. 192 pp., 78 figs.

Some of the latest thoughts about an uncertain subject.

Order ANOSTRACA. Fairy Shrimps

Although not an intertidal animal, the brine shrimp *Artemia* is well known to amateur aquarists, who hatch the eggs for fish food, and to visitors to the evaporation ponds of salt works, where it is associated with the unicellular green alga *Dunaliella,* upon which it feeds. Lately *Artemia* has become an important experimental animal.

KUENEN, D. J. 1939. Systematical and physiological notes on the brine shrimp, *Artemia.* Arch. Neerl. Zool. **3** (4): 365–449, 25 figs.

Based on studies of material hatched from eggs collected at Marina, near Monterey, and European material. The author suggests that the American species should be called *A. gracilis,* but recent opinion is that there is but the single species, *Artemia salina.*

LITTLEPAGE, JACK L., and MARILYNE N. MCGINLEY. 1965. A bibliography of the genus *Artemia* (*Artemia salina*), 1812–1962. San Francisco: San Francisco Aquarium Society, Special Pub. No. 1. 73 pp., mimeo.

Order DIPLOSTRACA

Suborder CLADOCERA

BAKER, HARRIET M. 1938. Studies in the Cladocera of Monterey Bay. Proc. Calif. Acad. Sci., (4) **23** (23): 311–65.

Subclass OSTRACODA

The Ostracoda are often numerous, and there are several species among seaweeds and bottom litter at the shore, but no attempt has been made to consider them. Unfortunately, there is no easily accessible reference, and no easy way into the literature.

In recent years, paleontologists have taken to calling these animals "ostracodes," but most zoologists refer to them as "ostracods." The Oxford Dictionary tells us that it should be "ostracode" from Greek οστρακώδης, meaning hard-shelled, whereas the Third Webster's omits "ostracode" entirely. Those who claim they are following classic tradition and purity should be aware that "ostracodes" is the singular form, and that "ostracode" is the neuter plural and is an adjective referring to the shell of a mollusk, at least as used by Aristotle (see *Progression of animals,* Harvard Classics, 1937, p. 536, l. 27). Were it not that some paleontologists consider this a matter of life and death, we would ignore it. Since they do claim classical authority, however, it is suggested that they reverse the usage and refer to an ostracod as an ostracodes, and to ostracods as ostracode, and let us go our way in peace. Either way, Aristotle would not have known what we were talking about.

BENSON, RICHARD H. 1959. Ecology of Recent ostracodes of the Todos Santos Bay region, Baja California, Mexico. Univ. Kans. Paleo. Contr., Arthropoda, Art. **1**, pp. 1–80, 11 pls., 20 figs.

A general account that may aid in identification of species at the southern end of our range. Fine maps, but uses overly detailed terminology for a general work.

——— 1966. Recent marine podocopid ostracodes. Oceanogr. Mar. Biol. Ann. Rev., **4**: 213–32.

"The purpose of this presentation is to acquaint those who work in the biological or historical aspects of marine science with the nature and significance of the paleontologically important living ostracodes. In a sense this is a report on the 'state of the art.' It is also a selective review, not over the last year's results, but over those of the last 100 years." But the extensive list of references does not include the titles of the papers; this is, unfortunately, apparently a policy of this series. The author takes J. Marvin Weller to task for lamenting that there is a dearth of new ideas in paleontology, and protests that, to the contrary, this indicates the soundness of the old ideas. Time stays, we go, I suppose.

LUCAS, V. Z. 1931. Some Ostracoda of the Vancouver Island region. Contr. Canad. Biol. and Fish., **7**: 397–416.

SKOGSBERG, TAGE. 1928. Studies on marine ostracods. Occ. Paps., Calif. Acad. Sci., **15**: 1–154.

——— 1950. Two new species of marine Ostracoda (Podocopa) from California. Proc. Calif. Acad. Sci., (4) **26** (14): 485–505, 4 pls.

Subclass COPEPODA

DUDLEY, PATRICIA L. 1966. Development and systematics of some Pacific marine symbiotic copepods: A study of the biology of the Notodelphyidae, associates of ascidians. Seattle: Univ. Wash. Press. 282 pp., 51 figs.

Comprehensive, exhaustive, and detailed. But somebody may need this information.

FRASER, J. H. 1936. The occurrence, ecology and life history of *Tigriopus fulvus* (Fisher). J. Mar. Biol. Assoc., **20**: 523–36, 5 figs.

——— 1936. The distribution of rock pool copepods according to tidal level. J. Anim. Ecol., **5**.

GOODING, RICHARD U. 1960. North and South American copepods of the genus Hemicyclops (Cyclopida: Clausidiidae). Proc. U.S. Nat. Mus., **112** (3434): 159–95, 10 figs.

Describes various species of copepods occurring on, or in the burrows of, *Callianassa* and *Upogebia*.

ILLG, PAUL L. 1949. A review of the copepod genus *Paranthessius* Claus. Proc. U.S. Nat. Mus., **99**: 391–428, 6 figs.

——— 1960. Marine copepods of the genus *Anthessius* from the northeastern Pacific Ocean. Pac. Sci., **14** (4): 337–72, 125 figs.

Most of these copepods are associated with mollusks.

JOHNSON, MARTIN W., and J. BENNET OLSON. 1948. The life history and biology of a marine harpacticoid copepod, *Tisbe furcata* (Baird). Biol. Bull., **95** (3): 320–32, 2 pls.

LANG, KARL. 1948. Monographie der Harpacticiden. Lund, Håkan Ohlssons Boktryckeri. 2 vols., 1683 pp., 605 figs., 371 distr. maps.

——— 1965. Copepoda Harpacticoidea from the California Pacific coast. K. Sven. Vet., **10** (2). 560 pp., 6 pls., 303 text figs.

Based principally on collections made at Dillon Beach and Pacific Grove.

LIGHT, S. F., and OLGA HARTMAN. 1936. A review of the genera *Clausidium* Kossmann and *Hemicyclops* Boeck (Copepoda) with the description of a new species from the northeast Pacific. Univ. Calif. Pub. Zool., **41**: 173–88.

MARSHALL, S. M., and A. P. ORR. 1955. The biology of a marine copepod *Calanus finmarchicus* (Gunnerus). Edinburgh: Oliver and Boyd. 188 pp., 63 figs.

Not, however, the last word, as several additional papers on this subject by these authors have since appeared.

MONK, C. R. 1941. Marine harpacticoid copepods from California. Trans. Amer. Micr. Soc., **60** (1): 75–99, 3 pls.

WILSON, C. B. 1935. Parasitic copepods from the Pacific coast. Amer. Midland Nat., **16**: 776–97.

FAHRENBACH, WOLF H. 1962. The biology of a harpacticoid copepod. La Cellule, **62** (3): 301–76, 9 pls., 4 text figs.

Originally described from *Halosaccion* at Moss Beach, this copepod is always found in or on red algae, often in the bladders of *Halosaccion*. It occurs from San Pedro to the Queen Charlotte Islands.

ZULUETA, ANTONIO DE. 1911. Los copépodos parásitos de los celentéreos. Mem. Real Soc. Esp. Hist. Nat., **7**. 58 pp.

HENDERSON, JEAN T. 1930. A new parasitic copepod (*Scolecimorpha huntsmani*, n. sp.). Contr. Canad. Biol. and Fish., **6**: 215–24.

MISTAKIDIS, M. 1949. A new variety of *Tigriopus lilljeborgii* Norman. Dove Mar. Lab. Rept., 3d ser., No. 10: 55–70, 3 pls.

Gives comparison tables of known species, including *T. californicus* (as *T. triangularis*); Lang (1948), p. 342, uses *T. californicus*, however.

VACQUIER, VICTOR D., and WILLIAM L. BELSER. 1965. Sex conversion induced by hydrostatic pressure in the marine copepod *Tigriopus californicus*. Science, **150** (3703): 1619–21.

Authors found that pressure applied during naupliar stage resulted in higher percentage of females. Anyhow, *Tigriopus* has become an experimental animal, and some attempts have been made to establish it as a sort of marine *Drosophila*. One problem with this is that *Drosophila* may be a little out of date as well.

Subclass CIRRIPEDIA

Order THORACICA. Barnacles

A comprehensive bibliography of barnacles would occupy a volume; we offer only a small sample. Barnacles, as components of communities and zones, occupy a place in many papers. Some of these are cited in the general section—see, for example, Connell's papers on predator-prey relationships as zonation control, etc.

BARNES, H. 1958. Regarding the southern limits of *Balanus balanoides* (L.). Oikos, **9** (2): 139–57, 9 figs.

This species occurs certainly as far south as Sitka, perhaps even farther south than Barnes suspects; in any event, his skepticism about the occurrence of *B. balanoides* at Hatteras was refuted by Wells, Wells, and Gray (see below).

——— 1959. Stomach contents and micro-feeding of some common cirripedes. Canad. J. Zool., **37**: 231–36.

BARNES, H., and MARGARET BARNES. 1958. Further observations on self-fertilization in *Chthamalus* sp. Ecology, **39** (3): 550.

——— 1959. Note on stimulation of cirripede nauplii. Oikos, **10** (1): 19–23.

——— 1959. The naupliar stages of *Balanus hesperius* Pilsbry. Canad. J. Zool., 37: 237–44, 4 figs.

——— 1959. The effect of temperature on the oxygen uptake and rate of development of the egg-masses of two common cirripedes, *Balanus balanoides* (L.) and *Pollicipes polymerus* J. B. Sowerby. Kieler Meeresf., **15** (2): 242–51, 2 figs.

——— 1959. Studies on the metabolism of cirripedes. The relation between body weight, oxygen uptake, and species habitat. Veröffent. Meeresf. Bremerhaven, **6** (2): 515–23, 2 figs.

BARNES, H., and E. S. REESE. 1959. Feeding in the pedunculate cirripede *Pollicipes polymerus* J. B. Sowerby. Proc. Zool. Soc. Lond., **132** (4): 569–85, 8 figs., 1 pl.

BERNARD, FRANCIS J., and CHARLES E. LANE. 1962. Early settlement and metamorphosis of the barnacle *Balanus amphitrite niveus*. J. Morphol., **110** (1): 19–40, 5 pls., 2 text figs.

BOHART, R. M. 1929. Observations on the attachment of *Balanus crenatus* Bruguière, found in the waters of Puget Sound. Amer. Nat., **43**: 353–61.

BROCH, H. 1922. Studies on Pacific cirripedes: No. 10 in Papers from Dr. Th. Mortensen's Pacific expedition, 1914–16. Vidensk. Medd. fra Dansk Naturh. Foren., **73**: 215–58, 77 text figs.

CORNWALL, I. E. 1925. Review of the Cirripedia of the coast of British Columbia. Contr. Canad. Biol. and Fish., n.s., **2**: 469–502.

——— 1927. A new species of barnacle from the coast of California. Ann. Mag. Nat. Hist., (10) **20**: 233–35.

——— 1936. On the nervous system of four British Columbian barnacles (one new species). J. Biol. Bd. Canad., **1**: 469–75.

——— 1955. The barnacles of British Columbia. B.C. Provincial Museum, Handbook No. 7, 69 pp., illus.

——— 1955. Cirripedia. Canadian Pacific Fauna. **10**: Arthropoda; 10e, Cirripedia. 49 pp., 40 figs.

——— 1956. Identifying fossil and Recent barnacles by the figures in the shell. J. Paleont., **30** (3): 646–51, 3 figs.

——— 1958. Identifying Recent and fossil barnacles. Canad. J. Zool., **36**: 79–89, 49 figs.

The patterns seen in sections of bits of shell are apparently specific, and serve as identifying characters.

Crisp, D. J., and A. J. Southward. 1961. Different types of cirral activity of barnacles. Phil. Trans. Roy. Soc. Lond., Ser. B, **243** (705): 271–308, 15 figs.

Fahrenbach, Wolf H. 1965. The micromorphology of some simple photoreceptors. Zeitschr. Zellf., **66**: 233–54, 13 figs.

Concerns *Balanus cariosus* and *B. amphitrite*.

Gwilliam, G. F. 1963. The mechanism of the shadow reflex in Cirripedia, I: Electrical activity in the supraesophagal ganglion and ocellar nerve. Biol. Bull., **125** (3): 470–85, 10 figs.

——— 1965. ——— II. Photoreceptor response, second order responses, and motor cell output. *Ibid.*, **129** (2): 244–56, 4 figs.

Henry, Dora Priaulx. 1940. Notes on some pedunculate barnacles from the North Pacific. Proc. U.S. Nat. Mus., **88**: 225–36.

——— 1940. The Cirripedia of Puget Sound, with a key to the species. Univ. Wash. Pub. Oceanogr., **4** (1): 1–48, 4 pls.

——— 1942. Studies on the sessile Cirripedia of the Pacific coast of North America. Univ. Wash. Pub. Oceanogr., **4** (3): 95–134, 4 pls., figs.

Heron-Allen, Edward. 1928. Barnacles in nature and myth. Oxford Univ. Press. 180 pp., illus.

"This volume . . . has no reason whatever for existence, it serves no useful purpose, and supplies no want, long-felt or otherwise" (from the Preface).

Herz, Ludwig E. 1933. The morphology of the later stages of *Balanus crenatus* Bruguière. Biol. Bull., **64**: 432–42.

Some data also on Monterey Bay littoral barnacles, whose larvae were found to be too delicate, from the aeration standpoint, for practical culture methods. *B. glandula* of the San Francisco Bay sloughs was found to breed only until June.

Pilsbry, H. A. 1907. The barnacles (Cirripedia) contained in the collections of the U.S. National Museum. Bull. U.S. Nat. Mus., **60**. 122 pp.

——— 1916. Sessile barnacles (Cirripedia) contained in the collections of the U.S. National Museum. Bull. U.S. Nat. Mus., **93**. 366 pp.

——— 1921. Barnacles of the San Juan Islands, Washington. Proc. U.S. Nat. Mus., **59**: 111–15.

Southward, A. J., and D. J. Crisp. 1954. Recent changes in the distribution of the intertidal barnacles *Chthamalus stellatus* Poli and *Balanus balanoides* L. in the British Isles. J. Anim. Ecol., **23** (1): 163–77, 7 figs.

——— 1956. Fluctuations in the distribution and abundance of intertidal barnacles. J. Mar. Biol. Assoc., U.K., **35**: 211–29, 1 fig.

——— 1965. Activity rhythms of barnacles in relation to respiration and feeding. J. Mar. Biol. Assoc., U.K., **45**: 161–85, 11 figs.

These are a few examples of the numerous papers by these authors that should be consulted by the serious cirripediologist.

WELLS, HARRY W., MARY JANE WELLS, and I. E. GRAY. 1960. On the southern limit of *Balanus balanoides* in the western Atlantic. Ecology, **41** (3): 578–80, 1 fig.

MONTEROSSO, B. 1930. Studi cirripedologici, VI: Reactions of *C. stellatus* under experimental conditions. Rend. Acc. Lincei, **9** (6): 501–5.

BARNES, HAROLD, and J. J. GONOR. 1958. Neurosecretory cells in the cirripede *Pollicipes polymerus* J. B. Sowerby. J. Mar. Res., **17**: 81–102, 15 figs.
HILGARD, GALEN HOWARD. 1960. A study of reproduction in the intertidal barnacle *Mitella polymerus*, in Monterey Bay, California. Biol. Bull., **119** (2): 169–88, 7 figs.
HOWARD, GALEN KENT, and HENRY C. SCOTT. 1959. Predaceous feeding in two common gooseneck barnacles. Science, **129** (3350): 717–18, 1 fig.

Order ACROTHORACICA. Burrowing Barnacles

TOMLINSON, JACK T. 1955. The morphology of an acrothoracican barnacle, *Trypetesa lateralis*. J. Morphol., **96** (1): 97–114, 4 pls.
This small burrowing barnacle lives in *Tegula* shells occupied by hermit crabs.

Order RHIZOCEPHALA

BOSCHMA, H. 1930–31. Rhizocephala. No. 55 in Papers from Dr. Th. Mortensen's Pacific expedition, 1914–16. Vidensk. Medd. fra Dansk Naturh. Foren., **89**: 297–380.
——— 1933. New species of Sacculinidae in the collection of the U.S. National Museum. Tijdschr. Ned. Dierk. Vereen., (3) **3** (4).
Including *Heterosaccus californicus*, sp. nov., on *Pugettia producta* from Santa Cruz, California.
——— 1934. The relationship between the Sacculinidae of the Pacific and their hosts. Proc. 5th Pac. Sci. Cong., **5**: 4195–97.
POTTS, F. A. 1912. *Mycetomorpha*, a new rhizocephalan (with a note on the sexual condition of *Sylon*). Zool. Jahrb., Abt. f. Syst., **33**: 575–94.
Parasitic on the shrimp *Crago communis* at Nanaimo. *Sylon* on *Spirontocaris, Pandalus*, etc., at Friday Harbor.
——— 1915. On the rhizocephalan genus *Thompsonia*. Carnegie Institution, Pub. Dept. Mar. Biol., **8**: 1–32.
REINHARD, EDWARD G. 1942. Studies on the life history and host-parasite relationship of *Peltogaster paguri*. Biol. Bull. **83** (3): 401–15.
——— 1944. Rhizocephalan parasites of hermit crabs from the northwest Pacific. J. Wash. Acad. Sci., **34** (2): 49–58.

——— 1950. An analysis of the effects of a sacculinid parasite on the external morphology of *Callinectes sapidus* Rathbun. Biol. Bull., **98** (3): 277–88.

See pp. 240–41; the effect of sacculinid parasitism suggests the peculiar phenomenon of sex determination in the echiurid *Bonellia*. In this worm, larvae that settle by themselves become females; if a larva settles on a female, however, it becomes a small, parasite-like male, which lies within the nephridium of its mate. There is some substance, hormone-like in action and secreted by the female, that inhibits the larva, although all young *Bonellias* are potentially female.

REISCHMAN, PLACIDUS G. 1959. Rhizocephala of the genus *Peltogasterella* from the coast of the State of Washington and the Bering Sea. Nederl. Akad. Wetensch. Proc., C., **62** (4): 409–35, 38 figs.

Loxothyeacus, Peltogaster, and/or other rhizocephalans increase in abundance with the increase in latitude. They are rare at Monterey, but abundant at Sitka, where *Lophopanopeus* and *Cancer oregonensis* are commonly infected.

Class MALACOSTRACA

Subclass LEPTOSTRACA

Order NEBALIACEA

CANNON, H. G. 1927. On the feeding mechanism of *Nebalia bipes.* Trans. Roy. Soc. Edin., **55**: 355–69.

CLARK, A. E. 1932. *Nebaliella caboti,* n. sp., with observations on other Nebaliacea. Trans. Roy. Soc. Canad., (3), Sec. 5, **26**: 217–35.

LAFOLLETTE, R. 1914. A *Nebalia* from Laguna Beach. J. Entom. and Zool., **6**: 204–6.

See the Clark (1932) reference. Identified as the Atlantic *Nebalia bipes* in former editions; although the females are almost inseparable from *N. bipes,* the males are so distinct that Clark considered a new genus justified. There may be other species, especially in the South.

Subclass PERACARIDA

Order SCHIZOPODA (obs.)

Euphausiacea and Mysidacea, comprising the obsolete order Schizopoda; the first is entirely pelagic, mostly on the high seas, and is therefore treated scantily here.

BANNER, ALBERT H. 1948–49. A taxonomic study of the Mysidacea and Euphausiacea (Crustacea) of the northeastern Pacific. Part I: Mysidacea from family Lophogastridae through tribe Erythropini. Trans. Roy. Canad. Inst. **26**: 345–99, 9 pls. (1948). Part II: Mysidacea, from tribe Mysini through subfamily Mysidellinae. *Ibid.,* **27**: 65–124, 7 pls. (1948). Part III: Euphausiacea. *Ibid.,* **28**: 1–62, 4 pls. (1950).

ESTERLY, C. O. 1914. The Schizopoda of the San Diego region. Univ. Calif. Pub. Zool., **13**: 1–20.

Taxonomy only, with no distribution or natural history; the bibliography cites Hansen, 1913, and Ortman, 1908, which need not be repeated here.

HANSEN, H. J. 1915. The Crustacea Euphausiacea of the U.S. National Museum. Proc. U.S. Nat. Mus., **48**: 59–114.

TATTERSALL, WALTER M. 1932. Contribution to a knowledge of the Mysidacea of California: I (La Jolla), and II (San Francisco). Univ. Calif. Pub. Zool., **37**: 301–47.

——— 1933. Euphausiacea and Mysidacea from western Canada. Contr. Canad. Biol. and Fish., **8**: 181–205.

Order MYSIDACEA

TATTERSALL, WALTER M. 1951. A review of the Mysidacea of the U.S. National Museum. U.S. Nat. Mus., Bull., **201**. 292 pp., 103 figs.

CANNON, H. G. 1927. On the feeding mechanism of a mysid crustacean. Trans. Roy. Soc. Edin., **55**: 219–54.

Order CUMACEA

CALMAN, W. T. 1912. The Crustacea of the order Cumacea in the collection of the U.S. National Museum. Proc. U.S. Nat. Mus., **41**: 603–76.

HART, JOSEPHINE F. L. 1931. Some Cumacea of the Vancouver Island region. Contr. Canad. Biol. and Fish., n.s., **6** (3): 25–40, 5 figs.

WIESER, WOLFGANG. 1956. Factors influencing the choice of substratum in *Cumella vulgaris* Hart (Crustacea: Cumacea). Limnol. and Oceanogr., **1** (4): 274–85, 4 figs.

ZIMMER, CARL. 1936. California Crustacea of the order Cumacea. Proc. U.S. Nat. Mus., **83**: 423–39.

None of these sometimes common but usually small (½ inch) and grotesque, shrimplike crustaceans are treated here. Two species are fairly common inshore in the region involved, and several extend northward from southeastern Alaska. One, *Colurostylis occidentalis* Calman, has just turned up at Corona del Mar.

Order CHELIFERA (Tanaidacea)

Small, chelate, isopod-like animals, especially common in mussel beds; usually treated with the isopods in systematic papers. *Leptochelia* is a genus commonly represented on this coast.

LANG, KARL. 1961. Further notes on *Pancolus californiensis* Richardson. Arkiv f. Zool., Ser. 2, **13** (30): 573–77, 2 figs., 1 pl.

MENZIES, ROBERT J. 1953. The apseudid Chelifera of the eastern tropical and north temperate Pacific Ocean. Bull. Mus. Comp. Zool., **107** (9): 443–96, 27 figs.

Order ISOPODA. Pill Bugs, etc.

HATCH, MELVILLE A. 1947. The Chelifera and Isopoda of Washington and adjacent regions. Univ. Wash. Pub. Biol., **10** (5): 155–274.

HOLTHUIS, L. B. 1949. The Isopoda and Tanaidacea of the Netherlands, including the description of a new species of *Limnoria*. Zool. Meddel. Leiden, **30** (12): 163–90, 4 figs.

Describes *Limnoria quadripunctata* from the Dutch coast, and points out that the "*Limnoria lignorum*" of the San Francisco Bay piling report is actually this species.

JOHNSON, MARTIN W. 1935. Seasonal migrations of the wood borer *Limnoria lignorum* (Rathke) at Friday Harbor, Washington. Biol. Bull., **69**: 427–38.

LANG, KARL. 1961. Contributions to the knowledge of the genus *Microcerberus* Karaman (Crustacea: Isopoda) with a description of a new species from the central California coast. Arkiv. Zool. Stockholm, Ser. 2, **13** (22): 493–510, 3 pls.

MALONEY, J. O. 1933. Two new species of isopod crustaceans from California. J. Wash. Acad. Sci., **23**: 144–47.

Synidotea macginitiei and *Pentidotea montereyensis*, Monterey Bay littoral.

Menzies, Robert J. 1950. Notes on California isopods of the genus *Armadilloniscus,* with the description of *Armadilloniscus coronacapitalis,* n. sp. Proc. Calif. Acad. Sci., (4) **26** (13): 467–81, 5 pls.

——— 1950. A remarkable new species of marine isopod, *Erichsonella crenulata,* n. sp., from Newport Bay, California. Bull. S. Calif. Acad. Sci., **49** (1): 29–35, 3 pls.

——— 1951. New marine isopods, chiefly from northern California, with notes on related forms. Proc. U.S. Nat. Mus., **101** (3273): 105–56, 33 figs.

——— 1951. A new species of *Limnoria* (Crustacea: Isopoda) from southern California. Bull. S. Calif. Acad. Sci., **50** (2): 86–88, 1 pl.

This is *Limnoria tripunctata,* which invades creosoted wood in southern California.

——— 1951. A new genus and new species of asellote isopod, *Caecijaera horvathi,* from Los Angeles–Long Beach Harbor. Amer. Mus. Novit., **1542.** 7 pp., 3 figs.

———, and J. Laurens Barnard. 1951. The isopodan genus *Iais* (Crustacea). Bull. S. Calif. Acad. Sci., **50** (3): 136–51, 9 pls.

Ondo, Yoshinori, and Syuiti Mori. 1956. Periodic behavior of the shore isopod *Megaligia exotica* (Roux), I: Observations under natural conditions. Jap. J. Ecol., **5** (4): 161–67, 5 figs.

Text in Japanese. From the adequately long summary, it appears that these rock lice travel down in lines toward the tide line in the morning, and go home in the late afternoon. They tend to go in groups according to age, the younger ones ahead of their elders, and follow established pathways. This does not seem to have been observed in our species, perhaps because no one has had the patience to watch them long enough.

Richardson, H. 1905. Isopods of North America. Bull. U.S. Nat. Mus., **54.** 727 pp.

——— 1909. Isopods collected in the Northwest Pacific. Proc. U.S. Nat. Mus., **37**: 75–129.

Stafford, B. E. 1912. Studies in Laguna Beach Isopoda. First Ann. Rept. Laguna Mar. Lab., pp. 18–33.

——— 1913. Studies in Laguna Beach Isopoda. Pomona J. Entom. and Zool., **5**: 161–72, 182–88.

Nierstrasz and Brender à Brandis. 1929. Epicaridea, I. No. 48 in Papers from Dr. Th. Mortensen's Pacific expedition, 1914–16. Vidensk. Medd. fra Dansk Naturh. Foren., **87**: 1–44.

Enright, J. T. 1965. Entrainment of a tidal rhythm. Science, **147** (3660): 864–67, 3 figs.

The endogenous activity of *Exocirolana chiltoni* related to tidal rhythm has been synchronized by simulating the wave action on the beach, which suggests that the stimulus for movement up and down the beach is mechanical in this case.

Hoestlandt, H. 1964. Examen comparé de races polychromatiques de Sphéromes (Crustacés Isopodes) des côtes atlantique européenne et pacifique américaine. Verh. Internat. Verein. Limnol., **15**: 371–78, 1 fig.

Color variations, including those of *Gnorimosphaeroma oregonensis.*

Menzies, Robert J. 1954. A review of the systematics and ecology of the genus *Exosphaeroma,* with the description of a new genus, a new species, and a new subspecies (Crustacea: Isopoda, Sphaeromidae). Amer. Mus. Novit., No. **1683.** 24 pp., 10 figs.

Riegel, J. A. 1959. Some aspects of osmoregulation in two species of sphaeromid isopod Crustacea. Biol. Bull., **116** (2): 272–84, 2 figs.

——— 1959. A revision of the sphaeromid genus *Gnorimosphaeroma* Menzies (Crustacea: Isopoda) on the basis of morphological, physiological, and ecological studies on two of its "subspecies." Biol. Bull., **117** (1): 154–62, 1 fig.

Author says that Menzies's subspecies ought to be species; both names of species and titles of the papers are overly long for a bibliographer's patience.

Idothea montereyensis Maloney . . . . §205

Lee, Welton L. 1966. Pigmentation of the marine isopod *Idothea montereyensis.* Comp. Biochem. Physiol., **18**: 17–36, 6 figs.

These isopods occur in green, brown, and red. They are basically yellowish; production of blue pigment makes them green, accumulation of red pigment turns them red, and the brown is transitional. It is therefore by different combinations of pigments that the isopods more or less match their backgrounds, and the abundance of pigment is related to diet.

——— 1966. Color change and ecology of the marine isopod *Idothea* (*Pentidotea*) *montereyensis* Maloney, 1933. Ecology, **47** (6): 930–41, 7 figs.

I. urotoma Stimpson (includes *I. rectilinea*) . . . . §§46, 205

Menzies, Robert J. 1950. The taxonomy, ecology, and distribution of northern California isopods of the genus *Idothea,* with the description of a new species. Wasmann J. Biol., **8** (2): 155–95, 10 pls., 3 figs.

I. (*Pentidotea*) *resecata* (Stimpson) . . . . §205

Menzies, Robert J., and Richard J. Waidzunas. 1948. Postembryonic growth changes in the isopod *Pentidotea resecata* (Stimpson), with remarks on their taxonomic significance. Biol. Bull., **95**: 107–13.

I. (*P.*) *schmitti* Menzies (formerly *P. whitei*) . . . . §172

I. (*P.*) *stenops* (Benedict) . . . . §172

I. (*P.*) *wosnesenski* (Brandt) . . . . §165

Ligia occidentalis Dana . . . . § 2

Armitage, Kenneth B. 1960. Chromatophore behavior in the isopod *Ligia occidentalis* Dana, 1855. Crustaceana, **1** (3): 193–207, 7 figs.

"One can infer that the functional significance of the chromatophore system is protective coloration."

Hilton, W. A. 1915. Early development of *Ligia.* Pomona J. Entom. and Zool., **7**: 211–27.

L. pallasii (Brandt) . . . . §162

Limnoria spp. . . . . §329

Three species of wood-boring *Limnoria* occur on the Pacific coast: *L. lignorum* from Alaska to Point Arena, *L. quadripunctata* from Tomales Bay to San Diego, and *L. tripunctata* from San Francisco Bay to San Quintín, Baja California. The last two species occur in San Francisco Bay; the species of the piling report, at least as illustrated, is *L. quadripunctata,* from the colder waters of the bay; *L. tripunctata* occurs in slightly warmer waters, and is capable of boring into creosoted timbers. All three species are widely distributed elsewhere in the world. In addition to the wood-boring *Limnoria,* there is a species on the Pacific Coast that acts as a parasite, boring into the living holdfasts of the large kelps.

Beckman, Carolyn, and Robert J. Menzies. 1960. The relationship of reproductive temperature and the geographical range of the marine wood-borer *Limnoria tripunctata.* Biol. Bull., **118** (1): 9–16, 2 figs.

Johnson, Martin W., and Robert J. Menzies. 1956. The migratory habits of the marine gribble *Limnoria tripunctata* Menzies in San Diego Harbor, California. Biol. Bull., **110** (1): 54–68, 5 figs.

Menzies, Robert J. 1954. The comparative biology of reproduction in the wood-boring isopod crustacean *Limnoria.* Bull. Mus. Comp. Zool., **112** (5): 363–88, 6 figs.

——— 1957. The marine borer family Limnoriidae (Crustacea: Isopoda). Bull. Mar. Sci. Gulf and Carib., **7** (2): 101–200, 42 figs.

Sphaeroma pentodon Richardson . . . . §336

BARROWS, A. L. 1919. The occurrence of a rock-boring isopod along the shore of San Francisco Bay, California. Univ. Calif. Pub. Zool., **19**: 299–316.

Tylos punctatus Holmes & Gay . . . . §184

Order AMPHIPODA. Beach Hoppers, Sand Fleas, Skeleton Shrimps, etc.

ALDERMAN, A. L. 1936. Some new and little-known amphipods of California. Univ. Calif. Pub. Zool., **41**: 53–74.

BARNARD, J. LAURENS. 1952. Some Amphipoda from central California. Wasmann J. Biol., **10** (1): 9–36, 9 pls.

——— 1954. Marine Amphipoda of Oregon. Corvallis: Ore. St. Univ. Press. 37 pp., 33 pls., 1 text fig.

——— 1965. Marine Amphipoda of the family Ampithoidae from southern California. Proc. U.S. Nat. Mus., **118** (3522): 1–46, 27 figs.

——— 1966. Benthic Amphipoda of Monterey Bay, California. Proc. U.S. Nat. Mus., **119** (3541): 1–41, 7 figs.

BOUSFIELD, E. L. 1957. Notes on the amphipod genus *Orchestoidea* on the Pacific coast of North America. Bull. S. Calif. Acad. Sci., **56** (3): 119–29, 2 pls.

——— 1959. New records of beach hoppers (Crustacea: Amphipoda) from the coast of California. Bull. Nat. Mus. Canad., **172**. 12 pp., 4 figs.

———, and W. L. KLAWE. 1963. *Orchestoidea gracilis,* a new beach hopper (Amphipoda: Talitridae) from Lower California, Mexico, with remarks on its luminescence. Bull. S. Calif. Acad. Sci. **62** (1): 1–8, 2 figs.

BOWERS, DARL E. 1963. Field identification of five species of Californian beach hoppers (Crustacea: Amphipoda). Pac. Sci., **17** (3): 315–20, 4 figs.

Identification by color patterns.

——— 1964. Natural history of two beach hoppers of the genus *Orchestoidea* (Crustacea: Amphipoda), with reference to their complemental distribution. Ecology, **45** (4): 677–96, 10 figs.

CHEVREUX, E., and L. FAGE. 1925. Amphipodes. Faune de France, **9**. 488 pp., 438 figs.

The most recent monograph on the group; since many species are cosmopolitan, many species of the Pacific coast are treated.

CRAWFORD, G. I. 1937. A review of the amphipod genus *Corophium,* with notes on the British species. J. Mar. Biol. Assoc., **21** (2): 589–630, 4 figs.

With keys to sections of the genus, including species occurring locally.

DOUGHERTY, ELLSWORTH C., and JOAN STEINBERG. 1953. Notes on the skeleton shrimps (Crustacea: Caprellidae) of California. Proc. Biol. Soc. Wash., **66**: 39–50.

ENRIGHT, J. T. 1963. The tidal rhythms of activity of a sand-beach amphipod. Zeitschr. vergl. Physiol., **40**: 276–313, 14 figs.

Concerns *Synchelidium* sp. from southern California beaches.

HOLMES, SAMUEL J. 1904. Amphipod crustaceans of the expedition. Harriman Alaska Exp., **10**: 233–46.

——— 1908. The Amphipoda collected by the U.S. Bureau of Fisheries steamer *Albatross* off the west coast of North America in 1903 and 1904, with descriptions of a new family and several new genera and species. Proc. U.S. Nat. Mus., **35** (1654): 489–543.

MILLS, ERIC L. 1959. Amphipod crustaceans of the Pacific coast of Canada, I: Family Atylidae. Bull. Nat. Mus. Canad. **172**: 13–33, 4 figs.

——— 1962. Amphipod crustaceans of the Pacific coast of Canada, II: Family Oedicerotidae. Natur. Hist. Paps. Nat. Mus. Canad., **15**. 21 pp., 6 figs.

——— 1965. The zoogeography of North Atlantic and North Pacific ampeliscid amphipod crustaceans. Syst. Zool., **14** (2): 119–30.

Shoemaker, C. R. 1916. Description of three new species of amphipods from southern California. Proc. Biol. Soc. Wash., **29**: 157–60.

——— 1925. XV: The Amphipoda collected . . . chiefly in the Gulf of California. Bull. Amer. Mus. Nat. Hist., **52**: 21–61.

——— 1926. Amphipods of the family Bateidae. Proc. U.S. Nat. Mus., **68**: 1–26.

——— 1934. Two new species of *Corophium* from the west coast of America. J. Wash. Acad. Sci., **24**: 356–60.

One dredged in Monterey Bay; the other may occur inshore from Peru to the Bering Sea.

——— 1938. Three new species of the amphipod genus *Ampithoë* from the west coast of America. J. Wash. Acad. Sci., **28** (1): 15–25, 4 text figs.

——— 1949. The amphipod genus *Corophium* on the west coast of America. J. Wash. Acad. Sci., **39** (2): 66–82, 8 figs.

——— 1952. A new species of commensal amphipod from a spiny lobster. Proc. U.S. Nat. Mus., **102** (3299): 231–33, 1 fig.

——— 1964. Seven new amphipods from the west coast of North America, with notes on some unusual species. Proc. U.S. Nat. Mus., **115** (3439): 391–430, 15 figs.

Stasek, Charles R. 1958. A new species of *Allogaussia* (Amphipoda: Lysianassidae) found living within the gastrovascular cavity of the sea anemone *Anthopleura elegantissima.* J. Wash. Acad. Sci., **48** (4): 119–26, 3 figs.

So far, *Allogaussia recondita* is found only in this anemone at Moss Beach, where it seems to be common. It was not found in *A. xanthogrammica,* but may invade this species under aquarium conditions. The amphipod wanders about the host with apparent impunity. We are also reminded that Moss Beach is a type locality for a number of interesting species.

Stout, V. R. 1912. Studies in Laguna Amphipoda, I. First Ann. Rept. Laguna Mar. Lab., pp. 134–49.

——— 1913. Studies in Laguna Amphipoda, II. Zool. Jahrb., Abt. f. Syst., **34**: 633–59.

Thorsteinson, Elsa D. 1941. New or noteworthy amphipods from the North Pacific coast. Univ. Wash. Pub. Oceanogr., **4** (2): 50–96.

SHOEMAKER, C. R. 1930. Description of two new amphipod crustaceans (*Talitridae*) from United States. J. Wash. Acad. Sci., **20**: 107–14.

O. californiana (Brandt) . . . . §185

O. corniculata Stout . . . . §154

ENRIGHT, J. T. 1961. Lunar orientation of *Orchestoidea corniculata* Stout (Amphipoda). Biol. Bull., **120** (2): 148–56, 4 figs.

They navigate up and down the beach by the position of the moon.

Polycheria antarctica (Stebbing) . . . . §100

The amphipods are one of the most abundant, and at the same time least-known, populations of our intertidal fauna. Revisionary work is still in progress; hence we are unable to do more than submit a list of common species.

These species build tubes of soft mud, frequently on pilings in harbors: *Jassa falcata* (Montagu), *Corophium acherusicum* (Costa), *C. insidiosum* Crawford, *C. spinicorne* Stimpson, and *Podocerus brasiliensis* (Dana). In bays or harbors, among hydroids, *Stenothoë marina* (Bate) is common. Also in bays, found in borings that it possibly enlarges from *Limnoria* borings, is *Chelura terebrans* Philipi.

Common intertidal species of the coast include the following: *Allorchestes angustus* Dana, *Amphilochus neapolitanus* Della Valle, *Ampithoë indentata* (Stout), *A. dalli* Shoemaker, *A. humeralis* Stimpson, *Anisogammarus pugettensis* (Dana), *Elasmopus antennatus* (Stout), *Ericthonius brasiliensis* (Dana), *Hyale frequens* (Stout) [possibly the *Hyale* sp. of our §153?], *Maera dubia* Calman, *M. simile* Stout, *Melita fresneli* (Audouin), *Paragrubia uncinata* (Stout), *Parallorchestes ochotensis* (Brandt).

Superorder EUCARIDA

Order EUPHAUSIACEA

Thysanoëssa gregaria G. O. Sars . . . . §330

According to Banner (1949, III), records for this species from off northern California to British Columbia actually refer to *T. longipes* Brandt; and he implies that *gregaria* may not occur here at all. See his paper for a consideration of other euphausiids commonly found near shore, especially along the coasts of Washington and British Columbia.

Order DECAPODA. Shrimps, Lobsters, and Crabs

In this, as in other appropriate sections of the Bibliography, references dealing with two or more major groups in the hierarchy of classification are treated under the highest general category, which requires that papers dealing with both anomurans and brachyurans be listed under Decapoda, and so on. Within the decapods, the limits of the suborders are uncertain because there are anomurous brachyurans and brachyurous anomurans, etc. We follow here, as a matter of convenience, the classification suggested by Fenner Chace in Waterman's *Physiology of Crustacea*. A truly "natural" classification of this complex group may never be attained.

BOOLOOTIAN, R. A., A. C. GIESE, A. FARMANFARMAIAN, and J. S. TUCKER. 1959. Reproductive cycles of five west coast crabs. Physiol. Zool., **32** (4): 213–20, 3 figs.

The cycles are based on observations of female crabs carrying egg masses, not on "breeding" in the barnyard sense. Perhaps it might be better to refer to this activity as "brooding," an activity undertaken by hens after "breeding." Certainly, a crab that may produce several successive egg masses after one copulation should not be said to "breed" every 30 days, and the inference that certain species may "breed" only in summer could be wide of the mark; "breeding" may have occurred in winter or spring. This semantic problem becomes a biological one when the authors suggest that reproduction in some of the species studied may be correlated with availability of food for the larval stages, and then go on to discuss seasons of "breeding." This work should be compared with that of Knudsen (cited below).

GLASSELL, STEVE A. 1935. New or little-known crabs from the Pacific coast of northern Mexico. Trans. San Diego Soc. Nat. Hist., **8** (14): 91–106.

——— 1938. New and obscure decapod Crustacea from the West American coasts. Trans. San Diego Soc. Nat. Hist., **8** (33): 411–54, 10 pls.

HART, JOSEPHINE F. L. 1930. Some decapods from the southeastern shores of Vancouver Island. Canad. Fld. Nat., **44** (5): 101–9.

Describes a new species of *Spirontocaris.*

——— 1940. Reptant decapod Crustacea from the west coasts of Vancouver and Queen Charlotte Islands, B.C. Canad. J. Res., D, **18**: 86–105.

KNUDSEN, JENS W. 1964. Observations on Brachyura and Anomura of Puget Sound. Pac. Sci. **18** (1): 3–33, 1 fig.

The temperature curves in this paper are based on almost weekly readings made by the author at Point Defiance on Puget Sound, five feet below the surface, for the appropriate periods. The various cycles of oogenesis, egg deposition (formation of berry), and hatching are clearly separated, as they are not in the similar but more preliminary paper by Boolootian, *et al.*; copulation, when observed, has been carefully recorded. Because of this difference in treatment, it is not possible to compare the reproductive cycles observed in Puget Sound with those observed for the same species at Monterey.

PETERS, HANS M. 1955. Die Winkgebärde von *Uca minuca* (Brachyura) in vergleichend-ethologischer, -ökologischer und -morphologischer-anatomischer Betrachtung. Zeitschr. Morph. u. Ökol. Tiere, **43**: 425–500, 51 figs.

Examples of the large and growing literature on behavior. Fiddler crabs are favorite subjects for this sort of thing. No one seems to have studied our own vanishing species adequately (vanishing in the sense that its habitat is vanishing).

RATHBUN, MARY J. 1904. Decapod crustaceans of the northwest coast of North America. Harriman Alaska Exp., **10**: 1–210.

Especially useful for northern *Spirontocaris.*

SCHMITT, WALDO L. 1921. Marine decapod Crustacea of California. Univ. Calif. Pub. Zool., **23**: 1–470.

WAY, W. F. 1917. Brachyura and crab-like Anomura of Friday Harbor, Washington. Pub. Puget Sd. Biol. Sta., **1**: 349–96.

Suborder NATANTIA. Shrimps and Prawns

BERKELEY, A. A. 1929. Commercial shrimps of British Columbia. Museum and Art Notes, **4** (3): 109–15.

BONNOT, PAUL. 1932. The California shrimp industry. Div. Fish and Game of Calif., Fish Bull. No. 38. 22 pp.

HART, JOSEPHINE F. L. 1964. Shrimps of the genus *Betaeus* on the Pacific coast of North America, with descriptions of three new species. Proc. U.S. Nat. Mus., **115** (3490): 431–66, 80 figs., 2 pls.

HYNES, FRANK W. 1929. Shrimp fishery of southeastern Alaska. Bur. Fish. Doc. **1052**. 18 pp.

NEEDLER, A. BERKELEY. 1934. Larvae of some British Columbia Hippolytidae. Contr. Canad. Biol. and Fish., **8**: 237–42.

Spirontocaris paludicola, S. brevirostris, etc.; *Hippolyte californiensis.*

NEWMAN, WILLIAM A. 1963. On the introduction of an edible oriental shrimp (Caridea: Palaemonidae) to San Francisco Bay. Crustaceana, **5** (2): 118–32, 3 figs.

ISRAEL, H. R. 1936. A contribution toward the life histories of two California shrimps. Calif., Fish Bull. No. 46, 28 pp.

Crangon californiensis (Holmes) .... §278

JOHNSON, MARTIN W., F. A. EVEREST, and R. W. YOUNG. 1947. The role of snapping shrimp (*Crangon* and *Synalpheus*) in the production of underwater noise in the sea. Biol. Bull., **93** (2): 122–38, 8 figs.

C. dentipes (Guèrin) .... §143

Hippolysmata californica Stimpson .... §134

Hippolyte californiensis Holmes .... §230

CHACE, FENNER A., JR. 1951. The grass shrimps of the genus *Hippolyte* from the west coast of North America. J. Wash. Acad. Sci. **41** (1): 35–39, 1 fig.

Restricts range of *H. californiensis* to Gulf of California to Bodega Bay; the species found from Puget Sound to Alaska is described as *H. clarki.*

Palaemon macrodactylus Rathbun .... §345

Pandalus danae Stimpson, and similar large, usually deep-water, shrimps ..... §283

BERKELEY, A. A. 1930. The post-embryonic development of the common pandalids of British Columbia. Contr. Canad. Biol. and Fish., n.s., **6** (6): 79–163.

Spirontocaris brevirostris (Dana) .... §150

S. *camtschatica* (Stimpson), and S. *cristata* (Stimpson) .... §278

S. *palpator* (Owen) .... §150

S. *paludicola* (Holmes) .... §278

S. *picta* (Stimpson) .... § 62

S. *prionota* (Stimpson) .... §150

S. *taylori* (Stimpson) .... §150

Suborder REPTANTIA

Section MACRURA

Superfamily SCHYLLARIDEA. Spiny Lobsters

Panulirus interruptus (Randall) .... §148

BACKUS, JOHN. 1960. Observations on the growth rate of the spiny lobster. Calif. Fish and Game, **46** (2): 177–81, 2 figs.

LINDBERG, ROBERT G. 1955. Growth, population dynamics, and field behavior in the spiny lobster, *Panulirus interruptus* (Randall). Univ. Calif. Pub. Zool., **59** (6): 157–248, 7 pls., 16 text figs.

WILSON, ROBERT C. 1948. A review of the southern California spiny lobster fishery. Calif. Fish and Game, **34** (2): 71–80, 6 figs.

Superfamily THALASSINIDEA. Ghost Shrimps and Mud Shrimps

Callianassa affinis Holmes .... § 57

C. californiensis Dana .... §292

MACGINITIE, G. E. 1934. The natural history of *Callianassa californiensis* Dana. Amer. Midland Nat., **15** (2): 166–77.

TOLLEFSON, H., and LOWELL D. MARRIAGE. 1949. The ghost shrimp fishery. Ore. Fish. Comm. Shellfish Invest. Progr. Rept. 16. 6 pp.

The fishing method is a portable pump, which forces the animals out of their burrows by pressure. They are sold as bait.

C. gigas Dana (= *C. longimana* Stimpson) .... §313

Upogebia pugettensis (Dana)....................................§313

MACGINITIE, G. E. 1930. The natural history of the mud shrimp *Upogebia pugettensis* (Dana). Ann. Mag. Nat. Hist., (10) **6**: 36–44.

Section ANOMURA
Hermit Crabs, Porcelain Crabs, Mole Crabs

GLASSELL, STEVE A. 1945. Four new species of North American crabs of the genus *Petrolisthes*. J. Wash. Acad. Sci., **35** (7): 223–29.

GLYNN, PETER W. 1961. The first recorded mass stranding of pelagic red crabs, *Pleuroncodes planipes*, at Monterey Bay, California, since 1859, with notes on their biology. Calif. Fish and Game, **47** (1): 97–101, 1 fig.

These bright red creatures caused quite a stir—all part, perhaps, of the perturbations of the ocean in recent years.

HART, JOSEPHINE F. L. 1937. Larval and adult stages of British Columbian Anomura. Canad. J. Res., D, **15**: 179–220.

JOHNSON, MARTIN W., and W. M. LEWIS. 1942. Pelagic larval stages of the sand crabs. Biol. Bull., **83** (1): 67–87.

KNIGHT, MARGARET D. 1966. The larval development of *Polyonyx quadrangulatus* Glassell and *Pachycheles rudis* Stimpson (Decapoda: Porcellanidae) cultured in the laboratory. Crustaceana, **10** (1): 75–97, 62 figs.

——— 1967. The larval development of the sand crab *Emerita rathbunae* Schmitt (Decapoda: Hippidae). Pac. Sci., **21** (1): 58–76, 47 figs.

MACKAY, DONALD C. G. 1932. Description of a new species of crab of the genus *Paralithodes*. Contr. Canad. Biol. and Fish., **7**: 335–40.

P. rostratus from 15 fathoms, near Prince Rupert, B.C.

MAKAROV, V. V. 1938. Anomura: Fauna of U.S.S.R. Crustacea, **10** (3). Translated from the Russian by the Israel Program for Scientific Translations, for the National Science Foundation, Washington, D.C., 1962. 278 pp., 113 figs., 5 pls.

Pagination of Russian original indicated in translation; for systematic purposes, this must, of course, be dated as of 1938. Includes North Pacific species in the old sense of Anomura (i.e., including *Thalassinidea*, etc.).

STEVENS, B. A. 1925. Hermit crabs of Friday Harbor, Washington. Pub. Puget Sd. Biol. Sta., **3**: 273–310.

——— 1928. *Callianassidae* from the west coast of North America. Pub. Puget Sd. Biol. Sta., **6**: 315 69.

Blepharipoda occidentalis Randall..................................§186

Cryptolithodes sitchensis Brandt...................................§132

Emerita analoga (Stimpson)...§186

EFFORD, IAN E. 1965. Aggregation in the sand crab *Emerita analoga* (Stimpson). J. Anim. Ecol., **34**: 63–75, 5 figs.

——— 1966. Feeding in the sand crab *Emerita analoga* (Stimpson) (Decapoda: Anomura). Crustaceana, **10** (2): 167–82, 9 figs., 1 pl.

——— 1967. Neoteny in sand crabs of the genus *Emerita*. (Anomura: Hippidae). Crustaceana, **13** (1): 81–93, 5 figs.

——— In process. Distribution of the sand crab *Emerita analoga* in the northeastern Pacific Ocean.

The small males of *E. analoga* are sexually precocious.

JOHNSON, MARTIN W. 1940. The correlation of water movements and dispersal of pelagic littoral animals, especially the sand crab *Emerita*. J. Mar. Res., **2** (3): 236–45, 4 figs.

KNOX, CAMERON, and RICHARD A. BOOLOOTIAN. 1963. Functional morphology of the external appendages of *Emerita analoga*. Bull. S. Calif. Acad. Sci., **62** (2): 45–68, 28 figs.

Although the similar paper by Snodgrass (1952) is not cited, this paper appears to have been organized on the same pattern. Obviously, Snodgrass's paper was not read by either author, or they might have investigated the interesting problem posed by the different structure of the fourth leg of the male in *Emerita analoga*, which was observed by Snodgrass. For that matter, a number of the papers cited were listed according to the same abridgements used in other references, so they apparently were not read either.

MacGinitie, G. E. 1938. Movements and mating habits of the sand crab, *Emerita analoga*. Amer. Midland Nat., **19**: 471–81.

Mead, H. T. 1917. Notes on the natural behavior of *Emerita analoga* (Stimpson). Univ. Calif. Pub. Zool., **16** (23): 431–38, 1 fig.

Snodgrass, R. E. 1952. The sand crab *Emerita talpoida* (Say) and some of its relatives. Smiths. Misc. Colls., **117** (8). 34 pp., 11 figs.

The structural anatomy, from the viewpoint of functional adaptations, in the various sand crabs, including *E. analoga*.

Weymouth, Frank W. 1919. Notes on the habits and use of the small sand crab (*E. analoga*). Calif. Fish and Game, **5**: 171–72.

———, and Richardson, C. H. 1912. Observations on the habits of the crustacean *Emerita analoga*. Smiths. Misc. Colls., **59** (7): 1–14, 1 pl.

Kurup, N. G. 1964. The intermolt cycle of an anomuran, *Petrolisthes cinctipes* Randall (Crustacea: Decapoda). Biol. Bull., **127** (1): 97–107, 19 figs.

Section BRACHYURA
The True Crabs: Shore Crabs, Pea Crabs, Spider Crabs, Edible Crabs, Fiddler Crabs, etc.

Edwards, E. 1966. Mating behaviour in the European edible crab (*Cancer pagurus* L.). Crustaceana, **10** (1): 23–30, 2 figs., 3 pls.

Notes attraction between sexes (ectohormones?). Compare this with the study by Snow and Nielsen under *Cancer magister*.

Garth, John S. 1958. Brachyura of the Pacific coast of America: Oxyrhyncha. Allan Hancock Pac. Exp., **21**. Part 1, text; Part 2, tables and plates.

———, and W. Stephenson. 1966. Brachyura of the Pacific coast of America: Portunidae. Allan Hancock Monogr. Mar. Biol., **1**. 154 pp., 12 pls.

Glassell, Steve A. 1933. Descriptions of five new species of Brachyura collected on the west coast of Mexico. Trans. San Diego Soc. Nat. Hist., **7** (28): 331–44.

HAIG, JANET. 1960. The Porcellanidae (Crustacea: Anomura) of the eastern Pacific. Allan Hancock Pac. Exp., **24**. 440 pp., 41 pls., 12 text figs., frontispiece.

HART, JOSEPHINE F. L. 1935. The larval development of British Columbia Brachyura, I. Canad. J. Res., **12**: 411–32, illus.

———. 1960. The larval development of British Columbia Brachyura, II: Majidae, Subfamily Oregoniinae. Canad. J. Zool., **38**: 539–46, 38 figs.

HOPKINS, THOMAS S., and THOMAS B. SCANLAND. 1964. The host relations of a pinnotherid crab, *Opisthopus transversus* Rathbun (Crustacea: Decapoda). Bull. S. Calif. Acad. Sci., **63** (4): 175–80, 2 figs.

KNUDSEN, JENS W. 1960. Reproduction, life history, and larval ecology of the California Xanthidae, the pebble crabs. Pac. Sci., **14** (1): 3–17, 4 figs.

MENZIES, ROBERT J. 1948. A revision of the brachyuran genus *Lophopanopeus*. Occ. Paps. Allan Hancock Found., **4**. 44 pp., 6 pls., 3 graphs.

MIR, ROBERT D. 1961. The external morphology of the first zoeal stages of the crabs *Cancer magister* (Dana), *Cancer antennarius* (Stimpson), *Cancer anthonyi* (Rathbun). Calif. Fish and Game, **47** (1): 103–11, 12 figs.

PEARCE, JACK B. 1962. Adaptation in symbiotic crabs of the family Pinnotheridae. Biologist, **45** (1–2): 11–15, 1 fig.

RATHBUN, MARY J. 1918. The grapsoid crabs of America. U.S. Nat. Mus. Bull., **97**. 461 pp., 161 pls., 172 figs.

——— 1925. The spider crabs of America. *Ibid.*, **129**. 613 pp., 283 pls., 153 figs.

——— 1930. The cancroid crabs of America of the families Euryalidae, Portunidae, Atelecyclidae, Cancridae, and Xanthidae. *Ibid.*, **152**. 609 pp., 230 pls., 85 figs.

——— 1937. The oxystomatous and allied crabs of America. *Ibid.*, **166**. 278 pp., 86 pls., 47 figs., endpiece.

WELLS, W. W. 1928. *Pinnotheridae* of Puget Sound. Pub. Puget Sd. Biol. Sta., **6**: 283–314.

——— 1940. Ecological studies on the Pinnotherid crabs of Puget Sound. Univ. Wash. Pub. Oceanogr., **2** (2): 19–50.

Similar to *C. magister*, but with more rounded and somewhat smaller, smoother carapace. Intertidal, especially from Dillon Beach northwards, and offshore in the South. Somehow overlooked in previous editions.

MACKAY, DONALD C. G. 1942. The Pacific edible crab, *Cancer magister*. Fish. Res. Bd. Canad. Bull., **62**. 32 pp.

———, and FRANK W. WEYMOUTH. 1935. The growth of the Pacific edible crab, *Cancer magister* Dana. J. Biol. Bd. Canad., **1**: 191–212.

POOLE, RICHARD L. 1966. A description of laboratory-reared zoeae of *Cancer magister* Dana, and megalopae taken under natural conditions (Decapoda: Brachyura). Crustaceana, **11** (1): 83–97, 7 figs.

SNOW, C. DALE, and JOHN R. NIELSEN. 1966. Pre-mating and mating behavior of the Dungeness crab (*Cancer magister* Dana). J. Fish. Res. Bd. Canad., **23** (9): 1319–23, 8 figs.

WALDRON, KENNETH D. 1958. The fishery and biology of the Dungeness crab (*Cancer magister* Dana) in Oregon waters. Fish. Comm. of Ore. Contr. **24**. 43 pp., 13 figs.

PEARCE, JACK B. 1966. The biology of the mussel crab *Fabia subquadrata*, from the waters of the San Juan Archipelago, Washington. Pac. Sci., **20** (1): 3–35, 7 figs.

DEHNEL, PAUL A., and DMITRY STONE. 1964. Osmoregulatory role of the antennary gland in two species of estuarine crabs. Biol. Bull., **126** (3): 354–72, 6 figs.

TODD, MARY-ELIZABETH, and PAUL A. DEHNEL. 1960. Effect of temperature and salinity on heat tolerance in two grapsoid crabs, *Hemigrapsus nudus* and *Hemigrapsus oregonensis*. Biol. Bull., **118** (1): 150–72, 7 figs.

SYMONS, P. E. K. 1964. Behavioral responses of the crab *Hemigrapsus oregonensis* to temperature, diurnal light variation, and food stimuli. Ecology, **45** (3): 580–91, 5 figs.

GROSS, WARREN J. 1959. The effect of osmotic stress on the ionic exchange of a shore crab. Biol. Bull., **116** (2): 248–57.

———, and LEE ANN MARSHALL. 1960. The influence of salinity on the magnesium and water fluxes of a crab. Biol. Bull., **119** (3): 440–53, 1 fig.
A neat diagram summarizes the findings.

HIATT, ROBERT M. 1948. The biology of the lined shore crab, *Pachygrapsus crassipes* Randall. Pac. Sci., **2** (3): 135–213, 2 pls., 18 figs.

GLASSELL, STEVE A. 1933. Notes on *Parapinnixa affinis* Holmes and its allies. Trans. San Diego Soc. Nat. Hist., **7** (27): 319–30.

PEARCE, JACK B. 1966. On *Pinnixa faba* and *Pinnixa littoralis* (Decapoda: Pinnotheridae) with the clam *Tresus capax* (Pelecypoda: Mactridae), in *Some contemporary studies in marine science*, pp. 565–89 (London, Allen and Unwin).
Tresus capax is infested by pinnixid crabs, whereas *T. nuttalli* is rarely so (at Bodega Bay, not at Humboldt Bay or Puget Sound).

SMITH, R. I. 1967. Osmotic regulation and adaptive reduction of water-permeability in a brackish-water crab, *Rhithropanopeus harrisi* (Brachyura: Xanthidae). Biol. Bull., **133** (3): 643–58, 5 figs.

CRANE, JOCELYN. 1941. Crabs of the genus *Uca* from the west coast of Central America. Zoologica, **26** (3): 145–208, 8 figs., 9 pls.
Uca crenulata, p. 198, etc.

DEMBROWSKI, JAN B. 1926. Notes on the behavior of the fiddler crab. Biol. Bull., **50**: 179–201.

Superorder HOPLOCARIDA (STOMATOPODA)

BIGELOW, R. P. 1894. Report upon the Crustacea of the order *Stomatopoda.* Proc. U.S. Nat. Mus., **17**: 489–550.

SCHMITT, WALDO L. 1940. The stomatopods of the west coast of America. Allan Hancock Pac. Exp., **5** (4): 129–225, 33 text figs.
Monographic for the region covered, chiefly in the Panamic area. Includes species occurring in California.

Subphylum **LABIATA**

Class CHILOPODA

Order GEOPHILOMORPHA

Some small centipedes are strongly halophilic, if not intertidal, and species have been described from many parts of the world. They occur in our region, but are often overlooked; and one species has been found at one of our favorite collecting localities.

CHAMBERLIN, RALPH V. 1960. A new marine centipede from the California littoral. Proc. Biol. Soc. Wash., **73**: 99–102, 2 figs.

CLOUDSLEY-THOMPSON, J. L. 1948. *Hydroschendyla submarina* (Grube) in Yorkshire: with an historical review of the marine Myriapoda. The Naturalist (London), Oct.–Dec., **1948**: 149–52.

——— 1951. Supplementary notes on Myriapoda. *Ibid.*, Jan.–Mar., **1951**: 16–17.

Attempts complete bibliography to 1951 of marine centipedes and millipedes.

Class INSECTA

Order Coleoptera

Intertidal insects are either rare or seldom sought after, but certain small beetles occur in the rocky intertidal, especially at Moss Beach and near Dillon Beach (where the editor found one associated with a sponge). Five genera, in four families, are represented on the Pacific coast: *Thalassotrechus* (Carabidae), *Liparocephalus*, *Diaulota* (Staphylinidae), *Endeodes* (Melyridae), and *Eurystethes* (Eurystethidae). Because of the considerable ecological interest of this group (despite which almost nothing is known of the life histories of these animals), we list the available literature. Perhaps this will stimulate someone to investigate them.

Blackwelder, R. E. 1932. The genus *Endeodes* Leconte (Coleoptera, Melyridae). Pan-Pac. Ent., **8** (3): 128–36.

Chamberlin, J. C., and G. F. Ferris. 1929. On *Liparocephalus* and allied genera. *Ibid.*, **5** (3): 137–43; **5** (4): 153–62.

Moore, Ian. 1956. A revision of the Pacific coast Phytosi with a review of the foreign genera (Coleoptera: Staphylinidae). Trans. San Diego Soc. Nat. Hist., **12** (7): 103–52, 4 pls.

——— 1956. Notes on some intertidal Coleoptera with descriptions of the early stages (Caribidae, Staphylinidae, Malachiidae). *Ibid.*, **12** (11): 207–30, 4 pls.

——— 1964. The Staphylinidae of the marine mud flats of southern California and northwestern Baja California (Coleoptera). *Ibid.*, **13** (12): 269–84, 4 figs.

Saunders, L. G. 1928. Some marine insects of the Pacific coast of Canada. Ann. Ent. Soc. Amer., **21** (4): 521–45.

Usinger, Robert L., ed. 1956. Aquatic insects of California. With keys to North American genera and California species. Berkeley: Univ. Calif. Press. 508 pp., illus.

Somewhere in this volume, most of the common insects encountered on the shore, in salt ponds and marshes, will be found.

Order Diptera

Many flies, of course, frequent the beaches; of these, the kelp flies, *Fucellia rufitibia* and *F. costalis*, which swarm around the masses of decaying kelp on the beach, are the most noticeable; at times they are so numerous as to make a visit to the shore (especially on a calm day) unpleasant. The first is a small fly with red tibiae; the second, somewhat larger, is predacious and may bite.

Smith, Leslie M., and Homer N. Lowe. 1948. The black gnats of California. Hilgardia, **18** (3): 157–83, 16 figs.

Describes the life history of the Bodega black gnat (*Holoconops kerteszi* Kieffer), which breeds in sand near the sea. This biting gnat is sometimes unpleasantly abundant at Dillon Beach, and is also known from Great Salt Lake and North Africa.

Wirth, W. W. 1949. A revision of the clunionine midges, with descriptions of a new genus and four new species (Diptera: Tendipedidae). Univ. Calif. Pub. Entom., **8** (4): 151–82, 7 figs.

These midges are often extremely abundant (especially during the winter months) on rocks in the intertidal. Some of the species are completely flightless, in some the males have functional wings, and in others both sexes are capable of flight. Immature stages are to be sought among *Enteromorpha* and similar green algae in the high intertidal.

——— 1952. The Heleidae of California. *Ibid.*, **9** (2): 95–266, 33 figs.

Monograph of the biting midges. Assigns *Holoconops* to *Leptoconops*. Discusses *Culicoides tristriatulus* (pp. 173–75) as a severe pest along parts of the Alaskan coast, especially at Valdez. It breeds in salt marshes in Alaska and northern California.

Phylum MOLLUSCA

The term mollusk is derived from Aristotle's word for the cuttlefish, and is considered to mean soft-bodied. Linnaeus grossly misunderstood this phylum, and he lumped several diverse groups, including echinoderms, with his Mollusca. Our modern concept of the phylum dates from Cuvier. All these naturalists understood the Mollusca as essentially soft-bodied animals, many of which produced external shells. They are indeed animals in which the processes of evolution are manifested chiefly by the rearrangements and specializations of their soft parts; they are "visceral beings." Those students who are aware of the true nature of mollusks and devote most of their time to studying the soft anatomy, especially in relation to function, are termed malacologists. Students to whom the principal appeal of mollusks is their shells are known as conchologists.

Since many more have responded to the esthetic and geometrical appeal of sea shells than to guts and gizzards, and can study them with little training in zoology or attention to the viscera, many well-meaning amateurs have managed to confuse the nomenclature with inadequately described species and misidentifications. Almost daily, some name has to be changed. Sometimes this is because of a legitimate reappraisal of zoological affinities that requires assignment to a different genus; but perhaps just as often it is because too many cooks have stirred the brew. This business is not, of course, a peculiarity of molluscan names; but because more people are interested in sea shells than in crabs or worms, the name changes and inadequate descriptions are more apparent.

Some of the most beautiful of mollusks have no shells, and the Pacific coast has a fair share of nudibranchs and tectibranchs, or sea slugs. Many of these are difficult to identify, since most of the critical characters are based on the internal anatomy. We can offer little encouragement to anyone who wants to know these animals intimately but does not care to dissect them and follow out the mysterious arrangements of their reproductive systems. Study of these animals on the basis of external appearances alone is the mark of the unreconstructed amateur. Often we are asked how to preserve nudibranchs so that they will not lose their color or shape; there is no way. However, color photography is now a universal and comparatively inexpensive hobby; good pictures have been taken with no more than a color Brownie camera.

Should anyone decide from the foregoing that the editor is hostile to conchologists, he hastens to state that he was born in the house next door to Henry Hemphill (an ardent conchologist and collector of the 1890's and later); and that the first live expert he met was probably Mrs. Ida Oldroyd, rustling among Hemphill's specimen cabinets. Henry Hemphill's demonstrations of natural variation, shells glued in patterns of fans and rosettes on blue cards with white trimmings, made an enduring impression on a young mind. But it is getting late in the day for reliance on shells and external characters alone; too late, in fact, for students of nudibranchs who look inside their lovely beasts only under duress.

In recent years, the study of Pacific Coast mollusks, by both malacologists and conchologists, has been encouraged by the establishment of *The Veliger*. Mollusks, as important components of the intertidal biota, have also attracted the interest of ecologists and students of behavior. Keeping up with all this literature is the task of Sisyphus; entire books can be overlooked, in spite of this computerized age.

ABBOTT, R. TUCKER. 1954. American seashells (rev. ed., 1956). New York: Van Nostrand. 541 pp., illus.
Also available in a condensed, paperback edition.

COWAN, I. MCT. 1964. New information on the distribution of marine Mollusca on the coast of British Columbia. Veliger, **7** (2): 110–13.

Penitella conradi extends from San Francisco north to Esperanza Inlet, boring in *Haliotis kamtschatikana*; *Opalia chacei* north from central Oregon to Queen Charlotte Strait; *Amphissa versicolor* north from Oregon to Queen Charlotte Strait.

DUGGAN, ELEANOR P. 1963. Report of non-indigenous marine shells collected in the state of Washington. Veliger, **6** (2): 112.

Notes appearance of *Nassarius fraterculus* (with oysters?) from Japan, and other exotic species.

GRANT, U. S., IV, and H. R. GALE. 1931. Catalogue of the marine Pliocene and Pleistocene Mollusca of California and adjacent regions. Mem. San Diego Soc. Nat. Hist., **1**: 1–1036.

A fairly definitive source for the terminology and taxonomy of extant local mollusks, despite its paleontological outlook.

HALL, CLARENCE A., JR. 1964. Shallow water marine climates and molluscan provinces. Ecology, **45** (2): 226–34, 6 figs.

HANNA, G. DALLAS. 1966. Introduced mollusks of western North America. Occ. Paps. Calif. Acad. Sci., **48**. 108 pp., 85 figs., 4 pls.

Considers terrestrial, freshwater, and marine introductions, immigrants, and strays.

KEEN, A. MYRA. 1937. An abridged checklist and bibliography of west American marine Mollusca. Stanford: Stanford Univ. Press. 88 pp.

——— 1958. Sea shells of tropical west America: Marine mollusks from Lower California to Colombia. Stanford: Stanford Univ. Press. 624 pp., text figs., 10 pls.

A magnificent book, although most of the species considered do not fall within our range. A companion volume for these northern climes would be useful.

——— 1962. Nomenclatural notes on some west American mollusks, with proposal of a new species name. Veliger, **4** (4): 178–80.

Scuttles dear old *Schizothaerus* in favor of *Tresus,* among other legalistic matters.

——— 1963. Marine molluscan genera of western North America: An illustrated key. Stanford: Stanford Univ. Press. 126 pp., illus.

Includes keys to all shelled genera, including chitons, some cephalopods, and appropriate opisthobranchs. Nudibranchs, of course, are not included. Nevertheless, a most useful guide.

———, and ALLYN G. SMITH. 1961. West American species of the bivalved gastropod genus *Berthelinia.* Proc. Calif. Acad. Sci., Ser. 4, **30** (2): 47–66, 33 figs., 1 col. pl.

KEEP, JOSIAH. 1935. West coast shells (revised by Joshua L. Baily, Jr.). Stanford: Stanford Univ. Press. 350 pp.

LA ROCQUE, AURELE. 1953. Catalogue of the Recent Mollusca of Canada. Bull. Nat. Mus. Canad., **129**. 406 pp.

A useful index to species of both Canadian coasts, as well as freshwater and terrestrial species. References to literature; type, locality, and range.

LEMCHE, HENNING, and KARL GEORG WINGSTRAND. 1959. The anatomy of *Neopilina galatheae* Lemche, 1957 (Mollusca: Tryblidiacea). Galathea Repts., **3**: 9–72, 56 pls.

A modern classic.

MOORE, RAYMOND C., ed. 1960. Treatise on invertebrate paleontology, Part I: Mollusca 1. Mollusca—General features. Scaphopoda. Amphineura. Monoplacophora. Gastropoda—General features. Archaeogastropoda and some (mainly Paleozoic) Caenogastropoda and Opisthobranchia. Geol. Soc. Amer. and Univ. Kansas Press. 351 pp.

Quite a title page. Useful for confirming genera, and as a general introduction.

MORTON, J. E. 1960. Molluscs: An introduction to their form and functions. New York: Harper Torchbooks. 232 pp. 23 figs.

Originally published 1958. Because of its availability as a paperback, this work inhabits many bookshelves. Unfortunately, it was rather loosely written and there are many errors of fact and emphasis. Too bad.

NEWELL, IRWIN M. 1948. Marine molluscan provinces of western North America: A critique and a new analysis. Proc. Amer. Phil. Soc., **92** (3): 155–66, 7 figs.
Criticism of Schenck and Keen (1936, 1937, 1940).

OLDROYD, IDA S. 1924. Marine shells of Puget Sound and vicinity. Pub. Puget Sd. Biol. Sta., **4**: 1–272.

——— 1924–27. Marine shells of the west coast of North America, Stanford Univ. Pub., Univ. Ser., Geol. Sci., I; II, Parts 1, 2, and 3.

PACKARD, E. L. 1918. Molluscan fauna from San Francisco Bay. Univ. Calif. Pub. Zool., **14**: 199–452.

PARIS, OSCAR H. 1960. Some quantitative aspects of predation by muricid snails on mussels in Washington Sound. Veliger, **2** (3): 41–47, 1 pl.
Suggests that limit and lower boundary are caused by predation.

QUAYLE, D. B. 1964. Distribution of introduced marine Mollusca in British Columbia waters. J. Fish. Res. Bd. Canad., **21** (5): 1155–81, 10 figs.

SCHENCK, HUBERT G., and A. MYRA KEEN. 1936. Marine molluscan provinces of western North America. Proc. Amer. Phil. Soc., **76** (6): 921–38.

——— 1937. An index method for comparing molluscan faunules. Proc. Amer. Phil. Soc., **77** (2): 161–82.

——— 1940. Biometric analysis of molluscan assemblages. Société biogeographique, **77** (2): 379–92, 2 pls.

SMITH, ALLYN G., and MACKENZIE GORDON, JR. 1948. The marine mollusks and brachiopods of Monterey Bay, California, and vicinity. Proc. Calif. Acad. Sci., (4) **26** (8): 147–245, 2 pls., 4 figs.
Comprehensive checklist, with descriptions of several new species and notes on ecology and distribution.

STASEK, CHARLES R. 1967. Autotomy in the Mollusca. Occ. Paps. Calif. Acad. Sci., **61**. 44 pp., 11 figs.

VALENTINE, JAMES W. 1966. Numerical analysis of marine molluscan ranges on the extratropical northeastern Pacific shelf. Limnol. and Oceanogr., **11** (2): 198–211, 7 figs.

WILBUR, KARL M., and C. M. YONGE, eds. 1964. Physiology of Mollusca, Vol. I. New York: Academic Press. 473 pp., illus.

——— 1966. Physiology of Mollusca, Vol. II. *Ibid.* 645 pp., illus.
Most of the information of interest to naturalists is in Volume I (ecology, reproduction, culture, etc.), but Volume II considers feeding, digestion, and ionic regulation. Serious students of Mollusca (i.e., malacologists) must have both volumes.

YONGE, C. M. 1947. The pallial organs in the aspidobranch Gastropoda and their evolution throughout the Mollusca. Phil. Trans. Roy. Soc. Lond., (B) **232** (591): 443–518, 1 pl., 40 figs.
A splendid paper by a master zoologist; by far the most interesting part of a mollusk is that which leaves the empty shell behind.

A comprehensive bibliography of Pacific coast mollusks is beyond our scope, although several of the works cited above contain useful bibliographies. We are still without a handy Pacific coast guide; Abbott's book concerns both coasts and omits many common Pacific coast species. The field book by Percy Morris is unsatisfactory, and we are still left with Josiah Keep's venerable *vade mecum*.

Class AMPHINEURA

Order POLYPLACOPHORA. Chitons (Sea Cradles)

BARNAWELL, EARL B. 1960. The carnivorous habit among the Polyplacophora. Veliger, **2** (4): 85–88, 1 pl.

——— 1960. *Mopalia hindsi recurians*, ssp. nov. (Amphineura). Veliger, **3** (2): 37–40, 1 pl.
From San Francisco Bay.

BERRY, S. STILLMAN. 1911. A new California chiton. Proc. Acad. Nat. Sci. Phil. for 1911: 487–92, 1 pl.

——— 1917 and 1919. Notes on west American chitons, I and II. Proc. Calif. Acad. Sci., (4) **7**: 229–48; and (4) **9**: 1–36.

——— 1922. Fossil chitons of western North America. Proc. Calif. Acad. Sci., (4) **11** (18): 399–526, 16 pls., text figs.

Of the 33 species treated, 25 are also Recent. There are detail illustrations, as well as geographic and bathymetric ranges, for these living species, making this as near as there is to a handbook of Pacific Recent species.

——— 1925. The species of Basiliochiton. Proc. Acad. Nat. Sci. Phil. for 1925: 23–29, 1 pl.

——— 1925. New or little-known southern California Lepidozoans. Proc. Malacol. Soc. Lond., **16** (5): 223–31, 1 pl.

——— 1927. Notes on some British Columbian chitons. Proc. Malacol. Soc. Lond., **17** (4): 159–64, 1 pl.

——— 1931. A redescription, under a new name, of a well-known Californian chiton. Proc. Malacol. Soc. Lond., **19** (5): 255–58, 1 pl.

——— 1946. A re-examination of the chiton *Stenoplax magdalenensis* (Hinds), with description of a new species. Proc. Malacol. Soc. Lond., **26** (6): 161–66, 2 pls.

DALL, W. H. 1919. Description of new species of chitons from the Pacific coast of North America. Proc. U.S. Nat. Mus., **55**: 499–516.

No illustrations. Short descriptions, including 17 intertidal species, mostly from southern California.

GIESE, A. C., J. S. TUCKER, and R. A. BOOLOOTIAN. 1959. Annual reproductive cycles of the chitons *Katharina tunicata* and *Mopalia hindsii*. Biol. Bull., **117** (1): 81–88, 3 figs.

HEATH, H. 1905. Breeding habits of chitons of the California coast. Zool. Anz., **29**: 390–93.

LOWENSTAM, HEINZ A. 1962. Magnetite in denticle capping in Recent chitons (Polyplacophora). Bull. Geol. Soc. Amer., **73**: 435–38, 1 pl.

This substance is hard enough to permit the abrasion of limestone (note the deep pits at La Jolla!) and might possibly be an aid in homing.

SMITH, ALLYN G. 1966. The larval development of chitons (Amphineura). Proc. Calif. Acad. Sci., **32** (15): 433–46, 11 figs.

Literature review.

THORPE, SPENCER R., JR. 1962. A preliminary report on spawning and related phenomena in California chitons. Veliger, **4** (4): 202–10, 1 fig.

Mostly about species of *Mopalia*.

YAKOVLEVA, A. M. 1966. Shell-bearing mollusks (Loricata) of the seas of the U.S.S.R. (*Pantsyrnye mollyuski morei SSSR*; 1952). Israel Program for Scientific Translations for the Smithsonian Institution. 127 pp., 11 pls.

YONGE, C. M. 1939. On the mantle cavity and its contained organs in the Loricata (Placophora). Quart. J. Micr. Sci., **81** (3): 367–90, 6 figs.

The term Loricata also applies to the group of crustaceans that includes the spiny lobsters, and to the crocodiles and alligators, as well as to the chitons. Slightly confusing.

OKUDA, SHIRO. 1947. Notes on the postlarval development of the giant chiton, *Cryptochiton stelleri* (Middendorff). J. Fac. Sci. Hokk. Imp. Univ. (6) Zool., **9** (3): 267–75, 14 figs.

TUCKER, JOHN S., and ARTHUR C. GIESE. 1959. Shell repair in chitons. Biol. Bull., **116** (2): 318–22, 1 fig.

Cyanoplax dentiens (Gould) . . . . §237

I. regularis (Carpenter) . . . . §128

Katharina tunicata (Wood) . . . . §177

GIESE, A. C., and M. A. HART. 1967. Seasonal changes in component indices and chemical composition in *Katharina tunicata*. J. Exper. Mar. Biol. and Ecol., **1** (1): 34–46, 7 figs.

Lepidochitona keepiana Berry . . . . §237

BERRY, S. STILLMAN. 1948. Two misunderstood west American chitons. Leaflets in Malacology, **1** (4): 13–15.

L. keepiana ranges from Monterey southward; *Cyanoplax dentiens* from Monterey northward. The two had been confused under the name *Lepidochitona dentiens.*

Lepidozona cooperi (Carpenter) . . . . §128

L. mertensi (Middendorff) . . . . §128

Mopalia ciliata (Sowerby) and *M. lignosa* (Gould) . . . . §128

M. muscosa (Gould) . . . . §221

Nuttallina californica (Reeve) . . . . §169

Placiphorella velata Carpenter . . . . §128

MCLEAN, JAMES H. 1962. Feeding behaviour of the chiton *Placiphorella*. Proc. Malacol. Soc. Lond., **35** (1): 23–26, 2 figs.

Eats live, moving prey (amphipods, shrimp, etc.), but may also subsist on algae. Flaps down on them.

Stenoplax conspicua (Carpenter) . . . . §127

S. heathiana Berry . . . . §127

Tonicella lineata (Wood) . . . . § 36

Class GASTROPODA

Snails, Limpets, Sea Hares, Nudibranchs, etc.

Subclass PROSOBRANCHIA

CROFTS, DORIS R. 1955. Muscle morphogenesis in primitive gastropods and its relation to torsion. Proc. Zool. Soc. Lond., **125** (3 & 4): 711–50, 30 figs.

EMERSON, DAVID N. 1965. Summer polysaccharide content in seven species of west coast intertidal prosobranch snails. Veliger, **8** (2): 62–66, 1 pl.

FRETTER, VERA, and ALASTAIR GRAHAM. 1962. British prosobranch molluscs. Their functional anatomy and ecology. London: Ray Society. 755 pp., 317 figs.

A monumental, exemplary, and indispensable work; a prime example of what a study of mollusks should be.

KEEN, A. MYRA, and C. L. DOTY. 1942. An annotated checklist of the gastropods of Cape Arago, Oregon. Ore. St. Monogr., Stud. in Zool., 3. 16 pp.

PAINE, ROBERT T. 1966. Function of labial spines, composition of diet, and size of certain marine gastropods. Veliger, **9** (1): 17–24, 2 figs.

Suggests that the spine is used to anchor predators while boring, or whatever.

STOHLER, RUDOLF. 1960. Fluctuations in mollusk populations after a red tide in the Estero de Punta Banda, Lower California, Mexico. Veliger, **3** (1): 23–28, 1 pl.

VOKES, H. E. 1936. The gastropod fauna of the intertidal zone at Moss Beach, San Mateo County. Nautilus, **50** (2): 46–50.

How many can still be found there after 30 years?

Order Archeogastropoda
Limpets, Abalones, Turbans

Suborder Fissurellacea. Keyhole Limpets, Abalones

Cox, Keith W. 1962. California abalones, Family Haliotidae. Calif. Dept. Fish and Game, Fish Bull. **118**. 133 pp., 61 figs., col. pls.
The standard reference for abalone fanciers.

Crofts, Doris R. 1929. *Haliotis*. Liverpool Mar. Biol. Comm. Mem., **29**: 1–174.

——— 1937. The development of *Haliotis tuberculata*, with special reference to organogenesis during torsion. Phil. Trans. Roy. Soc. Lond., (B), **228**: 219–68, 7 pls., 14 figs.

McLean, James H. 1966. A new genus of Fissurellidae and a new name for a misunderstood species of west American *Diodora*. L. A. County Mus. Contr. Sci., No. **100**. 8 pp., 4 figs.
A misunderstood name rather than species!

Pilson, Michael E. Q., 1965. Variation of hemocyanin concentration in the blood of four species of *Haliotis*. Biol. Bull., **128** (3): 459–72, 8 figs.

Margolin, Abe S. 1964. The mantle response of *Diodora aspera*. Anim. Behav., **12** (1): 187–94, 3 figs.

Carlisle, John G., Jr. 1945. The technique of inducing spawning in *Haliotis rufescens* Swainson. Science, n.s., **102**: 566.

Leighton, David L. 1961. Observations of the effect of diet on shell coloration in the red abalone, *Haliotis rufescens* Swainson. Veliger, **4** (1): 29–32, 1 pl.
When young abalone eat red algae, they are red; when they don't, they are not.

Lyons, Richard B. 1957. *Haliotis rufescens* at Sunset Bay, Oregon. Nautilus: **70** (4): 109–11.

Montgomery, David H. 1967. Responses of two haliotid gastropods (Mollusca), *Haliotis assimilis* and *Haliotis rufescens,* to the forcipulate asteroids (Echinodermata), *Pycnopodia helianthoides* and *Pisaster ochraceus.* Veliger, **9** (4): 359–68, 2 pls., 2 figs.
It would appear that abalone, with good reason, are more anxious to remove themselves from the vicinity of *Pycnopodia* than from that of *Pisaster*; but they obviously do not care to be eaten by either.

Suborder Patellacea. True Limpets

Frank, Peter W. 1965. Growth of three species of Acmaea. Veliger, **7** (3): 201–2.
A. digitalis may attain an age of 6 years; *A. paradigitalis,* 4 years.

Fritchman, Harry K., II. A study of the reproductive cycle in California Acmaeidae (Gastropoda). Part I, Veliger, **3** (3): 57–63, 1 pl. (1961). Part II, *ibid.* (4): 95–100, 3 pls. (1961). Part III, Veliger, **4** (1): 41–47, 6 pls. (1961). Part IV, *ibid.* (3): 134–40, 3 pls. (1962).

KESSEL, MARGARET M. 1964. Reproduction and larval development of *Acmaea testudinalis.* Biol. Bull., **127** (2): 294–303, 18 figs.

MARGOLIN, ABE S. 1964. A running response of *Acmaea* to seastars. Ecology, **45** (1): 191–93.

SHOTWELL, J. ARNOLD. 1950. Distribution of volume and relative linear measurement changes in *Acmaea,* the limpet. Ecology, **31** (1): 51–61, 11 figs.

Does not find that height of shell is correlated with zonal distribution.

——— 1950. The vertical zonation of *Acmaea,* the limpet. *Ibid.,* (4): 647–49, 3 figs.

Suggests desiccation as the primary factor in distribution of *Acmaea.*

TEST, A. R. (G.). 1945. Description of new species of *Acmaea.* Nautilus, **58** (3): 92–96.

——— 1945. Ecology of California *Acmaea.* Ecology, **26** (4): 395–405.

——— 1946. Speciation in limpets of the genus *Acmaea.* Contr. Lab. Vert. Zool., Univ. Mich., **31.** 24 pp.

Note on nomenclature: Various references to "Mrs. Grant" or "Grant (MS)" refer to her personal communications to E. F. Ricketts or to her unpublished thesis on the genus *Acmaea* on file in the University of California library. References to "Dr. A. R. (Grant) Test" are to the published work cited above.

TEST, FREDERICK H. 1945. Substrate and movements of the marine gastropod *Acmaea asmi.* Amer. Midland Nat., **33** (3): 791–93.

VILLEE, C. A., and T. C. GROODY. 1940. The behavior of limpets, with reference to their homing instinct. Amer. Midland Nat., **24**: 190–204.

Finds no evidence of homing in the strict sense. Bibliography includes Richardson (1934) and Wells (1917), Pacific references to the same subject.

YONGE, C. M. 1962. Ciliary currents in the mantle cavity of species of *Acmaea.* Veliger **4** (3): 119–23, 2 figs.

The 1948 ruling of the International Commission, concerning page priority, was withdrawn in favor of the first reviser in 1953; hence *A. cassis* is again *A. pelta.*

FRANK, PETER W. 1965. The biodemography of an intertidal snail population. Ecology, **46** (6): 831–44, 8 figs.

A distinct color variety of *A. limatula,* mistakenly called *A. scutum ochracea* Dall, occurs commonly at La Jolla and is one of the highest limpets.

SEAPY, ROGER R. 1966. Reproduction and growth in the file limpet, *Acmaea limatula* Carpenter, 1864. Veliger, **8** (4): 300–310, 7 figs.

SEGAL, EARL. 1956. Microgeographic variation as thermal acclimation in an intertidal mollusk. Biol. Bull., **111** (1): 129–52, 8 figs.

Both papers based on studies at Palos Verdes.

The puzzle of at least one of the intermediate forms of our limpets has been solved by Fritchman, who demonstrated that the "hybrid" between *A. digitalis* and *A. pelta* is actually a well-defined species. It occurs in the same region as *A. digitalis,* but lacks external ribbing and a massive brown spot at the interior apex.

FRITCHMAN, HARRY K., II. 1960. *Acmaea paradigitalis*, sp. nov. (Acmaeidae: Gastropoda). Veliger, **2** (3): 53–57, 4 pls.

A. pelta Eschscholtz (see note under *A. cassis pelta*) . . . . § 11

A. persona Eschscholtz . . . . § 11

A. scabra (Gould) . . . . § 11

HEWATT, WILLIS G. 1940. Observations on the homing limpet, *Acmaea scabra* Gould. Amer. Midland Nat., **24** (1): 205–8.

Finds evidence of homing.

Acmaea sp., similar to *A. scabra* . . . . § 25

A. scutum Eschscholtz, Plate Limpet . . . . § 25

A. triangularis (Carpenter) . . . . §274

Lottia gigantea Sowerby (*L. gigantea* [Gray] of authors), Owl Limpet . . . . § 7

ABBOTT, DONALD P. 1956. Water circulation in the mantle cavity of the owl limpet *Lottia gigantea* Gray. Nautilus, **69** (3): 79–87, 4 figs.

FISHER, W. K. 1904. The anatomy of *Lottia gigantea* Gray. Zool. Jahrb., **20** (1): 1–66.

Suborder TROCHACEA. Top Shells, Turbans

FRITCHMAN, HARRY K., II. 1965. The radulae of *Tegula* species from the west coast of North America and suggested intrageneric relationships. Veliger, **8** (1): 11–15, 10 figs.

T. gallina and *T. funebralis* appear related by radular type; other groups less coherent?

MCLEAN, JAMES H. 1964. New species of Recent and fossil west American aspidobranch gastropods. Veliger, **7** (2): 129–33, 1 fig., 1 pl.

States that common southern California *Tegula ligulata* really should be *T. mendella* n. sp., but codicil to paper says it was already named and should be referred to as *Tegula eiseni* Jordan—short-lived species! Los Angeles County to Magdalena Bay.

Astraea gibberosa (Dillwyn) (formerly *A. inequalis* [Martyn]) . . . . §117

A. undosa (Wood) . . . . §117

Calliostoma annulatum (Humphrey) . . . . §182

C. canaliculatum (Humphrey) . . . . §182

C. ligatum (Gould) (formerly *C. costatum* [Martyn]) . . . . §215

Norrisia norrisi (Sowerby) . . . . §136

Tegula brunnea (Philippi) . . . . § 29

T. eiseni Jordan (= *T. ligulata* [Menke]; see McLean, above) . . . . §137

T. funebralis (A. Adams) . . . . § 13

ABBOTT, DONALD P., *et al.*, eds. 1964. The biology of *Tegula funebralis* (A. Adams). Veliger, **6** (Supplement). 81 pp.

Consists of 19 papers on various aspects of *Tegula*, the results of an experimental course carried out at Hopkins Marine Station. The theme was this common snail; and each student worked on some aspect of the animal. Some of the papers are preliminary student reports; the main purpose in publishing them all in this manner was to demonstrate an interesting, and evidently successful, experiment in teaching biology at the seashore.

FRANK, PETER W. 1965. Shell growth in a natural population of the turban snail, *Tegula funebralis*. Growth, **29**: 395–403, 3 figs.

MCLEAN, JAMES H. 1962. Manometric measurements of respiratory activity in *Tegula funebralis*. Veliger, **4** (4): 191–93.

MERRIMAN, JEAN A. 1967. Systematic implication of radular structures of west coast species of *Tegula*. Veliger, **9** (4): 399–403, 2 pls., 2 figs.

Order MESOGASTROPODA
Periwinkles, Slipper Shells, Cowries, Moon Snails, Horn Shells, Heteropods, etc.

COLMAN, JOHN. 1932. A statistical test of the species concept in *Littorina.* Biol. Bull., **62** (3): 223–43, 11 figs.

DUERR, FREDERICK G. 1965. Survey of digenetic trematode parasitism in some prosobranch gastropods of the Cape Arago region, Oregon. Veliger, **8** (1): 42.
Littorina scutulata and *Olivella* parasitized; other common species not.

GHISELIN, MICHAEL T. 1964. Morphological and behavioral concealing adaptations of *Lamellaria stearnsii,* a marine prosobranch gastropod. Veliger, **6** (3): 123–24, 1 pl.
It looks like its prey and moves about at night.

JENSEN, AD. S. 1951. Do the Naticidae (Gastropoda: Prosobranchia) drill by chemical or by mechanical means? Vidensk. Medd. fra Dansk. Naturh. Foren., **113**: 251–61, 2 figs.
Favors boring by mechanical action with the radula, suggesting that acid is too weak to bore holes so rapidly.

HALL, R. P. 1925. Twinning in a mollusc. Science, **61**: 658.
As *Serpuloides vermicularis.*

COE, WESLEY R. 1948. Nutrition and sexuality in protandric gastropods of the genus *Crepidula.* Biol. Bull., **94** (2): 158–60.

——— 1949. Divergent methods of development in morphologically similar species of prosobranch gastropods. J. Morphol., **84** (2): 383–400, 10 figs.

CONKLIN, E. G. 1897. The embryology of *Crepidula.* J. Morphol., **13**: 1–226.

GOULD, HARLEY N. 1952. Studies on sex in the hermaphrodite mollusk *Crepidula plana,* IV: Internal factors influencing growth and sex development. J. Exp. Zool., **119** (1): 93–164, 16 figs., 1 pl.

MORITZ, C. E. 1939. Organogenesis in the gastropod *Crepidula adunca* Sowerby. Univ. Calif. Pub. Zool., **43** (11): 217–48, 22 figs.

ORTON, J. H. 1912. An account of the natural history of the slipper limpet (*Crepidula fornicata*). J. Mar. Biol. Assoc., n.s., **9**: 437–43. *And* The mode of feeding of the slipper limpet. *Ibid.*: 444–78.

MACGINITIE, NETTIE, and G. E. MACGINITIE. 1964. Habitat and breeding seasons of the shelf limpet *Crepidula norrisiarum* Williamson. Veliger, **7** (1): 34, 1 pl.
They say it breeds year-round; is found occasionally on the crab *Randallia bulligera.*

DARLING, STEPHEN D. 1965. Observations on the growth of *Cypraea spadicea.* Veliger, **8** (1): 14–15.
Does not grow much after maturity.

Order NEOGASTROPODA
Whelks, Rocksnails, Olives, Cones, etc.

HOWARD, FAYE B. 1962. Egg-laying in *Fusitriton oregonensis* (Redfield). Veliger, **4** (3): 160–68, 1 pl.
Nice photo of the process, and of rosette-like egg masses observed in June at about 57°N near Baranoff Island.

PHILPOTT, C. H. 1925. Observations on the early development of *Argobuccinum oregonensis*. Pub. Puget Sd. Biol. Sta., **3**: 369–81.

Ilyanassa obsoleta (Say) (*Nassa obsoleta*)..............................§340

DEMOND, JOAN. 1952. The Nassariidae of the west coast of North America between Cape San Lucas, Lower California, and Cape Flattery, Washington. Pac. Sci., **6** (4): 300–317, 2 pls.

DIMON, A. C. 1905. The mud snail *Nassa obsoleta*. Cold Spr. Harb. Monogr., **5**. 86 pp.

Nassarius fossatus (Gould) (*Nassa fossata, Alectrion fossatus*)..............§286

GORE, ROBERT H. 1966. Observations on the escape response in *Nassarius vibex* (Say) (Mollusca: Gastropoda). Bull. Mar. Sci., **16** (3): 423–34, 2 figs.
Our own behave the same way.

MACGINITIE, G. E. 1931. The egg-laying process of the gastropod *Alectrion fossatus* Gould. Ann. Mag. Nat. Hist., Ser. 10, **8**: 258–61.

N. tegula (Reeve) (*Alectrion tegula*)..............................§270

Ocenebra japonica (Dunker)..............................§340

Olivella baetica Carpenter (also known as *O. pedroana* and *O. intorta*).......§253

O. biplicata (Sowerby)..............................§253

STOHLER, RUDOLF. 1962. Preliminary report on growth studies in *Olivella biplicata*. Veliger, **4** (3): 150–51, 1 pl.

Opalia crenimarginata (Dall)..............................§ 64

ROBERTSON, ROBERT. 1963. Wentletraps (Epitoniidae) feeding on sea anemones and corals. Proc. Malacol. Soc. Lond., **35** (2, 3): 51–63, 3 pls.

THORSON, GUNNAR. 1957. Parasitism in the marine gastropod family Scalidae. Vidensk. Medd. fra Dansk Naturh. Foren., **119**: 55–58, 2 figs.
Reports on the situation at La Jolla.

Searlesia dira (Reeve)..............................§200

Thais canaliculata (Duclos), also known as *Purpura canaliculata*............§200

DALL, W. H. 1915. Notes on the species of the molluscan subgenus *Nucella*... northwest coast of America. Proc. U.S. Nat. Mus., **49**: 557–72.

T. emarginata (Deshayes)..............................§182

T. lamellosa (Gmelin)..............................§200

KINCAID, TREVOR. 1957. Local races and clines in the marine gastropod *Thais lamellosa* Gmelin: A population study. Seattle: The Calliostoma Co. [in Prof. Kincaid's garage]. 75 pp., 65 pls.

Urosalpinx cinereus (Say)..............................§286

FEDERIGHI, HENRY. 1931. Studies on the oyster drill (*Urosalpinx cinerea* Say). Bull. U.S. Bur. Fish., **47** (4): 85–115, 7 figs.

Subclass OPISTHOBRANCHIA

BARNES, H., and H. T. POWELL. 1954. *Onchidoris fusca* (Müller), a predator of barnacles. J. Anim. Ecol., **23** (2): 361–63, 1 pl.

GHISELIN, MICHAEL T. 1965. Reproductive function and the phylogeny of opisthobranch gastropods. Malacologia, **3** (3): 327–78, 7 figs.

HURST, ANNE. 1967. The egg masses and veligers of thirty northeast Pacific opisthobranchs. Veliger, **9** (3): 255–88, 31 figs., 13 pls.

LANCE, JAMES R. 1961. A distributional list of southern California opisthobranchs. Veliger, **4** (2): 64–69.

——— 1962. Two new opisthobranch mollusks from southern California. Veliger, **4** (3): 155–59, 8 figs., 1 pl.

Cadlina limbaughi, subtidal, San Diego area; *Phidiana pugnax*, intertidal, Pacific Grove to Point Loma.

——— 1962. A new *Stiliger* and a new *Corambella* (Mollusca: Opisthobranchia), from the northwestern [*sic*!] Pacific. Veliger, **5** (1): 33–38, 10 figs., 1 pl.

——— 1966. New distributional records of some northeastern Pacific Opisthobranchiata (Mollusca: Gastropoda), with descriptions of two new species. Veliger, **9** (1): 69–81, 12 figs.

MACFARLAND, FRANK M. 1966. Studies of opisthobranchiate mollusks of the Pacific coast of North America. Mem. Calif. Acad. Sci., **6**. 546 pp., 72 pls.

For years—for decades—we had awaited this great work; and finally, some 15 years after the author's death, it was published, according to the will of Mrs. MacFarland. There are many fine things in this book; but unfortunately we cannot say *ite missa est, nunc dimittis*, for many of the species proposed in this work had already been named by others in the interim, and it will take specialists as many more years to unscramble the complications. However, all this, and much more, has been set forth with marvelous tact and restraint by G. Dallas Hanna in the Introduction to the volume. For those who do not read prefaces, we can only say, *caveat emptor*.

MARCUS, ERNST. 1961. Opisthobranch mollusks from California. Veliger, **3** (Supplement): 1–85, 10 pls.

Provides definitive descriptions, with guts and gizzards, of most of the common species and some of the rare ones. Many sea slugs, alas, cannot be known from their pretty exteriors alone.

STEINBERG, JOAN E. 1961. Notes on the opisthobranchs of the west coast of North America, I: Nomenclatural changes in the order Nudibranchia (southern California). Veliger, **4** (2): 57–63.

——— 1963. Notes on the opisthobranchs of the west coast of North America, II: The order Cephalaspidea from San Diego to Vancouver Island. Veliger, **5** (3): 114–17.

——— 1963. Notes on the opisthobranchs of the west coast of North America, III: Further nomenclatural changes in the order Nudibranchia. Veliger, **6** (2): 63–67.

——— 1963. Notes on the opisthobranchs of the west coast of North America, IV: A distributional list of opisthobranchs from Point Conception to Vancouver Island. Veliger, **6** (2): 68–73.

Order CEPHALASPIDEA

Acteon punctocaelatus (Carpenter) (*Actaeon punctocoelatus* of authors) . . . . . §287

Aglaja diomedea (Bergh) . . . . not treated

This small, dark-brown to black slug frequents estuarine flats. It has been reported from Elkhorn Slough, and is the species that sporadically occurs in large numbers on the "High Clam Flats" of Dillon Beach. At the Dillon Beach locality, spawning occurs in July and August; eggs are deposited in pear-shaped masses on the flats. The large summer population of 1950 was reduced to a few scattered individuals by December. It also occurs in Puget Sound.

Bulla gouldiana (Pilsbry), also known as *Bullaria* and *Vesica* . . . . §287

BERRILL, N. J. 1931. The natural history of *Bulla hydatis* Linn. J. Mar. Biol. Assoc., **17**: 567–71.

Habits, development, and feeding mechanism of a form related to ours.

Haminoea vesicula (Gould) (also known as *Haminaea*) . . . . §274

Leonard, Ruth E. 1918. Early development of *Haminaea*. Pub. Puget Sd. Biol. Sta., **2**: 45–63.

Spicer, V. D. P. 1933. Report on a colony of *Haminoea* at Ballast Point, San Diego, Calif. Nautilus, **47**: 52–54.

Ecological notes, with data on ovipositing, of a new subspecies.

Navanax inermis (Cooper) . . . . §294

Paine, Robert T. 1963. Food recognition and predation on opisthobranchs by *Navanax inermis* (Gastropoda: Opisthobranchia). Veliger, **6** (1): 1–9, 1 fig., 1 pl.

Tracks them down by their mucus trails; eats *Hermissenda*.

——— 1965. Natural history, limiting factors and energetics of the opisthobranch *Navanax inermis*. Ecology, **46** (5): 603–19, 8 figs.

Philine sp. . . . . §274

Order Anaspidea

Beeman, Robert D. 1960. A new tectibranch, *Aplysia reticulopoda,* from the southern California coast. Bull. S. Calif. Acad. Sci., **59** (3): 144–52, 3 figs.

——— 1963. Notes on the California species of *Aplysia* (Gastropoda: Opisthobranchia). Veliger, **5** (4): 145–47.

Eales, N. B. 1960. Revision of the world species of *Aplysia* (Gastropoda: Opisthobranchia). Bull. Brit. Mus. (Nat. Hist.), Zool., **5** (10): 267–404, frontispiece, 61 figs.

Considers California species; *A. nettiae* Winkler is probably *A. californica*.

Aplysia californica Cooper (was *Tethys*) . . . . §104

Eales, N. B. 1921. *Aplysia*. Liverpool Mar. Biol. Com. Mem., **24**. 84 pp.

MacGinitie, G. E. 1934. The egg-laying activities of the sea hare *Tethys californicus* (Cooper). Biol. Bull., **67** (2): 300–303.

Phyllaplysia taylori Dall . . . . §274

McCauley, James E. 1960. The morphology of *Phyllaplysia zostericola,* new species. Proc. Calif. Acad. Sci., Ser. 4, **29** (16): 549–76, 6 figs.

Still the same species, to judge from MacFarland's account.

Order Sacoglossa

This order includes various interesting beasts that have usually escaped notice, although they are at times common enough. The famous bivalve sea slugs occur just south of our range; but they are within easy reach of many amateur naturalists these days.

Gonor, J. J. 1961. Notes on the biology of *Hermaeina smithi,* a sacoglossan opisthobranch from the west coast of North America. Veliger, **4** (2): 85–98, 13 figs.

Order Notaspidea

Tylodina fungina Gabb . . . . § 98

Order Nudibranchia

Agersborg, H. P. K. 1923. Notes on a new cladohepatic nudibranch from Friday Harbor, Washington. Nautilus, **36**: 133–38.

The dark brown *Olea hansineënsis,* from eelgrass.

Collier, Clinton L. 1963. A new member of the genus *Atagema* (Gastropoda: Nudibranchia), a genus new to the Pacific northeast. Veliger, **6** (2): 73–75, 5 figs.

Cook, Emily F. 1962. A study of food choices of two opisthobranchs, *Rostanga pulchra* McFarland [*sic*] and *Archidoris montereyensis* (Cooper). Veliger, **4** (4): 194–96, 4 figs.

Farmer, Wesley M., and Allan J. Sloan. 1964. A new opisthobranch mollusk from La Jolla, California. Veliger, **6** (3): 148–50, 2 figs., 1 pl.
Ancula lentiginosa.

Heath, Harold. 1917. The anatomy of an aeolid, *Chioraera dalli.* Proc. Acad. Nat. Sci. Phil., **1917**: 137–48.
From the southeastern Alaska littoral.

Hurst, Anne. 1966. A description of a new species of *Dirona* from the northeast Pacific. Veliger, **9** (1): 9–15, 7 figs., 1 pl.
This is an excellent example of how one of these animals should be described. The species is subtidal.

MacFarland, Frank M. 1906. Opisthobranchiate Mollusca from Monterey Bay. Bull. U.S. Bur. Fish., **25**: 109–51. Colored plates, etc.

——— 1912. The nudibranchiate family Dironidae. Zool. Jahrb., Supp. **25** (1): 515–36.

——— 1923. Morphology of the nudibranch genus *Hancockia.* J. Morphol., **38**: 65–92.

——— 1923. Acanthodoridae of the California coast. Nautilus, **39**: 1–27.

——— 1929. *Drepania,* a genus of nudibranchiate mollusks new to California. Proc. Calif. Acad. Sci., **18** (4): 485–96.

Marcus, Ernst. 1964. A new species of *Polycera* (Nudibranchia) from California. Nautilus, **77** (4): 128–31, 4 figs.
Polycera hedgpethi, from Tomales Bay; taken with *P. atra.*

O'Donoghue, C. H. 1926. A list of the nudibranchiate Mollusca recorded from the Pacific coast of North America. Trans. Roy. Canad. Inst., **15**: 199–247.

——— 1927. Notes on a collection of nudibranchs from Laguna Beach. Pomona J. Entom. and Zool., **19**: 77–117.

Guberlet, J. E. 1928. Observations on the spawning habits of *Melibe.* Pub. Puget Sd. Biol. Sta., **6**: 262–70.
Mentions the 1923 Agersborg morphological paper on this form.

MacFarland, Frank M., and C. H. O'Donoghue. 1929. A new species of *Corambe.* Proc. Calif. Acad. Sci., (4) **18**: 1–27.

Hermissenda crassicornis (Eschscholtz) ....§295

BÜRGIN, ULRIKE F. 1965. The color pattern of *Hermissenda crassicornis* (Eschscholtz, 1831). Veliger, **7** (4): 205–15, 9 figs.
Due partly to various arrangements of pigments, partly to food.

Hopkinsia rosacea MacFarland....................................§109

Rostanga pulchra MacFarland....................................§ 33

Triopha carpenteri (Stearns)....................................§ 34

T. maculata MacFarland ....§109

Subclass PULMONATA

Trimusculus reticulata (Sowerby) (formerly *Gadinia*)....................§147

HUBENDICK, BENGT. 1947. Phylogenie und Tiergeographie der Siphonariidae: Zur Kenntnis der Phylogenie in der Ordnung Basommatophora und des Ursprungs der Pulmonatengruppe. Zool. Bidr. Uppsala, **24**: 1–216, 107 figs.
Includes a discussion of the morphology of the Gadiniidae.

YONGE, C. M. 1960. Further observations on *Hipponix antiquatus*, with notes on North Pacific pulmonate limpets. Proc. Calif. Acad. Sci., Ser. 4, **31** (5): 111–19, 4 figs.

Class PELECYPODA or BIVALVIA
Clams, Cockles, Mussels, Oysters, Shipworms, etc.

ARMSTRONG, LEE R. 1965. Burrowing limitations in pelecypods. Veliger, **7** (3): 195–200, 4 figs.

BARRETT, ELINORE M. 1963. The California oyster industry. Calif. Dept. Fish and Game, Fish. Bull., **123**. 103 pp. 31 figs.

CAHN, A. R. 1949. Pearl culture in Japan. U.S. Fish and Wildl. Serv., Fish. Lflt. No. 357. 91 pp., 22 figs.
A complete account of the artificial propagation of pearls; the pearls, of course, are no less natural than the wild ones—in fact, they are a little better.

——— 1950. Oyster culture in Japan. *Ibid.* No. 383. 80 pp., 40 figs.
Also a complete account; both of these reports originally appeared as Reports 122 and and 134, Nat. Res. Section, Supreme Commander for Allied Powers, Tokyo.

COE, WESLEY R. 1941. Sexual phases in wood-boring mollusks. Biol. Bull., **81** (2): 168–76, 3 figs.

——— 1946. A resurgent population of the bay-mussel (*Mytilus edulis diegensis*). J. Morphol., **78** (1): 85–101, 2 pls.

——— 1948. Nutrition, environmental conditions, and growth of marine bivalve mollusks. J. Mar. Res., **7** (3): 586–601, 2 figs.

———, and DENIS L. FOX. 1944. Biology of the California sea mussel . . . III. Biol. Bull., **87**: 59–72.

FITCH, JOHN E. 1953. Common marine bivalves of California. Calif. Dept. Fish and Game, Fish. Bull., **90**. 102 pp., 63 figs.

FOX, DENIS L., ed. 1936. The habitat and food of the California sea mussel. Bull. Scripps Inst. Oceanogr., Tech. Ser., **4**: 1–64.

GLYNN, PETER W. 1964. *Musculus pygmaeus* spec. nov., a minute mytilid from the high intertidal zone at Monterey Bay, California. Veliger, **7** (2): 121–28, 2 figs. (one as Pl. 23).

GRAU, GILBERT. 1959. Pectinidae of the eastern Pacific. Allan Hancock Pac. Exp., **23**. 308 pp., 57 pls.

GUNTER, GORDON. 1950. The generic status of living oysters and the scientific name of the common American species. Amer. Midland Nat., **43** (2): 438–49.

Oysters with free larval development and estuarine habit should be referred to genus *Crassostrea* (incl. *gigas* and *virginica*); those with fertilization and development in the mantle cavity and more maritime habit are in *Ostrea* s. str. (incl. the European *O. edulis* and our *O. lurida*). There are also significant morphological differences.

HAAS, FRITZ. 1942. The habits of some west coast bivalves. Nautilus, **55** (4): 109–13.

HILL, C. L., and KOFOID, C. A. 1927. Marine borers on the Pacific coast. Final Report, S.F. Bay Mar. Piling Comm., 1–357.

KORRINGA, P. 1949. More light upon the problem of the oyster's nutrition? Bijdr. Dierk., **28**: 237–48.

——— 1951. On the nature and function of "chalky" deposits in the shell of *Ostrea edulis* Linnaeus. Proc. Calif. Acad. Sci., Ser. 4, **27** (5): 133–58, 2 figs.

——— 1951. The shell of *Ostrea edulis* as a habitat. Arch. Néerl. Zool., **10** (1): 32–152, 13 figs., tbls.

LOOSANOFF, VICTOR L., and HARRY C. DAVIS. 1963. Rearing of bivalve mollusks. Adv. Mar. Biol., **1**: 1–136, 43 figs.

———, HARRY C. DAVIS, and PAUL E. CHANLEY. 1966. Dimensions and shapes of larvae of some marine bivalve mollusks. Malacologia, **4** (2): 351–435, 61 figs.

MACGINITIE, G. E. 1941. On the method of feeding of four pelecypods. Biol. Bull., **80** (1): 18–25.

MARRIAGE, LOWELL D. 1954. The bay clams of Oregon: Their economic importance, relative abundance, and general distribution. Fish Comm. Ore., Contr. 20. 47 pp., 24 figs.

MAURER, DON. 1967. Filtering experiments on marine pelecypods from Tomales Bay, California. Veliger, **9** (3): 305–9.

——— 1967. Burial experiments on marine pelecypods from Tomales Bay, California. Veliger, **9** (4): 376–81.

QUAYLE, D. B. 1939. Notes on *Paphia bifurcata,* a new molluscan species from Ladysmith Harbor, B.C. Nautilus, **52** (4): 139.

——— 1941. The Japanese "Littleneck" clam accidentally introduced into British Columbia waters. Pac. Biol. Sta. Progr. Rept., **48**: 17–18.

——— 1943. Sex, gonad development, and seasonal gonad changes in *Paphia staminea* Conrad. J. Fish. Res. Bd. Canad., **6** (2); 140–51, 5 figs.

SOMMER, HERMANN, and KARL F. MEYER. 1935. Mussel poisoning. Calif. and West. Med., **42** (6): 423–26.

SOOT-RYEN, TRON. 1955. A report on the family Mytilidae (Pelecypoda). Allan Hancock Pac. Exp., **20** (1): 1–154, 78 figs., 10 pls.

STASEK, CHARLES R. 1963. Synopsis and discussion of the association of ctenidia and labial palps in the bivalved Mollusca. Veliger, **6** (2): 81–97, 5 figs.

STAUBER, LESLIE A. 1950. The problem of physiological species with special reference to oysters and oyster drills. Ecology, **31** (1): 109–18, 2 figs.

TOWNSLEY, P. M., and R. A. RICHY. 1965. Marine borer aldehyde oxidase. Canad. J. Zool., **43**: 1011–19, 5 figs.

TURNER, RUTH D. The family Pholadidae in the western North Atlantic and the eastern Pacific. Part I: Pholadinae. Johnsonia, **3** (33): 1–64, 34 pls. (1954). Part II: Martesiinae, Jouannetiinae, and Xylophaginae. *Ibid.*, (34): 65–160, 59 pls. (1955).

——— 1966. A survey and illustrated catalogue of the Teredinidae (Mollusca: Bivalvia). Boston: Mus. Comp. Zool. 265 pp., 25 figs., 64 pls.

YOCUM, H. B., and E. R. EDGE. 1929. The ecological distribution of the Pelecypoda of the Coos Bay region of Oregon. Nthwest. Sci., **5**: 65–71.

YONGE, C. M. 1939. The protobranchiate Mollusca: A functional interpretation of their structure and evolution. Phil. Trans. Roy. Soc. Lond., (B), **230**: 79–147, 39 figs., 1 pl.

——— 1949. On the structure and adaptations of the Tellinacea, deposit-feeding Eulamellibranchia. *Ibid.*, **234**: 29–76, 29 figs.

——— 1951. Studies on Pacific coast mollusks, I–III. Univ. Calif. Pub. Zool., **55** (6–8): 395–420, 13 figs.

Three species are considered: I, *Cryptomya californica*; II, *Platyodon cancellatus* (a rock borer); III, *Hinnites multirugosus* (*H. giganteus*).

——— 1960. Oysters. London: Collins (The New Naturalist series). 209 pp., 72 figs., 17 pls.

Order FILIBRANCHIA

Anomia peruviana d'Orbigny . . . . §217

Hinnites giganteus Gray . . . . §125

Lithophaga plumula Hanley . . . . §241

HODGKIN, NORMAN M. 1962. Limestone boring by the mytilid *Lithophaga*. Veliger, **4** (3): 123–29, 3 figs., 3 pls.

Favors chemical boring.

Mytilus californianus Conrad . . . . §158

CHEW, K. K., A. K. SPARKS, and S. C. KATKANSKY. 1964. First record of *Mytilicola orientalis* Mori in the California mussel *Mytilus californianus* Conrad. J. Fish. Res. Bd. Canad., **21** (1): 205–7, 1 fig.

In Humboldt Bay.

DEHNEL, PAUL A. 1956. Growth rates in latitudinally and vertically separated populations of *Mytilus californianus*. Biol. Bull., **110** (1) 43–53, 4 figs.

REISH, DONALD J. 1964. Discussion of the *Mytilus californianus* community on newly constructed rock jetties in southern California. Veliger, **7** (2): 95–101, 3 figs.

Colonization started with *Ulva*; *Mytilus* settled in spring.

RAO, K. PAMPAPATHI. 1953. Rate of water propulsion in *Mytilus californianus* as a function of latitude. Biol. Bull., **104** (2): 171–81, 5 figs.

——— 1954. Tidal rhythmicity of rate of water propulsion in *Mytilus*, and its modifiability by transplantation. Biol. Bull., **106** (3): 353–59, 5 figs.

STOHLER, R. 1930. Beitrag zur Kenntnis des Geschlechtszyklus von *Mytilus californianus* Conrad. Zool. Anz., **90**: 263–68.

See also the many papers by Fox *et al.*, and by Coe, summarized in Coe (1948), cited under Pelecypoda.

WHEDON, W. FOREST. 1936. Spawning habits of the mussel *Mytilus californianus* Conrad. Univ. Calif. Pub. Zool., **41** (5): 35–44, 1 pl.

YOUNG, R. T. 1941. The distribution of the mussel (*Mytilus californianus*) in relation to the salinity of its environment. Ecology, **22** (4): 379–86.

——— 1945. Stimulation of spawning in the mussel (*Mytilus californianus*). Ecology, **26** (1): 58–69.

Mytilus edulis L. . . . . §197

BAYNE, B. L. 1965. Growth and the delay of metamorphosis of the larvae of *Mytilus edulis* (L.). Ophelia, **2** (1): 1–47, 20 figs.

COE, WESLEY R. 1945. Nutrition and growth of the California bay-mussel (*Mytilus edulis diegensis*). J. Exp. Zool., **99** (1): 1–14, 2 figs.

CRAIG, G. Y., and A. HALLAM. 1963. Size-frequency and growth-ring analyses of *Mytilus edulis* and *Cardium edule*, and their palaeoecological significance. Paleontology, **6** (4): 731–50, 10 figs.

REISH, DONALD J. 1964. Studies on the *Mytilus edulis* community in Alamitos Bay, California, I: Development and destruction of the community. Veliger, **6** (3): 124–31, 5 figs.
Seasonal progression other than true succession occurs.

DAKIN, W. S. 1909. *Pecten.* Liverpool Mar. Biol. Comm. Mem., No. 17. 132 pp.
GUTSELL, J. S. 1930–31. Natural history of the bay scallop. Bull. U.S. Bur. Fish., **46**: 569–632.

Order EULAMELLIBRANCHIA

YONGE, C. M. 1967. Form, habit and evolution in the Chamidae (Bivalvia), with reference to conditions in the rudists (Hippuritacea). Phil. Trans. Roy Soc. Lond., (B), **252** (775): 49–105, 31 figs.
This study is based in large part on observations of California species at Pacific Grove.

WILLIAMS, WOODBRIDGE. 1949. The enigma of Mission Bay. Pac. Disc., **2** (Mar.–Apr.): 22–23.
A popular account, with excellent photographs, of a curious naked bivalve that looks like a nudibranch. Occurred in Mission Bay before progress invaded the place; may still be there. Reported as "rare" in the Monterey intertidal.

FRASER, C. MCLEAN. 1931. Notes on the ecology of the cockle. Trans. Roy. Soc. Canad., Sec. 5, **25**: 59–72.
TAYLOR, CLYDE C. 1960. Temperature, growth and mortality—The Pacific cockle. J. du Conseil, **26** (1): 117–24, 5 figs.
WEYMOUTH, FRANK W., and S. H. THOMPSON. 1931. The age and growth of the Pacific cockle (*Cardium corbis* Martyn). Bull. U.S. Bur. Fish., **46**: 633–41.

STEELE, E. N. 1964. The immigrant oyster (*Ostrea gigas*) now known as the Pacific oyster. Warren's Quick Print: Olympia, Washington. 179 pp.
By E. N. Steele, "Pioneer Olympia Oysterman," in cooperation with the Pacific Coast Oyster Growers Association. Although this book has the peculiarities to be expected of a work produced by an author who never wrote a book before and a printer who never published one before, it is, on the whole, informative and useful. There are some well-reproduced photographs of billboards (page 152A) testifying to the incompatibility of pulp mills and oysters, welcoming the passerby to eat oysters now that the pulp mill has closed down, etc.

Coe, Wesley R. 1932. Sexual phases in the American oyster (*Ostrea virginica*). Biol. Bull., **63**: 419–41.

Galtsoff, Paul S. 1964. The American oyster, *Crassostrea virginica* Gmelin. U.S. Fish and Wildl. Serv., Fish. Bull., **64**. 480 pp., 390 figs.

Dr. Galtsoff's summary of his life with oysters and of the work done on this much-studied animal by many others. It went immediately out of print; perhaps a new printing will be available, since it has been awarded an in-house prize.

Nelson, T. C. 1924. The attachment of oyster larvae. Biol. Bull., **46**: 143–51.

——— 1928. Relation of spawning of the oyster to temperature. Ecology, **9**: 145–54.

Johnson, Phyllis T. 1966. On *Donax* and other sandy-beach inhabitants. Veliger, **9** (1): 29–30.

Pohlo, Ross H. 1967. Aspects of the biology of *Donax gouldii* and a note on evolution in *Tellinacea* (Bivalvia). Veliger, **9** (3): 330–37, 4 figs.

Segerstråle, Sven G. 1965. Biotic factors affecting the vertical distribution and abundance of the bivalve *Macoma baltica* (L.) in the Baltic Sea. Botanica Gothobergensia, **3**: 195–204, 5 figs.

Pfitzenmeyer, Hayes T., and Carl N. Shuster. 1960. A partial bibliography of the soft-shell clam *Mya arenaria* L. Contr. No. 123, Maryland Dept. Res. and Educ., Chesapk. Biol. Lab.; also Inform. Ser., Pub. No. 4, Univ. Del. Mar. Labs. 29 pp. (processed).

——— 1965. Annual cycle of gametogenesis of the soft-shelled clam, *Mya arenaria*, at Solomons, Maryland. Chesapk. Sci., **6** (1): 52–59, 8 figs.

Bonnot, Paul. 1937. Setting and survival of spat of the Olympia oyster, *Ostrea lurida*, on upper and lower horizontal surfaces. Calif. Fish and Game, **23** (3): 224–28, 2 figs.

Coe, Wesley R. 1932. Development of the gonads, and the sequence of the sexual phases, in the California oyster. Bull. Scripps Inst. Oceanogr., Tech. Ser., **3**: 119–44.

Davis, Harry C. 1955. Mortality of Olympia oysters at low temperatures. Biol. Bull., **109** (3): 404–6.

Hopkins, A. E. 1935. Attachment of the larvae of the Olympia oyster. Ecology, **16**: 82–87.

——— 1937. Experimental observations on spawning, larval development, and setting in the Olympia oyster, *Ostrea lurida*. Bull. U.S. Bur. Fish., **48** (23): 439–503, 41 figs.

EVANS, JOHN W., and DAVID FISHER. 1966. A new species of *Penitella* (Family *Pholadidae*) from Coos Bay, Oregon. Veliger, **8** (4): 222–24, 1 fig., 1 pl.

Penitella turnerae. Scarce? A hybrid?

FRASER, C. McLEAN, and G. M. SMITH. 1928. Notes on the ecology of the littleneck clam, *Paphia staminea* Conrad. Trans. Roy. Soc. Canad., (4) **22**: 249–69.

See also the 1929 citation for *Saxidomus*.

FRASER, C. McLEAN. 1929. The spawning and free-swimming larval periods of *Saxidomus* and *Paphia*. Trans. Roy. Soc. Canad., (4) **23**: 195–98.

———, and G. M. SMITH. 1928. Notes on the ecology of the butter clam, *Saxidomus giganteus*. *Ibid*., **22**: 271–86.

FRASER, C. McLEAN. 1930. The razor clam ... of Graham Island, Queen Charlotte Group. Trans. Roy. Soc. Canad., (5) **24**: 141–54.

TAYLOR, CLYDE C. 1959. Temperature, growth and mortality—the Pacific razor clam. J. du Conseil, **25** (1): 93–100, 6 figs.

WEYMOUTH, FRANK W., and H. C. McMILLAN. 1931. Relative growth and mortality of the Pacific razor clam, *Siliqua patula,* and their bearing on the commercial fishery. Bull. U.S. Bur. Fish., **46**: 543–67.

Also the previous excellent report of Weymouth *et al.* (1925), and of Weymouth (1928).

YONGE, C. M. 1952. Studies on Pacific coast mollusks, IV: Observations on *Siliqua patula* Dixon and on evolution within the Solenidae. Univ. Calif. Pub. Zool., **55** (9): 421–38, 8 figs.

POHLO, ROSS H. 1963. Morphology and mode of burrowing in *Siliqua patula* and *Solen rosaceus* (Mollusca: Bivalvia). Veliger, **6** (2): 98–104, 6 figs.

DREW, G. A. 1907. Habits and movements of the razor clam *Ensis directus*. Biol. Bull., **12**: 127–40.

POHLO, ROSS H. 1966. A note on the feeding behavior in *Tagelus californianus* (Bivalvia: Tellinacea). Veliger, **8** (4): 225.

Not a deposit, but a suspension feeder.

STEPHENS, A. C. 1929. Notes on the rate of growth of *Tellina tenuis* da Costa in the Firth of Clyde. J. Mar. Biol. Assoc., **16**: 117–29.

Refers to previous work on natural history, 1928.

COE, WESLEY R. 1933. Sexual phases in *Teredo*. Biol. Bull., **65**: 283–303.

——— 1934. Sexual rhythm in *Teredo*. Science, **80**: 192.

GRAVE, B. H. 1928. Natural history of the shipworm *Teredo navalis* at Woods Hole, Mass. Biol. Bull., **55**: 260–82.

MILLER, R. C. 1924. The boring mechanism of *Teredo*. Univ. Calif. Pub. Zool., **26**: 41–80.

See also Turner's monograph on the Teredinidae, cited under the introduction to bivalves, above.

Tivela stultorum (Mawe), Pismo Clam.................................§190

BAXTER, JOHN L. 1961. Results of the 1955 to 1959 Pismo clam censuses. Calif. Fish and Game, **47** (2): 153–62, 3 figs.

The last clams to settle at the Morro census line, just north of the P.G.&E. outlet, did so in 1955, although there has been good recruitment in other localities since 1957. There may, of course, be no relation between these events.

COE, WESLEY R. 1949. Nutrition, growth, and sexuality of the Pismo clam (*Tivela stultorum*). J. Exp. Zool., **104** (1): 1–24, 4 figs.

FITCH, JOHN E. (1950). The Pismo clam. Calif. Fish and Game, **36** (3): 285–312, 13 figs.

——— 1965. A relatively unexploited population of Pismo clams, *Tivela stultorum* (Mawe, 1823) (Veneridae). Proc. Malacol. Soc. Lond. **36**: 309–12, Pl. 13.

HERRINGTON, W. C. 1930. The Pismo clam. Fish. Bull., Calif. Fish and Game, **18**: 1–69.

WEYMOUTH, FRANK W. 1923. Life history and growth of the Pismo clam. *Ibid.*, 7.

Transenella tantilla (Gould)...not treated

HANSEN, BENT. 1953. Brood protection and sex ratio of *Transenella tantilla* (Gould), a Pacific bivalve. Vidensk. Medd. fra Dansk. Naturh. Foren., **115**: 313–24, 4 figs.

A small clam of sand flats (confusable with *Gemma gemma*); of viviparous, protandrous habit.

Tresus nuttalli Conrad (formerly *Schizothaerus*!), Horse Clam ....§300

ADDICOTT, WARREN O. 1963. An unusual occurrence of *Tresus nuttalli* (Conrad, 1837) (Mollusca: Pelecypoda). Veliger, **5** (4): 143–45, 3 figs.

Found as a nestler in sandstone burrows of other bivalves.

PEARCE, JACK B. 1965. On the distribution of *Tresus nuttalli* and *Tresus capax* (Pelecypoda: Mactridae) in the waters of Puget Sound and the San Juan Archipelago. Veliger, **7** (3): 166–70, 1 fig., 1 pl.

POHLO, ROSS H. 1964. Ontogenetic changes of form and mode of life in *Tresus nuttalli*. Malacologia, **1** (3): 321–30, 6 figs.

SWAN, E. F., and J. H. FINUCANE. 1952. Observations on the genus *Schizothaerus*. Nautilus, **66** (1): 19–26, 3 pls.

Zirfaea pilsbryi Lowe (formerly known as *Z. gabbi* Tryon)................§305

LOWE, HOMER N. 1931. Notes on the west coast *Zirfaea*. Nautilus, **45**: 52–53.

Class CEPHALOPODA
Octopods or Octopuses (not Octopi, please!), Squids, Nautilus

The -pus of Octopus is from the Greek for foot, and is usually rendered -pod, or -poda in English; -pi is a false plural. By the same analogy, we should say cephalopi instead of cephalopods. One wonders how this got started.

AKIMUSHKIN, I. I. 1965. Cephalopods of the seas of the U.S.S.R. (Golovonogie mollyuski morei SSSR. Akad. Nauk., Inst. Okeanol., 1963). Israel Program for Scientific Translations, for the Smithsonian Institution. 233 pp., 60 figs.

BERRY, S. STILLMAN. 1910. Review of the cephalopods of western North America. Bull. U.S. Bur. Fish., **30**: 269–336.

——— 1913. Notes on some west American cephalopods. Proc. Acad. Nat. Sci. Phil., **1913**: 72–77, text figs.

FIELDS, W. GORDON. 1965. The structure, development, food relations, reproduction, and life history of the squid *Loligo opalescens* Berry. Calif. Dept. Fish and Game, Fish. Bull., **131**. 108 pp., 59 figs.

FOX, DENIS L. 1938. An illustrated note on the mating and egg-brooding habits of the two-spotted octopus. Trans. San Diego Soc. Nat. Hist., **7**: 31–34, 1 pl.

LANE, FRANK W. 1960. Kingdom of the octopus. New York: Sheridan House.

The 1962 paperback edition by Pyramid Books omits the extensive bibliography and some of the illustrations. The book is about cephalopods generally, and is excellent.

MCGOWAN, JOHN A. 1954. Observations on the sexual behavior and spawning of the squid *Loligo opalescens* at La Jolla, California. Calif. Fish and Game, **40** (1): 47–54, 5 figs.

PICKFORD, GRACE E., and BAYARD H. MCCONNAUGHEY. 1949. The *Octopus bimaculatus* problem. A study in sibling species. Bull. Bingh. Oceanogr. Coll., **12** (4). 66 pp., 28 figs.

A fine example of critical taxonomic study. Establishes two sympatric species of two-spotted octopus, differing principally in egg size and ecological habitat. *O. bimaculatus* is subtidal in rocky habitats; *O. bimaculoides* is low intertidal, especially on mud flats. They also differ in their assemblage of mesozoans! Both species are known only between Los Angeles and Ensenada. Study also indicates that the genus *Paroctopus* is invalid.

PILSON, M. E. Q., and P. B. TAYLOR. 1961. Hole drilling by octopus. Science, **134** (3487): 1366–68, 1 fig.

The octopus drills very small holes and injects venom from the salivary gland, which weakens the abalone, clam, or snail and allows the octopus to feed on it.

ROBSON, G. C. A monograph of the Recent Cephalopoda, Part I: Brit. Mus., 1929. 236 pp. Part II: Brit. Mus. (Nat. Hist.), 1932. 359 pp., 6 pls., 79 text figs.

SASAKI, M. 1929. A monograph of the dibranchiate cephalopods of the Japanese and adjacent waters. J. Fac. Agric. Hokk. Imp. Univ., **20** (Suppl.). 357 pp., 30 pls.

WELLS, M. J. 1962. Brain and behaviour in cephalopods. Stanford: Stanford Univ. Press. 171 pp., 53 figs.

Are cephalopods, after all these millions of years, at the end of their trail, lost in an evolutionary blind alley? It would seem so; in any event, they lack the equipment for colonizing fresh water or land (fortunately).

Octopus sp. .... §135

FISHER, W. K. 1923. Brooding habits of a cephalopod. Ann. Mag. Nat. Hist. (Ser. 9), **12**: 147–49.

——— 1925. On the habits of an octopus. *Ibid.* (Ser. 9), **15**: 411–14.

O. bimaculoides Pickford and McConnaughey .... §135

See Pickford and McConnaughey paper already cited. The *O. bimaculatus* of earlier editions, and of the MacGinities' book, is, of course, *O. bimaculoides*.

PETERSON, R. PRICE. 1959. The anatomy and histology of the reproductive systems of *Octopus bimaculoides*. J. Morphol., **104** (1): 61–88, 3 pls., 4 figs.

O. dofleini Wülker .... §135

JOHANSEN, KJELL, and ARTHUR W. MARTIN. 1962. Circulation in the cephalopod, *Octopus dofleini*. Comp. Biochem. and Physiol., **5** (3): 161–76, 16 figs.

WINKLER, LINDSAY R., and LAURENCE M. ASHLEY. 1954. The anatomy of the common octopus of northern Washington. Walla Walla Coll. Pub. Dept. Biol. Sci. and Biol. Sta., No. 10. 29 pp., 19 figs.

"The octopus, so well known as the 'devilfish,' has largely escaped anatomical research." The octopus shouldn't be blamed for that, I suppose. (Species indicated as *O. apollyon.*)

Phylum ECHINODERMA

BOOLOOTIAN, RICHARD A., ed. 1966. Physiology of Echinodermata. A collective effort by a group of experts. New York: Interscience Publishers of Wiley & Sons. 822 pp., illus.

As with all symposium volumes, some chapters are more satisfactory in matters of coverage and detail than others. As a whole, the book is a solid contribution (essential for serious students, etc.), and the editor is to be congratulated for having gotten it together, even if his own chapter seems to depend too much on unpublished and personally communicated references. The publishers are not to be congratulated on the price, but perhaps this has been gauged on the irreducible minimum of library sales needed to make the book break even before it becomes available for the Xerox machines.

BUSH, M. 1921. Revised key to the echinoderms of Friday Harbor, Washington. Pub. Puget Sd. Biol. Sta., 3: 65–77.

Includes description of a new ophiuran, *Amphiodia peloria*, very long-armed, in sand. Nomenclature obsolete.

CLARK, H. L. 1913. Echinoderms from Lower California. Bull. Amer. Mus. Nat. Hist., **32**: 185–236.

——— 1935. Some new echinoderms from California. Ann. Mag. Nat. Hist., Ser. 10, **15**: 120–29.

Describes a new sand dollar subspecies from 15 fathoms off Coronado and two new brittle stars from subtidal Corona del Mar, *Ophiacantha eurythra* and *Amphiodia psara*.

EKMAN, SVEN. 1946. Zur Verbreitungsgeschichte der Warmwasserechinodermen im stillen Ozean (Asteroidea, Ophiuroidea, Echinoidea). Nova Acta R. Soc. Sci. Upsaliensis, (4) **14** (2). 42 pp., 1 fig.

HILTON, W. A. 1918. Some echinoderms of Laguna Beach. J. Entom. and Zool., **10**: 78.

JOHNSON, MARTIN W., and LEILA T. JOHNSON. 1950. Early life history and larval development of some Puget Sound echinoderms. In *Studies honoring Trevor Kincaid*, pp. 73–153, 4 pls.

MOORE, RAYMOND C., ed. 1966. Treatise on invertebrate paleontology, Part U: Echinodermata 3. Volume 1, 366 pp., 271 figs.; Volume 2, 695 pp., 263 figs. Geol. Soc. Amer. and Univ. Kansas Press.

Volume 1 includes Asteroids, Ophiuroids, and Echinoids; Volume 2, Echinoids (concl.) and Holothuroids. Thus fossil representation of all groups commonly represented in our area is treated.

NICHOLS, DAVID. 1962. Echinoderms. London: Hutchinson Univ. Libr. 200 pp., 26 figs. (revised edition, 1966).

Well-written, concise introduction; inadequately illustrated.

ZIESENHENNE, FRED C. 1942. Some notes on the distribution records of little-known southern California echinoderms. Bull. S. Calif. Acad. Sci., **40** (3): 117–20.

Class ASTEROIDEA. Seastars, Starfish

FARMANFARMAIAN, A., A. C. GIESE, R. A. BOOLOOTIAN, and J. BENNETT. 1958. Annual reproductive cycles in four species of west coast starfishes. J. Exp. Biol., **138** (2): 355–67, 7 figs.

FISHER, W. K. 1911–30. Asteroidea of the North Pacific. Part 1, 1911; Part 2, 1928; Part 3, 1930. Bull. U.S. Nat. Mus., **76**.

O'DONOGHUE, C. H. 1924. On the summer migration of certain starfish in Departure Bay, British Columbia. Contr. Canad. Biol. and Fish., n.s., **1**: 455–72.

PAINE, V. L. 1929. The tube feet of starfish as autonomous organs. Amer. Nat., **43**: 517–29.

Astrometis sertulifera (Xantus) ....§ 72

JENNINGS, H. S. 1907. Behavior of the starfish *Asterias forreri* de Loriol. Univ. Calif. Pub. Zool., **4**: 53–185.

Astropecten armatus Gray ....§250

Dermasterias imbricata (Grube) ....§211

Evasterias troscheli (Stimpson) ....§207

CHRISTIANSEN, AAGE MØLLER. 1957. The feeding behavior of the sea star *Evasterias troescheli* (Stimpson). Limnol. and Oceanogr., **2** (3): 180–97, 15 figs.

Henricia leviuscula (Stimpson) ....§ 69

RASMUSSEN, B. 1965. On the taxonomy and biology of the North Atlantic species of the asteroid genus *Henricia* Gray. Medd. Danmarks Fish. Havunders., **4**: 157–213.

Henricia is not a carnivore, but a ciliary mucoid feeder.

Leptasterias hexactis (Stimpson) ....§§ 68, 210

CHIA FU-SHIANG. 1966a. Brooding behavior of a six-rayed starfish, *Leptasterias hexactis.* Biol. Bull., **130** (3): 304–15, 7 figs.

Description of how this small starfish broods the embryos by humping over the eggs.

——— 1966b. Systematics of the six-rayed sea star *Leptasterias* in the vicinity of San Juan Island, Washington. Syst. Zool., **15** (4): 300–306, 5 figs.

The forms under the names *Leptasterias hexactis* and *L. equalis,* which caused the late W. K. Fisher so much vexation, are convincingly demonstrated to be a single biological species, *Leptasterias hexactis.* It is suggested that all the North Pacific forms ascribed to this genus "may belong to one or only a few polytypic species."

L. pusilla Fisher ....§ 30

Linckia columbiae Gray ....§ 71

Orthasterias koehleri (de Loriol) ....§208

Patiria miniata (Brandt), formerly *Asterina miniata*..§ 26

ANDERSON, JOHN MAXWELL. 1959. Studies on the cardiac stomach of a starfish, *Patiria miniata* (Brandt). Biol. Bull., **117** (2): 185–201, 21 figs.

——— 1965. Studies on visceral regeneration in sea-stars. II. Regeneration of pyloric caeca in Asteriidae, with notes on the sources of cells in regenerating organs. Biol. Bull., **128** (1): 1–23, 36 figs.

MOORE, A. R. 1945. The individual in simpler forms. Ore. St. Monogr. Psychol., **2**. 143 pp., 13 pls., 9 figs. (See Ch. IV, pp. 61–79.)

NEWMAN, H. H. 1925. An experimental analysis of asymmetry in the starfish *Patiria miniata.* Biol. Bull., **49**: 111–38.

Pisaster brevispinus (Stimpson) f. *paucispinus* (Stimpson)..§209

Pisaster f. *brevispinus* occurs at Monterey only offshore, but at Crescent City it occurs on reefs. At Dillon Beach, it is often seen on the eelgrass flats.

P. giganteus (Stimpson)..§157

P. giganteus capitatus (Stimpson), the southern California subspecies........§142

P. ochraceus f. *ochraceus* (Brandt), chiefly on shores subject to wave impact. .§157

FEDER, HOWARD M. 1955. On the methods used by the starfish *Pisaster ochraceus* in opening three types of bivalve molluscs. Ecology, **36** (4): 764–67.

——— 1959. The food of the starfish *Pisaster ochraceus* along the California coast. *Ibid.*, **40** (4): 721–24, 2 figs.

The starfish may intrude its stomach through a very narrow crack and start digesting its victim long before the clam is "open," although brute force is also resorted to.

———, and REUBEN LASKER. 1964. Partial purification of a substance from starfish tube feet which elicits escape responses in gastropod molluscs. Life Sciences, **3** (9): 1047–51.

MAUZEY, KARL PERRY. 1966. Feeding behavior and reproductive cycles in *Pisaster ochraceus*. Biol. Bull., **131** (1): 127–44, 7 figs.
Feed little in winter, mostly in July and August; winter diet principally chitons, barnacles and limpets in summer.

QUAYLE, D. B. 1954. Growth of the purple seastar. Oyster Bull. Brit. Col. Dept. Fish., **5** (3): 11–13.

GREER, D. L. 1962. Studies on the embryology of *Pycnopodia helianthoides* (Brandt) Stimpson. Pac. Sci., **14**: 280–85.

Class OPHIUROIDEA
Brittle Stars, Serpent Stars

BUCHANAN, J. B., and J. D. WOODLEY. 1963. Extension and retraction of the tube-feet of ophiuroids. Nature, **197**: 616–17, 2 figs.
Notes great extensile capacity of the tube feet, but did not observe their use in feeding.

CLARK, A. H. 1921. A new ophiuran of the genus *Ophiopsila* from southern California. Proc. Biol. Soc. Wash., **34**: 109–10.

CLARK, H. L. 1911. North Pacific ophiurans. Bull. U.S. Nat. Mus., **75**: 1–302.

——— 1915. A remarkable new brittle star. J. Entom. and Zool., **75**: 64–66.
Ophiocryptus maculosus, from kelp holdfasts at Laguna Beach, 16 mm. disk, arms 9 mm. long.

MAGNUS, DIETRICH B. E. 1964. Gezeitenströmung und Nahrungsfiltation bei Ophiuren und Crinoiden. Helg. Wiss. Meeresunters., **10** (1–4): 104–17, 8 figs.

MAY, R. M. 1924. Ophiurans of Monterey Bay. Proc. Calif. Acad. Sci., (4) **13**: 261–303.

MCCLENDON, J. F. 1909. Ophiurans of the San Diego region. Univ. Calif. Pub. Zool., **6**: 33–64.

NIELSEN, EIGEL. 1932. Ophiurans from the Gulf of Panama, California, and the Strait of Georgia, No. 59 in Papers from Dr. Th. Mortensen's Pacific expedition, 1914–16. Vidensk. Medd. fra Dansk Naturh. Foren. Köbenhavn, **91**: 241–346.
Description of new species, including *Ophiocryptus granulosus* and *Amphichondrius granulosus* from La Jolla, with keys and taxonomic descriptions of all the ophiurans known from these regions. *Ophiactis simplex* (LeConte) and *Amphipholis squamata* (Della Chiaje) are reported from the La Jolla littoral.

STEPHENS, GROVER C., and RAGHUNATH A. VIRKAR. 1966. Uptake of organic material by aquatic invertebrates, IV: The influence of salinity on the uptake of amino acids by the brittle star *Ophiactis arenosa*. Biol. Bull., **131** (1): 172–85, 5 figs.
The uptake of glycine increases with reduction of salinity, but further decrease of salinity reduces uptake.

HILTON, W. A. 1918. The central nervous system of a long-armed serpent star. J. Entom. and Zool., **10**: 75.

Ophioplocus esmarcki Lyman...§ 45
Ophiopteris papillosa (Lyman) ....§124
Ophiothrix spiculata LeConte ....§123

With radial shields spiny. The related *O. rudis* Lyman, less common on the littoral, has bare radial shields and smooth spines.

Class ECHINOIDEA
Sea Urchins, Sand Dollars, and Heart Urchins

Clark, H. L. 1925. A catalogue of the recent sea urchins (*Echinoidea*). Br. Mus. 250 pp.

——— 1948. A report on the Echini of the warmer eastern Pacific, based on the collections of the *Velero III*. Allan Hancock Pac. Exp., **8** (5): 225–351, 37 pls., 3 figs.

Since most of central California species range well down Baja California, this report discusses five of the eight species we list below. A bibliography of Clark's echinoderm papers is included.

Durham, J. Wyatt. 1955. Classification of clypeasteroid echinoids. Univ. Calif. Pub. Geol. Sci., **31** (4): 73–198, 38 figs., 2 pls., frontispiece.

Grant, U. S., IV, and L. G. Hertlein. 1938. The west American Cenozoic Echinoidea. Pub. UCLA in Math. and Phys. Sci., **2**. 225 pp., 30 pls.

Although the emphasis is on paleontology, all the Recent Pacific coast species are also cited, with synonymy, references, and geographic range; often with descriptions, and sometimes with illustrations.

Kahl, M. E. 1950. Metabolism and cleavage of eggs of the sea urchin *Arbacia punctulata*: A review, 1932–1949. Biol. Bull., **98** (3): 175–217.

A summary of the work done on this Atlantic coast species, with bibliography.

Dendraster excentricus (Eschscholtz) ....§246
Echinarachnius parma (Lamarck) ....§246

Parker, G. H., and M. Van Alstyne. 1932. Locomotor organs of *Echinarachnius parma*. Biol. Bull., **62**: 195–200.

Lovenia cordiformis A. Ag...§247
Lytechinus anamesus A. Ag. and H. L. Clark.............................§269
L. pictus (Verrill) ....§218
Strongylocentrotus dröbachiensis (O. F. Müller).........................§212

Swan, Emery F. 1961. Some observations on the growth rate of sea urchins in the genus *Strongylocentrotus*. Biol. Bull., **120** (3): 420–27, 1 fig.

Mostly about *S. dröbachiensis*.

Weese, A. O. 1926. Food and digestive processes of *Strongylocentrotus dröbachiensis*. Pub. Puget Sd. Biol. Sta., **5**: 165–79.

S. franciscanus (A. Ag.) ....§ 73

Johnson, Martin W. 1930. Notes on the larval development of *Strongylocentrotus franciscanus*. Pub. Puget Sd. Biol. Sta., **7**: 401–11.

S. purpuratus (Stimpson) ....§174

The three species of *Strongylocentrotus* have been the subject of physiological and embryological experimentation for years; there are certainly dozens, probably hundreds of scattered papers relating to them. Loeb's classic work on parthenogenesis was done with Monterey Bay urchins. The most popular species for experimental purposes is *S. purpuratus*, and the species has become extremely scarce in certain easily accessible areas because of the harvesting by molecular and developmental biologists. Among those active in research on cellular and developmental processes based on the study of sea urchin eggs on the Pacific coast, and whose principal specimen is *S. purpuratus*, are W. E. Martin, D. Mazia, A. Tyler, and A. H. Whiteley. Perhaps this is the best and highest use of this material (to use the tax assessor's phrase); but the humble purple sea urchin is also of interest to ecologists, who prefer undisturbed populations. Furthermore, in certain small areas sea urchins have simply been trampled to bits by the numbers of people who like to walk in their haunts at low tide. We do not have enough space for more than a sample of the literature on this species; students should scan the back files and keep up with the *Biological Bulletin,* which is one of the major outlets for papers on this and other west coast echinoderms.

EBERT, THOMAS A. 1965. A technique for the individual marking of sea urchins. Ecology, **46** (1–2): 193–94, 2 figs.

——— 1967. Negative growth and longevity in the purple sea urchin *Strongylocentrotus purpuratus* (Stimpson). Science, **157** (3788): 557–58, 1 fig.

S. purpuratus may live ten years; larger animals may actually shrink a bit, possibly in unfavorable conditions.

——— 1967. Growth and repair of spines in the sea urchin *Strongylocentrotus purpuratus* (Stimpson). Biol. Bull., **133** (1): 141–49, 2 figs.

FARMANFARMAIAN, A., and A. C. GIESE. 1963. Thermal tolerance and acclimation in the western purple sea urchin, *Strongylocentrotus purpuratus.* Physiol. Zool., **36:** 237–43.

LASKER, REUBEN, and ARTHUR C. GIESE. 1954. Nutrition of the sea urchin *Strongylocentrotus purpuratus.* Biol. Bull., **106** (3): 328–40, 3 figs.

Class HOLOTHURIOIDEA. Sea Cucumbers, Sea Slugs

CLARK, H. L. 1924. Some holothurians from British Columbia. Canad. Fld. Nat., **38** (3): 54–57.

Eleven species from shore and from shallow dredgings.

DEICHMANN, E. 1938. New holothurians from the western coast of North America, and some remarks on the genus *Caudina.* Proc. N. Eng. Zool. Clb., **16:** 103–15, text figs.

HEDING, S. G. 1928. Synaptidae. No. 46 in Papers from Dr. Th. Mortensen's Pacific expedition, 1914–16. Vidensk. Medd. fra Dansk Naturh. Foren., **85:** 105–323.

Describes four new *Leptosynapta* and three *Chiridota* from La Jolla and Nanaimo.

WELLS, H. W. 1924. New species of *Cucumaria* from Monterey Bay. Ann. Mag. Nat. Hist., (9) **14:** 113–21.

COWLES, R. P. 1907. *Cucumaria curata* sp. nov. Johns Hopkins Univ. Circ., No. 195. 2 pp.

FILICE, FRANCIS P. 1950. A study of some variations in *Cucumaria curata* (Holothuroidea). Wasmann J. Biol., **8** (1): 39–48, 2 figs.

SMITH, EDMUND H. 1962. Studies of *Cucumaria curata* Cowles 1907. Pac. Nat., **3** (5): 233–46.

Describes brooding, etc. E. Deichmann thinks the species may be *C. lubrica.*

HALL, A. R. 1927. Histology of the retractor muscle of *Cucumaria miniata.* Pub. Puget Sd. Biol. Sta., **5:** 205–19.

GLYNN, PETER W. 1965. Active movements and other aspects of the biology of *Astichopus* and *Leptosynapta* (Holothuroidea). Biol. Bull., **129** (1): 106–27, 5 figs.

Lissothuria nutriens (H. L. Clark) . . . . §141

PAWSON, DAVID L. 1967. The psolid holothurian genus *Lissothuria*. Proc. U.S. Nat. Mus., **122** (3592): 1–17, 5 figs.
L. nutriens, pp. 6–8, not illustrated.

Molpadia arenicola (Stimpson) . . . . §264
Psolus chitonoides H. L. Clark. . . . . §226
Stichopus californicus (Stimpson) . . . . § 77

COURTNEY, W. D. 1927. Fertilization in *Stichopus californicus*. Pub. Puget Sd. Biol. Sta., **5**: 257–60.

S. parvimensis H. L. Clark. . . . . § 77
Thyone rubra H. L. Clark. . . . . §141

Phylum **CHORDATA**

Subphylum **UROCHORDATA (TUNICATA)**
Sea Squirts, Compound Ascidians, Tunicates

The large systematic literature on Pacific coast tunicates is catalogued in Van Name's monograph, and there is no need to cite earlier scattered papers here.

ABBOTT, DONALD P. 1953. Asexual reproduction in the colonial ascidian *Metandrocarpa taylori* Huntsman. Univ. Calif. Pub. Zool., **61** (1): 1–78, 5 figs., 14 pls.

BERRILL, N. J. 1950. The Tunicata, with an account of the British species. Ray Society. 354 pp., 120 figs.

MACGINITIE, G. E. 1939. The method of feeding of tunicates. Biol. Bull., **77**: 443–47.

MORGAN, T. H. 1941. Further experiments in cross- and self-fertilization of *Ciona* at Woods Hole and Corona del Mar. Biol. Bull., **80** (3): 338–53.

NEWBERRY, ANDREW TODD. 1965. Vascular structure associated with budding in the polystelid ascidian *Metandrocarpa taylori*. Ann. Soc. Roy. Zool. Belg., **95** (2): 57–74, 4 figs.

——— 1965. The structure of the circulatory apparatus of the test and its role in budding in the polystyelid ascidian *Metandrocarpa taylori* Huntsman. Mem. Acad. Roy. Belg. Sci., **16** (5). 57 pp., 42 figs. (plates outside of pagination!).

TRASON, WINONA B. 1963. The life cycle and affinities of the colonial ascidian *Pycnoclavella stanleyi*. Univ. Calif. Pub. Zool., **65** (4): 283–326, 125 figs.

VAN NAME, WILLARD G. 1945. The North and South American ascidians. Bull. Amer. Mus. Nat. Hist., **84**. 476 pp., 31 pls., text figs.

Amaroucium californicum Ritter and Forsyth. . . . . § 99

GRAVE, CASWELL. 1920. *Amaroucium pellucidum* forma *constellatum*, I: The activities and reactions of the tadpole larva. J. Exp. Zool., **30**: 239–57.
Natural history of an Atlantic form comparable to *A. californicum*.

Ascidia ceratodes (Huntsman) . . . . not treated

Phytoplankton is not unique in its erratic fluctuations; this ascidian "bloomed" in Tomales Bay in the early spring of 1950 and died off by fall. Not only were the animals abundant in all parts of the bay (including the intertidal), but specimens up to six inches long were seen. The range of this species is southern California to British Columbia.

VAN NAME, W. G. 1931. New North and South American ascidians. Bull. Amer. Mus. Nat. Hist., **61**: 207–25.

Subphylum **HEMICHORDATA (ENTEROPNEUSTA)**
Acorn or Tongue Worms, etc.

BARRINGTON, E. J. W. 1965. The biology of Hemichordata and Protochordata. San Francisco: Freeman (University Reviews in Biology). 176 pp., 82 figs.

Concerns balanoglossids, ascidians, and amphioxus.

BULLOCK, THEODORE H. 1945. The anatomical organization of the nervous system of *Enteropneusta*. Quart. J. Micr. Sci., **86** (1): 55–111, 7 pls., 1 fig.

DAVIS, B. N. 1908. The early life history of *Dolichoglossus pusillus* Ritter. Univ. Calif. Pub. Zool., **4** (30): 187–226, 5 pls.

HILTON, W. A. 1918. *Dolichoglossus pusillus* Ritter. Pomona J. Entom. and Zool., **10**. 76 pp.

RITTER, W. E. 1900. *Harrimania maculosa,* a new genus and species of *Enteropneusta* from Alaska. Papers from the Harriman Alaska Exp. II. Proc. Wash. Acad. Sci., **2**: 111–32.

Records a very common intertidal form, which should be noted here, since it seems to be a feature of under-rock collecting at Kodiak, and at Prince William Sound, Orca, and Valdez. This thick, dark acorn worm, up to six inches in length, does not burrow as do most *Enteropneusta,* but lies under stones after the fashion of holothurians. Other species are mentioned, to be named *Balanoglossus intermedia* and *B. californicus,* but these apparently remain MS species to this day.

——— 1902. The movements of the Enteropneusta and the mechanism by which they are accomplished. Biol. Bull., **3**: 255–61.

Two MS species mentioned: *D. pusillus,* described 1930 in Horst, below; and *B. occidentalis,* stated to be abundant in Puget Sound. The latter is apparently still a MS species.

———, and B. N. DAVIS. 1904. Studies on the ecology, morphology, and speciology of the young of some Enteropneusta of western North America. Univ. Calif. Pub. Zool., **1**: 171–210.

WILLEY, ARTHUR. 1931. *Glossabalanus berkeleyi,* a new enteropneust from the west coast. Trans. Roy. Soc. Canad., Ser. 3, **25** (5): 19–28.
Known from a single fragmentary specimen, 40 mm. × 6 mm.

Saccoglossus pusillus (Ritter) .... §267

HORST, C. J. VAN DER. 1930. Observations on some Enteropneusta. Papers from Dr. Th. Mortensen's Pacific expedition, 1914–16, II. Vedensk. Medd. fra Dansk Naturh. Foren., **87**: 135–200.
Ritter's description is on p. 154.

Subphylum CEPHALOCHORDATA. Lancelets

JORDAN, EVERMANN, and CLARK. 1930. As listed below under Pisces.
HUBBS, C. L. 1922. A list of the lancelets of the world. Occ. Paps. Mus. Zool., Univ. Mich., No. 105.

Branchiostoma californiense Andrews .... §266

Subphylum CRANIATA (VERTEBRATA)

Class PISCES. Fishes

Consideration of the fishes has been deliberately omitted, as was the case with the Protozoa, except where it was thought that mention of a few of the more obvious forms, especially those found under rocks and in the tide pools, would add to the interest or usefulness of this account.

ARORA, HARBANS L. 1948. Observations on the habits and early life history of the batrachoid fish *Porichthys notatus* Girard. Copeia, **1948** (2): 89–93, 1 pl.

BARNHART, P. S. 1936. Marine fishes of southern California. Univ. Calif. Press. 209 pp., 290 figs.

BOLIN, ROLF L. 1944. A review of the marine cottid fishes of California. Stanford Ichth. Bull., **3** (1). 135 pp., 40 figs.
With *toto* drawings of each species. Included are most of the common tide-pool fish. Descriptions of species rather than specimens, with practical keys.

BROWN, MARGARET E., ed. 1957. The physiology of fishes: I, Metabolism; II, Behavior. New York: Academic Press.

CARLISLE, JOHN G., JR., JACK W. SCHOTT, and NORMAN J. ABRAMSON. 1960. The barred surfperch (*Amphistichus argenteus* Agassiz) in southern California. Calif. Fish Bull., **109**. 79 pp., 40 figs., col. frontispiece.

CLEMENS, W. A., and G. V. WILBY. 1946. Fishes of the Pacific coast of Canada. Fish. Res. Bd. Canad. Bull., **68**. 368 pp., 253 figs.
An adequate and useful guide, insofar as tidepool fishes are concerned, from San Francisco, or possibly Monterey, northward.

HELLE, JOHN H., RICHARD S. WILLIAMSON, and JACK E. BAILEY. 1964. Intertidal ecology and life history of pink salmon at Olsen Creek, Prince William Sound, Alaska. U.S. Fish and Wildl. Serv., Spec. Scient. Rept. Fish. **483**. 26 pp., 20 figs., 3 appendix figs.
It would appear from this that *Oncorhynchus gorbuscha,* at least, might survive the steady encroachment of dams upon salmon streams. A large percentage (74%) of the run in some years spawns in gravel covered by high tide (up to 80% of the time) in the lower reaches of the stream, and development and hatching appear to be successful in this peculiar situation. Although there is a typical marine intertidal representation in the lower reaches, it is mostly of euryhaline organisms such as *Mytilus edulis, Hemigrapsus oregonensis,* "*Balanus* sp.," etc. Salinity within the gravel where the eggs are deposited is as high as 9.3 parts per thousand.

HUBBS, CARL L., and LAURA C. HUBBS. 1954. Data on the life history, variation,

ecology, and relationships of the kelp perch, *Brachyistius frenatus,* an embiotocid fish of the Californias. Calif. Fish and Game, **40** (2): 183–98, 3 figs.
Contra Tarp (1952), below.

HUBBS, CLARK. 1952. A contribution to the classification of the blennioid fishes of the family Clinidae, with a partial revision of the eastern Pacific forms. Stanford Ichth. Bull., **4** (2): 41–165, 64 figs.
Includes *toto* figures and descriptions of the various tide-pool blennies mentioned in this book.

JONES, ALBERT C. 1962. The biology of the euryhaline fish *Leptocottus armatus* Girard (Cottidae). Univ. Calif. Pub. Zool., **67** (4): 321–68, 10 figs.
Based on studies in Walker Creek at Tomales Bay.

JORDAN, D. S., and B. W. EVERMANN. 1896–1900. The fishes of north and middle America. Bull. U.S. Nat. Mus., **47**. 4 vols., 3313 pp., 392 pls.

———, B. W. EVERMANN, and H. W. CLARK. 1930. Checklist of the fishes and fish-like vertebrates of north and middle America. Rept. U.S. Comm. Fish. for 1928, Part II. 670 pp., etc.

MACGINITIE, G. E. 1939. The natural history of the blind goby *Typhlogobius californiensis* Steindachner. Amer. Midland Nat., **21** (2): 489–505, 2 pls.

MORRIS, ROBERT W. 1951. Early development of the cottid fish, *Clinocottus recalvus* (Greeley). Calif. Fish and Game, **37** (3): 281–300, 27 figs.
One of the common "tidepool johnnies" of the central California coast.

NOBLE, ELMER R. 1941. On distribution relationships between California tide-pool fishes and their myxosporidian (protozoan) parasites. J. Parasitol., **27** (5): 409–14, 1 pl.

PHILLIPS, JULIUS B. 1957. A review of the rockfishes of California. Calif. Fish Bull., **104**, 158 pp., 66 figs., col. frontispiece.

ROEDEL, PHIL M. 1953. Common ocean fishes of the California coast. Calif. Fish Bull., **91**. 184 pp., 175 figs., col. frontispiece.

———, and WM. ELLIS RIPLEY. 1950. California sharks and rays. Calif. Dept. Fish and Game, Fish Bull., **75**: 88 pp., 65 figs.
Man-eaters (*Carcharodon carcharias*) have been "rare" along our coast; but records as far north as Fort Bragg antedate 1959, the summer of the "warm" water and the sharks. Shark incidents can be expected to increase as the numbers of swimmers and divers increase. The best account so far of the sharks and the warm-water conditions is that by Harold Gilliam, in the "This World" magazine section of the San Francisco Chronicle for July 26, 1959.

SCHULTZ, L. P. 1936. Key to the fishes of Washington, Oregon, and closely adjacent regions. Univ. Wash. Pub. Biol., **2** (4): 103–228, 48 text figs.

TARP, FRED HARALD. 1952. A revision of the family Embiotocidae (the surf-perches). Calif. Fish Bull., **88**. 99 pp., 32 figs.

WALFORD, LIONEL A. 1937. Marine game fishes of the Pacific coast from Alaska to the equator. Berkeley: Univ. Calif. Press. 205 pp., map and frontispiece, 69 pls. (37 col.).

WILLIAMS, GEORGE C. 1957. Homing behavior of California rocky shore fishes. Univ. Calif. Pub. Zool., **59** (7): 249–84, 1 fig.

———, and DORIS C. WILLIAMS. 1955. Observations on the feeding habits of the opaleye, *Girella nigricans*. Calif. Fish and Game, **41** (3): 203–8, 2 figs.
The young of this species occur in tidepools in southern California.

YOUNG, PARKE H. 1963. The kelp bass (*Paralabrax clathratus*) and its fishery, 1947–1958. Calif. Fish Bull., **122**. 67 pp., 33 figs.

SCHULTZ, L. P., and A. C. DELACY. 1932. Nesting habits of . . . *A. purpurescens*. Science, Oct. 7, 1932, p. 311.

Gillichthys mirabilis Cooper, Sand Goby................................§266
Hypsypops rubicunda (Girard), Garibaldi..............................§149
Leuresthes tenuis (Ayres), Grunion....................................§192

OLSON, ANDREW C. 1950. Ground squirrels and horned larks as predators upon grunion eggs. Science, **36** (3): 323–27, 3 figs.

WALKER, BOYD W. 1952. A guide to the grunion. Science, **38** (4): 409–20, 6 figs.

Oligocottus maculosus Girard, Sculpin, Tidepool Johnny.................§149
Porichthys notatus Girard, Midshipman, Grunter.........................§240
Sicyogaster maeandrica (Girard), Clingfish.............................§ 51
Typhlogobius californiensis Steindachner, Blind Goby ....§ 58
Xiphister mucosus (Girard), Blenny....................................§ 51

Class AVES

General references on distribution:

GRINNELL, J., and A. H. MILLER. 1944. The distribution of the birds of California. Pac. Coast Avifauna, No. 27. 608 pages.
With excellent habitat notes.

MUNRO, J. A., and I. McT. COWAN. 1947. A review of the bird fauna of British Columbia. Brit. Colum. Provinc. Mus., Spec. Pub. No. 2: 1–285.

Selected references on ecological relations of birds of inshore waters and the intertidal zone:

BARTHOLOMEW, GEORGE A., JR. 1942. The fishing activities of double-crested cormorants on San Francisco Bay. Condor, **44**: 13–21.

——— 1943. The daily movements of cormorants on San Francisco Bay. Condor, **45**: 3–18.

BOND, R. M. 1942. Banding records of California brown pelicans. Condor, **44**: 116–21.

KENYON, K. W. 1949. Observations on behavior and populations of oyster-catchers in Lower California. Condor, **51**: 193–99.

LAWRENCE, G. E. 1950. The diving and feeding activity of the western grebe on the breeding grounds. Condor, **52**: 3–16.
Acmaea taken as prey on wintering waters.

McHUGH, J. L. 1950. Increasing abundance of albatrosses off the coast of California. Condor, **52**: 153–56.
See Bibliography for other recent papers on albatrosses.

McKERNAN, D. L., and VICTOR B. SCHEFFER. 1942. Unusual numbers of dead birds on the Washington coast. Condor, **44**: 264–66.

MARSHALL, J. T., JR. 1948. Ecologic races of song sparrows in the San Francisco Bay region. Condor, **50**: 193–215, 233–56.
Salt-marsh habitats described.

MILLER, A. H. 1943. Census of a colony of Caspian terns. Condor, **45**: 220–25.
On San Francisco Bay.

MOFFITT, J. 1941. Notes on the food of the California clapper rail. Condor, **43**: 270–73.

MURIE, O. J. 1940. Food habits of the northern bald eagle in the Aleutian Islands, Alaska. Condor, **42**: 198–202.

REEDER, W. G. 1951. Stomach analysis of a group of shorebirds. Condor, **53**: 43–45.
Records of prey taken at Sunset Beach and Point Mugu, California.

WEBSTER, J. D. 1941. Feeding habits of the black oyster-catcher. Condor, **43**: 175–80. See corrections, Condor, **53**: 54 (1951).

——— 1941. The breeding of the black oyster-catcher. Wilson Bull., **53**: 141–56. See corrections, Condor, **53**: 54 (1951).

WILLIAMS, L. 1942. Displays and sexual behavior of the Brandt cormorant. Condor, **44**: 85–104.

Class MAMMALIA

Man is not the only mammal who raids the intertidal for food; the raccoon (*Procyon lotor,* various subspecies) visits the shore at low tide from Puget Sound to Baja California, feeding principally on crabs and leaving his tracks along the sand. In Baja California, where the barren back country offers little nourishment, the coyote is a consistent intertidal feeder. Various mice, ground squirrels (see grunion), and possibly the mink also visit the shore.

As for the strictly marine mammals, the representatives of two mammalian orders (exclusive of the whales) occur along the shores of California. The Carnivora are represented by the sea otter, *Enhydra lutris nereis,* a marine member of the weasel family (Mustelidae), which feeds principally on subtidal abalone, sea urchins, and crabs. The sea otter, once almost banished from California shores, still exists in small, scattered herds along the less-frequented parts of the coast from Monterey to the Channel Islands; and hope has been expressed for its return in sufficient numbers to justify renewed harvesting for its valuable fur. (It might, however, compete with commercial abalone divers in that event.) Among the Pinnipedia (seals and sea lions), we most often see the Steller sea lion, *Eumetopias stelleri,* and the smaller California sea lion, *Zalophus californianus* (the trained seal of the circus), both members of the family *Otariidae,* or eared seals. Commercial fishermen make the perennial claim that these animals destroy valuable fish, especially salmon (a complaint first made in the 1870's), but investigations have failed to substantiate this, at least to the degree needed to justify wholesale slaughter of the animals. These animals, relatives of the famous Pribilof fur seal, have their principal breeding grounds or rookeries on the Channel Islands. The harbor seal, *Phoca vitulina,* is the only true seal likely to be seen by shorebound observers. Because it cannot turn its hind feet forward, it is clumsy on land, and cannot climb rocks, as do the sea lions. Small herds are to be seen in bays, where they "haul out" on sand bars; they are common in San Francisco and Tomales bays. The great sea elephant, *Mirounga angustirostris,* also a member of the true seals or Phocidae, was once reduced to a few small bands on the islands off Baja California; but it has recently been observed on the Channel Islands, and is now 15,000 strong. There is a small herd of them on Año Nuevo Island.

ANDREWS, ROY C. 1914. Monograph of the Pacific Cetacea, I: The California gray whale (*Rachianectes glaucus* Cope). Mem. Amer. Mus. Nat. Hist., n.s., **1** (5): 227–87, 22 figs., 9 pls.

The California gray whale, which calves in bays and lagoons, was once thought to be nearly extinct, but it is gradually returning; the present population is several thousand. Calving is now restricted to the lagoons of Baja California; but during the winter months, these whales migrate southward along the California coast, and may often be seen from shore.

BARTHOLOMEW, GEORGE A., JR. 1952. Reproductive and social behavior of the northern elephant seal. Univ. Calif. Pub. Zool., **47** (15): 369–472, 20 pls., 2 text figs.

Illustrated with excellent photographs, some of them recalling the famous Ninth Canto of Camões's *Lusiads*—although things were managed more gracefully, according to the Lusitanian bard:

Cahir se deixa aos pés do vencedor,
Que todo se desfaz em puro amor.

BOLIN, ROLF L. 1938. Reappearance of the southern sea otter along the California coast. J. Mammal., **19** (3): 301–6.

BONNOT, PAUL. 1951. The sea lions, seals, and sea otter of the California coast. Calif. Fish and Game, **37** (4): 371–89, 11 figs.

"In common with all organisms in a natural environment, the marine mammals were controlled by biological checks that maintained the populations in balance with all associated species. The advent of man into this orderly design demoralized it completely. The human animal is the most persistent and rapacious predator that has so far appeared on earth." P. 371.

BOOLOOTIAN, RICHARD A. 1961. The distribution of the California sea otter. Calif. Fish and Game, **47** (3): 287–92, 2 figs.

DAUGHERTY, ANITA E. 1965. Marine mammals of California. Sacramento: Dept. Fish and Game. 87 pp., illus.

A convenient pocket-sized handbook; well illustrated.

EBERT, EARL E. 1968. A food habits study of the southern sea otter, *Enhydra lutris nereis*. Calif. Fish and Game, **54** (1): 33–42. 1 fig.

GILMORE, RAYMOND M. 1961. The story of the gray whale. Privately printed. 17 pp., illus.

MATHISEN, OLE A., ROBERT T. BAADE, and RONALD J. LOPP. 1962. Breeding habits, growth and stomach contents of the Steller sea lion in Alaska. J. Mammal., **43** (4): 469–77, 3 figs.

NORRIS, KENNETH S., ed. 1966. Whales, dolphins, and porpoises. Berkeley: Univ. Calif. Press. 789 pp., illus.

The reports of a symposium; practically everyone active in studying cetaceans was in some way involved in this four-pound leviathan.

———, and JOHN R. PRESCOTT. 1961. Observations on Pacific cetaceans of Californian and Mexican waters. Univ. Calif. Pub. Zool., **63** (4): 291–402, 12 text figs., 15 pls.

Firsthand observations for the most part, especially of porpoises.

OGDEN, ADELE. 1941. The California sea otter trade, 1784–1848. Berkeley: Univ. Calif. Press. 251 pp., illus.

Primarily concerned with the economics of the trade, but with life history information, an excellent photo (opp. p. 146) of a sea-otter herd, and a color frontispiece.

ORR, ROBERT T., and THOMAS C. POULTER. 1965. The pinniped populaton of Año Nuevo Island, California. Proc. Calif. Acad. Sci., Ser. 4, **32** (13): 377–404, 12 figs.

This is the northern outpost of the elephant seal. Since access is moderately difficult because of tricky wave refraction between the island and the shore, this island will hopefully continue to be somewhat isolated from the mainland—except, of course, to devoted marine biologists who do not disturb things.

——— 1967. Some observations on reproduction, growth and social behavior in the Steller Sea Lion. Proc. Calif. Acad. Sci., Ser. 4, **35** (10): 193–226, 27 figs., 3 tables.

With good views and a map of Año Nuevo Island, and a fine photo of the junior author being accepted as a mother by a deserted pup.

PETERSON, RICHARD S., and GEORGE A. BARTHOLOMEW, JR. 1967. The California Sea Lion. Special Publication No. 1, American Society of Mammalogists. 79 pp., 26 figs.

SCAMMON, C. M. 1874. The marine mammals of the northwestern coast of North America, described and illustrated: Together with an account of the American whale-fishery. San Francisco: J. H. Carman. 319 pp., 27 pls.

This classic work, long an expensive collector's item, is being reproduced by Dover with a biographical sketch of Captain Scammon by Victor Scheffer.

SCHEFFER, VICTOR B. 1958. Seals, sea lions, and walruses: A review of the Pinnipedia. Stanford: Stanford Univ. Press. 179 pp., illus.

The tooth marks illustrated on Plate 32 are not those of a killer whale, but of a shark.

SCOFIELD, W. L. 1941. The sea otters of California did not reappear. Calif. Fish and Game, **27** (1): 35–38.

(They were there all the time.)

General Bibliography

WHEN BIBLIOGRAPHIES support text citations, they are, of course, a key to the source of information. Not every reference given here, however, has been directly quoted in the text; many of them have been included simply because they seem to belong here. In other words, this section is intended to be a bibliography in the sense of a guide to further reading, rather than simply a list of "references cited." For this edition I have added a section listing (and commenting on) all the books about seashore life of this coast that have come to my attention.

For the reader's convenience, this Bibliography has been subdivided into several general topic areas: I, ecological and faunistic accounts related to the Pacific coast; II, popular books and booklets about Pacific coast seashore life; III, papers of general scope on marine ecology and related matters; IV, texts and reference books; V, books on geology and paleontology; VI, popular books and texts about the ocean, published mostly since 1960. It should be apparent, of course, that these categories overlap substantially, with respect even to individual books.

This is not to be considered a complete list of every account describing marine invertebrates on the Pacific coast. Such a bibliography would fill an entire book—or several feet of computer tape. I know I have missed many things, but I hope that enough is included to enable the determined student to work into the literature and ferret them out. All I can say is that if I have missed some essential reference, it may be because the author did not send me a reprint; yet I do not wish it thought that I am soliciting reprints (a subject on which I am known to be an expert anyhow). Also, my present location is not exactly at the crossroads of culture, and many things escape notice in these days of abundant publication. I do thank all who have sent me their papers, and I only hope that I have remembered to include all those deserving mention.

I. Ecological and Faunistic Accounts Related to the Pacific Coast

ALEXANDER, JOSEPHINE. 1961. Grassroots oceanography. Pac. Disc., **14** (1): 18–23, illus.

Tomales Bay in its good old days.

ANDREWS, H. L. 1945. The kelp beds of the Monterey region. Ecology, **26**: 24–37.

BANDY, ORVILLE M., JAMES C. INGLE, JR., and JOHANNA M. RESIG. 1965. Foraminiferal trends: Hyperion outfall, California. Limnol. and Oceanogr., **10** (3): 314–32, 14 figs.

BARNARD, J. LAURENS, and OLGA HARTMAN. 1959. The sea bottom off Santa Barbara, California: Biomass and community structure. Pac. Nat., **1** (6). 16 pp., 7 figs.

BOUSFIELD, E. L. 1958. Ecological investigations on shore invertebrates of the Pacific coast of Canada, 1955. Ann. Report, Nat. Mus. Canad., 1955–56, Bull. **147**: 104–15, 2 pls.

CALCOFI, *see* Marine Research Committee.

CAPLAN, R. I. and RICHARD A. BOOLOOTIAN. 1967. Intertidal ecology of San Nicolas Island, Proc. Symposium Biol. California Islands, Santa Barbara Botanical Garden, pp. 203–17, 9 figs.

CAREY, A. G., W. G. PEARCY, and C. L. OSTERBERG. 1966. Artificial radionuclides in marine organisms in the northeast Pacific Ocean off Oregon. In *Disposal of radioactive wastes into seas, oceans and surface waters,* pp. 303–19. Vienna: International Atomic Energy Authority.

COE, WESLEY R. 1932. Season of attachment and rate of growth of sedentary marine organisms. Bull. Scripps Inst. Oceanogr. Tech Ser., **3**: 37–86.

———, and W. E. ALLEN. 1937. Growth of sedentary marine organisms on experimental blocks and plates for nine successive years at the pier of the Scripps Institution of Oceanography. Bull. Scripps Inst. Oceanogr. Tech. Ser., **4**: 101–36.

CONNELL, JOSEPH H. 1963. Territorial behavior and dispersion in some marine invertebrates. Res. on Population Ecology, **5** (2): 87–101, 3 figs., 2 pls.

DITSWORTH, GEORGE R. 1966. Environmental factors in coastal and estuarine waters. Bibliographic Ser., Vol. I: Coast of Oregon. U.S. Dept. Int., Fed. Water Pollution Contr. Admin. Pub. WP-20-2. 61 pp.

DOTY, MAXWELL S. 1946. Critical tide factors that are correlated with the vertical distribution of marine algae and other organisms along the Pacific coast. Ecology, **27** (4): 315–28, 6 figs.

DRUEHL, LOUIS D. 1967. Vertical distributions of some benthic marine algae in a British Columbia inlet, as related to some environmental factors. J. Fish. Res. Bd. Canad., **24** (1): 33–46, 7 figs.

FILICE, FRANCIS P. 1954. An ecological survey of the Castro Creek area in San Pablo Bay. Wasmann J. Biol., **12** (1): 1–24, 2 figs.

——— 1954. A study of some factors affecting the bottom fauna of a portion of the San Francisco Bay estuary. Wasmann J. Biol., **12** (3): 257–92, 7 figs.

——— 1958. Invertebrates from the estuarine portion of San Francisco Bay and some factors influencing their distributions. Wasmann J. Biol., **16** (2): 159–211, 3 figs.

——— 1959. The effect of wastes on the distribution of bottom invertebrates in the San Francisco Bay estuary. Wasmann J. Biol., **17** (1): 1–17, 1 fig.

FOLSOM, THEODORE H., and JOHN H. HARVEY. 1957. Comparison of some natural radiations received by selected organisms. N.A.S. Pub. **551**: 28–33, 2 figs.

FRASER, C. MCLEAN. 1932. A comparison of the marine fauna of the Nanaimo region with that of the San Juan archipelago. Trans. Roy. Soc. Canad., Ser. 3, **26** (Sec. 5): 49–70.

——— 1942. The collecting of marine zoological materials in B.C. waters. Canad. Fld. Nat., **56**: 115–20.

FROLANDER, HERBERT F. 1964. Biological and chemical features of tidal estuaries. J. Water Poll. Contr. Fed., August 1964: 1037–48, 7 figs.

Concerns Yaquina Bay, Oregon.

GISLÈN, TORSTEN. 1943–44. Physiological and ecological investigations concerning the littoral of the northern Pacific. Sections 1–4. Lunds Univ. Ars. N F Avd. 2, **39** (5) and **40** (8): 63 and 91 pp., pls.

Comparisons between the fauna and the environmental factors of Pacific Grove, California, and Misaki, Japan, in the same latitude. Statement of the ecological associations, with correlations of temperature, wave-shock, etc., for the coasts of California and northern Baja California. Extensive information not elsewhere available.

GLYNN, PETER W. 1966. Community composition, structure, and interrelationships in the marine intertidal *Endocladia muricata–Balanus glandula* association in Monterey Bay, California. Beaufortia, **12** (148): 1–198, 76 figs.

Monumental, comprehensive, and exhaustive.

GRAHAM, H. W., and HELEN GAY. 1945. Season of attachment and growth of sedentary marine organisms at Oakland, California. Ecology, **26**: 375–86.

GRIER, MARY C. 1941. Oceanography of the North Pacific Ocean, Bering Sea, and Bering Strait: A contribution towards a bibliography. Libr. Ser., Univ. Wash. Pub., **2**. 290 pp.

Includes zoology, botany, and explorations.

GRINNELL, JOSEPH, and JEAN M. LINSDALE. 1936. Vertebrate animals of Point Lobos Reserve, 1934–35. Carnegie Inst. Wash. Pub., **481**. 159 pp., 39 pls.

Contains considerable information about marine mammals and shore birds, with excellent photographs.

GURJANOVA, EUPRAXIE. 1966. Comparative research of biology of the littoral in the far eastern seas. Proc. Ninth Pac. Sci. Congr., **19** (Zoology): 75–86.

HALL, CLARENCE A., JR. 1964. Shallow water marine climates and molluscan provinces. Ecology, **45** (2): 226–34, 6 figs.

HAMBY, ROBERT J. 1964. Drift bottle studies at Bodega Head, California. Pac. Mar. Sta., Res. Repts. (unnumbered). 30 pp., 37 figs.

HARTMAN, OLGA. 1960. The benthonic fauna of southern California in shallow depths and possible effects of wastes on the marine biota. In *Waste Disposal in the Marine Environment*, pp. 57–81. New York: Pergamon Press. 7 figs.

——— 1966. Quantitative survey of the benthos of San Pedro Basin, Southern California, Part II: Final results and conclusions. Allan Hancock Pac. Exp., **19** (2): 185–456, 2 maps, 13 pls.

HARTMAN, OLGA, and J. LAURENS BARNARD. 1958. The benthic fauna of the deep basins off southern California, Part I. Allan Hancock Pac. Exp., **22** (1): 1–67, 1 cht., 2 pls.

——— 1960. Benthic fauna . . . , Part II. *Ibid.* (2): 69–297, 19 pls., map.

HEWATT, W. G. 1935. Ecological succession in the *Mytilus californianus* habitat, as observed in Monterey Bay, California. Ecology, **16**: 244–51.

——— 1937. Ecological studies on selected marine intertidal communities of Monterey Bay, California. Amer. Midland Nat. **18** (2): 161–206.

——— 1938. Notes on the breeding seasons of the rocky beach fauna of Monterey Bay, California. Proc. Calif. Acad. Sci., (4) **23** (19): 283–88.

——— 1946. Marine ecological studies on Santa Cruz Island, California. Ecol. Monogr., **16**: 185–210.

A survey involving more than 500 species, mostly intertidal.

HOLM-HANSEN, O., J. D. H. STRICKLAND, and P. M. WILLIAMS. 1966. A detailed analysis of biologically important substances in a profile off southern California. Limnol. and Oceanogr., **11** (4): 548–61, 6 figs.

HUBBS, CARL L. 1948. Changes in the fish fauna of western North America correlated with changes in ocean temperature. J. Mar. Res., **7** (3): 459–82, 6 figs.

———, and GUNNAR I. RODEN. 1964. Oceanography and marine life along the Pacific coast. Handbook of Middle American Indians, Univ. Texas, **1**: 143–86, 26 figs.

Concerns mostly the regions south of our limits.

ISAACS, JOHN D. 1961. Capacity of the oceans. Science and Technology, prototype issue: 38–43, illus.

——— 1965. Larval sardine and anchovy relationships. CalCOFI Repts., **10**: 102–40, 75 figs.

JOHNSON, RALPH G. 1962. Mode of formation of marine fossil assemblages of the Pleistocene Millerton Formation of California. Bull. Geol. Soc. Amer., **73**: 113–30, 1 fig.

——— 1965. Temperature variation in the infaunal environment of a sand flat. Limnol. and Oceanogr., **10** (1): 114–20, 7 figs.

——— 1965. Pelecypod death assemblages in Tomales Bay, California. J. Paleont., **39** (1): 80–85, 1 fig.

——— 1967. Salinity of interstitial water in a sandy beach. Limnol. and Oceanogr., **12** (1): 1–7, 1 fig.

——— 1967. The vertical distribution of the infauna of a sand flat. Ecology, **48** (4): 571–78, 3 figs.

Polychaetes, at least on Tomales Bay flats where surface is wet, do not migrate down with low tide but may be randomly distributed with reference to tidal level.

JONES, GILBERT F. 1965. The distribution and abundance of subtidal benthic mollusca on the mainland shelf of southern California. Malacologia, **2** (1): 43–68, 5 figs.

JONES, MEREDITH L. 1961. A quantitative evaluation of the benthic fauna off Point Richmond, California. Univ. Calif. Pub. Zool., **67** (3): 219–320, 30 figs.

KELLEY, D. W. (compiler). 1966. Ecological studies of the Sacramento–San Joaquin estuary, Part I: Zooplankton, zoobenthos, and fishes of San Pablo and Suisun Bays, zooplankton and zoobenthos of the delta. Calif. Dept. Fish and Game, Fish Bull., **133**. 133 pp., illus.

LIMBAUGH, CONRAD. 1961. Cleaning symbiosis. Sci. Amer., **205** (2) (August 1961): 42–49, illus.

MACGINITIE, G. E. 1935. Ecological aspects of a California marine estuary. Amer. Midland Nat. **16** (5): 629–765, 20 figs.

Life in Elkhorn Slough before progress improved the place. In this work, little faith is placed in the mapping of communities and concurrent attempts to divide them into zones. Whether from the influence of this paper or from other factors, quantitative population studies have been scarce on this coast. When the papers of Shelford and his students, the work of Hewatt, and the unpublished work of Pitelka and Paulson (see our pp. 374–76) have been cited, that is about all we have. One of the most important deterrents to this work, of course, is the bewildering complexity of our fauna and the taxonomic uncertainty in which so many of the species are involved.

——— 1939. Littoral marine communities. Amer. Midl. Nat., **21** (1): 28–55.

MCLEAN, JAMES H. 1962. Sublittoral ecology of kelp beds of the open coast area near Carmel, California. Biol. Bull., **122** (1): 95–114, 2 figs.

MARINE RESEARCH COMMITTEE. California Cooperative Sardine Research Program, Progress Reports, 1950–57, ed. T. A. Manar. Sacramento: Calif. State Printing Office.

———, Sardine Research Program, Progress Report VII (January 1960). Symposium: The changing Pacific Ocean in 1957 and 1958, eds. Oscar E. Sette and John D. Isaacs, pp. 13–217.

———, Sardine Research Program, Progress Report VIII (January 1961). Symposium on Fisheries Oceanography, ed. Maurice Blackburn, pp. 19–130.

Marx, Wesley. 1966. The frail ocean. New York: Coward McCann. 248 pp.

The author lives in southern California, and the larger part of this book concerns problems along California shores and in San Francisco Bay. See also Manwell and Baker, 1967, Moore, 1967 (in the section on selected extralimital papers), and George R. Stewart's *Not so rich as you think* (New York: Houghton-Mifflin, 1968) for further discussions of the critical problem of environmental deterioration.

Murphy, Garth I. 1961. Oceanography and variations in the Pacific sardine population. CalCOFI Repts., **8**: 55–64, 5 figs.

——— 1966. Population biology of the Pacific sardine (*Sardinops caerulea*). Proc. Calif. Acad. Sci., (4) **34** (1): 1–84, 17 figs.

Neushul, M. 1967. Studies of subtidal marine vegetation in western Washington. Ecology, **48** (1): 83–94, 9 figs.

North, Wheeler J. 1967. Integration of environmental conditions by a marine organism, pp. 195–222 in *Pollution and marine ecology*, ed. Theodore A. Olson and Frederick J. Burgess. New York: Interscience Pub. of Wiley & Sons. 14 figs.

Concerning *Macrocystis*.

Pavlovskii, E. N., ed. 1966. Atlas of the invertebrates of the far eastern seas of the U.S.S.R. Israel Program for Scientific Translations. 457 pp., 66 pls.

English version of a book first published in 1955; available from U.S. Department of Commerce Clearinghouse for Scientific and Technical Information, TT 66–51067. Actually, there was little need to translate this, since the text is simple and telegraphic, and the pictures are the main thing. Some of them are better in this than in the original, some worse. Many North Pacific species are included, and it is especially useful for workers in the Bering Sea.

Pequegnat, Willis E. 1961. New world for marine biologists. Nat. Hist., April 1961: 8–17, illus.

——— 1961. Life in the scuba zone. Nat. Hist., May 1961: 46–55, illus.

Popular account of studies with the aqualung in the Newport–Corona del Mar area.

——— 1963. Population dynamics in a sublittoral epifauna. Pac. Sci., **17** (4): 424–30, 1 fig.

——— 1966. The epifauna of a California siltstone reef. Ecology, **45** (2): 272–83, 2 figs.

Less popular accounts of the same experience.

——— 1968. Distribution of epifaunal biomass on a sublittoral rock-reef. Pacific Science, **22** (1): 37–40, 1 fig.

Radovich, John. 1961. Relationships of some marine organisms of the northeast Pacific to water temperatures, particularly during 1957 through 1959. Calif. Dept. Fish and Game, Fish Bull., **112**. 62 pp., 11 figs.

Reid, Joseph L., Jr. 1960. Oceanography of the northeastern Pacific Ocean during the last ten years. CalCOFI Repts., **7**: 77–90, 18 figs.

———, Gunnar I. Roden, and John G. Wyllie. 1958. Studies of the California current system. CalCOFI Repts., **6**: 28–57, 22 figs.

Reid, Joseph L., Jr., and Richard A. Schwartzlose. 1962. Direct measurements of the Davidson Current off central California. J. Geophys. Res., **67** (6): 2491–97, 7 figs.

"The driving force for the Davidson Current is not the local winds but some larger-scale aspect."

Reish, Donald J. 1961. A study of benthic fauna in a recently constructed boat harbor in southern California. Ecology, **42** (1): 84–91, 3 figs.

Rigg, George B., and R. C. Miller. 1949. Intertidal plant and animal zonation in the vicinity of Neah Bay, Washington. Proc. Calif. Acad. Sci., (4) **26** (10): 323–51, 8 figs.

The authors recognize four zones: splash, upper tidal, lower tidal, and demersal, corresponding to the 1–4 system of this book.

RODEN, GUNNAR I. 1961. On seasonal temperature and salinity variations along the west coast of the United States and Canada. CalCOFI Repts., **8**: 95–119, 20 figs.

SCHEER, BRADLEY T. 1945. The development of marine fouling communities. Biol. Bull., **89** (1): 103–21, 4 figs.

The timetable of establishment in Newport Bay, Calif.

SETTE, OSCAR E., and JOHN D. ISAACS, eds. 1960. The changing Pacific Ocean in 1957 and 1958. CalCOFI Repts., **7**: 13–217, 178 figs.

SHELFORD, V. E., and E. D. TOWLER. 1925. Animal communities of the San Juan Channel and adjacent waters. Pub. Puget Sd. Biol. Sta., **5**: 33–74.

Application of community principles to Puget Sound waters; only the animals are considered. Since our standards of appraising communities have changed, it is difficult to recognize some of these communities now.

SHELFORD, V. E., *et al.* 1935. Some marine biotic communities of the Pacific coast of North America. Ecol. Monogr., **5**: 251–345.

STEVENSON, R. E., ed. 1959. Oceanographic survey of the continental shelf area of southern California. State Water Pollution Control Board, Sacramento; Pub. **20**. 559 pp., illus. (mimeo).

STEPHENSON, T. A., and ANNE STEPHENSON. 1961. Life between tide-marks in North America. IVa: Vancouver Island, I. J. Ecol., **49** (1): 1–29, 9 figs., 13 pls. IVb, Vancouver Island, II. *Ibid.*, (2): 227–43, 2 figs., 2 pls.

STOHLER, R. 1930. Gewichtsverhältnisse bei gewissen marinen Evertebraten. Zool. Anz., **91**: 149–55.

Tables of fresh weight, skeletal weight, and in some cases, weight of water and "meat," etc., are given for such common Pacific forms as *Pisaster, Patiria* (as *Asterina*), *Pycnopodia,* urchins, *Acmaea,* chitons, snails, *Mytilus,* etc.

VOGL, RICHARD J. 1966. Salt marsh vegetation of upper Newport Bay, California. Ecology, **47** (1): 80–87, 3 figs.

WIESER, W. 1959. The effect of grain size on the distribution of small invertebrates inhabiting the beaches of Puget Sound. Limnol. and Oceanogr., **4** (2): 181–94, 10 figs.

WYRTKI, KLAUS. 1965. The annual and semiannual variation of sea surface temperature in the North Pacific Ocean. Limnol. and Oceanogr., **10** (3): 307–13, 6 figs.

ZOBELL, CLAUDE E., and CATHERINE B. FELTHAM. 1942. The bacterial flora of a marine mud flat as an ecological factor. Ecology, **23** (1): 69–78, 1 fig.

Concerns the bacterial flora of Mission Bay.

II. Books and Booklets about Pacific Coast Seashore Life

In this section of the Bibliography, we are concerned with books intended primarily for the lay public (perhaps because there are so few others), both young and old, about seashore life on the Pacific coast. Some of these are books about both Atlantic and Pacific coasts, and most of them have been published on the Atlantic seaboard. Appropriately enough, however, our first book on Pacific coast seashore life was set, printed, and bound in San Francisco—in those happy days when San Francisco really had hopes of becoming a printers' town. This was Josiah Keep's little book for school children, *Shells and Sea-life* (1901). Today, no publisher flourishes actually in San Francisco, although the Freeman firms, both senior and junior, claim the city as their home. The work is done elsewhere, however, and San Francisco is the publisher's town only by courtesy. It is a source of embarrassment to anyone from the Bay Area (there is only one Bay Area on the Pacific coast) to acknowledge that outside of the university presses, the most active publisher on the coast for general books of indigenous origin, including at least two seashore items, is Binfords & Mort of Portland. The only explanation for this seems to be that Portland has not learned it really doesn't amount to much, or that Binfords & Mort must receive assistance from its authors in paying the bills. As for that great urban sprawl beneath the inversion layer, the L.A.–San Diego aggregation of automobiles and tract housing, its chief contribution to publishing is paperbacks of the sort that arouse the ire of police sergeants and Sunday-school teachers and provoke intense boredom on the part of jurors and most sensible people over 21. There are two ambitious publishers in the Bay Area who may join the honorable company of publishers of seashore books, although at this time, one is a publisher of choo-choo books, for the most part, and the other specializes in garden handbooks and travel guides.

Only books of general scope have been included in this section; those on seaweeds and seashells have already been listed. We lack a good treatment of our seashells, or, for that matter, of seaweeds along the entire coast. *Between Pacific Tides* is not intended to be a handbook of the sort that will make it easy for a person who has had no biology at any level to identify a given animal or plant. It is, of course, possible to identify most of the common animals with this book, but not by the means that are usual with wildflower guides (by color) or bird handbooks (by wing pattern, etc.). A guide using these field marks is really impossible, and one must inevitably look at illustrations and hope to find something that resembles the specimen at hand.

The oldest general guide to seashore invertebrates is that by Johnson and Snook, published 40 years ago and still useful despite the antiquated names. It has just been reprinted. The next in age, Guberlet's *Animals of the Seashore,* first appeared in 1936; and although it is currently in print, it is still much the same as it was 30 years ago, except that the printing process has made some of the illustrations muddier than they were to begin with. We are critically in need of something between this book and Light's more professionally oriented *Manual for the Pacific Coast,* but no one seems to have the time or the energy to do the job for the entire coast. Yet interest in seashores and in poking around in tidepools has increased, at least in proportion to the population increase, and there is no dearth of pretty picture books to stimulate even more interest. The latest of these is William H. Amos's *The Life of the Seashore,* which has many fine color scenes of Pacific coast shores, as well as Atlantic scenes.

AMOS, WILLIAM HOPKINS. 1958. Life in Pacific tidepools. Garden City, N.Y.: Doubleday. 45 pp., illus., paper.

The illustrations are 30 brightly colored stamps from photographs, printed separately and pasted in designated places by the reader. Most of them were taken at Pacific Grove by George Waters. This booklet is one of a series "prepared in cooperation with the National Audubon Society."

ARNOLD, AUGUSTA FOOTE. 1916. The sea-beach at ebb-tide: A guide to the study of the seaweeds and the lower animal life found between tide-marks. New York: Century. 490 pp., text figs., 85 pls.

The lady who compiled this book (which includes the fauna and flora of both coasts of North America) must have been a remarkable person. The book is essentially sound, although the nomenclature is out of date. However, the treatment is predominantly Atlantic coast.

BRAUN, ERNEST, and VINSON BROWN. 1966. Exploring Pacific coast tide pools. Healdsburg, Calif.: Naturegraph. 56 pp., 134 figs., 40 col. pls.

"This book seeks to show the thrill and adventure of the sea and its life along our Pacific shores wherever the great waves and the rocks meet. In the surge and crash of the salt water there is not only a music of sound, but one also of feeling and understanding how life lives in incredible diversity among the tidepools, in the crevices, among the seaweeds, and under the rocks." Too bad the pictures are not as highly colored as the prose, for despite the claim that they are in "full color," they are reproduced by a cheap off-color process. Obviously intended for the more emotive of nature lovers, with the usual offerings of synthetic common names.

CARL, G. CLIFFORD. 1963. Guide to marine life of British Columbia. Handbook No. 21, British Columbia Provincial Museum, Victoria. 135 pp., illus., paper.

The illustrations are excellent line and wash drawings (b&w). Most of the first 99 pages concern mammals and birds, so only the most conspicuous and/or common invertebrates are treated in the remaining pages. The book concludes with a somewhat mysterious reference to something called *Cadborosaurus*, a serpent-like creature whose total length depends "upon circumstances or condition of observer." There are two useful notes, one on how to make kelp pickle (apparently for human consumption), the other on how to preserve starfish and sea urchins. This book was listed at 50 cents (Canadian) in 1963.

CASTENHOLTZ, R. W. 1967. Stability and stresses in intertidal populations, pp. 15–28 in *Pollution and marine ecology*, ed. Theodore A. Olson and Frederick J. Burgess. New York: Interscience Pub. of Wiley & Sons. 7 figs.

CLARKE, WILLIAM D., and MICHAEL NEUSHUL. 1967. Subtidal ecology of the southern California coast. *Ibid.*, pp. 29–42, 6 figs.

CLEMONS, ELIZABETH. 1964. Tidepools and beaches. Illustrated by Joe Gault. New York: Knopf. 84 pp.

Introduces the common animals and plants of both the Atlantic and Pacific coasts, with clear sharp drawings and scientific names. Written with a matter-of-fact, no-nonsense approach that more writers of such juveniles should emulate.

FLORA, CHARLES J., and EUGENE FAIRBANKS. 1966. The sound and the sea: A guide to Northwestern neritic invertebrate zoology. Bellingham, Wash.: Pioneer Printing Co. 455 pp., 33 figs.

Although the format is similar to Guberlet's, more species are included, and the primary emphasis is on the fauna of Puget Sound. A number of the synonyms given are apparently perpetuated typographical errors, which should be forgotten. Most of the photographs are adequate, but there are occasional very dead animals. This is the second edition of a work that appears to be used primarily for instruction at Bellingham.

GUBERLET, MURIEL LEWIN. 1936. Animals of the seashore. Portland, Ore.: Metropolitan Press. 412 pp., illus.

——— 1942. The seashore parade. With illustrations by Jan Ogden. Lancaster, Pa.: Jaques Cattell. 197 pp., text figs., col. pls.

Standard kiddie stuff, although I do not care for the bold poster aspect of the color plates.

——— 1946. Hermie's trailer house. Illustrations by Marjorie Kincaid Illman. Lancaster, Pa.: Jaques Cattell. 32 pp., illus.

The illustrations are in perfect accord with the text; fortunately the book has long since fallen out of print.

——— 1962. Animals of the seashore. Revised and enlarged edition. Portland, Ore.: Binfords & Mort. 450 pp., illus.

There is also a 1949 edition of this book, which has somehow survived the silent treatment from the professionals. The 1962 edition is subtitled "Descriptive guide to seashore invertebrates of the Pacific coast," which is stretching verity somewhat. Perhaps a third of the photographs are useless for identification (the latest edition has most of the illustrations of the first, printed to a larger size but with a muddier process). The 1962 ediiton has a well-intended glossary that only betrays the author's lack of zoological knowledge. The author is also responsible for a book on seaweeds at ebb tide.

HEDGPETH, JOEL W. 1961. Common seashore life of southern California. Illustrated by Sam Hinton, edited by Vinson Brown. Healdsburg, Calif.: Naturegraph. 65 pp., illus. (b&w, col. frontispiece). Volume 1, Naturegraph Ocean Guidebooks.

The first 34 pages of this were prepared as a guide to be presented to the assembled multitude at the great La Jolla symposium, "Perspectives in Marine Biology," but because of delays in illustrating, the work was not ready in time. The editor decided this was not quite enough, and added the remaining 30 pages of keys and thumbnail drawings (by Carol Lyness). Thus are pots kept boiling.

——— 1962. Introduction to seashore life of the San Francisco Bay region and the coast of northern California. Illustrated by the author and Lynn Rudy. Berkeley: Univ. Calif. Press. Natural History Guides, 9. 136 pp., b&w illus., 8 col. pls.

I am pleased to be informed that this is selling well.

HOLLING, HOLLING CLANCY. 1957. Pagoo. Illustrated by the author and Lucille Webster Holling. Boston: Houghton Mifflin. 87 pp., illus. (b&w and col.).

Most children's books (or "juveniles," as the trade calls them) are ephemeral and soon go out of print. This one, on the life of a hermit crab (Pagoo for *Pagurus*), has survived for several years and still shows signs of going strong. This may be due in part to the superb illustrations, which convey a vivid feeling of the sweep and surge of the sea along the shore. The various animals illustrated along the margins and in the color plates are all southern California species. There is an inevitable tinge of the pathetic fallacy in the text, but it is difficult to avoid this as one identifies oneself with the perilous life of a hermit crab; and it is certainly less upsetting to the old pros in their dank laboratories than are the anthropomorphic conversations encountered in another juvenile on the life of a hermit crab. We are glad to see the book still in print.

HUNTINGTON, HARRIET E. 1941. Let's go to the seashore. Illustrated with photographs by the author. New York: Doubleday. Junior Books.

The photographs are excellent; the text is reasonable, although set as free verse. Not specifically for the Pacific coast, and the illustrations are not identified by common name, although all of them are obviously from somewhere south of Monterey.

HYLANDER, CLARENCE J. 1950. Sea and shore. New York: Macmillan. 242 pp., illus.

The publishers indicate that this book is intended for ages 10–14; also, it concerns plants and animals of both Atlantic and Pacific coasts. Like all such ambivalent books, it gives more space to the Atlantic coast.

JOHNSON, MYRTLE ELIZABETH, and HARRY JAMES SNOOK. 1927. Seashore animals of the Pacific coast. New York: Macmillan. 659 pp., 700 text figs., 11 col. pls.

This book is a remarkable achievement for its time, and although much of the nomenclature has fallen into obsolescence, the book is still useful and sought after. Indeed, a paperback reprint, without change, has been published by Dover. It is certain that copies of the original edition will continue to be in demand; many of them are annotated (who, we wonder, now has Wesley Coe's well-annotated copy that was "borrowed" from his office at Scripps many years ago?). The original publication was made possible by the generosity of Ellen Browning Scripps; yet within a few years after publication (around 1932–33), the publishers had remaindered the stock, and the book went out of print.

KEEP, JOSIAH. 1901. Shells and sea-life. San Francisco: Whitaker and Ray. 200 pp., b&w illus., col. frontispiece.

This is the first book about Pacific coast shore life; half of it naturally concerns mollusks, and the rest is about other creatures, including sea lions, birds, and the harvest of the sea. There are even some poems, but Professor Keep was much better at prose (much better in fact than several more recent authors who have written for the schoolbook trade).

LIGHT, S. F. 1941. Laboratory and field text in invertebrate zoology. Berkeley: Associated Students Store. 232 pp., illus., paper.

——— 1954. Intertidal invertebrates of the central California coast. S. F. Light's *Laboratory and field text in invertebrate zoology*, revised by Ralph I. Smith, Frank A. Pitelka, Donald P. Abbott, and Frances M. Wessner, with the assistance of many other contributors. Berkeley: Univ. Calif. Press. 446 pp., 133 figs.

The revised edition, posthumously produced thanks principally to the efforts of R. I. Smith, is the standard *vade mecum* for those who wish to identify some of their creatures. As a group effort, it has some parts that are better than others and some that are out of date; but it is still the best we have. Users are warned that the keys are not strictly applicable very far south or north of the coast between the Russian River and Carmel.

MACGINITIE, G. E., and NETTIE MACGINITIE. 1949. Natural history of marine animals. New York: McGraw-Hill. 473 pp., 282 figs. Second edition, 1968, 523 pp.

For many years the MacGinities inhabited the laboratory at Corona del Mar, domesticating all sorts of unlikely marine invertebrates in the aquaria there. Much of this book concerns these animals, in the aquaria and in the nearby environment. George MacGinitie is responsible for making *Urechis* famous, through his classic studies (along with the late W. K. Fisher) of the ecology and feeding habits of this strange worm. The second edition is almost the same as the first, with a few name changes and some replaced illustrations; the printing is clearer. New material has been added in the form of notes to the text in an appendix. Most of this information is based on observations at Point Barrow or during European trips.

MILLER, MARGARET. 1942. Along our coast: Stories of sea creatures of the Pacific Ocean. New York: Dodd, Mead. 60 pp., illus.

The author, having had no college course in biology and lacking the advantage of being married to a professor of the subject, overcame these handicaps to produce a rather good book on various sea creatures, mostly around La Jolla and Scripps in its halcyon days. The photographs are old-style, still stuff, in striking contrast to the contemporaneous work of Huntington. But the text is better.

RAE, JESS CAMPBELL. 1941. Beach magic. Illustrated by Margaret Newton. Portland: Binfords & Mort. 48 unnumbered pp.

A book of verses about things on the seashore, mostly animals. Some of them are rather apt, and one begins to think that perhaps the author is hiding some professional knowledge of biology until the verse about the sand crab:

If it is doggone lonesome being unlike your kin,
Remember variations make a specie win;

But we should have known all along that there are no editors in Portland. Still, we rather like the opening verse:

Could mortal leave upon the strand,
When life has ebbed away,
A thing as perfect as this shell,
He would not dread decay.

ROBINSON, W. W., and IRENE ROBINSON. 1942. At the seashore. Pictures by Irene Robinson. New York: Macmillan. 38 pp., charcoal and pastel illus.

A large picture book with dialogue attributed to the human characters rather than to the starfish; from the illustrations and the mention of abalone and sea lions, it is obvious that this book was written and illustrated on the California coast; perhaps the publishers wished to conceal this fact to improve Atlantic coast sales.

SLOAN, TAY. 1961. Wonders of the Pacific shore. Photographs and text by Tay Sloan; drawings by Lura Karlsson. New York: Columbia Record Club, Panorama Colorslide Nature and Science Program. 45 pp.; front pocket with 25 transparencies; back pocket with 6″ 33⅓ r.p.m. record narrated by Hans Conreid.

According to the title page, the photographs were taken at Pacific Marine Station, and Joel W. Hedgpeth assisted in preparation of the manuscript and photographs. True enough; not mentioned is the intervention of some Madison Avenue wordsmith, who undid some painful assistance and sneaked back into the text the egregious error that the high tides are springs and the low tides are neaps, as well as a few other infelicities. Many of the photographs are excellent; the drawings on the margins are a bit too free and easy. Use of this book requires a special projector and a phonograph. The whole Panorama Project seems to be a gimmick that did not quite come off. The book itself has just been republished, but I have not seen a copy as this goes to press.

SMITH, DICK, and FRANK VAN SCHAICK. 1964. Beachwalker's guide. A pictorial introduction to marine life of the Pacific coast. Santa Barbara, Calif.: McNally & Loftin. Distributed by Lane Books, Menlo Park, Calif. 63 pp., illus., paper.

If taken as a guide to what may be found dead on the beach, this book might be acceptable. However, the drawings are very poorly done, and the photographs are reproduced by coarse screen that does not do some of them justice. Nevertheless, the thing seems to be doing well; copies are on the magazine stands of supermarkets all along the coast and in the racks at various tourist traps featuring deadly octopuses, sea lions, and so forth along Highway 101. It is a pity something better could not be managed for this trade.

SMITH, LYNWOOD. 1962. Common seashore life of the Pacific Northwest. Illustrated by Constance Gabriel Schulte. Healdsburg, Calif.: Naturegraph. 66 pp., b&w illus., col. frontispiece.

The drawings are artistic—if you like the scratchy pen and ink style—but not always useful for identification. The text is reasonably informative, and somewhat better than Guberlet; but it must still be said that the Puget Sound area, with its marvelously rich fauna, lacks an adequate handbook. This small book is used in teachers' courses—for want of anything better.

THOMPSON, BOB, ed. 1966. Sunset beachcombers' guide to the Pacific coast. Menlo Park, Calif.: Lane Books. 112 pp., illus.

Although not strictly speaking a book about seashore life, there are items here and there about clamming, sea lions, tides, waves, and the like, as well as hints on interesting places to go.

TIERNEY, ROBERT J., JOSEPH W. ULMER, LEONARD J. WAXDECK, HARRIS N. FOSTER, and JOHN R. ECKENROAD. 1966. Exploring tidal life along the Pacific coast. 64 pp., illus.

This is an illustrated key to common seashore plants and animals of the Bay area, designed by a group of high school teachers for persons having no background in biology. Available from Tidepool Associates, 70 Halkin Lane, Berkeley, Calif. 94708, at $0.85.

VESSEL, MATTHEW F., and HERBERT H. WONG. 1965. Seashore life of our Pacific coast. Illustrated by Joseph Capozio. Palo Alto: Fearon. 60 pp., illus.

We understand that the authors hoped that this book would be adopted for textbook use at elementary levels, as those by Josiah Keep and Harrington Wells were in years before. It did get on some sort of list. It falls short of its predecessors, however, for this purpose. It is unfortunate that the illustrator did not go forth and obtain original models instead of ransacking all the available sources—from which he has glibly copied, down to the same bends in the worms, the same shapes of sea slugs, etc. There are some odd errors in the text that have escaped editorial hands, as well as inappropriate application of East Coast popular names, such as laver and bladderwrack.

WAKEMAN, NORMAN HAMMOND. 1961. Wonders of the world between the tides. New York: Dodd, Mead. 63 pp., illustrated with photographs by the author.

Some of the photographs appear to have been converted from movie footage and are not as sharp as they ought to be. A standard sort of juvenile treatment, lacking the precision to be expected of a biology instructor. One of the photographs shows *Pleuroncodes* on an uninterested anemone in an aquarium: "The tuna crab falls easy prey to the sea anemone. Tuna crabs are fed upon by tuna in the open sea," etc.

WELLS, HARRINGTON. 1942. Seashore life. Sacramento: Calif. St. Dept. Educ. 271 pp., illus.

The successor to Keep's amiable little volume for primary grades. A much more detailed and serious treatment than apparently considered appropriate in our more enlightened day, when we must write for the intellectual level of essentially illiterate teachers instead of for the children.

WILLIS, ELIZABETH BAYLEY. 1938. Little bay creatures. Illustrated by M. Erckenbrack Hennessy. Portland, Ore.: Binfords & Mort., 89 pp.

Page 43:

" 'Please, please go away,' wailed a small voice almost under his feet. 'You almost stepped on me. Please let me alone. I am laying my eggs in this pool.'

" 'Why, Butter Clam, I beg your pardon. I didn't know that you were in the pool. Are you of all people laying eggs? What a surprise! I had forgotten that there were clam eggs,' replied the Starfish."

One may also wonder why some professional artists prefer to copy photographs instead of actual specimens; in this book the famous left-handed *Polinices* of Guberlet's *Animals of the Seashore* is rather sloppily copied (p. 41). The color plates seem to have been done without reference to nature. The book is still in print (unfortunately).

III. Selected Papers on Marine Ecology and Related Matters

ALLEE, W. C. 1923. Studies in marine ecology, I and II. Biol. Bull., **44**: 157–253.

One of the few comprehensive treatments of a restricted area in this country. Although concerned with the fauna of Woods Hole, it pictures a state of affairs applicable anywhere.

ALLEN, E. J. 1899. On the fauna and bottom deposits near the thirty-fathom line. J. Mar. Biol. Assoc., n.s., **5**: 365–542.

Although dealing exclusively with dredged forms, this lengthy and excellent account is still valuable for general ecological information, factors of distribution, interrelations, etc. Type of bottom is considered an important factor.

AX, PETER. 1966. Die Bedeutung der interstitiellen Sandfauna für allgemeine Probleme der Systematik, Ökologie und Biologie. Veröff. Inst. Meeresf. Bremerhaven, **11**: 15–65, 39 figs.

BALLENTINE, W. J. 1961. A biologically defined exposure scale for the comparative description of rocky shores. Field Studies, **1** (3). 19 pp., 5 figs.

BASSINDALE, R., and R. B. CLARK. 1960. The Gann Flat, Dale: Studies on the ecology of a muddy beach. Field Studies, **1** (2). 22 pp., 12 figs.

BASSINDALE, R., F. J. EBLING, J. A. KITCHING, and R. D. PURCHON. 1948. The ecology of the Lough Ine rapids, with special reference to water currents, I: Introduction and hydrography. J. Ecol., **36** (2): 305–22, 9 figs., 1 pl., cht.

See also Ebling, *et al.*

BENNETT, ISOBEL, and ELIZABETH C. POPE. 1953. Intertidal zonation of the exposed rocky shores of Victoria, together with a rearrangement of the biogeographical provinces of temperate Australian shores. Austr. J. Mar. and Freshw. Res., **4** (1): 105–59, 5 figs., 6 pls.

——— 1960. Intertidal zonation of the exposed rocky shores of Tasmania, and its relationship with the rest of Australia. Austr. J. Mar. and Freshw. Res., **11** (2): 181–221, 3 figs., 7 pls.

BEVERIDGE, W. A., and V. J. CHAPMAN. 1950. The zonation of marine algae at Piha, New Zealand, in relation to the tidal factor (Studies in intertidal zonation, 2). Pac. Sci., **4** (3): 188–201, 14 figs.

See also Dellow, Chapman, below.

BLACKER, R. W. 1957. Benthic animals as indicators of hydrographic conditions and climatic change in Svalbard waters. Min. Agr., Fish and Food., Fishery Invest., Ser. 2, **20** (10). 49 pp., 30 figs., 20 foldout tbls.

BOLIN, ROLF L. 1949. The linear distribution of intertidal organisms and its effect on their evolutionary potential. Proc. XIII Int. Congr. Zool.: 459–60.

CABIOCH, LOUIS. 1961. Étude de la repartition des peuplements benthiques au large de Roscoff. Cah. Biol. Mar., **2**: 1–40, foldout cht.

CASPERS, H. 1950. Der Biozönose und Biotopbegriff vom Blickpunkt der Marinen und Limnischen Synökologie. Biol. Zentralb., **69** (1/2): 43–63.

——— 1950. Die Lebensgemeinschaft der Helgoländer Austernbank. Helgol. Wiss. Meeresunters., **3**: 119–69, 15 figs.

——— 1951. Biozönotische Untersuchungen über die Strandarthropoden im bulgarischen Küstenbereich des Schwarzen Meeres. Hydrobiologica, **3** (2): 131–93, 15 figs.

——— 1951. Quantitative Untersuchungen über die Bodentierwelt des Schwarzen Meeres im bulgarischen Küstenbereich. Arch. Hydrobiol., **45**: 1–192, 66 figs., 7 pls.

A useful series of papers. The third item is noteworthy as a contribution to the study of sandy shores, which are too often neglected by ecologists.

CHAMBOST, L. 1928. Essai sur la région littorale dans les environs de Salammbô. Bull. Sta. Oceanogr. Salammbô, **8**. 28 pp., 7 figs., 1 cht.

Missed in the various comprehensive bibliographies, but worth citing as a reference to a sandy, tideless environment.

CHAPMAN, GARTH. 1958. The hydrostatic skeleton in the invertebrates. Biol. Revs., **33** (3): 338–71, 5 figs.

CHAPMAN, V. J. 1946. Marine algal ecology. Bot. Rev., **12**: 628–72.

Literature review.

——— 1950. The marine algal communities of Stanmore Bay, New Zealand (Studies in intertidal zonation, 1). Pac. Sci., **4** (1): 63–68, 3 figs.

The three papers so far published in this series (Beveridge and Chapman, and Dellow) are principally concerned with ecology of algae.

——— 1957. Marine algal ecology. Bot. Rev., **23** (5): 320–50, 2 figs.

COLMAN, JOHN. 1933. The nature of the intertidal zonation of plants and animals. J. Mar. Biol. Assoc., n.s., **18**: 435–76.

Establishes close correlation between critical distributional horizons and tidal marks.

——— 1943. Some intertidal enigmas. Proc. Linn. Soc. Lond., **1941–42**, (3): 232–34.

Poses some critical questions about intertidal zonation, which still remain to be answered.

CONNELL, JOSEPH H. 1961. Effects of competition, predation by *Thais lapillus*, and other factors on natural populations of the barnacle *Balanus balanoides*. Ecol. Monogr., **31**: 61–104, 22 figs.

——— 1961. The influence of interspecific competition and other factors on the distribution of the barnacle *Chthamalus stellatus*. Ecology, **42** (4): 710–23.

——— 1963. Territorial behavior and dispersion in some marine invertebrates. Jap. J. Pop. Ecol., **5** (2): 87–101, 3 figs.

Craig, Gordon Y., and G. Oertel. 1966. Deterministic models of living and fossil populations of animals. Quart. J. Geol. Soc. Lond., **122** (3): 315–55, 18 figs.

Dahl, Erik. 1948. On the smaller Arthropoda of marine algae, especially in the polyhaline waters off the Swedish west coast. Unders. Öresund, **35**. 193 pp., 42 figs.

Another of these excellent, comprehensive Scandinavian papers (cf. Gislèn, below), beautifully printed, which are the product of a well-known fauna and a sound tradition in marine ecology.

Dakin, W. J., Isobel Bennett, and Elizabeth Pope. 1948. A study of certain aspects of the ecology of the intertidal zone of the New South Wales coast. Austr. J. Scient. Res., (B), **1** (2): 176–230, 9 pls., 3 figs.

Excellently illustrated description and analysis of the Australian coast, with lists of species characteristic of the zones recognized. Of special interest is the information about the marine spider *Desis crosslandi,* which lives in the low intertidal.

Davenport, Demorest. 1966. The experimental analysis of behavior in symbioses. In *Symbiosis,* Vol. 1, pp. 381–429. New York: Academic Press. 10 figs.

Dellow, Vivienne. 1950. Intertidal ecology at Narrow Neck Reef, New Zealand (Studies in intertidal zonation, 3). Pac. Sci., **4** (4): 355–74, 13 figs.

——— 1955. Marine algal ecology of the Hauraki Gulf, New Zealand. Trans. Roy. Soc. N.Z., **83** (1): 1–91, 11 figs., 4 pls.

———, and R. Morrison Cassie. 1955. Littoral zonation in two caves in the Auckland district. *Ibid.*, (2): 321–31, 7 figs.

Dexter, Ralph W. 1947. The marine communities of a tidal inlet at Cape Ann, Massachusetts: A study in bioecology. Ecol. Monogr., **17**: 261–94, 17 figs.

Descriptive population ecology, with biomes and faciations, and coaction diagrams.

Donaldson, Lauren R., and Richard F. Foster. 1957. Effects of radiation on aquatic organisms. N.A.S. Pub. **551**: 96–102.

Doty, Maxwell S. 1957. Rocky intertidal surfaces. Ch. 18 in *Treatise on marine ecology and paleoecology, Vol. I: Ecology.* Memoir, **67** (1), Geol. Soc. Amer., pp. 535–85. 18 figs., foldout cht.

Ebling, F. J., J. A. Kitching, R. D. Purchon, and R. Bassindale. 1948. The ecology of the Lough Ine rapids, with special reference to water currents, 2: The fauna of the *Saccorhiza* canopy. J. Anim. Ecol., **17** (2): 223–44, 14 figs.

The two papers on Lough Ine rapids describe a situation similar to that in the surge channels and narrow passages of the Puget Sound region, where comparable studies would be of interest. See also Bassindale, *et al.,* 1948.

Einarsson, Hermann. 1941. Survey of the benthonic animal communities of Faxa Bay, Iceland. Meddel. Komm. Danmarks Fisk. og Havunders. (Fiske.), **11** (1): 46 pp., 19 figs., 6 tbls.

Endean, R., R. Kenny, and W. Stephenson. 1956. The ecology and distribution of intertidal organisms on the rocky shores of the Queensland mainland. Austr. J. Mar. and Freshw. Res., **7** (1): 88–146, 13 figs., 7 pls.

——— 1956. The ecology and distribution of intertidal organisms on certain islands off the Queensland coast. Austr. J. Mar. and Freshw. Res., **7** (3): 317–42, 8 figs., 4 pls.

Ercegovic, A. 1934. Wellengang und Lithophytenzone an der Ostadriatischen Küste. Acta Adriatica (Split, Jugoslavia), **3**: 1–20.

Stresses the significance of the wave (i.e., surf and astronomical or barometric tide) as a primary oceanographic factor, and emphasizes the importance of wave shock as a formative and segregating factor for lithophytic organisms.

Evans, R. G. 1947. The intertidal ecology of Cardigan Bay. J. Ecol., **34**: 273–309.

——— 1947a. The intertidal ecology of selected localities in the Plymouth neighborhood. J. Mar. Biol. Assoc., **27**: 173–218.

These two papers are the beginning of a series planned to describe, in Stephenson's manner, the coasts of England.

Fager, E. W. Communities of organisms, in *The sea,* ed. M. N. Hill, Vol. 2, pp. 415–37. Wiley & Sons, New York.

Folsom, Theodore H., and John H. Harley. 1957. Comparison of some natural radiations received by selected organisms. N.A.S. Pub. **551**: 28–33, 2 figs.

Gislèn, Torsten. 1930. Epibioses of the Gullmar Fjord, II. Kristin. Zool. Sta. 1877–1927, No. 4: 1–380.

Sweden, west coast. Recognizes and enumerates the plant-animal communities in a tideless region, starting from above mean sea level and working down. This "marine sociology" would seem to be the most correct system of ecological analysis yet devised, but one unfortunately not adapted to a popular treatise such as our book, and restricted furthermore to areas better known taxonomically and physically than the Pacific coast of North America.

——— 1931. A survey of the marine associations in the Misaki district. J. Fac. Sci. Imp. Univ. Tokyo, Sec. 4, **2**: 398–444.

Glynne-Williams, J., and J. Hobart. 1952. Studies on the crevice fauna of a selected shore in Anglesey. Proc. Zool. Soc. Lond., **122** (3): 797–824, 5 figs., 1 pl.

Grave, B. H. 1933. Rate of growth, age at sexual maturity, and duration of life in sessile organisms at Woods Hole. Biol. Bull., **65**: 375–86.

Finds that many prolific animals have a life-span of less than one year and produce several generations during the summer, which accounts for the rapid rehabilitation of depleted areas.

Guiler, Eric R. 1950. The intertidal ecology of Tasmania. Pap. and Proc. Roy. Soc. Tasman., **1949**: 135–201, 30 figs., 2 pls.

——— 1951. The intertidal ecology of Pipe Clay Lagoon. Pap. and Proc. Roy. Soc. Tasman., **1950**: 29–52, 8 figs., 2 pls.

——— 1951. Notes on the intertidal ecology of the Freycinet Peninsula. *Ibid.*: 53–70, 4 figs., 3 pls.

Author suggests that conditions in Tasmania are more comparable to those of the South African coast than of New South Wales.

——— 1959. Intertidal belt-forming species on the rocky coasts of northern Chile. Pap. and Proc. Roy. Soc. Tasman., **93**: 33–58, 2 pls. (dated; Amen!).

——— 1959. The intertidal ecology of the Montemar area, Chile. *Ibid.*: 165–83, 3 pls. (also dated).

——— 1960. The intertidal zone-forming species on rocky shores of the east Australian coast. J. Ecol., **48** (1): 1–28.

See Bennett and Pope (1960) for comments.

Gurjanova, Eupraxie. 1958. Comparative research of biology of the littoral in the Far Eastern seas. Proc. Ninth Pac. Sci. Congr., **19** (Zoology): 75–86.

———, J. Y. Liu, O. A. Scarlato, P. V. Uschakov, Wu Bao-ling, and Tsi Chung-yen. 1958. A short report on the intertidal zone of the Shantung Peninsula (Yellow Sea). Bull. Inst. Mar. Biol., Acad. Sinica, **1** (2): 1–43, 3 figs., 7 tbls. (pp. 1–21 in Chinese; pp. 24–43 in Russian; titles in English and Chinese).

Perhaps there is something to be said for Esperanto after all, although we understand Swahili is more expressive.

Harding, J. P., and Norman Tebble, eds. 1963. Speciation in the sea. Systematics Assoc., Pub. **3**. 199 pp., illus.

Includes essays on speciation in intertidal animals, *Fucus,* lugworms, etc.

Hedgpeth, Joel W. 1947. Fishers of the Murex (Notes for a bibliography of marine natural history). Isis, **37** (107–8): 27–32.

——— 1957. Concepts of marine ecology. Ch. 3 in *Treatise on marine ecology and paleoecology, Vol. I: Ecology.* Memoir, **67** (1), Geol. Soc. Amer., pp. 29–52, 7 figs.

Heegaard, P. 1944. The bottom fauna of Praestø Fjord. Folia Geographica Danica, **3** (3): 59–81, 7 figs., 5 tbls.

Hodgkin, E. P. 1960. Patterns of life on rocky shores. J. Roy. Soc. West. Austr., **43** (2): 35–43, 2 text figs., 2 pls.

HOSHIAI, TAKAO. Synecological study on intertidal communities. I: The zonation of intertidal animal community with special reference to the interspecific relation. Bull. Mar. Biol. Sta. Asamushi, **9** (1): 27–33, 1 fig. (1958). II: On the interrelation between the Hijikia fusiforme zone and the *Mytilus edulis* zone. *Ibid.*, (3): 123–26, 1 fig. (1959). VI: A synecological study on the intertidal zonation of the Asamushi coastal area with special reference to its re-formation. *Ibid.*, **12** (2–3): 93–126, 17 figs. (1964).

HUNTSMAN, A. G. 1918. The vertical distribution of certain intertidal animals. Trans. Roy. Soc. Canad., Ser. 3, **12** (Sec. 4): 53–60.

HUTCHINS, LOUIS W. 1947. The bases for temperature zonation in geographical distribution. Ecol. Monogr., **17**: 325–35, 8 figs.

Recognizes four basic types of zonation for marine organisms in the Northern Hemisphere based on temperature differences. Because of the importance of this paper, we consider a summary in this place inadvisable. After twenty years, it is a classic in the literature of marine ecology.

HUTCHINSON, G. EVELYN. 1948. Circular causal systems in ecology. Ann. N.Y. Acad. Sci., **50** (4): 221–46, 4 figs.

JOHNSON, D. S., and H. H. YORK. 1915. The relation of plants to tide levels. Carnegie Inst. Wash. Pub. **206**. 162 pp.

Establishes the great importance of the tidal-exposure factor in the distribution of shore algae, emphasizing the fact that plants occur in zones through the operation of this factor, and that "the vertical range of a plant common to two localities with different ranges of tide will be found exactly proportional in each place to the local range of tide."

JOHNSON, MARTIN W. 1948. Sound as a tool in marine ecology, from data on biological noises and the deep scattering layer. J. Mar. Res., **7** (3): 443–58, 4 figs.

One of the interesting and unexpected products of wartime research was the study of underwater noises made by shrimp and fish. This paper summarizes and lists recent papers on the subject.

JONES, N. S. 1950. Marine bottom communities. Biol. Revs., **25**: 283–313.

Summary of the research and thought on bottom communities, especially in European waters, with a comprehensive list of literature.

JØRGENSEN, C. BARKER. 1955. Quantitative aspects of filter feeding in invertebrates. Biol. Revs., **30**: 391–454, 13 figs.

KETCHUM, BOSTWICK H. 1960. Oceanographic research required in support of radioactive waste disposal. In *Disposal of radioactive wastes, II* (Monaco conference), pp. 283–91.

KINNE, OTTO, and AURICH KINNE, eds. 1964. Fourth marine biological symposium. Helg. Wiss. Meeresunters., **10** (1–4). 476 pp.

Principally North Sea marine biology, but with sections on general ecology and production.

KNOX, G. A. 1953. The intertidal ecology of Taylor's Mistake, Banks Peninsula. Trans. Roy. Soc. N.Z., **81** (2): 189–220, 8 figs., 2 pls.

——— 1963. The biogeography and intertidal ecology of the Australian coasts. Oceanogr. Mar. Biol., Ann. Rev., **1**: 341–404, 5 figs.

Summary and review paper.

KORRINGA, P. 1947. Relations between the moon and periodicity in the breeding of marine animals. Ecol. Monogr., **17**: 347–81, 5 figs.

Review paper, with discussion of typical species that show periodism, especially oysters and polychaetes.

KRUMHOLZ, LOUIS A., EDWARD D. GOLDBERG, and HOWARD BOROUGHS. 1957. Ecological factors involved in the uptake, accumulation, and loss of radionuclides by aquatic organisms. N.A.S. Pub. **551**: 69–79.

KUSSAKIN, O. G. 1961. Nekotorye zakonomernosti raspedeleniya fauny i flory v osushnoi zone yuzhnykh Kuril'skikh Ostrovov. (Certain controlling factors of the occurrence of fauna and flora of the spray zone of the southern Kurile Islands). Issl. Dalnev. Morei, **7**: 312–43, 3 figs.

LAWSON, G. W. 1954. Rocky shore zonation on the Gold Coast. J. Ecol., **44** (1): 153–70, 6 figs., 1 pl.

——— 1955. Rocky shore zonation in the British Cameroons. J. West. Afr. Sci. Assoc., **1** (2): 78–88, 4 figs., 1 pl.

——— 1957. Some features of the intertidal ecology of Sierra Leone. J. West. Afr. Sci. Assoc., **3** (2): 166–74, 3 figs.

——— 1957. Seasonal variation of intertidal zonation on the coast of Ghana in relation to tidal factors. J. Ecol., **45**: 831–60, 9 figs.

LEWIS, JOHN R. 1953. The ecology of rocky shores around Anglesey. Proc. Zool. Soc. Lond., **123** (III): 481–549, 15 figs., 4 pls.

——— 1954. The ecology of exposed rocky shores of Caithness. Trans. Roy. Soc. Edinb., **62** (III, 17): 695–723, 6 figs., 4 pls.

——— 1955. The mode of occurrence of the universal intertidal zones in Great Britain: With a comment by T. A. and Anne Stephenson, J. Ecol., **43** (1): 270–90, 4 figs.

——— 1957. Intertidal communities of the northern and western coasts of Scotland. Trans. Roy. Soc. Edinb., **63** (I, 10): 185–220, 8 figs., 2 pls.

——— 1964. The ecology of rocky shores. London: English Universities Press. 323 pp., 85 figs., 40 pls.

An exhaustively thorough treatment, zone by zone, of the shores of the British Isles.

———, and H. T. POWELL. 1960. Aspects of the intertidal ecology of rocky shores in Argyll, Scotland: I and II. Trans. Roy. Soc. Edinb., **64** (3–4): 45–100, 9 figs., 4 pls.

LONGHURST, ALAN R. 1958. An ecological survey of the West African marine benthos. Colonial Office, Fishery Pub. **11**. 102 pp., 11 figs.

——— 1964. A review of the present situation in benthic synecology. Bull. Inst. Oceanogr., **63** (1317). 54 pp.

Most of the work is yet to be done.

MANWELL, CLYDE, and C. M. ANN BAKER. 1967. Oil and detergent pollution: Past, present, politics, and prospects. J. Devon Trust Nature Conservation, Supplement, July 1967: 39–72.

At times somewhat intemperate, especially about goings on in the Establishment, this is nevertheless a forceful account of the *Torrey Canyon* affair, of which the last word will not be said for many years.

MARSH, LOISETTE M., and E. P. HODGKIN. 1962. A survey of the fauna and flora of rocky shores of Carnac Island, Western Australia. West. Austr. Nat., **8**: 62–72, 8 figs., 1 pl.

MOKIEVSKII (MOKYEVSKY), O. B. 1953. K faune litorali Okhotskogo morya (The littoral fauna of the Sea of Okhotsk). Trudy Inst. Okeanol., **7**: 167–97, 2 figs.

MOKYEVSKY, O. B. 1960. Geographical zonation of marine littoral types. Limnol. and Oceanogr., **5** (4): 389–96, 1 fig.

——— 1966. Some principles of marine biogeography applied to the intertidal zone. Abstr. 2nd Internat. Oceanogr. Conf.: 257–58.

MOLANDER, A. 1928. Animal communities on soft bottom areas in the Gullmar Fjord. Kristin. Zool. Sta., 1877–1927, No. 2: 1–90.

MOORE, N. W. 1967. A synopsis of the pesticide problem. Advances in Ecological Research, **4**: 75–129, 5 figs.

NELSON, THURLOW C. 1947. Some conditions from the land in determining conditions of life in the sea. Ecol. Monogr., **17**: 337–46, 7 figs.

ODUM, HOWARD T., JOHN E. CANTLON, and LOUIS S. KORNICKER. 1960. An organizational hierarchy postulate for the interpretation of species-individual distributions, species entropy, ecosystem evolution, and the meaning of a species-variety index. Ecology, **41** (2): 395–99, 3 figs.

Life is all logarithms, or *Om mane padme om.*

ORTON, J. H. 1920. Sea temperature, breeding and distribution in marine animals. J. Mar. Biol. Assoc., n.s., **12**: 339–66.

PEARSE, A. S., H. J. HUMM, and G. W. WHARTON. 1942. Ecology of sand beaches at Beaufort, North Carolina. Ecol. Monogr., **12**: 135–90, 24 figs.

California beaches are actually much richer in species-mass than the Beaufort beach, but little has been done with them except studies of the Pismo clam, *Emerita,* and the worm *Thoracophelia* (q.v.).

PÉRÈS, J. M., and J. PICARD. 1958. Manuel de bionomie benthique de la Mer Méditerranée. Trav. sta. mar. d'Endoume, **23** (14): 7–122, 8 figs.

PETERSEN, C. G. JOH. 1918. The sea bottom and its production of fishfood: A survey of the work done in connection with valuation of the Danish waters from 1883–1917. Rept. Dan. Biol. Sta. to Bd. of Agric. 62 pp., 10 pls., map.

This is the summary paper of the classic work of Petersen and his colleagues. The first two installments, under the title of *Valuation of the Sea,* appeared as reports 20 (1911) and 21 (1913) of the Danish Biological Station. Although the efficacy of Petersen's sampling methods has been criticized, the work remains the foundation of all subsequent work in marine population ecology. Of Petersen, Allee *et al.* (p. 51) say: "It is not always recognized that this man is among the great in the history of ecology and hydrobiology."

PLESSIS, YVES. 1961. Ecologie de l'estran rocheux du Calvados: étude des biocenoses et récherches expérimentales. Ann. Inst. Oceanogr., **38** (3): 233–323, 8 figs.

PRENANT, MARCEL. 1960. Études écologiques sur les sables intercôtidaux, I: Questions de méthode granulométrique; Application à trois anses de la Baie de Quiberon. Cah. Biol. Mar., **1**: 295–340, 10 figs.

REID, D. M. 1932. Salinity interchange between sea water in sand, and overflowing fresh water at low tide, II. J. Mar. Biol. Assoc., **18**: 299–306.

RIEDL, RUPERT. 1962. Probleme und methoden der Erforschung des litoralen Benthos. Verh. Deutsch. Zool. Gesellsch., **1962**: 505–67, 36 figs.

——— 1964. Lo studio del litorale marino in rapporto alla moderna biologia. Atti sem. studi biol. **1**: 3–30, 15 figs.

——— 1964. Die Erscheinungen der Wasserbewegung und ihre Wirkung auf Sedentarier im mediterranen Felslitoral. Helgol. Wiss. Meeresunters., **10** (1–4): 155–86, 13 figs.

——— 1964. 100 Jahre Litoralgliederung seit Josef Lorenz, neue und vergessene Gesichtspunkte. Int. Rev. ges. Hydrobiol., **49** (2): 281–305.

——— 1966. Biologie der Meereshohlen: Topographie, faunistik, und Ökologie eines unterseeischen Lebensraumes. Eine Monographie. Hamburg: Paul Farey. 636 pp., 328 figs., 16 col. pls.

A magnificent, exhaustive treatise on undersea caves, grottoes, potholes, and crevices, in Teutonic detail; superbly illustrated. Though there may be a tendency to make a general statement about each particular case, the book does demonstrate what an intelligent and determined (and evidently tireless!) diver can do with the opportunity to observe instead of spearing fish and photographing his girl friend. Unfortunately these caves are rare in our waters, and much more dangerous because of heavy surf.

RÉNAUD-DEBYSER, JEANNE. 1963. Récherches écologiques sur la faune interstitielle des sables: Bassin d'Arcachon, île de Bimini, Bahamas. Vie et Milieu, suppl. **15**. 157 pp., 6 pls., 72 figs.

ROUNSEFELL, GEORGE A., and WALTER R. NELSON. 1966. Red-tide research summarized to 1964, including an annotated bibliography. U.S. Fish and Wildl. Ser., Spec. Sci. Rept. Fisheries No. **535**. 85 pp.

SANDERS, H. L., R. R. HESSLER, and G. R. HAMPSON. 1965. An introduction to the study of deep-sea benthic faunal assemblages along the Gay Head–Bermuda transect. Deep-sea Res., **12**: 845–67, 8 figs.

SCHUSTER, REINHARDT. 1962. Das marine Litoral als Lebensraum terrestrischer Kleinarthropoden. Int. Rev. ges. Hydrobiol., **47** (3): 359–412, 12 figs.

SOURIE, R. 1954. Contribution à l'étude écologique des côtes rocheuses du Sénégal. Mem. l'Inst. Fran. Afr. Noire, **38**. 342 pp., 45 figs., 23 pls.

SOUTHWARD, A. J. 1953. The ecology of some rocky shores in the south of the Isle of Man. Proc. and Trans. Liverpool Biol. Soc., **59**: 1–50, 17 figs.

——— 1958. The zonation of plants and animals on rocky seashores. Biol. Revs., **33**: 137–77, 6 figs.

———, and J. H. Orton. 1954. The effects of wave-action on the distribution and numbers of the commoner plants and animals living on the Plymouth breakwater. J. Mar. Biol. Assoc., **33**: 1–19, 7 figs.

Spassky, N. N. 1961. Litoral yugo-vostochnogo poberezh'ya Kamchatki (Littoral of the southeastern shore of Kamchatka). Issl. Dalnev. Morei, **7**: 261–311.

Stephen, A. C. 1929–30. Studies on the Scottish marine fauna; The fauna of the sandy and muddy areas of the tidal zone. Trans. Roy. Soc. Edinb., **56**: 291–306, and 521–35.

Zonation and optimum areas per species.

Stephenson, T. A. 1947. The constitution of the intertidal fauna and flora of South Africa. Ann. Natal Mus., **11** (2): 207–324.

Part III of the report of a ten-year survey. Bibliography.

Stephenson, T. A., and Anne Stephenson. 1949. The universal features of zonation between tidemarks on rocky coasts. J. Ecol., **37** (2): 289–305, 4 figs., 1 pl.

——— 1950. Life between tidemarks in North America. Endeavour, **9** (33). 3 pp., 4 col. pls.

——— 1950. Life between tidemarks in North America, I: The Florida Keys. J. Ecol., **38** (2): 354–402, 10 figs., 7 pls.

——— 1952. Life between tidemarks in North America, II: Northern Florida and the Carolinas. J. Ecol., **40** (1): 1–49, 9 figs., 6 pls.

——— 1954. Life between tidemarks in North America, IIIa: Nova Scotia and Prince Edward Island: Description of the region. J. Ecol., **42** (1): 14–45, 6 figs., 8 pls.; IIIb: Nova Scotia and Prince Edward Island: The geographical features of the region. *Ibid.*: 46–70.

——— 1961. Life between tidemarks in North America, IVa: Vancouver Island, I. J. Ecol., **49** (1): 1–29, 9 figs., 3 pls.; IVb: Vancouver Island, II. *Ibid.*, (2): 227–43, 2 figs., 2 pls.

Stephenson, W., R. Endean, and Isobel Bennett. 1958. An ecological survey of the marine fauna of Low Isles, Queensland. Austr. J. Mar. and Freshw. Res., **9** (2): 261–318, 2 figs., 12 pls.

Stephenson, W., and R. B. Searles. 1960. Experimental studies on the ecology of intertidal environments at Heron Island, I: Exclusion of fish from beach rock. Austr. J. Mar. and Freshw. Res., **11** (2): 241–67, 4 figs., 3 pls.

Thompson, T. E., *et al.* 1966. Contributions to the biology of the Inner Farne. Trans. Nat. Hist. Soc. Northumberland, Durham, and Newcastle-upon-Tyne, **15** (5): 197–225, 6 figs., 3 pls.

Good example of a group study of a small area.

Thorson, Gunnar. 1946. Reproduction and larval development of Danish marine bottom invertebrates, with special reference to the planktonic larvae in the Sound (Öresund). Meddel. Komm. Danmarks Fisk. og Havunders. (Plankton), **4** (1). 523 pp., 198 figs.

——— 1950. Reproductive and larval ecology of marine bottom invertebrates. Biol. Revs., **25**: 1–45, 6 figs.

Correlates development of nonpelagic larvae with conditions unfavorable to larval feeding, i.e., short seasons of plankton production and low temperatures. An excellent review paper.

——— 1966. Some factors influencing the recruitment and establishment of marine benthic communities. Neth. J. Sea Res., **3** (2): 267–93.

Tokioka, Takasi. 1953. Invertebrate fauna of the intertidal zone of the Tokara Islands, I: Introductory notes, with the outline of the shore and the fauna. Pub. Seto Mar. Biol. Lab., **3** (2): 123–38, 6 figs., 3 pls.

——— 1963. Supposed effects of the cold weather of the winter 1962–63 upon the intertidal fauna in the vicinity of Seto. Pub. Seto Mar. Biol. Lab., **11** (2): 415–24, 2 figs., 3 pls.

VATOVA, ARISTOCLE. 1940. Le zoocenosi della Laguna Veneta. Thalassia, 3 (10). 28 pp., 10 pls.

——— 1943. Le zoocenosi dell'Alto Adriatico presso Rovigno e loro variazioni nello spazio e nel tempo. Thalassia, **5** (6). 61 pp., 1 pl.

——— 1949. Caratteri di alcune facies bentoniche della Laguna Veneta. Nova Thalassia, **1** (4). 14 pp., 1 pl.
The population ecology of the canals and lagoon of Venice.

VAUGHAN, T. W. 1940. Ecology of modern marine organisms, with reference to paleogeography. Bull. Geol. Soc. America, **51**: 433–68.

VERRILL, A. E., and S. I. SMITH. 1871–72. Report on the invertebrate animals of Vineyard Sound and adjacent waters. Rept. U.S. Fish. Comm., **1871–72**: 295–778.
Out of date, but still surprisingly useful in principle and excellent reading. They divide the fauna into animals predominantly inhabiting (a) bays and sounds, (b) estuaries, harbors, etc., and (c) ocean shores (comparable to the classification we employ). Each region is subdivided according to type of bottom—sand, mud, rock, etc.

WEBB, J. E., *et al.* 1958. The ecology of Lagos Lagoon. Phil. Trans. Roy. Soc. Lond., Ser. B, **241** (683): 307–419, 7 pls.
A series of papers by Webb and others on the lagoons of the Guinea coast, lagoon deposits, and *Branchiostoma nigeriense* Webb.

WHITTEN, H. L., HILDA F. ROSENE, and JOEL W. HEDGPETH. 1950. The invertebrate fauna of Texas coast jetties: A preliminary survey. Pub. Inst. Mar. Sci. Univ. Tex., **1** (2): 53–87, 4 figs., 1 pl.
Describes the fauna of a little-known environment, peculiar in that it is artificially introduced upon a sandy littoral.

WIESER, W. 1960. Meeresökologie. Fortschr. Zool., **12**: 336–78.
A general review.

WILSON, DOUGLAS P. 1949. The decline of *Zostera marina* L. at Salcombe and its effects on the shore. J. Mar. Biol. Assoc., **28**: 395–412, 1 fig., 4 pls.

——— 1951. A biological difference between natural sea waters. J. Mar. Biol. Assoc., **30**: 1–20, 2 pls.
Water from two different localities gave very different results in rearing certain marine invertebrates, but lacked any perceptible chemical difference.

——— 1952. The influence of the nature of the substratum on the metamorphosis of the larvae of marine animals, especially the larvae of *Ophelia bicornis* Savigny. Ann. l'Inst. Oceanogr., **27** (2): 49–156, 2 figs.

WOMERSLEY, H. B. S. 1947. The marine algae of Kangaroo Island, I: A general account of the algal ecology. Trans. Roy. Soc. S. Austr., **71** (2): 228–52, 5 pls., 5 figs.

WOMERSLEY, H. B. S., and S. J. EDMONDS. 1952. Marine coastal zonation in southern Australia in relation to a general scheme of classification. J. Ecol., **40** (1): 84–90, 1 fig.
The authors disagree with the Stephensonian terminology in some details.

——— 1958. A general account of the intertidal ecology of south Australian coasts. Austr. J. Mar. & Freshw. Res., **9** (2): 217–60, 2 figs., 12 pls.

YONGE, C. M. 1948. Bottom fauna of the sea. Research, **13**: 589–95, 3 figs.
Excellent nontechnical review.

ZENKEVITCH, L. 1961. Certain quantitative characteristics of the pelagic and bottom life of the ocean. In Sears, ed. (q.v., Sect. VI), pp. 323–34. 6 figs.

IV. Texts and Reference Books

ALLEE, W. C., A. E. EMERSON, O. PARK, T. PARK, and K. P. SCHMIDT. 1949. Principles of animal ecology. Philadelphia: Saunders. 837 pp., 263 figs.
This milestone (in weight as well as content) of ecology is a mine of information that

brings together a great mass of recent literature and digests it (pardon the metaphor) for the reader, with varying success. An indispensable reference, it nevertheless leaves much to be desired in regard to marine ecology, an indication, perhaps, of the comparative youth of this branch of the science. Nevertheless, its historical introduction provides the best short summary of the development of marine ecology, and the book contains much of the tangential information on terrestrial and lacustrine problems that must be scanned by the marine ecologist. Perhaps it should be pointed out that ecology, as understood by these authors, is not the practice of describing the phenomena of natural history in an esoteric terminology (the old-fashioned definition of ecology as "scientific natural history"), but the analytical study of the dynamics and interrelationships of groups of organisms in what Ed Ricketts was fond of calling the "holistic" manner.

Ax, Peter. 1960. Die Entdeckung neuer Organisationstypen im Tierreich. Wittenberg-Lutherstadt: A. Ziemsen. 116 pp., 88 figs.

A concise account of most of the recent zoological oddities (except the bivalve gastropod); brings together information not available outside of special journals.

Bainbridge, R., G. C. Evans, and O. Rackham, eds. 1966. Light as an ecological factor. Brit. Ecol. Soc. Symp., 6. New York: Wiley & Sons. 452 pp., illus.

Several articles concern marine organisms and the aspects of light and its measurement in the sea. See, e.g., N. Millot, "The enigmatic echinoids"; F. Evans, "The role of light in the zonation of periwinkles"; and T. Levring, "Submarine light and algal shore zonation."

Barnes, Robert D. 1963. Invertebrate zoology. Philadelphia: Saunders. 632 pp.

Since its appearance, this has been a successful text; and up to 1967 it was the only adequate one-volume text available for invertebrate courses. Since then some formidable competition has appeared (see Meglitsch). A revised edition appeared in 1968, with added chapters on insects and parasitic groups (omitted from the first edition). The author has taken kindly to suggestions, and has gone through the text rather thoroughly, with the result that the first edition is effectively superseded.

Barrington, E. J. W. 1967. Invertebrate structure and function. Boston: Houghton Mifflin. 549 pp., illus.

An original, essential text, exactly what its title implies, which should at least make zoologists of a lot of previously dull fellows. "My underlying theme is a self-evident one: that the business of animals is to stay alive until they have reproduced themselves, and that the business of zoologists is to try to understand how they do it." Amen.

Brown, F. A., Jr., ed. 1950. Selected invertebrate types. New York: Wiley & Sons. 597 pp., 234 figs.

A dissection and reference guide to eastern types; static and disconnected in treatment, but with an excellent bibliography of recent summary papers dealing with morphology and anatomy.

Buchsbaum, Ralph. 1948. Animals without backbones: An introduction to the Invertebrata. Revised ed. Univ. Chi. Press. 405 pp., illus.

Now a standard introductory text, for advanced high schools or retarded colleges (if used beyond the freshman year). Many of the photos are of Pacific coast subjects.

Bullock, Theodore Holmes, and G. Adrian Horridge. 1965. Structure and function in the nervous systems of invertebrates. San Francisco and London: Freeman. Vol. I, pp. 1–798; Vol. II, pp. 801–1719. Illus.

This is, of course, more than a cornerstone; it is an entire edifice.

Bullough, W. S. 1950. Practical invertebrate anatomy. London: Macmillan. 463 pp., 168 figs.

Similar to the above, but the work of a single author, with many standard forms reexamined and some forms not usually included in a one-volume text.

Buzzati-Traverso, A. A. 1958. Perspectives in marine biology. Berkeley: Univ. Calif. Press. 621 pp., illus.

The "minutes" of a famous symposium held in the languid air of La Jolla so many years ago.

Calman, W. T. 1949. The classification of animals. New York: Wiley & Sons. 54 pp.

This small volume, one of the Methuen monographs, is a concise introduction to the practice and problems of taxonomy, written by an accomplished taxonomist.

Carter, G. S. 1948. A general zoology of the invertebrates. Rev. ed. London: Sidgwick and Jackson. 509 pp., 172 figs.

A readable textbook of invertebrate physiology.

CARTHY, J. D. 1958. An introduction to the behavior of invertebrates. New York: Macmillan. 380 pp., 4 pls., 148 figs.

Mostly about terrestrial invertebrates; good comprehensive bibliography.

CENTRE NATIONAL DE LA RÉCHERCHE SCIENTIFIQUE. 1952. Colloques internationaux, XXXIII: Écologie. Paris. 582 pp., illus.

The papers given at the international colloquium on ecology at Paris, 1950; also published in Parts 2–7 of *Année Biologique,* Vol. 27, 1951. A good summary of recent thought and work in ecology, especially on the nature of communities and the biocenosis.

CRISP, D. J., ed. 1964. Grazing in terrestrial and marine environments. A symposium of the British Ecological Society, Blackwell, Oxford. 322 pp., illus.

Barnacles, limpets, and nudibranchs are among the subjects considered.

DOUGHERTY, ELLSWORTH C., ed. (in collaboration with Zoe Norwood Brown, Earl D. Hanson, and Willard D. Hartman). 1963. The lower Metazoa: Comparative biology and phylogeny. Berkeley: Univ. Calif. Press. 478 pp., text figs.

The proceedings (in part) of an international symposium held at Asilomar in 1960. Most of this is zoologists' zoology. Other results of this symposium are the series of papers by Karling on California Turbellaria and Lang's monograph on California Harpactoidea; both of these are based on collections made by the authors during their stay in California before and after the symposium.

DOWDESWELL, W. H. 1961. Animal ecology. New York: Harper Torchbooks. 209 pp., 45 text figs., 16 pls.

A reprint of the 1959 edition (Methuen). A good introductory treatment. (For sixth form in Britain; suitable for junior college or university freshmen in the U.S.)

EALES, N. B. 1949. The littoral fauna of Great Britain: A handbook for collectors, 2d ed. Cambridge Univ. Press. 305 pp., frontispiece, 24 pls.

An annotated manual, with keys and diagnostic drawings for most of the British shore animals. Bibliographies at the beginning of each chapter. We quote, as being even more applicable to conditions in America, some of the remarks in the Foreword by Stanley Kemp: "One of the chief difficulties which lies in the way of a student who is beginning work on the marine fauna of the British coasts is the identification of the specimens which he collects. The information he wants is as a rule to be found in large monographs, or in papers scattered through long series of zoological journals; and these will almost always include descriptions of very many species that he is never likely to obtain . . . . Only those who have worked on systematic zoology can realize the vast amount of labour which this book has entailed . . . . Continental countries are much ahead of us in the production of such handbooks as this." There is an appendix, with practical hints and a glossary.

EDMONDSON, C. H. 1946. Reef and shore fauna of Hawaii. Bernice P. Bishop Mus., Spec. Pub., **22**. 381 pp., 223 figs.

A guide to the Hawaiian fauna, provided with short keys; now obsolete and being revised.

EDMONDSON, W. T. 1966. Marine biology III. Proceedings of the Third Interdisciplinary Conference, N.Y. Acad. Sci. 313 pp., illus.

This conference was held in 1964; it concerned the ecology of invertebrates, which means that almost everything and anything was discussed.

EKMAN, SVEN. 1935. Tiergeographie des Meeres. Leipzig: Akad. Verlags. M. B. H. 542 pp.

——— 1953. Zoogeography of the sea. London: Sidgwick and Jackson. 417 pp., 121 figs.

Although the English edition is completely revised and brought up to date in many respects, serious biogeographers must still refer to the more detailed German edition.

GAYEVSKAYA, N. S., ed. 1948. Opredelitel' fauny i flory severnykh morei S.S.S.R. (Manual of the fauna and flora of the northern seas of the U.S.S.R.) Moscow State Publishing House. 740 pp., 136 pls., 77 text figs.

A well-illustrated "handbook" (it is about the size of a telephone directory!) of arctic and high boreal marine plants and invertebrates. Some of the species included occur within our range. This useful work—for those who read Russian—could well serve as a model for the sort of guide we need for our own shores.

HADZI, J. 1963. The evolution of the Metazoa. New York: Macmillan. 499 pp., 62 figs.

A contentious, overlong, somewhat tedious and overrated book. It belongs to that school of phylogeny that declares speculations to be facts, and from them argues the truth of unprovable theories.

HARDY, ALISTER C. The open sea: Its natural history. Boston: Houghton-Mifflin.

Part I: The world of plankton. 335 pp., 103 text figs.; 24 b&w, 24 col. pls. (1956). Part II: Fish and fisheries. 322 pp., 114 text figs.; 32 b&w, 16 col. pls. (1959).

A splendid book, written for layman and professional. If you ever had a college education, and a book like this is beyond you, you have wasted four years. However, it is doubtful that such persons get this deep into a bibliography.

HEDGPETH, JOEL W., ed. 1957. Treatise on marine ecology and paleoecology, Vol. I: Ecology. Memoir, **67** (1). Geol. Soc. Amer. 1296 pp., illus.

In addition to the summary chapters by various authorities, there are extensive annotated bibliographies.

HESSE, RICHARD. Translated and prepared by W. C. Allee and Karl P. Schmidt. 1937. Ecological animal geography. New York: Wiley & Sons. 597 pp., 135 figs. 2d ed., 1951. 715 pp., 142 figs.

Marine communities are extensively considered, both for themselves and in connection with environmental limitations; methodology illuminated by the modern technic of holistic thinking.

HICKMAN, CLEVELAND P. 1967. Biology of the invertebrates. St. Louis: Mosby. 673 pp., 596 figs.

A rather pedestrian text, which says little about biology as such. Of mollusks: "Endless evolutionary trial and error on the part of the shell has evolved structural changes in keeping with the diverse ecologic pattern of the group." Concerning the internal anatomy of some arthropods (not including copepods, sacculinids, pycnogonids, etc.): "The arthropods are all provided with organ systems characteristic of higher forms." There are somewhat better texts.

HYMAN, LIBBIE H. 1940–59. The invertebrates. New York: McGraw-Hill. I, Protozoa through Ctenophora (1940). 726 pp., 221 text figs. II, Platyhelminthes and Rhynchocoela: The acoelomate Bilateria (1951). 550 pp., 208 figs. III, Acanthocephala, Aschelminthes, and Entoprocta: The pseudocoelomate Bilateria (1952). 572 pp., 223 text figs. IV, Echinodermata: The coelomate Bilateria (1955). 763 pp., 280 figs. V. Smaller coelomate groups: Chaetognatha, Hemichordata, Pogonophora, Phoronida, Ectoprocta, Brachipoda, Sipunculida: The coelomate Bilateria (1959). 783 pp., 240 text figs. VI, Mollusca I: Aplacophora, Polyplacophora, Monoplacophora, Gastropoda: The coelomate Bilateria (1967). 792 pp., 249 figs.

JØRGENSEN, C. BARKER. 1966. Biology of suspension feeding. New York: Pergamon. 357 pp., illus. (sparsely).

A fine summary of this important aspect of biology in the sea.

KAESTNER, ALFRED. 1967. Invertebrate zoology, Vol. I. Translated and adapted by Herbert W. Levi and Lorna R. Levi. New York: Interscience Pub. of John Wiley & Sons. 597 pp., illus.

This is intended to fill the gap between a one-volume text and a more comprehensive monographic series. The work will take up four volumes in English. Teutonically thorough, it will indeed supply many details, and should stand as a fine example of a style of zoology becoming rare in our midst.

LAUFF, GEORGE H. 1967. Estuaries. Amer. Assoc. Adv. Sci., Pub. **83**. 757 pp., illus.

The results (at last!) of a week-long meeting of specialists from various parts of the world down near the Marshes of Glynn in 1964. The most comprehensive approach to the complex problems of estuaries yet to appear in print.

MACFAYDEN, A. 1957. Animal ecology, aims and methods. London: Pitman. 264 pp., illus.

Still a useful guide to methods and approaches.

MEGLITSCH, PAUL A. 1967. Invertebrate zoology. London: Oxford Univ. Press. 961 pp., illus.

One of the better comprehensive one-volume invertebrate texts available. Written with somewhat more zoological sophistication than its current rivals. The illustrations are clearer, and the recent literature has been adequately ransacked. This is the best of the systematically oriented texts; combined with Barrington (see above) it provides a sound beginning for all who would be zoologists.

MOORE, HILARY B. 1958. Marine ecology. New York: Wiley & Sons. 493 pp., illus.

Based principally on British sources, and too detailed for elementary use, this book is nevertheless a convenient introduction to the back files of the *Journal of the Marine Biological Association.*

NATIONAL ACADEMY OF SCIENCES. 1957. The effects of atomic radiation on oceanography and fisheries. NAC-NRC Pub. **551**. 137 pp., illus.

A revision is in prospect.

NICOL, J. A. COLIN. 1960. The biology of marine animals, 2d ed. New York: Wiley. & Sons. 699 pp., illus.

This is an expert summary of a vast amount of work on the physiology of marine animals (including fish); as "biology," however, it overlooks, for the most part, the rather essential phenomenon of reproduction. In spite of this omission, no practicing marine biologist can be without this book.

OPPENHEIMER, CARL H. 1966. Marine biology II: Proceedings of the Second International Interdisciplinary Conference. N.Y. Acad. Sci. 369 pp., illus.

The theme of these discussions (in 1962) was phytoplankton, with various ecological asides.

POLIKARPOV, G. G. 1966. Radioecology of aquatic organisms: The accumulation and biological effect of radioactive substances. New York: Reinhold. 314 pp., 34 figs.

POTTS, W. T. W., and GWYNETH PARRY. 1964. Osmotic and ionic regulation in animals. New York: Macmillan. 423 pp., illus.

Brings Krogh up to date.

PROSSER, C. LADD, and FRANK R. BROWN. 1961. Comparative animal physiology, 2d ed. Philadelphia: Saunders. 688 pp., illus.

Revision of an excellent comprehensive monograph, with comprehensive bibliographies. Unfortunately, the practice of offering paraphrases of content, rather than actual titles of references, has been continued in this new edition.

RAY, DIXY LEE., ed. 1959. Marine boring and fouling organisms. Seattle: Univ. Wash. Press. 536 pp. + 7 pp., illus.

About shipworms and gribbles and their boring ways; nevertheless, it was a stimulating and pleasant symposium, held, appropriately enough, in the woods at Friday Harbor.

REMANE, ADOLF, and C. SCHLIEPER. 1958. Die biologie des Brackwassers: Die Binnengewässer, Bd. 22. Stuttgart. 348 pp., 139 figs., 43 tbls., 5 chts.

A solidly Germanic treatment; a worthy candidate for translation.

SCHEER, BRADLEY T. 1948. Comparative physiology. New York: Wiley & Sons. 563 pp., 68 figs.

Somewhat more readable than the Prosser text, but sometimes too general.

SCHENK, EDWARD T., and JOHN H. MCMASTERS. 1948. Procedure in taxonomy. Rev. ed., enlarged and in part rewritten by A. Myra Keen and Siemon William Muller. Stanford: Stanford Univ. Press. 93 pp.

Because so much of the invertebrate reconnaissance work here on the Pacific coast is constantly involved in taxonomic vexations, it seems well to cite here this manual of procedure.

SOCIÉTÉ DE BIOGÉOGRAPHIQUE. 1940. Contribution à l'étude de la répartition actuelle et passée des organismes dans la zone neritique. Paris: Lechevalier. 434 pp.

Memoir VII of the Société de Biogéographique; a symposium volume containing papers on Atlantic, Mediterranean, and Australian intertidal ecology, and a paper by Schenck and Keen on molluscan assemblages.

SOUTHWOOD, T. R. E. 1966. Ecological methods: With particular reference to the study of insect populations. London: Methuen. 391 pp., 101 figs.

"This volume aims to provide a handbook of ecological methods pertinent for the study of animals. Emphasis is placed on those most relevant to work on insects and other non-microscopic invertebrates of terrestrial and aquatic environments, but it is believed that the principles and general techniques will be found of value in studies on vertebrates and marine animals." (p. xv). Excellent; one of the best handbooks for procedures in modern ecology; how to keep up with the numbers game.

STÜMPKE, HARALD. 1961. Bau und Leben der Rhinogradentia. Stuttgart: Fischer. 83 pp., 12 figs., 15 tbls. Translated 1967 by Leigh Chadwick as *The snouters: form and life of the rhinogrades.* Garden City: Natural History Press. 92 pp., 15 pls., 12 figs.

Discusses the fascinating patterns of evolutionary radiation in an order of mammals now unfortunately exterminated by the destruction of the obscure Pacific islands in which they were endemic. Many of the species were marine.

TRESSLER, DONALD K., and JAMES MCW. LEMON. 1951. Marine products of commerce: Their acquisition, handling, biological aspects of the science and technology of their preparation and preservation, 2d ed. New York: Reinhold. 782 pp., illus.

First published in 1923, this has long been considered a standard reference work. Many of the chapters were satisfactorily brought up to date; but the sections on mussels, abalones, crabs, and the like appear to have been revised by the office boy. The section on abalones is a miserable joke, and the dangerous statement of the 1923 edition—"Mussels are best from December to July; they therefore may be used as a substitute for oysters during the early summer"—has been repeated without making it clear which species or coast is meant (see p. 676).

WILLIAMS, C. B. 1964. Patterns in the balance of nature: And related problems in quantitative ecology. New York: Academic Press. 324 pp., 126 figs.

The problems discussed are also good examples of methodology.

WILMOTH, JAMES H. 1967. Biology of Invertebrata. Englewood Cliffs, N.J.: Prentice-Hall. 465 pp., illus.

Still another text for 1967. Better than Hickman, but not as good as Meglitsch. Unfortunately, it is illustrated with poorly processed copies from divers sources—via xerography, that modern tyrant of literate society. Also, the labeling of figures is often inadequate. The English usage is at times unfortunate, but again not as unfortunate as Hickman's.

V. Geology and Paleontology

ADEGOKE, OLUWAFEYISOLA S. 1967. Earliest Tertiary West American species of Platyodon and Penitella. Proc. Calif. Acad. Sci., Ser. 4, **35** (1): 1–22, 26 figs.

AGER, DEREK V. 1963. Principles of paleoecology. An introduction to the study of how and where animals and plants lived in the past. New York: McGraw-Hill. 371 pp., illus.

An introductory text, with some useful viewpoints for the neontologist.

ANDERSON, CHARLES A., *et al.* 1950. 1940 E. W. Scripps cruise to the Gulf of California. Geol. Soc. Amer., Mem., **43**.

A collection of independently paged papers concerning geology, paleontology, and oceanography of the Gulf of California, by various authors, with an excellent bathymetric map. A suitable companion piece to *Sea of Cortez.*

ANDERSON, F. M. 1938. Lower Cretaceous deposits in California and Oregon. Geol. Soc. Amer. Spec. Pap. **19**. 339 pp., 84 pls. (Pl. 84 is a map of principal beds.)

The fossil record along the Pacific coast is spotty. It jumps from the amazing Cambrian beds of the Burgess shale in British Columbia to the Lower Cretaceous, followed by a rich Tertiary. The Cambrian has not been treated in a single monograph, but can be sampled in the many papers by Walcott, published principally in the Smithsonian Miscellaneous Series.

CAMP, CHARLES L., and G. DALLAS HANNA. 1937. Methods in paleontology. Berkeley: Univ. Calif. Press. 153 pp., 58 figs.

A useful manual for handling fossils.

DAETWYLER, CALVIN C. 1966. Marine geology of Tomales Bay, central California. Pac. Mar. Sta., Res. Rpt. **6**. 169 pp., 40 figs.

Mainly concerning the subsurface structure of Tomales Bay.

EHLEN, JUDI. 1967. Geology of state parks near Cape Arago, Coos County, Oregon. Ore Bin, **29** (4): 61–82, 12 figs.

Good descriptive account of one of Oregon's principal research and field trip areas; locale of University of Oregon's Institute of Marine Biology.

FAIRBRIDGE, RHODES W. 1961. Eustatic changes in sea level. In *Physics and chemistry of the earth,* **4**: 99–185. New York: Pergamon Press. 15 figs.

One would almost hope these changes were a bit more rapid; many problems would be solved by a rise in sea level to the fourth deck of the freeway plexus back of the Los Angeles city hall.

GLEN, WILLIAM. 1959. Pliocene and Lower Pleistocene fauna of the western part of the San Francisco Peninsula. Univ. Calif. Pub. Geol. Sci., **36** (2): 147–98, 3 pls., 5 text figs.

IMBRIE, JOHN, and NORMAN NEWELL. 1964. Approaches to paleoecology. New York: Wiley & Sons. 432 pp., illus.

A symposium volume; see the paper by Johnson, "Community approach to paleoecology," based in part on studies in Tomales Bay.

JENKINS, OLAF P. 1943. Geologic formations and economic development of the oil and gas fields of California. Calif. Div. Mines., Bull. **118**. 773 pp.

Some detailed accounts of principal beaches.

———, ed. 1951. Geologic guidebook of the San Francisco Bay counties: History, landscape, geology, fossils, minerals, industry, and routes to travel. Calif. Div. Mines Bull., **154**. 392 pp., illus.

An excellent compilation of information, including a chapter on the Farallons and material on the central California sea coast. Here and there the touch of a biologist would have improved the text, but one cannot have everything in this world.

JOHNSON, D. W. 1919. Shore processes and shoreline development. New York: Wiley & Sons. 584 pp.

A comprehensive source book with much ecologically applicable information; includes more data on waves, and is more easily understandable for the general reader, than anything else we have found. Johnson's classification of shorelines as emergent and submergent is oversimplified, but despite its signs of age, this book still remains the best available treatment of coastal geomorphology, and has recently been reprinted by Stechert. The plates are not well reproduced in the photo-reprint edition.

KEEN, A. MYRA, and HERDIS BENTSON. 1944. Checklist of California Tertiary marine mollusca. Geol. Soc. Amer. Spec. Pap., **56**. 280 pp.

KUENEN, P. H. 1950. Marine geology. New York: Wiley & Sons. 568 pp., 246 figs., 2 chts.

With special emphasis on the Malay Archipelago. Contains an excellent chapter on coral reefs.

KULM, L. D., and JOHN V. BYRNE. 1966. Sedimentary response to hydrography in an Oregon estuary. Mar. Geol., **4**: 85–118, 14 figs.

Yaquina Bay.

LADD, HARRY S., ed. 1957. Treatise on marine ecology and paleoecology, II: Paleoecology. Memoir, **67** (2), Geol. Soc. Amer. 1077 pp. illus.

Bibliographies in this volume emphasize groups with hard parts.

MENARD, H. W. 1964. Marine geology of the Pacific. New York: McGraw-Hill. 271 pp., illus.

About the deep ocean basin. "The only synthesis. . . . of any entire ocean basin," say the publishers.

NORTH, WILLIAM B., and JOHN V. BYRNE. 1965. Coastal landslides of northern Oregon. Ore Bin, **27** (11): 217–41, 16 figs.

RATHBUN, MARY J. 1926. The fossil stalk-eyed Crustacea of the Pacific slope of North America. U.S. Nat. Mus., Bull. **138**. 149 pp., 39 pls.

SCHÄFER, WILHELM. 1962. *Aktuo-Paläontologie nach Studien der Nordsee.* Frankfurt am Main: Waldemar Kramer. 666 pp., 277 figs., 37 pls.

This book is about a part of the North Sea coast and its living and recently dead organisms, as an introduction to the potential past of the paleontologist. It is based on the philosophy that the present is the key to the past. Each major section of the book ends with a summary or review statement in English, a useful feature of which quite a few people who should consult the book are apparently unaware.

SCHENCK, HUBERT G., and A. MYRA KEEN. 1950. California fossils for the field geologist. Stanford: Stanford Univ. Press. 88 pp., incl. 56 plates. Photolith, spiral-bound.

A handy guide, principally to mollusks. Intended to be used in the field, it needs a stronger binding to wear well.

SHEPARD, FRANCIS P. 1948. Submarine geology. New York: Harper. 348 pp., 105 figs., 1 cht.

With special reference to continental shelves and slopes, and to submarine canyons, hence complementary to the Kuenen text.

——— 1964. The earth beneath the sea. New York: Atheneum **56**. 275 pp., 113 figs., paper.

Concerns the geology of the vast extent of land under the oceans, with perhaps more reference to the Pacific. Suitable for textbook use, as well as general information for those satisfied only by lots of details instead of flowery generalities.

———, and K. O. EMERY. 1941. Submarine topography off the California coast: Canyons and tectonic interpretation. Geol. Soc. Amer. Spec. Pap., **31**. 171 pp., 42 figs., 4 chts.

The charts that accompany this monograph show the bottom formations along the California coast in layered colors, and are handsome examples of the cartographer's art.

VALENTINE, JAMES W. 1961. Paleoecologic molluscan geography of the Californian Pleistocene. Univ. Calif. Pub. Geol. Sci., **34** (97): 309–442, 16 figs.

Concerns southern California localities.

WEAVER, CHARLES E. 1937. Tertiary stratigraphy of western Washington and northwestern Oregon. Univ. Wash. Pub. Geol., **4**. 266 pp., 15 pls.

With good maps of the beds.

——— 1943. Paleontology of the marine Tertiary formations of Oregon and Washington. Univ. Wash. Pub. Geol., **5**. 788 pp., 104 pls.

Monographic treatment of larger invertebrate fossils.

VI. Books about the Ocean, Texts, etc. (Mostly since 1960)

AMOS, WILLIAM HOPKINS. 1966. The life of the seashore: Our living world of nature. New York: McGraw-Hill. 231 pp., illus.

Most of the illustrations are in color, and include many Pacific scenes and creatures. An excellent popular introduction.

VON ARX, WILLIAM S. 1962. An introduction to physical oceanography. Reading, Mass.: Addison-Wesley. 422 pp., illus.

Illustrated with diagrams and halftones, including some of historical interest. A good introduction to the subject for students with mathematical ability; it should be accessible to high school students as a demonstration of the need to study mathematics for oceanography, although its treatment is not deeply mathematical.

BARNES, HAROLD, ed. 1966. Some contemporary studies in marine science: A collection of original scientific papers presented to Dr. S. M. Marshall, O.B.E., F.R.S., in recognition of her contribution with the late Dr. A. P. Orr to marine biological progress. London: Allen and Unwin. 716 pp., illus.

BASCOM, WILLARD N. 1961. A hole in the bottom of the sea: The story of the Mohole Project. Garden City, N.Y.: Doubleday. 352 pp., illus.

Written by the man who first supervised the ill-fated Mohole Project; perhaps after we get through with Vietnam we will try again.

——— 1964. Waves and beaches: The dynamics of the ocean surface. Garden City, N.Y.: Doubleday. Anchor Books, S 34. 267 pp., 77 figs., 24 pls.

An essential book for anyone who spends any time along the shore; a simple straightforward discussion of waves, tides, and how the sea interacts with the land. There is no hardcover edition.

BURTON, MAURICE. 1960. Under the sea. New York: Watts. 256 pp., illus.

A literary rehash of information about the sea and about animals therein (mostly). The same material is covered in dozens of other books.

CAIDIN, MARTIN. 1964. Hydrospace. New York: Dutton. 320 pp., illus.

This is essentially about the gadgetry and committee thinking about oceanography.

CARRINGTON, RICHARD. 1960. A biography of the sea: The story of the world ocean, its animal and plant populations, and its influence on human history. New York: Basic Books. 286 pp., illus.

A popular treatment, with more emphasis on life in the sea than is found in Cowen's book.

CARSON, RACHEL L. 1951. The sea around us. New York: Oxford Univ. Press. 230 pp.

Primarily a well-done popularization of oceanography, with little reference to the life of the seas, this deservedly best-selling book has made oceanography famous. An earlier book by Miss Carson, *Under the sea wind,* has been reissued by Oxford University Press; it describes, in slightly more tinted language, the life of birds, fish, and eels on the eastern coast. Unfortunately, the glossary was not corrected, and the life histories are more imaginative than necessary.

——— 1955. The edge of the sea. Boston: Houghton Mifflin. 276 pp., illus.

Primarily about seashore life of the Atlantic and Gulf coasts, now widely available as a paperback.

COKER, R. E. 1962. This great and wide sea. New York: Harper Torchbooks, 551. 325 pp., 23 figs., 32 pls.

Widely used in schools; the original hardcover edition appeared in 1947, and the book is beginning to show its age, although it does provide more information on biology in the seas than most other general books available in paperback.

COUSTEAU, JACQUES-YVES, with JAMES DUGAN. 1963. The living sea. New York: Harper & Row. 325 pp., illus. (incl. col. pls.).

The latest installment of Captain Cousteau's adventures in oceanography; the reader will learn what Cousteau and his divers are up to, but not much about the hard core of oceanography. However, this romantic, episodal approach reaches some students that are not stimulated by more matter-of-fact approaches. Librarians with limited budgets should realize that books like this cannot fill the need for reference material (there is no index).

COWEN, ROBERT C. 1960. Frontiers of the sea. The story of oceanographic exploration. Garden City, N.Y.: Doubleday. 307 pp., illus.

One of the better general accounts of oceanography, written in clear readable style. The emphasis is more on physical and geological aspects than on biology.

CROMIE, WILLIAM J. 1962. Exploring the secrets of the sea. New York: Prentice-Hall. 300 pp., illus.

——— 1966. The living world of the sea. New York: Prentice-Hall. 343 pp., illus.

The first of these is similar in scope to other books whose authors' names begin with C: Cowen, Carrington, *et al.* The second book is an improvement on Burton's. However, Mr. Cromie is primarily a journalist, and is sometimes a bit imprecise with his words.

DAKIN, WILLIAM J. 1952. Australian seashores: A guide for beach-lovers, the naturalist, the shore fisherman, and the student. Sydney: Angus and Robertson. 372 pp., 23 figs., 99 pls.

A fine book, obviously a labor of love; seen through the press by his students Isobel Bennett and Elizabeth Pope.

DARWIN, CHARLES. 1962. The structure and distribution of coral reefs. Foreword by H. W. Menard. Berkeley: Univ. Calif. Press. 214 pp., 3 pls.

Reprinted from the first paperback edition (1851!), with a foreword saying that Darwin is still essentially correct. A brief summary of Darwin's hypothesis appears in almost every general book about the ocean, earth science textbook, etc., and has been authenticated by *Life* magazine. He could ask for no more.

DAUGHERTY, CHARLES MICHAEL. 1961. Searchers of the sea: Pioneers in oceanography. New York: Viking. 160 pp., illus.

For "older boys and girls"; an episodic and biographical history of oceanography.

DEFANT, ALBERT. 1958. Ebb and flow: The tides of earth, air, and water. Ann Arbor: Univ. Mich. Press. 121 pp.

Available in both cloth and paperback editions. A good introduction to tides.

——— 1961. Physical oceanography. New York: Pergamon. 2 vols.

Formidable; for dedicated oceanographers.

DOUGLAS, JOHN SCOTT. 1952. The story of the oceans. New York: Dodd, Mead. 315 pp.

Somewhat more lurid in spots than *The sea around us,* and with more about animals, this book seems to have been written primarily "for older boys." Since it is dedicated to that eminent diver and abalone ecologist, Thomas Delmer Reviea (see *Fortnight,* Jan. 21, 1952, pp. 12–13), we wonder who is responsible for the idea that he would "be squeezed by tremendous water pressure" if the octopus ripped open his suit with its horny beak at "sixty feet below the sparkling blue waters of the mid-California coast (p. 181)." Such a gruesome scene is oddly out of place in a book that is, on the whole, surprisingly accurate in its popularization of zoology and in its general statements.

ENGEL, LEONARD (and the editors of *Life*). 1961. The sea. New York: Time, Inc. Life Nature Library. 190 pp., illus.

A surprisingly good popular summary; one of the best volumes in this series. This text should be suitable for any intelligent child who has learned to read (by phonics or any other system), but nevertheless the book has been simplified for school use by the Silver-Burdett editors. At times it is too simple; e.g., "Most diatoms are yellow or brown; other algae may be green or red." Sometimes the abridgment is unnecessary, as well as misleading:

Original: "The poisonous land animals have their counterparts in the sea. Dozens of fish, including the ones pictured here, have hollow spines on their bodies, which can inject venom into an enemy. Humans touching these fish often receive stings that result in excruciating pain, severe shock, and even death."

Silver-Burdett version: "Animals that are just as dangerous as the rattlesnake make their homes in the sea. Many fish have poisonous needles hidden in their bodies. If an unlucky swimmer touches one of them, the poison is then shot into his body. This sting can be extremely painful; it can even cause death."

The Silver-Burdett version does have most of the illustrations of the standard edition, but it can hardly be said that the text has been improved by the educators.

FAIRBRIDGE, RHODES W. 1966. The encyclopedia of oceanography. New York: Reinhold. 1021 pp., illus.

Heretofore, much of this information has been scattered through various textbooks of oceanography and standard reference works; this volume brings it all conveniently together. Generously illustrated; one of the best of this publisher's series. (The series on biology is rather uneven; for some reason, it omits almost all of the phylum Coelenterata, as well as many other items.)

FLATTELY, F. W., and C. L. WALTON. 1922. The biology of the seashore. London: Macmillan. 336 pp., 16 pls., 21 figs.

An old classic, still valuable.

GASKELL, T. F. 1960. Under the deep oceans: Twentieth-century voyages of discovery. London: Eyre & Spottiswoode. 240 pp.

A well-written book, mostly about geological matters.

GORSKY, BERNARD. 1957. Vastness of the sea: Adventure in the mysterious depths. Boston: Little, Brown. 305 pp., illus. (sparsely).

Translated from the French. A typical example of the diary-style revelations of scuba diving. There are dozens of these books, all more or less alike. They have about as much to do with oceanography as books on big game hunting have to do with the natural history of Africa.

HARDY, SIR ALISTER. 1965. The open sea: Its natural history. Boston: Houghton Mifflin. 335 + 322 pp., illus. (Part I, *The world of plankton,* and Part II, *Fish and fisheries,* are bound together as one volume).

This is one of the great books about the sea written in our generation. Although British in orientation, it is the best layman's introduction to marine biology that we have.

——— 1967. Great waters: A voyage of natural history to study whales, plankton and the waters of the Southern Ocean. New York: Harper and Row. 542 pp., photographs, drawings, col. pls.

This is far more than a memory of the *Discovery* forty years before; it is the story of a great effort in oceanography, how the research has grown, and what it means. A marvelous book.

HILL, M. N., ed. 1963. The sea: Ideas and observations on progress in the study of the seas. New York: Interscience Pub. of Wiley & Sons. Vol. 1, Physical oceanography (1962), 864 pp. Vol. 2, Composition of sea water; Comparative and descriptive oceanography (1963), 554 pp. Vol. 3, The earth beneath the sea; History (1963), 963 pp.

These three expensive volumes are essentially a sort of journal, with many contributors, and a fourth volume may eventually appear. The original purpose was to bring Sverdrup, Johnson, and Fleming up to date. It is somewhat leaner in biology than in physical aspects.

HULL, SEABROOK. 1964. The bountiful sea. Englewood Cliffs, N.J.: Prentice-Hall. 340 pp., illus.

The author is an active figure on the fringe of the Washington Establishment of oceanography, and his approach is primarily based on anecdotal fallout from committee meetings and a preoccupation with getting something done. There are many signs of hasty writing, and the book cannot be recommended for its style. Caidin's book, which has somewhat the same coverage, is better and costs a dollar less.

IDYLL, C. P. 1964. Abyss—the deep sea and the creatures that live in it. New York: Crowell. 396 pp., illus.

Since the author is a professional biologist, his work is somewhat better than that of the journalists.

KING, CUCHLAINE A. M. 1965. An introduction to oceanography. New York: McGraw-Hill. 337 pp., illus.

A general introduction, intended for those without a background in physics, mathematics, and biology. Good for high school senior-level reading, although the avoidance of mathematical treatment may lull some to a false sense of security.

LONG, E. JOHN. 1964. Ocean sciences. Annapolis: U.S. Naval Institute. 304 pp., illus.

This is a collection of articles, of varying degrees of coverage, by the important figures in oceanography, mostly in various government agencies in Washington. They all belong to several committees and change chairs at different meetings, and are too busy to write about more than what they heard at the last meeting. Most of the articles appear to have been prepared for the higher levels of administration—or for admirals—so if you want to be informed enough to make a political speech, this may suffice. There is a glossary and an index.

MARSHALL, N. B. 1954. Aspects of deep sea biology. London: Hutchinson. 380 pp., text figs, col. pls.

An authoritative treatment but not overly technical. In the United States, this book has been available under the imprint of the Philosophical Library, which imported the sheets and doubled the price. We continue to watch Marboro's catalogues hopefully.

MERO, JOHN L. 1965. The mineral resources of the sea. New York: Elsevier. 312 pp., illus.

What one engineer thinks we might expect in the way of phosphates, manganese, gold, diamonds, etc. from the sea.

MILLER, ROBERT C. 1966. The sea. New York: Random House. 316 pp., 81 col. pls., b&w illus.

Superbly illustrated and printed, this is essentially a library table book. The text is well written, but intended to be of general application. It is the most handsome of the current crop of books about the sea.

MINER, ROY WALDO. 1950. Field book of seashore life. New York: Putnam. 888 pp., 251 pls. (24 in color).

Compact, concise guide to Atlantic coast invertebrates from New England to Beaufort, North Carolina. In preparation for 20 years, it supplants all previous Atlantic guides, although some of the nomenclature is not the most recent.

MOORE, HILARY B. 1958. Marine ecology. New York: Wiley & Sons. 493 pp., illus.

Based principally on British sources, and too detailed for elementary use, this book is nevertheless a convenient introduction to the back files of the *Journal of the Marine Biological Association.*

MURRAY, JOHN, and JOHAN HJORT. 1912. The depths of the ocean. London: Macmillan. 821 pp., 575 figs., 9 pls.

A monumental source book, especially for the North Atlantic. There is a recent facsimile reprint.

NEUMANN, GERHARD, and WILLARD J. PIERSON, JR. 1966. Principles of physical oceanography. Englewood Cliffs, N.J.: Prentice-Hall. 545 pp., illus.

An advanced, highly mathematical text.

OMMANNEY, F. D. 1949. The ocean. Oxford Univ. Press (Home University Library). 238 pp., illus.

This is a successor to the volume of this title by Sir John Murray, which is now rare even in secondhand bookstores. Bookstores, too, are now rare; the once friendly haunts of Los Angeles's 6th Street have been usurped by a wilderness of parking lots, and only two remain.

PICCARD, JACQUES, and ROBERT S. DIETZ. 1961. Seven miles down: The story of the bathyscaphe *Trieste.* New York: Putnam. 249 pp., illus.

All about the bathyscaphe and its deepest dive. This sort of thing is great for inspiring young hopefuls to take up oceanography; it can only be hoped that teachers can supply a few grains of salt or drops of cool water.

PICKARD, GEORGE L. 1964. Descriptive physical oceanography: An introduction. New York: Macmillan. A Pergamon Press Book. 199 pp., 311 figs.

This is probably the most scholarly physical oceanography treatment available at comparatively low price; some background in physics will help the reader. In price and physical structure, this book is something more than a paperback; but it is not a durably bound "hardcover" book, being bound in stiff cardboard.

RAYMONT, JOHN E. G. 1963. Plankton and productivity in the oceans. London: Pergamon; New York: Macmillan. 660 pp., illus.

About the plankton and other life of the sea, mostly in the North Atlantic.

RILEY, J. P., and G. SKIRROW. 1965–66. Chemical oceanography. New York: Academic Press.

A specialized research treatise, with many contributors.

RUSSELL, F. S., and C. M. YONGE. 1936. The seas: Our knowledge of life in the sea and how it is gained. Rev. ed. London: Frederick Warne. 379 pp., 127 pls.

Still in print and apparently going strong after all these years.

RUSSELL, R. C. H., and D. H. MACMILLAN. 1953. Waves and tides. London: Hutchinson; New York, Philosophical Library. 348 pp., illus.

Some knowledge of what an equation is supposed to mean will be of help, but tides and waves are sort of mathematical anyhow.

SEARS, MARY, ed. 1961. Oceanography: Invited lectures presented at the International Oceanographic Congress held in New York, 31 August–12 September 1959. Amer. Assoc. Adv. Sci., Wash., D.C. Pub. 67. 654 pp., illus.

Several of the recent popular books on oceanography, e.g., Cowen and Cromie, have mined material from this book.

SOUTHWARD, A. J. 1965. Life on the seashore. London: Heinemann; Cambridge: Harvard Univ. Press. 153 pp., 50 text figs., 8 b&w pls.

The animals and plants are British, but the principles are universal. The most valuable part of this book is its list of suggestions for observations and experiments on the shore. These will be especially useful for instructors who have had inadequate exposure to the scientific method, thanks to the stupidities of our teacher training programs.

STEINBECK, JOHN, and EDWARD F. RICKETTS. 1941. Sea of Cortez: A leisurely journal of travel and research. New York: Viking. 598 pp., 40 pls. (8 in color).

The first half of the book is a popular account of an informal expedition into the Gulf of California by small boat, the object of which was to survey the shore animals of that inaccessible and still little-known region. The latter half of the book is strictly technical, devoted mostly to lists, and consists of an annotated catalogue of the 550 species encountered, a resumé of the literature (with a bibliography of some 479 titles), and a summary of our knowledge concerning the natural history of the Panamic faunal province. All these together comprise materials for a source book of the marine invertebrates of this region.

SVERDRUP, H. U., MARTIN W. JOHNSON, and RICHARD H. FLEMING. 1942. The oceans: Their physics, chemistry, and general biology. Englewood Cliffs, N.J.: Prentice-Hall. 1087 pp., illus.

This now hardy and classic textbook has been kept in print without substantial change for more than 20 years. Oceanography has become so complex that its like will probably never be approached again, and the more recent texts have all been more or less oriented to particular phases of oceanography.

TRICKER, R. A. R. 1965. Bores, breakers, waves and wakes. New York: Elsevier. 250 pp., illus.

Very well illustrated, including good color plates.

WIEGEL, ROBERT L. 1964. Oceanographical engineering. Englewood Cliffs, N.J.: Prentice-Hall. 532 pp., illus.

Concerns problems of structures on the seashore and in harbors (in relation to the action of waves, tides, and currents), and related matters. At this time, oceanographic engineering is a field in which the demand exceeds the supply.

WILSON, DOUGLAS P. 1935. Life of the shore and shallow sea. London: Nicholson and Watson. 150 pp., 8 text figs., col. frontispiece, 52 pls. (7 × 9½ inches).

The finest book of the sort that has come to our attention until Yonge's. The 128 photographic illustrations of the living animals, reproduced generally very well by the plate-gravure process, are little short of superb. Treatment is ecological and semipopular. Sublittoral species are included. Southern England. (2d ed., 1951, 213 pp., 44 pls., 10 figs.)

——— 1947. They live in the sea. London: Collins. 128 pp., illus.

For younger readers; illustrated with 90 fine photographs.

YONGE, C. M. 1944. British marine life. London: Collins. 48 pp., 26 figs., 8 color pls.

A volume of the Britain in Pictures Series, illustrated with plates taken from old seashore books and monographs. The text is a précis of marine biology in Britain.

——— 1949. The seashore. London: Collins. 311 pp., 88 figs., 32 b&w, 40 col. pls.

This is the finest book of this genre yet published; a *sine qua non* for the bookshelf of all who go to the shore. Our own shores are no less beautiful than those in the color photographs of this book, but our printers work shorter hours and charge more for their time. Most of the photographs are by Douglas P. Wilson. If the reader finds it hard to choose between Yonge and Wilson, we can only recommend that he buy both books.

ZENKEVITCH, L. 1963. Biology of the seas of the U.S.S.R. London: Allen & Unwin; New York: Wiley & Sons. 955 pp., illus.

Translated from the Russian; a good account of what is in the oceans around Russia, including the Bering Sea and parts of the North Pacific.

Glossary

Abdomen. That part of an animal that houses the hind gut or intestines, when body regions may clearly be recognized, as in arthropods or vertebrates (but not in mollusks or echinoderms).

Adductor muscle. A muscle that brings a part toward the center or axis in vertebrates; in clams, ostracods, and nebaliids the muscle that closes the two valves.

Ambulacral groove. A groove down the lower side of each ray on a sea-star, along which the tube feet are arranged; associated with the radial canal. *See* water-vascular system.

Aperture. Usually an opening, like the opening of a snail shell; in bryozoans the term is used to describe an area of the zooecium that is covered with chitinous membrane.

Autotomy. The process of self-amputation of extremities, which are later regenerated; autotomy is especially characteristic of crabs and echinoderms.

Avicularium (pl. **avicularia**). A modified zooecium of a bryozoan, a tweezerlike structure resembling a bird's beak; serves to keep unwelcome objects and guests off.

Biomass. The amount of living material in a given area or volume of habitat.

Biome. A major climax community of plants and animals associated with a stable environmental life zone or region, e.g. tundra or grasslands.

Bursae. Pouches in an animal's body.

Byssus. The strands of horny, plastic material secreted by a special gland in mussels and many clams; lost by most species in early youth. The byssal hairs or filaments secure a mussel to the rock.

Cerata (singular, **ceras**). Fingerlike processes on the back of an eolid nudibranch, usually containing diverticula of the digestive gland (liver); in those species feeding on coelenterates the cerata may be "loaded" with nematocysts.

Cilia. Microscopic, hairlike vibratile projections found in many animals and consisting of a closed tube with two central filaments and nine peripheral; cilia are used for swimming and feeding, and for moving material in tubes or along channels.

Circumboreal. Describes a species or condition found around the world in the cold temperate region below the Arctic.

Climax or **climax development.** The stage of development at which an ecological unit has stabilized and may be expected to remain much the same.

Comb jelly. A common name for ctenophores; the "combs" referred to are small rows of cilia that are fused at the bases and move as a unit.

Commensal. An organism that lives with and shares the food of its host organism. There is generally no harm to the host, except for the increased competition for available food, but the commensal generally gains a great deal of protection or mobility.

Coppinia mass. Clusters of gonangia found on the stems of individual zooids in a hydroid colony.

Cypris. The second larval stage of a barnacle; resembles an ostracod of the genus *Cypris,* with a bivalve carapace, adductor muscle, and six thoracic appendages.

Ecdysis. The shedding of an outer covering; molting.

Ecosystem. A structure of organisms and environment, primarily arranged in terms of flow of energy through the system.

Eltonian pyramid (for Charles Elton). The concept of the food chain based on numbers and/or mass; the greatest mass is that of the primary plant food; a lesser mass of herbivores is above; and the carnivores are at the top of the pyramid.

Estuary. A section at the mouth of a river where saltwater tides mix with the freshwater river current, consequently a region of variable salinity.

Exopodite. The outer of two main branches in a crustacean appendage.

Faciation. A subdivision of an ecological association in which two or more, but not all, of the associated forms are dominant. This sort of terminology is falling into disuse.

Food chain. The plants and/or animals upon which a given organism depends; a large carnivore depends on herbivores, which depend on plants, and so on; when considered in terms of numbers of each food type, it is also referred to as the Eltonian pyramid.

Foramen. A small opening or orifice.

Gonangium. A structure in which gonads are produced; the modified theca protecting the gonad (blastostyle) of calyptoblast hydroids.

Gonophore. The reproductive zooid of a hydroid colony.

Hermaphrodite. An organism with both male and female organs in one individual.

Heteronereis. The specialized reproductive stage of some polychaete worms.

Holdfast. A branched, tangled modification of the stipe that attaches larger algae to rocks or substrate.

Holothurian. A sea cucumber.

Hydranth. A nutritive, or feeding, polyp in a hydroid colony.

Hydrothecae. Cuplike structures that enclose the individual polyps in a hydroid colony.

Krill. Planktonic crustaceans; the term is now restricted to the euphausiids, especially abundant in the Southern Ocean, where they are the principal food of baleen whales and penguins.

Lophophore. A coiled, looped, or circular structure consisting of a central member bearing ciliated tentacles that accumulate food and convey it to the mouth; somewhat different kinds occur in bryozoans, brachiopods, and phoronids.

Madreporite. A hard, perforated plate that filters water entering the water-vascular system.

Mantle. An outer integument of mollusks and brachiopods; the mantle covers the animal and secretes the shell, if a shell is present.

Manubrium. The stalk or fingerlike process of many coelenterates on which the mouth is situated.

Medusa. A free-swimming, usually bell-shaped coelenterate of the sexual generation; a jellyfish.

Megalops. The second larval form of crabs, characterized by stalked eyes and a well-developed abdomen.

Molting. The process of casting off an exoskeleton to allow for further growth.

Nauplius. The free larval stage of crustaceans, characterized by having three pairs of appendages, a single median eye, and a blind digestive tract. It is found in barnacles, copepods, ostracods, and fairy shrimp (e.g. *Artemia*), but not in the so-called "higher Crustacea" (Malacostraca), except for the krill and the penaeid prawns.

Neap tide. That part of the lunar tidal cycle with the least ranges of highs and lows, associated with the quarters of the moon.

Nematocyst. Specialized microscopic stinging structures that are characteristic of coelenterates, consisting of a capsule and a long, eversible thread, often armed with spines. They are not "cells."

Nephridia. Excretory organs, usually of invertebrates, consisting of collecting funnel, sac or flame cells, and a tube or system of tubes leading to the exterior. Often function in maintaining internal salt concentration in estuarine animals.

Neritic. Pertaining to the shallow sea or near land.

Operculum. A structure used to close the opening of a shell or tube.

Osculum (pl. **oscula**). The large, exhalant (outgoing) opening of a sponge.

Ostium. The small, inhalant opening of a sponge.

Ovigerous. Bearing eggs.

Pedicellariae. Small grasping defensive or cleaning structures in some echinoderms; a pedicellaria generally has two or three pincer arms (sometimes five) on a stalk; some are provided with poison glands.

Peduncle. Stalk for the attachment of an animal or an organ, often flexible.

Pelagic. Pertaining to the open ocean.

Pharynx. The tubular section of the alimentary tract that is immediately behind the mouth.

Phytoplankton. Microscopic floating algae and other autotrophs of the plankton.

Plankton. A collective term for the drifting forms of plant and animal life found in the upper layers of the sea.

Planula. A ovoid, ciliated larval type found in many coelenterates.

Polychaete. A member of the group of predominantly marine annelids characterized by possessing bundles of chaetae on lateral extensions (parapodia) of the body segments; bristleworms.

Polyp. The individual, asexual member of a hydroid or coral colony, or a sea anemone. Usually it is attached at the base but there are modified floating polyps, e.g. *Velella* and some anemones. Also used in the literary sense for octopus.

Polypide. In bryozoans the "soft parts," especially the lophophore, of the individual zooid, often used for the entire individual except the case (zooecium).

Primary productivity. The initial fixation of carbon by photosynthetic plants; the beginning of the food chain.

Proboscis (proper pl. **proboscides**). A general term for some sort of anterior process, usually capable of expansion and retraction; in an invertebrate it may be the anterior part of the pharynx including the mouth and an armature of teeth or other hard parts (as in polychaetes and gastropods), or the entire anterior part of the animal (as in sipunculids), or a structure separate from the digestive system, completely eversible (as in nemerteans) or expansile and retractile (as in some echiurids). It is usually controlled by a combination of hydrostatic and muscular action.

Protandric. Referring to a hermaphroditic animal that is first male and later female; some hermaphrodites may alternate sexually several times.

Radula. A ribbon or bandlike structure with rows of chitinous teeth; characteristic of chitons and gastropods.

Rostrum. A beaklike spine or process projecting forward between the eyes in crustaceans.

Seta (setae). Very small hairlike bristles, usually arising from a socket or special base; in Arthropoda and Annelida (as chaetae).

Siphon. A tubelike structure for taking in or discharging water; found in mollusks and ascidians.

Spat. Larval clams and oysters at the time of settling.

Spring tide. The tide cycle of greatest range, that is with the highest high and lowest low, associated with the full and new moons or the period of maximum declination.

Stipe. The stemlike part of the algal thallus or plant; in the large kelps the stipe may be many feet in length.

Strobila. The "stack of saucers" stage in some scyphozoans; the saucers later break off to become individual medusae.

Telson. The last segment of the abdomen in crustaceans.

Tentacle. A flexible, elongated structure for grasping and feeling.

Test. The hard outer shell of a sea urchin or sand dollar; also, the outer sheath, or tunic, of a tunicate.

Thallus. The entire individual seaweed or alga, including the flattened sheet, or blade, and the narrow stipe (q.v.).

Theca. A cup or sheath protecting a sessile animal, especially the individual of a hydroid colony.

Trochophore. A larva with apical tuft of cilia, complete digestive tract and protonephridal excretory system, propelled by bands of cilia around the inverted pear-shaped or nearly spherical body; characteristic of annelids and related phyla.

Tube feet. The hollow, rubbery appendages used for locomotion and feeding in echinoderms; the tube feet derive their strength partly from muscles and partly from hydraulic action controlled by the water-vascular system.

Tubercules. Small, rounded projections.

Tunic. The outer cuticle that covers an ascidian.

Valve. Any single section or plate of an invertebrate shell.

Veliger. The larval stage of bivalves and gastropods, developed from the trochophore by extension of the ciliary girdle to form swimming lobes; usually with a shell.

Water-vascular system. A system of canals in echinoderms, consisting basically of a ring canal around the mouth, five radial canals and a system of lateral canals to the podia (tube feet), and a stone canal open to the exterior through the madreporite. The system operates partly by ciliary, partly by muscular action; in the seastars the tube feet are expanded or contracted by moving water in and out with the aid of muscular ampullae. In crinoids and some ophiurians the tube feet are the principal feeding structures, passing food along the ambulacral groove to the mouth.

Zooea. The larval stage of crabs and some shrimp, having seven pairs of appendages and compound eyes.

Zooecium. The outer membranous or calcareous case of the individual member of a bryozoan colony, housing the zooid (consisting of the polypide and its basal part, the cystid).

Zooplankton. Animal members of the plankton, such as copepods, arrow worms, small medusae, etc.

GENERAL INDEX

General Index

ALL ANIMALS mentioned in the text and Systematic Index are arranged alphabetically in this General Index, which has been greatly expanded for this edition. Both common names and scientific names are listed. To aid those with a limited knowledge of zoological classification, each index entry for the common name of an animal or group of animals ("crab," e.g.) includes, in parentheses, the lowest-level taxonomic heading in the Systematic Index that embraces all of the animals referred to by that common name; most such entries also list the names of all genera into which these animals fall, to facilitate text reference. Authors cited in the text, as well as all authors whose works are listed in the Systematic Index and the General Bibliography, are indexed; the reader will thus be able to locate all of the books and papers of any one author. The Glossary is not indexed, except for a few terms also treated extensively in the text.

C

D

F

G

M

P

Q

R

S

T

tide pool habitat, 162–69

Notes

Notes

Notes

Notes

Notes

Notes

Notes

Notes

Notes